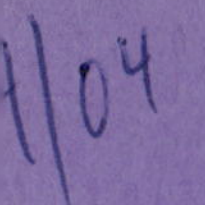

The Museum of
Broadcast Communications

Encyclopedia of Radio

The Museum of Broadcast Communications

Encyclopedia of Radio

VOLUME 2

F–N

Editor
CHRISTOPHER H. STERLING

Consulting Editor
MICHAEL C. KEITH

FITZROY DEARBORN
AN IMPRINT OF THE TAYLOR & FRANCIS GROUP

NEW YORK • LONDON

Published in 2004 by
Fitzroy Dearborn
An imprint of the Taylor and Francis Group
29 West 35th Street
New York, NY 10001

Published in Great Britain by
Fitzroy Dearborn
An imprint of the Taylor and Francis Group
11 New Fetter Lane
London EC4P 4EE

10 9 8 7 6 5 4 3 2 1

Library of Congress Cataloging-in-Publication Data

The Museum of Broadcast Communications encyclopedia of radio / editor,
Christopher H. Sterling ; consulting editor, Michael Keith.
p. cm.
ISBN 1-57958-249-4 (set : alk. paper) -- ISBN 1-57958-431-4 (vol. 1 :
alk. paper) -- ISBN 1-57958-432-2 (vol. 2 : alk. paper) -- ISBN
1-57958-452-7 (vol. 3 : alk. paper)
1. Radio--Encyclopedias. 2. Radio programs--Encyclopedias. I.
Sterling, Christopher H., 1943- II. Museum of Broadcast Communications.
III. Title.
TK6544.M84 2004
384.54'03--dc22

2003015683

First published in the USA and UK 2004

Typeset by Andrea Rosenberg
Printed by Edwards Brothers
Cover design by Peter Aristedes, Chicago Advertising and Design, Chicago, Illinois

Cover photos: (from upper left): Guglielmo Marconi; *Let's Pretend;* Aimee Semple McPherson; Casey Kasem; Mary Margaret McBride; A. Atwater Kent; Larry King; Kay Kyser; Rush Limbaugh; Huey Long

CONTENTS

ADVISERS

Stanley R. Alten
Syracuse University

Frank J. Chorba
Washburn University

Lynn A. Christian
International Radio Consultant

Ed Cohen
Clear Channel Broadcasters

Norman Corwin
Radio Playwright

Susan J. Douglas
University of Michigan

James E. Fletcher
University of Georgia

Robert S. Fortner
Calvin College

Stan Freberg
Radio Producer

Donald G. Godfrey
Arizona State University

Marty Halperin
Pacific Pioneer Broadcasters

Gordon H. Hastings
Broadcasters' Foundation

Robert Henabery
Radio Producer

Robert L. Hilliard
Emerson College

Michele Hilmes
University of Wisconsin, Madison

Chuck Howell
Library of American Broadcasting
University of Maryland

Stanley Hubbard
Hubbard Communications

John D. Jackson
Concordia University

Jack Mitchell
University of Wisconsin, Madison

Graham Mytton
BBC World Service (ret.)

Horace Newcomb
Director, Peabody Awards
University of Georgia

Peter B. Orlik
Central Michigan University

Ed Shane
Shane Media Services

Marlin R. Taylor
XM Satellite Radio

CONTRIBUTORS

Michael H. Adams
Alan B. Albarran
Pierre Albert
Craig Allen
Steven D. Anderson
Larry Appelbaum
Edd Applegate
Sousan Arafeh
John S. Armstrong
Philip J. Auter
Robert K. Avery
Glenda R. Balas
Mary Christine Banwart
Warren Bareiss
Ray Barfield
Kyle S. Barnett
Douglas L. Battema
Mary E. Beadle
Christine Becker
Johnny Beerling
Alan Bell
Louise Benjamin
ElDean Bennett
Marvin Bensman
Jerome S. Berg
Rosemary Bergeron
William L. Bird, Jr.
Howard Blue
A. Joseph Borrell
Douglas A. Boyd
John Bradford
L. Clare Bratten
Mark Braun
Jack Brown
Michael Brown
Robert J. Brown
Donald R. Browne
John H. Bryant
Joseph G. Buchman
Karen S. Buzzard
Paul Brian Campbell
Dom Caristi
Ginger Rudeseal Carter
Dixon H. Chandler II
Frank J. Chorba
Lynn A. Christian
Claudia Clark
Kathleen Collins
Jerry Condra
Harold N. Cones
Bryan Cornell
Elizabeth Cox
Steve Craig
Tim Crook
Marie Cusson
Keri Davies
E. Alvin Davis
J.M. Dempsey
Corley Dennison
Neil Denslow
Steven Dick
John D.H. Downing
Pamela K. Doyle
Christina S. Drale
Susan Tyler Eastman
Bob Edwards
Kathryn Smoot Egan
Lyombe Eko
Sandra L. Ellis
Ralph Engelman
Erika Engstrom
Stuart L. Esrock
Charles Feldman
Michel Filion
Howard Fink
Seth Finn
Robert G. Finney
Margaret Finucane
James E. Fletcher
Corey Flintoff
Joe S. Foote
Robert C. Fordan
Robert S. Fortner
James C. Foust
Ralph Frasca
James A. Freeman
Elfriede Fürsich
Charles F. Ganzert
Ronald Garay
Philipp Gassert
Judith Gerber
Norman Gilliland

Donald G. Godfrey
Douglas Gomery
Jim Grubbs
Joanne Gula
Paul F. Gullifor
Linwood A. Hagin
Donna L. Halper
Tona J. Hangen
Margot Hardenbergh
Jeffrey D. Harman
Dorinda Hartmann
Gordon H. Hastings
Joy Elizabeth Hayes
John Allen Hendricks
Alexandra Hendriks
Ariana Hernandez-Reguant
Robert L. Hilliard
Jim Hilliker
Michele Hilmes
John Hochheimer
Jack Holgate
Herbert H. Howard
Chuck Howell
Kevin Howley
W.A. Kelly Huff
Peter E. Hunn
John D. Jackson
Randy Jacobs
Glen M. Johnson
Phylis Johnson
Sara Jones
Lynda Lee Kaid
Stephen A. Kallis, Jr.
Steve Kang
Michael C. Keith
Ari Kelman
Colum Kenny
John Michael Kittross
Frederica P. Kushner
Philip J. Lane
Matthew Lasar
Laurie Thomas Lee
Renée Legris
Frederic A. Leigh
Lawrence W. Lichty
Lucy A. Liggett
Val E. Limburg
Robert Henry Lochte
Jason Loviglio
Gregory Ferrell Lowe
Christopher Lucas
Mike Mashon
Marilyn J. Matelski
Peter E. Mayeux
Dennis W. Mazzocco
Thomas A. McCain
Jeffrey M. McCall
David McCartney
Tom McCourt
Brad McCoy
Allison McCracken
Drew O. McDaniel
Michael A. McGregor
Robert McKenzie
Elizabeth McLeod
Mike Meeske
Fritz Messere
Colin Miller
Toby Miller
Bruce Mims
Jack Minkow
Jack Mitchell
Jason Mittell
Barbara Moore
Matthew Murray
Graham Mytton
Gregory D. Newton
Greg Nielsen
D'Arcy John Oaks
William F. O'Connor
Cary O'Dell
Robert M. Ogles
Ryota Ono
Peter B. Orlik
Pierre-C. Pagé
Brian T. Pauling
Manjunath Pendakur
Douglas K. Penisten
Stephen D. Perry
Patricia Phalen
Steven Phipps
Joseph R. Piasek
Gregory G. Pitts
Mark Poindexter
Tim Pollard
Robert F. Potter
Alf Pratte
Patricia Joyner Priest
Dennis Randolph
Lawrence N. Redd
David E. Reese
Patton B. Reighard

Andre Richte
Edward A. Riedinger
Terry A. Robertson
Melinda B. Robins
América Rodríguez
Eric W. Rothenbuhler
Richard Rudin
Joseph A. Russomanno
Anne Sanderlin
Erica Scharrer
Steven R. Scherer
Karl Schmid
Clair Schulz
Ed Shane
Pam Shane
Mitchell Shapiro
Jason T. Siegel
Ron Simon
B.R. Smith
Ruth Bayard Smith
Lynn Spangler
David R. Spencer
David Spiceland
Laurie R. Squire
Michael Stamm
Christopher H. Sterling
Will Straw
Michael Streissguth
Mary Kay Switzer
Rick Sykes
Marlin R. Taylor
Matt Taylor
Herbert A. Terry
Richard Tiner
Regis Tucci
David E. Tucker
Don Rodney Vaughan
Mary Vipond
Randall Vogt
Ira Wagman
Andrew Walker
Peter Wallace
Jennifer Hyland Wang
Richard Ward
Mary Ann Watson
Brian West
Gilbert A. Williams
Sonja Williams
Wenmouth Williams, Jr.
Roger Wilmut
Stephen M. Winzenburg
Richard Wolff
Roosevelt "Rick" Wright, Jr.
Edgar B. Wycoff
Thimios Zaharopoulos

LIST OF ENTRIES

Volume 2

F

Fadiman, Clifton 1904–1999

U.S. Writer, Editor, and Radio Emcee

Clifton Fadiman came to radio not from newspapers, vaudeville, or Hollywood but from the outwardly quieter world of book and periodical writing, editing, and publishing. As the master of ceremonies of *Information, Please* and other programs between 1938 and the mid-1950s, he used his knowing tone, witty repartee, and fondness for punning to become an appealing advocate of culture, learning, and civilized conversation for the World War II and postwar generations.

Fadiman learned the pleasures of knowledge at an early age. His older brother Edwin taught him to read when he was four, and before his teen years he was absorbing Milton, Homer, Dante, and other classic writers. To support himself during high school and college, he pieced together many jobs, beginning with mixing sodas in his father's Brooklyn drug store and then simultaneously reporting, selling ads, distributing copies, and otherwise helping with his brother's Long Island newspaper. He became a book reviewer for *The Nation* at age 17, and during his Columbia University days he was a ship's chandler, a bookseller, and a paid breaker-in of wealthier students' smoking pipes. By the time he finished his A.B. degree in 1925, he had gained direct experience of popular taste in many fields.

After two years of high school teaching, Fadiman joined Simon and Schuster in 1927, and as general editor there from 1929 to 1935 he made a number of shrewd publication choices that produced best-sellers. At the same time, he lectured at the People's Institute of New York and participated in many public forums, one of which would have a direct bearing on his radio career several years later. In 1934, when he assumed the editorship of *The New Yorker*'s book review page, he also had his first taste of sustained radio work as an on-air book reviewer for WJZ, but that stint lasted only six months. His best radio days were still to come.

In 1938 Dan Golenpaul, a creator of informational radio programs, was brooding over conventional quiz shows, which regularly dragged audience members to the microphone and exposed the shallowness of their knowledge. Golenpaul outlined a fresh approach: invite the public to send in questions to test a panel of experts. In choosing a master of ceremonies, Golenpaul recalled Clifton Fadiman's crisp contribution to a New School for Social Research radio forum on modern literature a few years earlier. Invited to lead the new quiz, Fadiman teasingly framed the questions for a panel gathered to record an audition disc. After some network doubts about public interest, *Information, Please* (titled after telephone operators' then-customary greeting) was first heard on 17 May 1938 on National Broadcasting Company (NBC) Blue. An unexpected hit, it rose to an estimated peak listenership of more than 9 million during its decade on the air and made Clifton Fadiman a popular icon of the intellectual establishment.

While producer Golenpaul battled sponsors' intrusions and brushed aside network directives, Fadiman tweaked the chemistry of the panel, which included newspaper columnist Franklin P. Adams, sportswriter and naturalist John Kiernan, and pianist Oscar Levant. The fourth chair, like the third one when Levant was absent in alternate weeks, was reserved for guest panelists from a wide range of performers, authors, statesmen, and athletes. Fadiman introduced the questions in a tone of mock menace, offering (as a 1941 review put it) an "ingratiating personality, with its intriguing dash of affable arrogance." Seldom missing an opening for a pun, he called Othello's killing of Desdemona an instance of "smother love," and when correspondent John Gunther correctly identified Reza Pahlavi as Iran's head of state, Fadiman pressed the question with "Are you shah?" while Gunther counterpunned, "Sultanly." Years after the program left the air, Fadiman confided that the questions and answers were only "an armature on which to build a sculpture of genuine conversation."

Clifton Fadiman fronted other programs, too. He and composer-conductor Morton Gould led the win-the-war

Clifton Fadiman in 1938
Courtesy AP/Wide World Photos

entertainment show *Keep 'Em Rolling* on Mutual in 1941–42, and from 1954 to 1956 he hosted NBC's sustaining *Conversation* with author-publisher Bennett Cerf and panel guests. In 1955 he was named a "communicator" for NBC's weekend program *Monitor.* His television contributions included *This Is Show Business* and a short-lived attempt to bring *Information, Please* to the newer medium.

After the mid-1950s, Fadiman largely focused on literary interests. He had been a board member and reviewer for the Book-of-the-Month Club since 1944, and he remained on its board of directors for more than 50 years. Begun in 1949, his "Party of One" column appeared in *Holiday* magazine for a decade. He especially valued his essay on children's literature for the *Encyclopaedia Britannica,* and he continued to assemble anthologies and to write books and essays urging the public to know the pleasures of books, wine, and mathematics.

This man of wit and prodigious memory was genially satirized as the *Duffy's Tavern* habitué Clifton Finnegan, who seemed almost too stupid to breathe. In the intellectually gritty 1950s, Dwight MacDonald more somberly dismissed Fadiman as a "midcult" peddler of learning to unwashed masses, and more recently critic John Leonard bemoaned Fadiman's "philistine" failure to appreciate William Faulkner's novels. Several generations of readers have been grateful, however, for Fadiman's invitations to learning in *Reading I've Liked* and *The Lifetime Reading Plan,* and listeners to *Information, Please* and *Conversation* discovered that knowledge and wit could be both gratifying and greatly entertaining.

Although the arc of Fadiman's career began and ended in writing and publishing, the middle span made a notable contribution to radio's upward aspirations. In fact, he was most valuable to radio precisely because he was not *from* radio, and for those who wished to condemn broadcasting as merely a noisy, empty-minded enterprise, Clifton Fadiman remained a hard nut to crack.

RAY BARFIELD

Clifton Fadiman. Born in Brooklyn, New York, 15 May 1904. A.B. degree, Columbia University, 1925. Taught at Ethical Culture High School, 1925–27; editor at Simon and Schuster, 1927–35; book editor, *New Yorker* magazine, New York City, 1933–43; debuted on radio doing book reviews, 1934; noted as moderator of *Information, Please,* 1938–48; Book-of-the-Month co-founder and review panelist, 1944–99; member, Board of Editors, Encyclopaedia Britannica, Chicago, 1959–99; Regents Lecturer, University of California, Los Angeles, 1967. Recipient: Award for distinguished service to American literature for radio program, *Information, Please!*, 1940; Clarence Day Award, American Library Association, 1969; Dorothy C. McKenzie Award, Southern California Council on Literature for Children and Young People, 1986; National Book Award for distinguished contribution to American letters, 1993. Died on Sanibel Island, Florida, 20 June 1999.

Radio Series

1938–48 *Information, Please*
1941–42 *Keep 'Em Rolling*
1944–45 *Words at War*
1949 *This is Broadway*
1955 *Monitor*
1954–56 *Conversation*

Television

This Is Show Business, 1949–54, 1956; *Information, Please,* 1952; *First Edition,* 1983–84, 1986

Selected Publications

Reading I've Liked: A Personal Selection, 1941
Party of One, 1955
Any Number Can Play, 1957
The Lifetime Reading Plan, 1960
Enter, Conversing, 1962

The New Lifetime Reading Plan (with John S. Major), 4th edition, 1999

Further Reading

Ashley, Sally, *F.P.A.: The Life and Times of Franklin Pierce Adams,* New York: Beaufort Books, 1986

Fadiman, Anne, *Ex Libris: Confessions of a Common Reader,* New York: Farrar Straus and Giroux, 1998; London: Allen Lane, 1999

Kieran, John, *Not under Oath: Recollections and Reflections,* Boston: Houghton Mifflin, 1964

Fairness Doctrine

Controversial Issue Broadcasting Policy

Until 1987 (with related parts lasting until 2000), the Federal Communications Commission (FCC) adhered to a series of policy guidelines collectively called the *fairness doctrine.* These guidelines encouraged stations to cover issues of public controversy, and to provide a variety of points of view on those issues. While they lasted, the policies were among the most controversial of all FCC program regulations.

Origin

A station licensee's duty to present diverse views on public issues was first declared by the Federal Radio Commission in 1928. A dozen years later, however, the FCC reversed direction when it strongly criticized a station for its practice of editorializing. In its 1941 *Mayflower* decision, the FCC concluded that with limited frequencies available for broadcasting, the public interest could not be well served by dedication of a broadcast facility to the support of its own partisan ends. In line with the *Mayflower* decision, broadcasters began to prohibit the sale of commercial time to deal with controversial issues—a policy that also helped them financially since such ads would only serve to anger some listeners and other advertisers.

In 1949 the FCC reversed itself, reconfirming that while stations have an obligation to cover controversial issues of public importance they now could (but did not have to) editorialize. When WHKC in Columbus, Ohio, refused to sell airtime to a labor union, the FCC stated that the station must be sensitive to the problems of public concern in the community and make sufficient time available on a nondiscriminatory basis. The commission concluded that radio stations have the "responsibility for determining the specific program material to be broadcast over their stations." Therefore, they were required to devote broadcast time to "issues of interest in the community served by their stations and [ensure] that such programs be designed so that the public has a reasonable opportunity to hear different opposing positions on the public issues of interest and importance in the community."

To nail down the proposed new policy on editorializing, the FCC held hearings on the matter. From the hearings came a 1949 statement, *In the Matter of Editorializing by Broadcast Licensees*, which placed two primary obligations on the broadcasters. What would become known later as the "fairness doctrine" required broadcasters (1) to cover controversial issues of public importance, and (2) to provide a reasonable opportunity for the presentation of contrasting viewpoints on those issues.

Development

A decade later, in 1959, Congress entered the fray. Legislators amended Section 315 (the political "equal opportunity" section of the Communications Act) to limit the applicability of the requirement to four types of news programs. At the same time, they made more concrete the broadcaster's responsibility to afford reasonable opportunity for the discussion of conflicting views on issues of public importance. This added phrase would cause considerable legal confusion in the future.

By 1967 the FCC had extended what was now commonly referred to as "the fairness doctrine" to include broadcast advertising of cigarettes, reasoning that because smoking was a controversial health issue, broadcasters were therefore required to provide contrasting viewpoints (This provision lasted until cigarette advertising was removed from the air entirely in the early 1970s.)

Complaining about yet another extension of the fairness doctrine, broadcasters asked what other types of program or advertising might trigger fairness doctrine concerns. To clarify the scope of their doctrine, the FCC instituted a wide-ranging inquiry into the fairness doctrine and its efficacy. As one result, the commission created three "contingent rights of tence, and hard cash was always in short supply. In addition, farmers generally lived at great distances from stations and needed to buy more expensive receivers to get satisfactory reception. And despite New Deal rural electrification efforts, many regions still lacked electricity, forcing use of battery-powered radios. By the 1940 census, 92 percent of urban U.S. homes reported owning radios, but only 70 percent of rural farm homes did. The situation among rural nonwhites was far worse. The chronic poverty among minority farmers meant that as late as 1940, only 20 percent owned radios.

Farm Radio and Country Music

Early listener response convinced broadcasters that rural and urban audiences differed considerably in their musical tastes. Farmers, it was believed, much preferred what was then called "hillbilly" music. This style was based on the folk songs commonly performed in rural areas, usually by one or two musicians playing simple stringed instruments. As the need grew for more programming to attract and hold the farm audience, several large stations developed live musical variety shows with a distinctly rural flavor. *National Barn Dance*, from Chicago's WLS, and *The Grand Ole Opry*, from WSM in Nashville, were two of the earliest and most successful.

By providing an audience for budding new performers and a ready market for their records, farm radio music shows played an essential role in the development of country music. Record companies began providing free or low-cost performers in exchange for the promotional value of having their stars heard on radio broadcasts. The fact that Nashville was the home of the powerful WSM and its immensely popular *Grand Ole Opry* was a decisive factor in that city's becoming the country music capital of the world.

The National Association of Farm Broadcasters

Radio stations soon recognized the need for specialized broadcast personnel to produce agricultural news and information programming. The position of station "farm director" was generally filled by someone who knew farming well and who could dedicate full attention to researching and reporting on agricultural issues. Often, male farm directors were assisted by women who were delegated the duties of reporting on rural home economics and hosting homemaker-oriented programs of interviews, recipes, and household hints. Today, farm broadcasters are often graduates of specialized university programs in agricultural journalism, and, although men still dominate the field, the role of female broadcasters has broadened considerably.

In the 1940s farm directors from several stations met and formed what would eventually be known as the National Association of Farm Broadcasters (NAFB). Today, the NAFB is farm broadcasting's major trade organization, offering members a news service, sales and marketing assistance, and farm audience research.

Farm Radio in the Television Age

The coming of television meant changes throughout the radio industry. Many of the powerful big-city radio stations no longer found it profitable to target rural audiences, and farm radio programming increasingly became the province of the growing number of lower-powered regional or local stations serving rural areas. At the same time, agricultural news and information programs began to appear on many local television stations that served farm audiences.

Yet for a number of reasons, farm radio has remained a viable medium. The low cost and portability of modern radio receivers means today's farm families can own several sets and listen wherever they happen to be. Radios installed in trucks, tractors, and other farm vehicles can accompany farmers throughout the workday. Timely weather forecasts and market reports remain just as important to farmers today as they were in the early days of radio. At the same time, the relatively low cost of operating a local radio station means that farm broadcasters can stay profitable even while appealing to a relatively narrow audience. In fact, it is just this characteristic that attracts agricultural advertisers, who can zero in on their target audience at a relatively low cost. These characteristics mean that farm radio will continue to flourish.

STEVE CRAIG

See also Trade Associations; WHA and Wisconsin Public Radio

Further Reading

Baker, John Chester, *Farm Broadcasting: The First Sixty Years*, Ames: Iowa State University Press, 1981

Evans, James F., *Prairie Farmer and WLS: The Burridge D. Butler Years*, Urbana: University of Illinois Press, 1969

Frost, S.E., Jr., *Education's Own Stations: The History of Broadcast Licenses Issued to Educational Institutions*, Chicago: University of Chicago Press, 1937

Malone, Bill C., *Country Music, U.S.A.*, Austin: University of Texas Press, 1968; revised edition, 1985

National Association of Farm Broadcasters: About NAFB, "History of NAFB," <www.nafb.com/news.cfm>

Rural Radio Listening: A Study of Program Preferences in Rural Areas of the U.S., New York: Rural Research Institute, 1951

Smulyan, Susan, *Selling Radio: The Commercialization of American Broadcasting, 1920–1934*, Washington, D.C.: Smithsonian Institution Press, 1994

U.S. Department of Agriculture, Bureau of Agricultural Economics, "Attitudes of Rural People Toward Radio Service: A Nation-Wide Service of Farm and Small-Town People," Washington, D.C.: United States Department of Agriculture, January 1946

Wik, Reynold M., "The USDA and the Development of Radio in Rural America," *Agricultural History* 62, no. 2 (1988)

Faulk, John Henry 1913–1990

U.S. Radio Humorist

When his developing radio career was cut short in the mid-1950s because of political blacklisting, Texas-born humorist John Henry Faulk decided to fight back. He undertook a six-year legal battle and eventually beat the right-wing blacklisters who had controlled network assignments of radio and television creative personnel for years.

Early Years

John Henry Faulk (he always used his full name) was born in Austin, Texas, just before World War I. He grew up in a mixed-race neighborhood and was encouraged by his liberal parents to treat everyone alike. His own subsequent liberal politics were partially formed by his parents, both leftist political activists. Working and attending classes part-time, Faulk entered the University of Texas in 1929 but took a decade to earn his bachelor's degree in English. While in school he became fascinated with Texas and American folklore of all kinds. Just over a year later, with a Works Progress Administration research grant, he earned a master's degree with an emphasis in American folklore. He stayed on in Austin to teach English courses at the university for two more years. Despite the military build-up in the early 1940s, the army initially turned him down for health reasons, and Faulk served with the merchant marine and later the Red Cross. The army finally accepted him for stateside service as a medic in 1944.

All through this period, Faulk honed his already impressive storytelling abilities. Overhearing him at a New York party, a Columbia Broadcasting System (CBS) official offered him a program slot of his own. Faulk hosted *Johnny's Front Porch* on the CBS radio network in 1945–46. He moved on to various on-air positions with stations WOV and WPAT in New Jersey from 1946 to 1948. He returned to CBS to host the weekday hour-long daily *John Henry Faulk Show* on the network's flagship station, WCBS in New York, starting in 1951. His good-natured Texas humor and characters gathered a growing listening audience and attracted appreciative advertisers.

Blacklisted and a Landmark Suit

By the mid-1950s, as his radio career developed, Faulk had became more concerned about the communist-baiting approach of the officers of the New York local chapter of the American Federation of Radio and Television Artists (AFTRA) performer's union, of which he was a member. For several years union officials had actively cooperated with (and some were officers in) AWARE, Inc., one of the blacklisting organizations that "investigated" the political backgrounds of performers and writers. Along with CBS journalist Charles Collingwood and comedian Orson Bean, Faulk ran for the AFTRA board and was one of 27 "middle of the road slate" members elected to the 35-member board. He was subsequently elected second vice president, taking office in January 1956. The new AFTRA officers promised to cut ties with the blacklisters.

Nursing their wounds, the former union officials used their newsletter to attack Faulk for his own political views and associations, including thinly veiled accusations that, based on his political activity (he had backed Henry Wallace in the 1948 presidential campaign, for example), he at least had strongly communist leanings. In the political climate of the times, with "McCarthyism" in full flower, this was a potentially damning indictment that often cost the accused his or her job. As the attackers expected and as Faulk feared, their classic tactic led to advertiser nervousness about Faulk, and he began to lose commercial advertisers for his program. Although the station initially stood by its employee, diminishing advertiser support and income raised concern, though his contract was renewed in December 1956.

Faced with the loss of his livelihood for a series of false accusations, Faulk decided to take on his tormenters, although

John Henry Faulk
Courtesy AP/Wide World Photos

he knew it would be a difficult and expensive process. With financial support from CBS newsman Edward R. Murrow, among others, Faulk filed a libel suit on 26 June 1956 against AWARE and its officers. But the overall situation turned bad—the "middle of the road slate" lost ground in AFTRA and became a minority again. Though his program continued for a few more months, Faulk was fired by registered letter from CBS while he was on vacation in July 1957.

AWARE managed to delay trial on the libel suit for some five years through a variety of legal maneuvers. In the meantime, Faulk could not obtain work in the broadcast or entertainment fields and had to rely on friends and supporters for the financial means to survive, let alone pay for the ongoing case. The trial of AWARE and its officers finally got under way in the spring of 1962 and lasted for 11 weeks. On 28 June 1962 the New York State jury awarded $3.5 million in damages to Faulk (the largest libel judgment to that point), though this was later reduced to $550,000 by an appeals court. Faulk eventually received about $75,000 for his years of effort and deprivation. Through his lawsuit, he had made an important point about civil liberties that was widely reported, AWARE had declared bankruptcy, and the general public now knew about—and was increasingly appalled by—blacklisting.

Later Life

Faulk returned to Austin in 1968 and lived in Texas for the rest of his life. For a time he operated a small advertising agency. His blacklisting case became known to a new generation in October 1975 when CBS broadcast a two-hour made-for-television docudrama movie based on Faulk's story. William Devane played Faulk, and George C. Scott played attorney Louis Nizer.

From 1975 to the early 1980s, Faulk played a homespun character on the nationally syndicated *Hee-Haw* television variety program. During the 1980s Faulk wrote and produced two one-man plays. In both *Deep in the Heart* and *Pear Orchard, Texas,* he portrayed characters with the best of human instincts but exhibiting the worst of cultural prejudices. He also lectured widely on college campuses, largely about civil liberties and freedom of expression, drawing on his own case. He died of cancer in Austin in 1990, aged 76.

The central branch of the public library in Austin is named after him, as are the John Henry Faulk Awards given out annually since 1986 by the Texas Storytelling Association for the state's best storyteller. The Center for American History at the University of Texas at Austin (which holds Faulk's professional and legal papers) sponsors an annual John Henry Faulk Conference on the First Amendment.

CHRISTOPHER H. STERLING

See also American Federation of Television and Radio Artists; Blacklisting; Collingwood, Charles; Columbia Broadcasting System; Murrow, Edward R.

John Henry Faulk. Born in Austin, Texas, 21 August 1913. Attended the University of Texas, BA in English, 1939; Masters degree in Folklore, 1940; doctoral work, 1940–42. Served with the Merchant Marine, 1942–43; Red Cross in Cairo, Egypt, 1943–44; U.S. Army, psychiatric social worker, 1944–46. Hosted *Johnny's Front Porch* on CBS radio network, 1945–46; held various on-air positions, WOV, New York City, 1946–47, and WPAT, Paterson, New Jersey, 1948; hosted *The John Henry Faulk Show* on WCBS in New York, 1951–57; elected vice-president, New York local, American Federation of Television and Radio Artists (AFTRA), 1955; filed a libel suit against AWARE, Incorporated, 1956; was fired by CBS and blacklisted for five years; won the largest libel judgment in history, 1962; eventually returned to making motion pictures, wrote a one-man play and acted on television. Died in Austin, Texas, 9 April 1990.

Radio Series

1945–46 *Johnny's Front Porch*
1951–57 *The John Henry Faulk Show*

Television

Leave It to the Girls, 1954; *The Morning Show,* 1955; *Fear on Trial* (writer), 1975; *Hee Haw,* 1975–83; *Adam,* 1983

Film

All the Way Home, 1963; *The Best Man,* 1964; *Texas Chain Saw Massacre,* 1974; *Lovin' Molly,* 1974; *Leadbelly,* 1976

Stage
Pear Orchard, Texas, 1970s

Selected Publications
Fear on Trial, 1964
The Uncensored John Henry Faulk, 1985

Further Reading
Burton, Michael C., *John Henry Faulk: The Making of a Liberated Mind: A Biography,* Austin, Texas: Eakin Press, 1993
Carleton, Don E., *Red Scare! Right-Wing Hysteria, Fifties Fanaticism, and Their Legacy in Texas,* Austin: Texas Monthly Press, 1985
Cogley, John, *Report on Blacklisting,* 2 vols., New York: Fund for the Republic, 1956 (see especially vol. 2, *Radio-Television*)
Foley, Karen Sue, *The Political Blacklist in the Broadcast Industry: The Decade of the 1950s,* New York: Arno Press, 1979
Nizer, Louis, *The Jury Returns,* Garden City, New York: Doubleday, 1966
O'Connor, John J., "TV: CBS Dramatizes John Faulk's 'Fear on Trial,'" *New York Times* (2 October 1975)
Talbot, David, and Barbara Zheutlin, "The Password Is Blacklist: George C. Scott, CBS, and John Henry Faulk: The Way We Really Were," *Crawdaddy* (October 1975)
Wakefield, Dan, "Disc Jockey Fights the Blacklist," *The Nation* (20 December 1958)

Federal Communications Commission

The Federal Communications Commission (FCC) is the federal agency charged with regulating broadcasting and other electronic communications media in the United States; it licenses stations to operate in the "public interest, convenience, and necessity." The FCC was created by Congress in 1934 to succeed the Federal Radio Commission. It is an independent federal agency established by the Communications Act of 1934 to regulate domestic interstate and international electronic communication, both wired and wireless.

Because the FCC was established as an independent federal agency, the "checks and balances" on it are not the same as they would be for administrative agencies (such as the Food and Drug Administration), which answer directly to the U.S. president. Although the president selects the chair and the commissioners (who must be approved by the Senate), the president does not have the authority to remove commissioners during their terms. The FCC is much more beholden to Congress, which controls not only appropriations but also the commission's very existence. Since 1981, the FCC is no longer a permanent agency but instead must be reauthorized by Congress every two years. Therefore, Congress' influence over the FCC has increased significantly during the 1980s and 1990s.

The FCC has a dual role: on the one hand, it makes rules and regulations to carry out the Communications Act, but it also serves as a judicial body, hearing appeals of its decisions. As a quasi-judicial agency, the FCC has the duty of both making rules and also serving as an adjudicator in cases dealing with rules violations and challenges. While the FCC has the responsibility for making, policing, and judging the rules, its decisions are subject to court review. For example, the FCC created a rule requiring regular station identification. It also enforces the rule by asking stations whether they have adhered to it. In cases where stations have been found to violate the rule, the FCC must decide what punishment, if any, to apply. If the offending station challenges the decision, it appeals the judgment to the FCC. In this role, the FCC serves as the equivalent of a federal district court. FCC decisions that are upheld in appeal can then be challenged by appealing directly to the Federal Court of Appeals for the Washington, D.C. Circuit.

Commissioners

The FCC has five commissioners (reduced from seven in 1983), one of whom serves as the chair. Members are appointed by the U.S. president and approved by the Senate. The term of office for commissioners is five years, and they may serve multiple terms. No more than three commissioners from one political party may serve simultaneously.

Although the FCC chair has the same one vote as any of the other four commissioners, the chair has a greater ability to influence the direction of the FCC. The chair's role in selecting issues to pursue sets an agenda for the commission. During the 1970s and 1980s, for example, the FCC adopted a more deregulatory approach, eliminating a number of rules and streamlining the radio license renewal process. Under the leadership of chairmen Richard Wiley, Charles Ferris, and Mark Fowler, the

FCC revisited its responsibilities under the concept of the public interest, adopting the philosophy that the public interest is best served by allowing marketplace forces to function. In the 1990s Chairman Reed Hunt decided to investigate the possibilities of high-definition television and formed a task force to study it. Chairman William Kennard pursued the possibility of adding low-power FM stations to the radio band to provide increased opportunities for disenfranchised members of society to be heard. Although no chair has been successful in pursuing all of his interests, each has had the opportunity to set the commission's, and thus to a certain extent the nation's, communications policy agenda.

The majority of commissioners over the years have been lawyers (the last engineer commissioner retired in 1963). FCC commissioners are creatures of politics and as such are often more versed in politics than in technology. A number of commissioners have had no technological background prior to joining the commission. They count on their staff advisers and employees to provide them with the necessary background information. Fewer than half the commissioners have served their full five-year terms. When they leave the commission, they frequently join communications companies or legal firms providing consulting services. Commission staff members often find themselves dealing with former commissioners.

The personality of the FCC changes over time, based on the various personalities of the commissioners who serve and the political climate of the period. Space does not permit a listing of all former FCC commissioners and their contributions, but a few should be noted. Frieda B. Hennock, the first woman appointed to the commission in 1948, served during the critical period of the television "freeze" (1948–52). During those four years, the FCC stopped licensing new TV stations while it decided the issues of color TV, UHF versus VHF transmission, and channel allocation policies. Benjamin L. Hooks, the first African-American commissioner, was appointed in 1972 and worked diligently for the enforcement of equal employment opportunities. Henry Rivera was the first Hispanic commissioner, appointed in 1981. Robert E. Lee has the distinction of having served longer than any other commissioner to date, from 1953 until 1981. William Kennard was the first African American chairman when appointed in 1993.

FCC commissioners have been perceived alternately as pro- and anti-broadcasting. FCC chairmen James L. Fly (appointed 1939) and Newton N. Minow (appointed 1961) were public interest advocates who raised the ire of many broadcasters during their respective terms. Fly chaired the FCC during the forced sale of the NBC "Blue" Network. Minow is perhaps best remembered for referring to television as a "vast wasteland" in a speech to the National Association of Broadcasters.

On the other hand, James Quello joined the commission in 1974 after retiring from his position as vice president and general manager of station WJR in Detroit. Quello's Senate confirmation hearings lasted longer than any other commissioner's because a number of public interest groups, fearing that he would be too favorable to broadcast interests (Quello replaced public interest advocate Nicholas Johnson), opposed his nomination. In spite of the lengthy process, Quello was overwhelmingly approved by the Senate. While Quello served as interim FCC chair, *Broadcasting* magazine called him "the broadcasters' chairman." In spite of his strong support of broadcast interests, Quello was critical of indecency on radio.

Staff

Although the FCC has only five commissioners, there are nearly 2,000 staff members in dozens of different departments, including a dozen field offices across the U.S. The five FCC commissioners must officially hold a public meeting at least once a month. The business of the commission is largely conducted as items circulate among the commissioners in between meetings, and at the staff level in the operating bureaus. The FCC is divided administratively into a number of offices and bureaus, to which the bulk of the commission's work is delegated. The six major operating FCC bureaus are Consumer and Governmental Affairs, Enforcement, International, Media, Wireless Telecommunications, and Wireline Competition. Most licensing and regulatory activity undertaken in the name of the FCC occurs at this level.

Of greatest concern to broadcasters is the Media Bureau (previously called the Broadcast Bureau), which regulates AM, FM, and television broadcast stations and related facilities. It assigns frequencies and call letters to stations and designates operating power and sign-on and sign-off times. It also assigns stations in each service within the allocated frequency bands with specific locations, frequencies, and powers. It regulates existing stations, ensuring that stations operate in accordance with rules and in accordance with the technical provisions of their authorizations.

The Media Bureau has five divisions: Audio, Video, Policy, Industry Analysis, and Engineering. The Audio Division receives and evaluates approximately 5,500 applications per year for the nation's approximately 14,000 AM, FM commercial, FM noncommercial educational, and FM translator and booster stations. These applications include station modification applications, applications for new stations, assignment or transfer applications, license applications, and renewal applications.

Since the 1996 Telecommunications Act, radio and television stations are licensed for eight years (into the 1980s, the standard term was only three). Licensees are obligated to comply with statutes, rules, and policies relating to program content, such as identifying sponsors of material that is broadcast. The bureau ensures that licensees make available equal opportunities for use of broadcast facilities by political candidates or opposing political candidates, station identification, and identification of recorded programs or program segments. Licensees who have violated FCC rules, and most especially licensees who have "misrepresented" themselves, are subject to such sanctions as forfeitures (fines), short-term renewals, or (rarely) license revocation.

Other FCC offices are established by the commission as organizational tools for handling the tremendous variety of commission tasks. The current FCC offices are Administrative Law Judges, Communication Business Opportunities, Engineering and Technology, General Counsel, Inspector General, Legislative Affairs, Managing Director, Media Relations, Plans and Policy, and Workplace Diversity.

In its attempt to carry out the wishes of Congress as expressed in the Communications Act, the FCC creates or deletes rules. Proposals for new rules come from a variety of sources both within and outside the commission. To begin assessing a new service or proposal on which it has little information, the commission may issue a Notice of Inquiry (NOI) seeking advice and answers to specific questions. If the FCC plans a new rule, it first issues a Notice of Proposed Rule Making (NPRM). The NPRM gives formal notice that the FCC is considering a rule and provides a required length of time during which the commission must allow public comments. After sufficient comment and discussion, new rules are adopted in FCC Reports and Orders and are published in its own official report series, the *FCC Record*, as well as in the government's daily *Federal Register*. The rules are collected annually into Title 47 of the Code of Federal Regulations.

Licensing

Without a doubt, the greatest power of the FCC is its power to license or not license a station. Since its inception, the commission has modified its rules and procedures for determining the manner in which it grants licenses. In keeping with the directive of the Communications Act, the FCC has always licensed stations to operate "in the public interest, convenience, and necessity." The specific manner in which it has made determinations about individual licensees, however, has undergone changes. For decades, the FCC would examine all applicants for a license to determine which would best serve the public interest, considering a range of characteristics, including preferences for local ownership, minority ownership, experienced ownership, and the character of the owners. Even a station that had been in operation for years faced the prospect of losing its license to a superior applicant in a renewal proceeding. The FCC had to conduct time-consuming and expensive comparative hearings whenever two or more applicants sought the same frequency.

The FCC and the broadcast industry have long been critical of the comparative process, which was largely eliminated by the 1996 Telecommunications Act. Now, existing stations have an expectation of license renewal, provided they have not violated important commission rules. The FCC does not allow a new challenger to draw an existing station that has served the public interest into a comparative process. To avoid comparative hearings in the case of new stations, the FCC in 1999 adopted an auction process similar to what it had used earlier for other nonbroadcast frequencies. Rather than spend months (sometimes years) in comparative hearings, the FCC instead determines whether applicants meet minimum eligibility requirements. If they do, the applicants can then bid for new, available stations or purchase existing outlets.

The FCC does not monitor the broadcasts of radio and television stations. It relies on information provided to it by broadcasters, competitors, and audiences. Determinations about whether a station deserves to have its license renewed are based on documents filed by the station, any public comments the commission receives about that station, and challenges to the renewal by interested parties.

The commission is able to enforce regulations primarily through the *threat* of license action. The majority of license renewals are granted with no disciplinary action by the FCC. However, should the commission find rules violations by a licensee, its actions can range from a letter admonishing a station to fines of up to $250,000 and a short-term renewal of the license, and even to the revocation of a license. It is the threat of this action that keeps broadcasters in compliance.

The FCC has the ultimate authority to revoke a station's license or deny its renewal, although that action has rarely been taken. In more than 60 years, the FCC has taken this action only 147 times: an average of fewer than three per year out of thousands of license renewals. More than one-third of those revocations and nonrenewals were due to misrepresentations to the commission. Although such cases have been infrequent, the FCC has acted severely in cases where licensees have intentionally lied.

In 1998, the FCC revoked the license for Bay City, Texas, station KFCC because the owner "engaged in a pattern of outright falsehoods, evasiveness, and deception." Chameleon Radio Corporation had been awarded a license to serve Bay City, but it attempted to move the transmitter closer to Houston, to the extent that Bay City would not even be served by the station. The FCC cited the station for repeated misrepresentations and lack of candor. Chameleon attempted to argue that the merit of its programming should protect it from revocation. The FCC responded that "meritorious programming does not mitigate serious deliberate misconduct such as misrepresentation." Also that year, licenses were revoked for seven stations in Missouri and Indiana that were owned by Michael Rice through three different corporations. Rice was convicted of 12 felonies involving sexual assaults of children. The FCC was prepared to allow the corporations to continue their ownership of the stations after their assertion that Rice would not be involved in station operation. The commission revoked the licenses on a finding that Rice was in fact involved in station operation and that "misrepresentations and lack of candor regarding his role at the stations" was cause for revocation.

Despite all the powers given to the FCC, the 1934 Communications Act specifically states that the commission may not censor the content of broadcasts. According to Section 326, "Nothing in this Act shall be understood or construed to give this Commission the power of censorship over the radio communications or signals transmitted by any radio station." Nevertheless, the FCC's reprimands and fines of stations for broadcasting indecent material at inappropriate times have been upheld by the Supreme Court. The FCC acknowledges its obligation to stay out of content decisions in most areas. The commission has declined to base license decisions on the proposed format of a radio station, and it no longer stipulates the amount of time stations should devote to public service announcements.

Spectrum Management

The FCC also has the important responsibility of managing all users (except the federal government) of the electromagnetic spectrum. This involves two processes. First, the FCC must allocate different uses for different parts of the spectrum in an efficient way that does not create interference. This is followed by the allotment of specific frequencies or channels to particular areas. Licensing of the allotments follows. (AM does not have allotments.) There have been two significant spectrum reallocations affecting radio. In 1945 the FCC moved FM's allocation from 42 to 50 megahertz up to its current location at 88 to 108 megahertz, more than doubling the available channels but rendering the existing FM receivers and transmitters obsolete. In 1979 the upper end of the AM band moved from 1605 kilohertz to 1705 kilohertz to accommodate ten additional AM channels—a change that took years to implement (the first stations shifted to the new higher frequencies only moved in the 1990s).

A second part of maintaining spectrum efficiency is approving equipment that uses the electromagnetic spectrum. The FCC must be certain that all devices emitting electromagnetic signals do so within their prescribed limits. "Type acceptance" is the FCC process of approving equipment that emits electromagnetic radio waves. In most cases, the evidence for type acceptance is provided by the equipment's manufacturer, who provides a written application with a complete technical description of the product and a test report showing compliance with the technical requirements. An FCC identification number can be found on the backs of telephones, walkie-talkies, pagers, and even microwave ovens. The number does not imply that the FCC has inspected that particular unit but rather that the product meets certain FCC minimal standards designed to avoid interference.

Policy is sometimes the result of *in*action by the FCC. For years, the commission was faced with what to do about AM stereo. Rather than choosing a standard from among competing applicants, the FCC adopted in 1982 a marketplace philosophy to allow a winner to emerge from marketplace decisions rather than their own. Many broadcasters believe the introduction of AM stereo was negatively affected by the commission's "decision not to decide." Some stations were reluctant to invest in AM stereo equipment, fearing they might select the losing standard and then lose thousands of dollars. Radio receiver manufacturers had the same concern. The FCC's decision "not to decide" virtually killed any chance for AM stereo.

DOM CARISTI

See also Blue Book; Communications Act of 1934; Deregulation of Radio; Equal Time Rule; Fairness Doctrine; Federal Radio Commission; Frequency Allocation; Licensing; Public Interest, Convenience, or Necessity; Regulation; Stereo; Telecommunications Act of 1996; United States Congress and Radio

Further Reading

Cole, Barry G., and Mal Oettinger, *Reluctant Regulators: The FCC and The Broadcast Audience,* Reading, Massachusetts: Addison-Wesley, 1978

Federal Communications Commission website, <www.fcc.gov>

Flannery, Gerald V., editor, *Commissioners of the FCC 1927–1994,* Lanham, Maryland: University Press of America, 1994

Hilliard, Robert L., *The Federal Communications Commission: A Primer,* Boston: Focal Press, 1991

Kahn, Frank J., editor, *Documents of American Broadcasting,* 4th edition, Englewood Cliffs, New Jersey: Prentice-Hall, 1984

Krasnow, Erwin G., Lawrence D. Longley, and Herbert A. Terry, *The Politics of Broadcast Regulation,* New York: St. Martin's Press, 1973; 3rd edition, 1982

Ray, William B., *FCC: The Ups and Downs of Radio-TV Regulation,* Ames: Iowa State University Press, 1990

Federal Radio Commission

Predecessor to the Federal Communications Commission

The first agency created in the United States specifically to license and regulate radio, the Federal Radio Commission (FRC) existed from 1927 to 1934. During that time, the FRC was responsible not only for licensing radio stations in the United States, but also for laying the regulatory foundation that to some extent still exists today.

Origins

The United States passed the Radio Act in 1912, which provided authorization for licensing of radio stations. It was not until 1920, however, that the secretary of commerce began to exercise this authority. Congress had decided to empower the secretary of commerce because radio was seen as interstate commerce: thus, the Department of Commerce was the logical choice to handle the authorization of these new radio stations.

It did not take long to realize that radio licensing and regulation was a much larger task than could be handled by the secretary of commerce in addition to his regular duties. At the First National Radio Conference in Washington, D.C., in 1922, government officials and amateur and commercial radio operators met to discuss problems facing the infant industry. A technical committee's report resulted in the introduction of legislation in 1923, but it never got out of Senate committee.

At the Second National Radio Conference in 1923, many of the same problems were revisited, most notably concerns about interference. Because interference was greatest in places with the most transmitters (population centers) and less of a problem in remote areas, a recommendation emerged from the conference that the nation be divided into zones and that different rules be established for different zones based on their own specific needs. Again, legislation was introduced into Congress, but it never advanced beyond committee. A third national conference in 1924 produced no legislation either. Finally, the Fourth National Radio Conference in 1925 produced the proposals that would lead to the Radio Act of 1927.

During all this time, the secretary of commerce continued to license radio stations. The difficulty came when secretary Herbert Hoover attempted to *not* license a station. In 1923 Intercity Radio applied for a renewal of its license. The request was denied based on a determination that there was no longer spectrum space available. Intercity appealed the case, and the Federal Circuit Court of Appeals ruled in their favor. According to the court, the 1912 Act authorized the secretary of commerce to grant licenses: it did not grant the authority to deny them. In the court's judgment, the secretary of commerce had to accommodate applicants by finding them spectrum space that would cause the least amount of interference.

The secretary's authority was further undermined in a 1926 court decision. Zenith Radio had been operating a station in Denver with a prescribed allocation limiting it to two hours of broadcasts per week. The station challenged the authorization by using other frequencies not specifically allocated to it. When secretary of commerce Herbert Hoover filed suit against Zenith, the court ruled that Zenith was within its rights, citing a section of the Radio Act that allowed stations to use "other sending wave lengths." For all intents and purposes, such a decision authorized any licensed station to operate on virtually any frequency in addition to the one allocated.

The result of the two court cases was devastating to the secretary of commerce's assumed authority to regulate radio stations. Following the decision in the Zenith case, the secretary sent a letter to the U.S. Attorney General requesting an opinion on the authority vested in the secretary of commerce by the Radio Act. The attorney general's response only reinforced what the courts had already decided. The interpretation of the Radio Act of 1912 was that licensed stations could use virtually any frequency they wanted, at whatever times and using whatever transmitting power they wanted. With the secretary's authority eviscerated, President Calvin Coolidge asked Congress to create new legislation. With the president's endorsement, legislation that had been proposed as a result of the 1925 National Radio Conference moved swiftly through Congress, and the Radio Act of 1927 was signed into law by President Coolidge on 23 February 1927.

Initial Tasks

Central to the 1927 Act was the establishment of the FRC. Instead of having one person overseeing licensing and regulation of radio, a commission of five people appointed by the president and approved by the Senate would handle the duties. Perhaps influenced by the thinking that different geographic regions would have different needs, commissioners were to represent those different regions. When it passed the Radio Act of 1927, Congress naively believed that the FRC would need to act for only one year to straighten out all the confusion in radio regulation and licensing. After that, things would be well enough established that the agency could serve only as an appellate board for actions by the secretary of commerce. W. Jefferson Davis wrote in the *Virginia Law Review* that after the first year, "the Secretary of Commerce will handle most of the problems that arise, and the Commission will probably function only occasionally." Clearly, that was not to be the case. The FRC's authority was made permanent in 1930 and was extended until 1934, when it was replaced by the Federal Communications Commission (FCC).

The FRC got off to what was at best a shaky start, as Congress had not financed it. The FRC came into existence without an appropriation from Congress. The original commissioners were required to do their own clerical work, and engineers had to be "borrowed" from other agencies for several years. Furthermore, Congress confirmed only three of the five nominees from President Coolidge in the first year. Two of the five appointees died during the Commission's first year. Only one of the original FRC members, Judge Eugene O. Sykes, was still serving just two years later. In spite of these difficulties, the FRC was able to dramatically advance the nation's radio regulatory policy in its eight years of existence, at a crucial time for the development of commercial radio. Even the publication *Radio News* recognized the enormity of the FRC's task and the efficiency with which it worked. In November 1929, the publication stated,

> Not in the history of federal bureaus has any commission ever been called upon to perform so great a task in so short a span of time. The already overloaded departments of the federal government could not have treated with radio problems on this scale without a great increase in personnel and what would have been tantamount to the setting up of a radio commission within the department to which it might have been assigned. By its segregation and absolute independence, radio has been regulated and its major problems have been solved without handicapping any other federal bureau ("Public Interest, Convenience and Necessity," in *Radio News* [November 1929]).

The Radio Act of 1927 empowered the FRC with the authority and regulatory discretion that the secretary of commerce had lacked under the Radio Act of 1912. The new act specifically granted the FRC authority to license stations for a limited period of time; to designate specific frequencies, power, and times of operation of stations; to conduct hearings and serve as a quasi-judicial body; and to deny a license or to revoke an existing license. All of these powers had been denied the secretary of commerce.

Legal Challenges

It did not take long for the FRC's enforcement authority to be legally challenged. Technical Radio Laboratory sought a license renewal for its station in Midland Park, New Jersey, and was denied by the FRC because there was not adequate spectrum available. Whereas Intercity had successfully challenged denial of its renewal request, Technical Radio Laboratory was not so fortunate, because the courts found that the FRC was within its authority. Dozens of other cases would follow, with similar results. Congress had taken the appropriate steps to provide the FRC with licensing authority.

According to the Radio Act, the FRC's jurisdiction was based on Congress' authority to regulate interstate commerce. Thus, radio stations that did not transmit across a state line might not have to be regulated by them. In theory, at least, radio stations involved only in intrastate commerce rather than interstate commerce were not subject to federal jurisdiction. Just such an assertion was made in the case of *United States v Gregg*. An unlicensed Houston station challenged the FRC's authority, claiming that its signal was not interstate commerce. The court accepted the FRC's argument that it had to have authority over all transmissions, even those that did not cross state lines, because otherwise it could not control the interference that might affect other, regulated stations. Leery of raising questions of states' rights, Congress had avoided the issue in the Radio Act of 1927. In 1933 the court extended FRC authority to cover all radio transmissions.

For the most part, the FRC fared quite well in challenges to its regulatory authority. To be sure, there were cases that the FRC did not win, but it certainly won many more than it lost. The Commission's ability to exercise regulatory authority over the broadcast spectrum became greater with each legal decision. In two highly visible cases, the FRC was able to deny license renewals for stations that had not acted in the public interest. In 1931 the court upheld the FRC in denying a renewal to Dr. John R. Brinkley's station, KFKB (Brinkley had been using his station to prescribe medications). Brinkley claimed that the FRC had no authority to censor him. The court held that the FRC was not engaging in censorship by examining the record to determine if a licensee had acted in the public interest. The following year, Dr. Robert Schuler of Trinity Methodist Church appealed the denial of his station license, KGEF, claiming that his free speech rights had been violated. The court rejected that argument on the premise that the Commission was not preventing Schuler from making his vitriolic comments. As with Brinkley, the FRC could use his past record as an indication of how he would serve the public interest.

Defining the Public Interest

Included in the Radio Act of 1927 was the stipulation that the FRC would act "as public convenience, interest, or necessity requires." Likewise, stations were to be licensed to serve the same public interest, convenience, and necessity. Congress borrowed the language from other legislation regulating public utilities. This vague directive from Congress served as the guiding philosophy for the FRC. In one action in 1928, the FRC denied 62 license renewals and modified the operations of dozens more. A month later, the FRC issued a statement to explain its interpretation of the public interest standard and how it was to be applied. As the FRC pointed out in its statement, "no attempt is made anywhere in the act to define the term 'public interest, convenience, or necessity,' nor is any illustration given of its proper application" (FRC annual report, 166 [1928]). While asserting that a specific definition was neither possible nor desirable, the FRC set forth some general principles regarding the public interest. Perhaps most illustrative of all is the concluding sentence from the FRC's statement: "The emphasis must be first and foremost on the interest, the convenience, and the necessity of the listening public, and not on the interest, convenience, or necessity of the individual broadcaster or the advertiser." The language mirrors the sentiment of Rep. Wallace H. White, cosponsor of the bill that became the Radio Act of 1927, who said:

> We have reached the definite conclusion that the right of our people to enjoy this means of communication can be preserved only by the repudiation of the idea underlying the 1912 law that anyone who will may transmit—and by the assertion in its stead of the doctrine that the right of the public to service is superior to the right of any individual to use the ether (67th Cong. Rec. 5479 [1926]).

The FRC determined that the public interest is served by having a "substantial band of frequencies set aside for the exclusive use of broadcasting stations and the radio listening public." It also adopted the general premise that the greatest good is served by minimizing interference. It follows from this that denying licenses in order to prevent interference, detrimental though it may be to some prospective broadcasters, serves the greatest good. The commission stated its intent to use the past record of a licensee to determine whether that station is deserving of a license. The reliability of a station's transmissions were also to be considered. Stations that could not be relied upon to transmit at regularly scheduled, announced times or whose transmission frequencies wandered around the spectrum did not serve the public interest. More than 70 years later, these interpretations of serving the public interest are still considered valid.

Also included in the 1928 statement on the public interest was the principle that stations should operate at different classes of service in order to ensure that there would be some stations serving larger geographic areas, while other stations served only small communities. This coincided with Congress' view, which had been stated earlier that same year when the Radio Act of 1927 was amended by the Davis Amendment in 1928. In addition to extending the FRC's authorization beyond its original year, the amendment directed the FRC to devise a scheme for providing equitable radio services in all zones of the country. One week following the FRC's statement on the public interest, it issued a plan for providing different classes of radio service. Each of the nation's zones would have an equal number of channels assigned as clear channels, regional channels, and local channels. Clear channels were designated to be high-power stations audible at a distance, whereas regional and local channels had decreasing coverage areas. Eight clear channels, seven regional channels, and six local channels were assigned to each of the five zones. The FRC's basic concept of clear, regional, and local channels remained in force more than six decades later.

One of the public-interest principles established by the FRC that did not continue with the FCC was the concept that licenses should not be provided to stations that offer services that duplicate those already available. The FRC stated that simply playing phonograph records on the air does not provide the listening public with anything that it cannot otherwise obtain. The FRC would have maintained a licensing scheme that would compare the programming intentions of the applicants. It can also be inferred that the FRC did not favor licensing stations whose formats mirrored those of stations already in the community. Based on today's radio business, that policy clearly did not survive

DOM CARISTI

See also Brinkley, John R.; Censorship; Clear Channels; Controversial Issues; First Amendment and Radio; Frequency Allocation; Hoover, Herbert; Licensing; Localism in Radio; Public Interest, Convenience, or Necessity; Regulation; United States Congress and Radio; White, Wallace H.; Wireless Acts of 1910 and 1912/Radio Acts of 1912 and 1927

Further Reading

Benjamin, Louise M., "In the Public Interest: The Radio Act and Actions of the Federal Radio Commission," Chapter 5 in *Freedom of the Air and the Public Interest,* Carbondale: Southern Illinois University Press, 2001

Bensman, Marvin R., "Regulation Under the Act of 1927," Chapter 5 in *The Beginning of Broadcast Regulation in the Twentieth Century,* Jefferson, North Carolina: McFarland, 2000

Davis, Stephen Brooks, *The Law of Radio Communication,* New York: McGraw-Hill, 1927

Federal Radio Commission, "Annual Report," Washington, D.C.: Government Printing Office, 1927–33; reprint, New York: Arno Press, 1971

Goodman, Mark, and Mark Gring, "The Ideological Fight Over Creation of the Federal Radio Commission in 1927," *Journalism History* 26 (Autumn 2000)

Schmeckebier, Laurence F., *The Federal Radio Commission: Its History, Activities and Organization,* Washington, D.C.: Brookings Institution, 1932

Slotten, Hugh R., "Radio Engineers, the Federal Radio Commission, and the Social Shaping of Broadcast Technology," chapter 2 in *Radio and Television Regulation: Broadcast Technology in the United States, 1920–60,* Baltimore, Maryland: Johns Hopkins University Press, 2000

U.S. House of Representatives, Committee on Merchant Marine and Fisheries, *Federal Radio Commission: Hearings,* 70th Cong., 2nd Sess. (January–February 1929)

Female Radio Personalities and Disk Jockeys

Radio's female pioneers, although limited by the social conventions of the times, were quite successful in the early days of radio. Since radio's inception, female personalities have sometimes been limited by legal, economic, and social constraints, as well as by their own perceptions of what is expected of them by listeners and the industry. As radio became more established, women had less formalized input into programming decisions, yet they have had a significant impact on the radio industry. Indeed, a congressional study in the early 1990s concluded that women-owned radio stations were 20 percent more likely to air women's programming than male-owned stations were and 30 percent more likely than non-minority-owned stations were (see Halonen, 1992).

Donna Halper (2001) is one of the few radio historians to track the lives of the early female pioneers: Eunice Randall, Emilie Sturtevant, Marie Zimmerman, Eleanor Poehler, and Halloween Martin. Randall made her radio debut in 1918 and became one of the nation's first female announcers. Sturtevant was one of Boston's first radio programmers in the 1920s. Zimmerman and Poehler were two of the first female radio station managers. The first woman to command Chicago morning radio in the 1920s was Halloween Martin, long before programmers realized the importance of that time slot.

Soap and Sisterhood

By the 1940s, female talk show hosts became popular role models for their female listeners, who sought advice on children, relationships, and detergents, as well as a bit of celebrity gossip and companionship. Television, ironically, seemed to supply an abundant pool of female hosts to the radio networks. While many female listeners tuned to Mary Margaret McBride and Kate Smith, others enjoyed the popular soap operas. The success of female radio personalities was often evaluated in terms of their ability to move merchandise.

By 1954, as the industry's reliance on the radio networks declined, female disc jockeys (although merely a handful across the nation) began to redefine their role in radio. Martha Jean "The Queen" Steinberg was one of the first women in the nation to make the leap from being hostess of a homemaker show to being a rhythm-and-blues disc jockey on Memphis' legendary WDIA-AM. Also across Memphis airwaves, WHER-AM became the first all-girl radio station in the U.S. in 1955. Many of the disc jockeys remained at the station through 1972. WHER was the brainchild of station owner and record producer Sam Phillips. Apart from these exceptions, however, radio disc jockeys were predominantly white and male. With the emergence of rock and roll, these men attempted to out-shock their competitors in the quest for the most listeners, and in the process, they gained power and control over music selection and programming. Women faded behind the scenes for the next decade.

Sexpot Radio

In the mid-1960s, New York's WNEW-FM created "Sexpot Radio," an all-female lineup of disc jockeys. From this experiment, which lasted only 18 months, Alison Steele became known as the "Nightbird." Her legendary sultry voice captivated night audiences as she read poetic and Biblical verses in between music and interviews with rock stars. Steele, who died in 1995, was a member of the Rock and Roll Hall of Fame and the first woman to receive *Billboard Magazine*'s FM Personality of the Year award in 1976. Unlike Steele, many women entered radio in the 1970s as part-time reporters and weekend announcers—after women's groups pressured the Federal Communications Commission (FCC) to revise its affirmative action policies. With free-form FM radio giving way to increasingly segmented formatting, women were often defined, not by their individuality, but within the context of a male-dominated morning show.

Before the turn of the decade, another radio station would once again provide new opportunities for women. Like WHER, it had a female staff and male owner. Connecticut's

WOMN-AM debuted in January 1979. It aired several news stories about women and played numerous female artists throughout the day. The short-lived effort paved the way for less stereotypical advertisements and created a demand for more female artists by radio stations across the nation.

Strong Talk in the 1980s

Music-oriented stations decreased their news and public service commitments and increased the number of songs played per hour. Still, there were signs that "chick" talk would become a valuable commodity in the industry in the years ahead. By the late 1970s, Sally Jessy Raphael had established herself as a popular radio talk show host, broadcasting her advice on love, family, and relationships to a late-night audience at a time when talk radio began to boom. Concurrently, television talk shows began to attract large female audiences away from radio. With every station playing the same songs, personality once again became an integral part of the programming mix.

Strong female personalities, such as Washington, D.C.'s Robin Breedon, soared in their ratings, past their white male competitors. With a decade of television broadcast experience and a degree from Howard University, Breedon became the number-one Arbitron-ranked morning personality in D.C. She proved that many listeners were seeking personalities with compassion and community commitment. *The Washington Post* referred to her as the "Queen of Radio." During her ten-year radio stint, she won seven Emmy nominations and two American Women in Radio and Television National Awards; then-Mayor Sharon Pratt-Kelly even proclaimed a day named in her honor. Breedon left radio in 1998.

By 1988 Howard Stern was on his way to becoming a national icon. Robin Quivers, Stern's articulate sidekick, became a dominant part of what was becoming known as "shock radio." As early as the 1980s, many women entered radio as sidekicks who typically read the morning headlines and provided a laugh track for male-dominated morning shows. In some cases, these women also became the targets of sexist jokes told by male hosts.

The Rise of the Shockette Jock

The real revolution was happening in Rhode Island. Carolyn Fox, Providence's number-one afternoon personality, spouted her liberal views on everything from sex to politics—before Howard Stern even gained national prominence. Fox paved the way for a number of women, such as San Francisco's Darian O'Toole, Austin's Sara Trexler, Denver's Caroline Corley, and Detroit's Kelly Walker, who became the new shock jocks of the 1990s. Many of these women began their careers as sidekicks or night personalities. Trexler, who began her career in 1986, was selected as *Billboard Magazine*'s Small Market Local Air Personality of the Year in 1999. Emulating their male predecessors, many of these shock jockettes featured the same type of locker-room humor, but from a female perspective.

Karin Begin (a.k.a. Darian O'Toole) has been billed as America's First Shockette. Although Carolyn Fox is her predecessor by more than a decade, the press and the radio industry has portrayed Begin as an American trailblazer for female radio broadcasters. Begin was born and raised in Nova Scotia and worked at a number of small Canadian markets before landing in the United States.

The turning point in her radio career came when she met Program Director Shawn Kelly, a big Howard Stern fan, who encouraged Begin to seek opportunities in the United States. Some of her early gigs included on-air stints in New Jersey, Philadelphia, Baltimore, and Sacramento. In San Francisco, she would become known as Darian O'Toole—the "Caustic Canadian Swamp Witch." Her morning show skyrocketed to number one in San Francisco, climbing from 23rd to 1st place in the market in only three years. She left San Francisco's KBIG in 1997 to take her show to New York. In September 1999 she returned to San Francisco to work at KSAN.

Alternate Models for Female Personalities

The ability to move merchandise, whether soap or soda, will always remain an essential part of commercial radio, for obvious reasons. With every product sold on radio, early female broadcasters knew that their bargaining power would increase tenfold. Some commercial gimmicks to market the female experience, such as the rise and fall of "Sexpot Radio" or the all-women radio station WOMN, have quickly failed within the past 40 years. On the other hand, the Seattle-based syndicated nighttime personality Delilah, a 25-year veteran, is taking a new spin on an old formula—a mix of advice and inspiration to a predominantly female audience, with listeners calling in to her show from more than 200 affiliates around the nation. Reminiscent of Casey Kasem's long-distance dedications, but a bit more personal, she is very much like Mary Margaret McBride in her desire to chat about love, family, and relationships. In October 1998, she announced her pregnancy and promised to share her experience with her listeners. Many American listeners seem comforted by female radio personalities who symbolize traditional family values. Dr. Laura Schlessinger, a controversial conservative talk show host in the 1990s, starts her weekday show by saying "I'm my kid's mom." In the final analysis, there has never been one personality style that has worked for all female broadcasters; rather, the means of success has been their ability to connect to the listeners—both men and women—in some unique way.

PHYLIS JOHNSON

See also American Women in Radio and Television; Association for Women in Communication; McBride, Mary Margaret; WHER; Women in Radio

Further Reading

Baehr, Helen, and Ann Gray, editors, *Turning It On: A Reader in Women and Media,* London and New York: Arnold, 1996

Broadcast Archive: Radio History on the Web, <www.oldradio.com>

Carter, Kevin, and Marc Schiffman, "More Women Talk the Shock Talk: Different Limits Apply to Female Jocks," *Billboard* 110, no. 3 (1998)

Douglas, J.C., "The Darian Deal," *Halifax Daily News* (4 October 1998)

Gaar, Gillian G., *She's a Rebel: The History of Women in Rock and Roll,* Seattle: Seal Press, 1992

Halonen, Doug, "Court Axes FCC's Gender Policy," *Electronic Media* (24 February 1992)

Halper, Donna L., *Invisible Stars: A Social History of Women in American Broadcasting,* Armonk, New York: M.E. Sharpe, 2001

Hinckley, David, "For O'Toole, Old Saw 'Try Again' Applies," *New York Daily News* (3 August 1998)

Smith, Wes, *The Pied Pipers of Rock 'n' Roll: Radio Deejays of the 50s and 60s,* Marietta, Georgia: Longstreet Press, 1989

Unesco, *Women and Media Decision-Making: The Invisible Barriers,* Paris: Unesco, 1987

Fessenden, Reginald 1866–1932

Canadian Electrical Engineer and Wireless Inventor

Reginald A. Fessenden was a seminal figure in the development of wireless telephony technology, the first important North American inventor to experiment with the wireless telephone, the immediate precursor of radio. Born in Canada, he undertook his most important inventive work in the United States, including the development of continuous-wave transmission and heterodyne principals (he was granted 229 U.S. patents from 1891 to 1936), and he conducted what was probably the world's first broadcast.

Origins

Fessenden was born in 1866 in what would become the Canadian province of Quebec, the first of four sons of an Episcopalian minister. His parents supported his drive for education, and he excelled in school, especially in math and science, earning an invitation from his father's alma mater, Bishop's College, to teach math and languages, although because of doing so, he never completed his own degree work. While teaching, he became increasingly fascinated with the scientific journals of the time, focusing especially on developments in electricity. He continued his teaching at a secondary education institute in Bermuda for another year or so.

There were several key turning points in Fessenden's professional life, and the first came in 1886 when he decided to leave secondary (high school) teaching and become involved more directly in the field of electrical engineering, in which he was largely self-taught. He began as a field tester with Thomas Edison's company, which was then wiring the streets of New York City. Within a year he was working in power engineering at Edison's New Jersey laboratory, a post he held for nearly three years, from 1887 to 1890. Here he learned by observation the importance of patents and the scientific method—but apparently not the importance of the market or of the process of successful innovation. He was increasingly attracted to the study of Hertzean waves, reading laboratory journals in his spare time. Financial problems at the labs led to layoffs of many workers, including Fessenden. His practical experience in electricity continued as an electrical assistant at the Westinghouse subsidiary, the United States Company, in 1890. Just a year later, in 1891, he joined the Stanley Company of Pittsfield, Massachusetts.

Though he lacked academic credentials, Fessenden had by the early 1890s already published in respected journals. Based on his record, he was named a professor of electrical engineering at Purdue University in 1892, an institution then striving to develop a reputation in this field. After a year at Purdue, in 1893 Fessenden accepted an offer to move to the University of Western Pennsylvania (later the University of Pittsburgh) to occupy a new chair of electrical engineering. Half of his salary was paid by Westinghouse, which hoped to make use of Fessenden for its own research needs (he undertook incandescent

U.S. President Franklin D. Roosevelt moments before his fireside chat on 12 March 1933
Courtesy AP/Wide World Photos

Roosevelt was the first president to make extensive and continuous use of modern electronic means to speak directly to his constituents. This capability has contributed to the increased power of the executive branch of government. Roosevelt's use of radio allowed him to influence the national agenda and to counter opposing newspaper editorials. His radio addresses played a major role in his popular image, in his being elected four times to the presidency of the United States, and in the success of his efforts to lead the nation through the years of economic depression and world war.

B.R. SMITH

See also Politics and Radio; United States Presidency and Radio

Further Reading

Becker, Samuel I., "Presidential Power: The Influence of Broadcasting," *Quarterly Journal of Speech* (February 1961)

Braden, Waldo W., and Earnest Brandenburg, "Roosevelt's Fireside Chats," *Speech Monographs* 22, no. 5 (November 1955)

Brandenburg, Earnest, "The Preparation of Franklin D. Roosevelt's Speeches," *Quarterly Journal of Speech* 25 (1949)

Brown, Robert J., "The Radio President," in *Manipulating the Ether: The Power of Broadcast Radio in Thirties America*, by Brown, Jefferson, North Carolina: McFarland, 1998

Chester, Edward W., *Radio, Television, and American Politics,* New York: Sheed and Ward, 1969

Roosevelt, Franklin D., *FDR's Fireside Chats,* edited by Russell D. Buhite and David W. Levy, Norman: University of Oklahoma Press, 1992

Rosenman, Samuel Irving, *Working with Roosevelt,* New York: Harper, 1952

Schlesinger, Arthur Meier, *The Imperial Presidency,* Boston: Houghton Mifflin, 1973

Smith, B.R., "FDR's Use of Radio in the War Years," *Journal of Radio Studies* 4 (1997)

Winfield, Betty, *FDR and the News Media,* Urbana: University of Illinois Press, 1990

First Amendment and Radio

The First Amendment (1791) to the U.S. Constitution provides, in part, that Congress shall make no law abridging the freedom of speech or of the press. Yet in spite of this proscription, there exist a number of regulations that limit free expression on radio. The very fact that stations must be licensed is a restriction that would be considered clearly unconstitutional if it were applied to print media. On the other hand, the courts have stated that radio broadcasting is entitled to First Amendment protection. The amount of protection is less than that enjoyed by print media, but it is still significant.

The fact that the First Amendment protections extend to radio as well as the press was made clear by a 1948 Supreme Court decision, *United States v Paramount Pictures,* which stated, "We have no doubt that moving pictures, like newspapers and radio, are included in the press whose freedom is guaranteed by the First Amendment." In order to best understand what free expression rights are due to radio, one needs to examine the rationale for regulating radio. Courts and legal scholars have provided a variety of arguments for regulation that fit into four general categories: scarcity of broadcast frequencies, the broadcast spectrum as a public resource, the need to alleviate interference, and the pervasiveness and power of the broadcast media.

Scarcity

In a 1984 decision, the Supreme Court stated, "The fundamental distinguishing characteristic of the new medium of broadcasting that, in our view, has required some adjustment in First Amendment analysis is that broadcast frequencies are a scarce resource that must be portioned out among applicants." This scarcity rationale has undergone a number of attacks in recent years with the proliferation of media, but in fact it is still considered a valid regulatory rationale. Although the number of radio stations (as well as the number of most other media outlets) has increased, courts continue to accept a scarcity rationale. The reason for this is that scarcity does not depend on the number of existing media outlets but rather on the determination of whether a new applicant stands a good chance of entry to the market. A vast number of existing media outlets implies that diversity exists, not that scarcity has been eliminated. If new applicants want to obtain station licenses and are unable to do so, that implies scarcity. Scarcity is a function of the number of people desiring a station to the number of stations available. As long as applicants exceed available frequencies, scarcity exists. As the Supreme Court noted in its *Red Lion* decision in 1969, "When there are substantially more individuals who want to broadcast than there are frequencies to allocate, it is idle to posit an unabridgeable First Amendment right to broadcast comparable to the right of every individual to speak, write, or publish." This is why there will never be scarcity for newspaper publishers, no matter how many newspapers are published in the United States. In theory, at least, any American can start a newspaper (at least there is no legal restriction). The same is not true for starting a radio station.

In 1943 the Supreme Court supported the notion that scarcity entitled the Federal Communications Commission (FCC) to make judgments about who would best serve the public interest. In *National Broadcasting Company v United States,* the Court rejected the argument that chain broadcasting rules were a violation of the First Amendment:

> If that be so, it would follow that every person whose application for a license to operate a station is denied by the Commission is thereby denied his constitutional right of free speech. Freedom of utterance is abridged to many who wish to use the limited facilities of radio. Unlike other modes of expression, radio is not inherently available to all. That is its unique characteristic, and that is why, unlike other modes of expression, it is subject to governmental regulation. Because it cannot be used by all, some who wish to use it must be denied. But Con-

> gress did not authorize the Commission to choose among applicants upon the basis of their political, economic or social views, or upon any other capricious basis. . . . The licensing system established by Congress in the Communications Act of 1934 was a proper exercise of its authority over commerce. The standard it provided for the licensing of stations was the "public interest, convenience, or necessity." Denial of a station license on that ground, if valid under the Act, is not a denial of free speech.

Public Resource

In 1962 President John F. Kennedy referred to the broadcast spectrum as a "critical natural resource." The federal government typically regulates the use of natural resources to ensure that they are not damaged and that their use is in the public interest. Viewing the spectrum as a public resource results in a philosophy that views users of the public resource as public trustees, who as such can be expected to act according to the dictates of those allowing them to use the resource. The government could have adopted other models for rationing spectrum, but it didn't. The assumption is that those who use the public resource have some degree of public service obligation.

A good example of this requirement is the demand, found in Section 312 of the Communications Act, that broadcasters provide reasonable access to candidates for federal office. This affirmative obligation on broadcasters, which would be unconstitutional if applied to print media, can only be justified under a public-resource rationale. In the 1981 Supreme Court Decision *Columbia Broadcasting System v Federal Communications Commission,* the Court wrote that such a rule "represents an effort by Congress to assure that an important resource—the airwaves—will be used in the public interest. [The rule] properly balances the First Amendment rights of federal candidates, the public, and broadcasters."

Interference

Undoubtedly the oldest of the regulatory rationales is the assertion that the government must regulate the broadcast spectrum in order to prevent interference. This was provided as rationale for the passage of the Radio Acts in 1912 and 1927. Failure to limit interference would result in a "cacophony" in which no one would be heard. Thus, the government exercises its authority to limit the free speech of some so that others might be heard. Some might contend that interference and scarcity are actually the same rationale, when in fact they are different. Their connection in broadcast contexts is understandable, because the spectrum is subject to both scarcity and interference. It is possible, however, to have interference when there is no physical scarcity. It is interesting to note that the Supreme Court's 1969 *Red Lion* decision quoted a 1945 print media case involving the Associated Press when it stated "the right of free speech . . . does not embrace a right to snuff out the free speech of others." Clearly, there can be interference without scarcity. In *Red Lion,* the Court stated:

> When two people converse face to face, both should not speak at once if either is to be clearly understood. But the range of the human voice is so limited that there could be meaningful communications if half the people in the United States were talking and the other half listening. Just as clearly, half the people might publish and the other half read. But the reach of radio signals is incomparably greater than the range of human voice and the problem of interference is a massive reality. The lack of know-how and equipment may keep many from the air, but only a tiny fraction of those with resources and intelligence can hope to communicate by radio at the same time if intelligible communication is to be had, even if the entire radio spectrum is utilized in the present state of commercially acceptable technology.

The Supreme Court justified broadcast regulation, in part at least, because of broadcasting's unique physical characteristics.

Pervasiveness and Power

Perhaps most controversial of all the rationales, this claim asserts that broadcast media should be regulated because of the media's unique role in the lives of Americans. In the famous *Pacifica* case (dealing with George Carlin's "Seven Dirty Words" monologue), the Supreme Court stated that "the broadcast media have established a uniquely pervasive presence in the lives of all Americans." Yet no one would attempt to assert that a small-town radio station has more pervasiveness and power than, say, *The New York Times*. Perhaps a more appropriate term for the Court to have used would have been *invasive* rather than *pervasive*. The Court seemed to be influenced by the fact that radio transmissions come into the privacy of one's home and automobile and are instantly available to children, unlike newspapers, which wait outside our homes for us to collect them and are unreadable by children still too young to read. In *Columbia Broadcasting System v Democratic National Committee* in 1973, the Court stated a concern dating back to the 1920s that radio's audience is in a sense "captive" because it cannot simply ignore the messages sent by broadcasters.

It is this rationale that supports limits on broadcast indecency. FCC rules that restrict the use of indecent language during certain hours of the broadcast day (6 A.M. to 10 P.M.) are based on the premise that the audience will consist of a number of minors who should not be subjected to indecent language. Allowing the restriction of indecent material on the air is a recognition of broadcasting's pervasive nature.

First Amendment Protections

In spite of the regulations that do exist, radio is not without First Amendment rights. Section 326 of the Communications Act specifically states:

> Nothing in this Act shall be understood or construed to give the Commission the power of censorship over the radio communications or signals transmitted by any radio station, and no regulation or condition shall be promulgated or fixed by the Commission which shall interfere with the right of free speech by means of radio communication.

Although some might contend that the Section 326 provision is rendered either superfluous by the First Amendment or invalid by rules such as those limiting indecency, the courts have continued to support the general principle that the FCC may not censor broadcasts.

Radio stations also have the right to decide who uses their facilities. The Supreme Court has unequivocally stated that the need to serve the public interest does not require that broadcasters provide access for individuals or organizations. Those who would like to present their positions on public issues have ample opportunity to do so without a government requirement that stations afford them airtime.

The Supreme Court has suggested that the balance between the First Amendment rights of broadcasters and the need for government regulation is not static and that changing conditions might warrant a change in the balance between the two. In the 1973 decision *Columbia Broadcasting System v Democratic National Committee,* the Court stated, "the history of the Communications Act and the activities of the Commission over a period of 40 years reflect a continuing search for means to achieve reasonable regulation compatible with the First Amendment rights of the public and the licensees." Eleven years later, in *Federal Communications Commission v League of Women Voters,* the Supreme Court made it even more clear that regulatory rationales were open to review and revision. In two rather significant footnotes, the Court signaled its willingness to accept a regulatory scheme that was less demanding of broadcasters. In addressing the scarcity rationale, the Court wrote, "We are not prepared, however, to reconsider our longstanding approach *without some signal from Congress or the FCC* that technological developments have advanced so far that some revision of the system of broadcast regulation may be required" (emphasis added). Although the Court was not prepared to lay the fairness doctrine or the scarcity rationale to rest, it opened the door for others to do so. After a series of legal actions, the FCC did in fact eliminate the fairness doctrine.

The entire concept of treating broadcast differently from print media has been challenged for some time, but the practice continues. Modifications have been made, and radio has significantly fewer regulations today than it had prior to the deregulation movement that began in the 1970s. Nonetheless, some would assert that the changing nature of mass media will make it more difficult to have different regulatory schemes based on modes of transmission. With media converging as they are, will regulatory policies that treat media differently based on modes of transmission be able to survive? In an era in which both newspaper and radio messages can reach their audience via the internet, should one be regulated differently from the other? These are questions that have been posed for decades, yet our regulatory policy remains essentially unchanged. Broadcast media are subject to regulation based on the four regulatory rationales stated above, while the print media are largely unregulated. The amount of regulation that will be tolerated is subject to the balancing engaged in by the Supreme Court, but radio (along with television) continues to be subject to regulation.

DOM CARISTI

See also Communications Act of 1934; Federal Communications Commission; Federal Radio Commission; Frequency Allocation; Network Monopoly Probe; Public Interest, Convenience, or Necessity; Red Lion Case; Seven Dirty Words Case; United States Congress and Radio; United States Supreme Court and Radio

Further Reading

Associated Press v United States, 326 U.S. 1 (1945)

Bittner, John R., *Broadcast Law and Regulation,* Englewood Cliffs, New Jersey: Prentice Hall, 1982; 2nd edition, as *Law and Regulation of Electronic Media,* 1994

Caristi, Dom, *Expanding Free Expression in the Marketplace: Broadcasting and the Public Forum,* New York: Quorum Books, 1992

Carter, T. Barton, Marc A. Franklin, and Jay B. Wright, *The First Amendment and the Fifth Estate: Regulation of Electronic Mass Media,* 5th edition, New York: Foundation Press, 1999

Columbia Broadcasting System v Democratic National Committee, 412 U.S. 94 (1973)

Emery, Walter B., *Broadcasting and Government: Responsibilities and Regulations,* East Lansing: Michigan State University Press, 1961; revised edition, 1971

Federal Communications Commission v League of Women Voters of California, 468 U.S. 364 (1984)

Federal Communications Commission v Pacifica Foundation, 438 U.S. 726 (1978)

Ginsburg, Douglas H., *Regulation of Broadcasting: Law and Policy towards Radio, Television, and Cable Communications,* St. Paul, Minnesota: West, 1979; 2nd edition, as *Regulation of the Electronic Mass Media: Law and Policy for Radio, Television, Cable, and the New Video Technologies,* by Ginsburg, Michael Botein, and Mark D. Director, 1991

National Broadcasting Company v United States, 319 U.S. 190 (1943)

Powe, Lucas A. Scot, *American Broadcasting and the First Amendment,* Berkeley: University of California Press, 1987

Red Lion Broadcasting Company v Federal Communications Commission, 395 U.S. 367 (1969)

United States v Paramount Pictures, 334 U.S. 131 (1948)

Fleming, Sir John Ambrose 1849–1945

British Electrical Engineer; Inventor of the Vacuum Tube

Sir John Ambrose Fleming led an active scientific life. His career covered the time from James Clerk Maxwell to the advent of electronic television. Fleming has been described as the scientific and technical link between Maxwell and Guglielmo Marconi. He was an outstanding teacher and highly successful popular scientific lecturer. He published more than 100 important papers on his discoveries. Fleming is best known for the thermionic vacuum tube or valve, the first electron tube that could change alternating current, such as a radio wave, to pulsating, one-way flow direct current. Fleming's diode improved radio reception and was a forerunner of the triode tube developed by Lee de Forest. Although the transistor eventually replaced Fleming's valve, his valve remained an important component of radios for nearly three decades and was used in the early days of computers and television.

Fleming began his study of electricity and mathematics under James Clerk Maxwell in the new Cavendish Laboratory at St. John's College in Cambridge. During his studies there, Fleming worked on improving the Carey Foster Bridge, a method for measuring the difference between two nearly equal resistances in electrical conduction. Fleming's improvement made the measuring device faster and more accurate. Maxwell labeled the device "Fleming's banjo" because of the measuring wire's circular shape.

After receiving his doctor of science degree in 1880, Fleming worked as a consultant for private industry. His consulting work resulted in many new methods and instruments for measuring high-frequency currents and new transformer designs. Fleming was a primary contributor to the development of electrical generator stations and distribution networks for several companies, including the London National Company, the Edison Telephone and Electric Light Companies, and the Swan Lamp Factory. For both the Swan and Edison companies, Fleming lent his expertise to photometry and helped develop the large-bulb incandescent lamp that used an aged filament as the light source.

In 1899 the Marconi Wireless Telegraph Company hired Fleming as a scientific adviser to help design the Poldhu wireless station in Cornwall, England. This was the largest wireless station in England and the source of the first transatlantic wireless telegraph transmission in 1901.

Fleming's most important contribution to electrical engineering was his vacuum tube, widely used in both radiotelegraphy and radiotelephony. In 1888 Thomas Edison announced his "Edison effect," which described how electronic particles were emitted from a hot electric lamp filament. Fleming had repeated Edison's findings in 1899 but had found little practical use for Edison's discovery.

In 1904, however, while searching for a more efficient and reliable detector of weak electrical currents, Fleming was inspired to make a new lamp, or valve, that would have a hot filament and an insulated plate sealed inside a high vacuum tube. When a current was passed through the carbon or metal filament, the rarefied air between the hot filament and the cold plate filled with electrons and became a conductor of electricity. He found that the electrons would travel only when the plate was attached to the positive terminal of a generator and that the plate would attract the negatively charged electrons. Fleming also noticed that this flow of electrons was in only one direction, from the hot filament to the cold metal plate, and not in the reverse direction. Alternating current would enter the device, but direct current would leave. Fleming had converted alternating-current radio signals into weak direct-current signals that could be heard with a telephone receiver. This was a major advance in radio technology. Fleming called his discovery a thermionic valve or tube, because it acted much like a check valve, which allows fluids to flow in only one direction. Eventually, the device was labeled "Fleming's valve." It provided the first truly reliable method to measure high-frequency radio waves. Fleming patented his valve in 1904. This discovery revolutionized radio telegraphy communication technology. The vacuum tube was the foundation of electronics until the 1960s, when solid-state technology was developed, replacing vacuum tubes in most electronic devices.

In 1906 Lee de Forest added a third element to Fleming's diode valve, thus effectively separating the high-frequency circuit from that of the filament, making amplification of radio signals possible. Litigation of the de Forest and Fleming patents continued for years. Court decisions in 1916 tied most companies into knots. As the United States entered World War I, the navy offered to indemnify all manufacturers of radio apparatus for the armed forces against any resulting patent infringement suits. This pooling of all patents enabled manufacturers to produce modern equipment without fear of lawsuits. Patent disputes between de Forest and Fleming were not fully resolved until after American Telephone and Telegraph (AT&T) bought de Forest's Audion patent, the Radio Corporation of America (RCA) acquired rights to the Fleming valve, and AT&T and RCA entered into a cross-licensing agreement in 1920.

Throughout his long career, Fleming lectured often at University College, the Royal Institution, and the Royal Society of Arts. He published extensively and presented several important research papers at learned societies' conferences. After Fleming read a paper on the need for an authoritative body for electrical standards for the burgeoning electric lighting industry, the Board of Trade Laboratory and eventually the National Physical Laboratory were established in Great Britain. Fleming is credited with developing a direct-reading potentiometer, set to read current and potential directly in amperes and volts, and with encouraging R.E.B. Crompton to put it on the market in a practical form.

Fleming placed his long scientific career into perspective when he wrote that in comparing the last half of the 19th century and the first third of the 20th century, there was an enormous increase in practical technical achievement, despite the diminished confidence we now have in the validity of our theoretical explanations of natural phenomena.

His peers regarded Fleming very highly because of his extraordinary devotion to his work. He never lost sight of the potential for wireless. He wrote that

> radiotelegraphy has not only given to mankind a superlatively beneficial means of communication, but has also opened up for discussion physical and cosmical problems of profound interest. . . . We are only at the very beginning of this evolution, yet it has already completely revolutionised the practical side of wireless telegraphy, as well as telephony (Fleming, 1921).

PETER E. MAYEUX

See also De Forest, Lee; Early Wireless; Marconi, Guglielmo; Maxwell, James Clerk

John Ambrose Fleming. Born in Lancaster, England, 29 November 1849. Attended University College, London, B.S., 1870; Royal School of Mines, South Kensington, 1872–74; St. John's College, Cambridge, Doctor of Science degree, 1880; worked as clerk in stockbrokerage firm, 1868–70; science master, Rosall School and Cheltenham College, 1872–74; chair of mathematics and physics, University College, Nottingham, 1881; consultant, Edison Telephone and Electric Light Company, Swan Lamp Factory, and London National Company, 1882; elected fellow, St. John's, 1883; professor, electrical technology (engineering), University College, 1885–1926; scientific advisor, Marconi Wireless Telegraph Company, 1899; developed the Fleming valve, 1904; appointed professor emeritus, University College, 1926; president, Television Society of London, 1930–45; received Royal Society fellowship, 1892; Hughes Gold Medal, Royal Society of London, 1910; Albert Medal, Royal Society of Arts, 1921; Faraday Medal, British Institution of Electrical Engineers, 1928; knighthood, 1929; Duddell Medal, Physical Society, 1931; Gold Medal, Institute of Radio Engineers (U.S.), 1933; Franklin Medal, Franklin Institute (U.S.), 1935; Kelvin Medal, 1935; elected honorary fellow, Cambridge, 1927; received honorary degree of D.Eng., Liverpool University, 1928. Died in Sidmouth, Devon, England, 18 April 1945.

Selected Publications

Alternate Current Transformer in Theory and Practice, 1888, 1895
Elementary Manual of Radiotelegraphy and Radiotelephony for Students and Operators, 1908
Principles of Electric Wave Telegraphy and Telephony, 1906
Propagation of Electric Currents in Telephone and Telegraph Conductors, 1911
Thermionic Valve and its Developments in Radiotelegraphy and Telephony, 1919
Wonders of Wireless Telegraphy Explained in Simple Terms for the Non-Technical Reader, 1914, 1919
Fifty Years of Electricity; The Memories of an Electrical Engineer, 1921
Memories of a Scientific Life, 1934

Further Reading

Aitken, Hugh G.J., *Syntony and Spark: The Origins of Radio,* New York: Wiley, 1976
Bondyopadhyay, P.K., "Fleming and Marconi: The Cooperation of the Century," *The Radioscientist and Bulletin* 5 (June 1994)
MacGregor-Morris, John Turner, *The Inventor of the Valve: A Biography of Sir Ambrose Fleming,* London: Television Society, 1954

MacGregor-Morris, John Turner, "Fleming, Sir (John) Ambrose," in *Dictionary of National Biography, 1941–1950*, edited by Leopold George Wickham Legg and Edgar Trevor Williams, London and New York: Oxford University Press, 1959

Sterling, Christopher H., and John M. Kittross, *Stay Tuned: A History of American Broadcasting*, 3rd edition, Mahwah, New Jersey: Lawrence Erlbaum, 2002

Tyne, Gerald F.J., *Saga of the Vacuum Tube*, edited by Diana D. Menkes, Indianapolis, Indiana: Sams, 1977

Flywheel, Shyster, and Flywheel

Radio Comedy Program

Although much of their fame rests on the dozen films they made between 1929 and 1950, the Marx Brothers, working together and as solo performers, enjoyed a measure of success in radio and later television broadcasting. The National Broadcasting Company's (NBC) weekly comedy *Flywheel, Shyster, and Flywheel* was the first network radio program to feature the Marx Brothers. Or, more accurately, it featured two of the four-member comedy team: Groucho and Chico. The remaining brothers—Harpo's silent clown and Zeppo's straight man—were less suitable for radio. Despite the fact that only 26 episodes were produced between November 1932 and May 1933, *Flywheel, Shyster, and Flywheel* opened up new avenues for the Marx Brothers' comic genius.

Flywheel, Shyster, and Flywheel's origins are typical of many programs produced for American radio during the early 1930s. Following on the heels of its rival's success with the *Texaco Fire Chief Program*, the Standard Oil Company sought a vehicle to promote its new product line: Esso gasoline and Essolube motor oil. Working with its advertising agency, McCann-Erickson, Standard Oil agreed to sponsor a weekly variety program called *Five Star Theater*. Every night of the week featured a different program: detective stories, dramas, musicals, and comedies. As Michael Barson notes in the introduction to his edited collection of the program's scripts, "the jewel of the enterprise was Monday night's entry, *Beagle, Shyster, and Beagle, Attorneys at Law*," which featured Groucho as Waldorf T. Beagle, a wisecracking ambulance chaser, and Chico as his incompetent assistant, Emmanuel Ravelli. Indeed, with four successful feature films to their credit—*The Coconuts, Animal Crackers, Monkey Business*, and *Horsefeathers*—landing even half of the Marx Brothers was quite a coup for Standard Oil.

Beagle, Shyster, and Beagle debuted on 28 November 1932 over the NBC Blue network. Although audience reaction is difficult to gauge, at least one listener, a New York attorney named Beagle, was not amused. Anxious to avoid a lawsuit, the network changed the name of Groucho's character to Flywheel and promptly altered the program's title accordingly. Not surprisingly, the scripts for *Flywheel, Shyster, and Flywheel* are characteristic of the Marx Brothers' penchant for rapid-fire one-liners, puns, putdowns, and malapropisms. And as in their movies, on radio the Marx Brothers had little regard for the rule of law or high society: few cherished American values or institutions were spared a "Marxist" skewering. For example, at the end of one episode, Flywheel (Groucho) advises a would-be philanthropist, "Instead of leaving half of your money to your children and the other half to the orphanage, why not leave your children to the orphanage . . . and the million to me?"

What is most significant about these scripts (the original programs were not recorded, but the majority of the show's transcripts survive in the Library of Congress) is their relationship to the Marx Brothers' film work. In some instances, entire routines from earlier films were reworked for *Flywheel, Shyster, and Flywheel*. For example, some episodes featured plot lines and dialogue taken from the Broadway hit and subsequent film *Animal Crackers*. Even the name of Chico's character, Emmanuel Ravelli, came directly from this film. Several scenes from *Monkey Business* found their way into episodes of the radio program as well. On the other hand, a number of *Flywheel, Shyster, and Flywheel* scripts foreshadowed the Marx Brothers' later film work. Of particular interest are early drafts of now archetypal routines and dialogue from the Marx Brothers' classic *Duck Soup*. The film's infamous trial sequence owes much of its funny business to a *Flywheel, Shyster, and Flywheel* script, as does Chico's hilarious recitation on his difficulties as a spy: "Monday I shadow your wife. Tuesday I go to the ball game—she don't show up. Wednesday she go to the ball game—I don't show up. Thursday was a doubleheader. We both no show up. Friday it rain all day—there'sa no ball game, so I go fishing." The name of Groucho's character, Waldorf T. Flywheel, would be recycled some years later in the 1941 film *The Big Store*.

The need for this recycling of old gags and testing of new material is understandable. Along with their writers, Nat Perrin and Alan Sheekman, Groucho and Chico soon grew tired of traveling cross-country from Hollywood to New York to do a weekly radio program. In fact, in January 1933, *Flywheel, Shyster, and Flywheel* took the then unprecedented step of relocating its broadcast from WJZ in New York to Hollywood for a time. Still, the time constraints facing both writers and performers undoubtedly contributed to their willingness to borrow from established routines while refining others. By the middle of 1933, however, it was a moot point. *Flywheel, Shyster, and Flywheel* was taken off the air. Although its ratings were quite respectable, considering the less-than-desirable airtime of 7:30 P.M., the sponsors were disappointed with the show's performance.

Throughout the 1930s and 1940s, Groucho and Chico returned to the airwaves in various guises. In 1934 they were hired by the Columbia Broadcasting System (CBS) to spoof the latest news in a short-lived program called *The Marx of Time*. Both Groucho and Chico struck out on their own as well. Chico made a number of radio appearances as a musical accompanist and band leader, and Groucho served as host for programs such as *Pabst Blue Ribbon Town*. During the war years, the Marx Brothers, including Harpo, made guest appearances on the Armed Forces Radio Service. Of special note, however, is Groucho's role as the judge in Norman Corwin's fanciful courtroom drama from 1945, *The Undecided Molecule*. Groucho's true calling on radio came in 1947 as a quiz show host on *You Bet Your Life*. Curiously, this popular program shuffled between the radio networks before finding a permanent home on NBC.

In an odd but telling postscript, the British Broadcasting Corporation (BBC) began airing recreations of *Flywheel, Shyster, and Flywheel* in 1990. The programs proved quite popular with British audiences and have subsequently been picked up for broadcast in the United States through National Public Radio (NPR).

KEVIN HOWLEY

See also Comedy; You Bet Your Life

Cast

Waldorf T. Flywheel	Groucho Marx
Emmanuel Ravelli	Chico Marx

BBC Cast

Waldorf T. Flywheel	Michael Roberts
Emmanuel Ravelli	Frank Lazarus

Writers

Nat Perin, Arthur Sheekman, Tom McKnight, and George Oppenheimer

Programming History

NBC Blue	28 November 1932–22 May 1933
BBC	1990–92 (19 Episodes)

Further Reading

Adamson, Joe, *Groucho, Harpo, Chico, and Sometimes Zeppo: A History of the Marx Brothers and a Satire of the Rest of the World,* New York: Simon and Schuster, 1973

Barson, Michael, editor, *Flywheel, Shyster, and Flywheel: The Marx Brothers' Lost Radio Show,* New York: Pantheon, 1988

Marx, Groucho, *Groucho and Me,* New York: Geis, and London: Gollancz, 1959

Mitchell, Glenn, *The Marx Brothers Encyclopedia,* London: Batsford, 1996

FM Radio

Frequency modulation (FM) radio, more usually called VHF radio outside the United States, began with experiments in the 1920s and 1930s, expanded to commercial operation in the 1940s, declined to stagnation in the 1950s in the face of competition from television, resumed growth in the 1960s, and rose to dominance of American radio listening by the late 1970s. This entry focuses first on the basics of FM broadcasting and then explores the development of the service in the United States, where it was first invented and developed; finally, this essay turns to selective brief coverage of FM outside the U.S.

FM Basics

FM transmitters modulate a carrier wave signal's frequency rather than its amplitude. That is, the power output remains the same at all times, but the carrier wave frequency changes in relation to the information (music or talk programs, e.g.)

transmitted. Electronic static (most of which is amplitude modulated) may flow with but cannot attach to FM waves, which allows the desired FM signal information to be separated from most interference by special circuits in the receiver.

Because U.S. FM channels are each 200 kilohertz wide (allowing a wide frequency swing), a high-quality sound image is transmitted (up to 15,000 cycles per second—almost three times the frequency response of AM signals and close to the 20,000-cycle limit of human hearing), usually in multiplexed stereo. The cost for this sound quality is paid for in spectrum—each FM station takes up 20 times the spectrum of a single AM station, although only a portion is used for actual signal transmission, with the remainder serving to protect signals of adjacent stations. FM radio in the United States is allocated to the very high frequencies (VHF), occupying 100 channels of 200 kilohertz each between 88 and 108 megahertz. Each FM channel accommodates hundreds of stations—there are more than 7,000 on the air at the beginning of the 21st century.

VHF transmissions follow line-of-sight paths from antenna to receiver, and thus FM transmitters (or television stations, which use neighboring frequencies) are limited in their coverage to usually not more than 40 to 60 miles, depending on terrain and antenna height. That limitation is balanced by the lack of the medium wave interference that AM radio has, which is caused by signals arriving from ground waves or sky waves at slightly different times because of the distances covered.

Experimental Development (to 1940)

No one person "invented" FM radio—indeed, the man most credited with developing the system, Edwin Howard Armstrong, readily conceded that point. The first patents concerning an FM transmission system were granted to Cornelius Ehret of Philadelphia in 1905, probably the first such patents in the world. Scattered mentions of FM in subsequent years focused on its negative aspects, suggesting that, based on what was then known, FM would not be a useful broadcast medium. Still, technical work continued, and more than two dozen patents had been granted to various inventors and companies by 1928. Much of the impetus behind research into FM work was the search for a solution to the frustrating interference problem with AM radio. By the late 1920s, it was clear that simply using more AM transmitter power would not overcome static, which made AM unlistenable in electrical storms. Something new was needed.

From 1928 to 1933, Edwin Armstrong, a wealthy radio inventor then on Columbia University's physics faculty, focused on trying to utilize FM in a viable broadcast transmission system. Rather than working with narrow bands as had others before him, Armstrong's key breakthrough was to use far wider channels, eventually 20 times wider than those used by AM. The frequency could then modulate over about 150 kilohertz (though it normally used far less), leaving 25-kilohertz sidebands to prevent interference with adjacent channels. This allowed for greatly improved frequency response, or sound quality. Armstrong incorporated various circuits to allow precise tuning of the wide channels while at the same time eliminating most static and interference. Armstrong applied for the first of his four basic FM patents in 1930 and for the last in 1933; all four were granted late in 1933.

From 1934 to 1941, Armstrong further developed and demonstrated FM, working toward Federal Communications Commission (FCC) approval of a commercial system. After a number of long-distance tests (successfully sending signals up to 70 miles with only 2,000 watts of power) in cooperation with the Radio Corporation of America (RCA), Armstrong announced his system to the press early in 1935. A more formal demonstration to a meeting of the Institute of Radio Engineers later that year (and the published paper that resulted) marked the beginning of active FM innovation.

Resistance to the FM idea began to develop at about this time, usually growing out of the competing interests of two other broadcast services. Owners of AM stations, including the major networks, were concerned about the new technology that might totally replace their existing system. And companies already investing heavily in television research, especially RCA, thought that the new video service should receive priority in allocations and industry investment. FM was seen by some as merely a secondary audio service, albeit a far better one technically.

In July 1936 Armstrong obtained permission from the FCC to construct the world's first full-scale FM station in Alpine, New Jersey, across the Hudson River from New York City. After a technical hearing, the FCC provided initial allocations for FM and television (among other services), granting the fledgling FM technology's backers the right to experiment on 13 channels scattered across three widely separated parts of the spectrum—26, 43, and 117 megahertz. Early in 1939 the allocation was expanded to 75 channels located more conveniently between 41 and 44 megahertz. In the meantime, Armstrong's experimental station—the world's first FM transmitter—had gone on the air as W2XMN with low-power tests in April 1938.

Developing further experimentation but also looking toward commercial FM operations, the New England–based Yankee Network began to build two large transmitters in 1938–39. General Electric built two low-power FM transmitters at the same time, and the National Broadcasting Company's (NBC) experimental station began operating in January 1940. The first FM station west of the Alleghenies began transmission tests in Milwaukee just a few days later. Transmitters for most of these operations came from Radio Engineering Laboratories. Receivers were first manufactured by General

Electric in 1939, with other companies joining in the next year; however, most FM sets cost a good deal more than their AM counterparts.

Early Operations (1940–45)

The FCC became the arena for a 1940 battle over whether or not to authorize commercial FM service, and if so, on how many channels and with what relationship to developing television. In March 1940, more than a week of hearings were held to air the industry's conflicting views over the merits of FM and television. On 20 May 1940, the commission released its decision allowing the inception of commercial FM operation as of 1 January 1941; the decision allocated 40 channels on the VHF band (42–50 megahertz), reserving the lowest five channels for noncommercial applicants. Final technical rules were issued a month later. The first 15 commercial station construction permits were issued on 31 October 1940.

As the new year dawned, 18 commercial and 2 educational stations aired (compared to more than 800 AM stations at the time). The first commercial license was granted to W47NV, affiliated with AM station WSM in Nashville. The first West Coast station, a Don Lee network outlet, went on the air in September 1941. FM outlets briefly used unique call signs that combined the letters used with AM stations with numbers indicating the channel used (e.g., W55M in Milwaukee broadcast on 45.5 megahertz). This system was replaced with normal four-letter call signs in mid-1943.

By the end of 1941, and after the United States had entered World War II, the FCC reported 67 commercial station authorizations, with another 43 applications pending. About 30 of the former were actually on the air. Wartime priorities forced the end of further license grants and limited construction material availability after March 1942. By the end of October 1942, 37 stations were in operation, plus an additional 8 outlets still devoted to experiments. But construction materials and replacement parts were increasingly difficult to find, and some owners turned back their authorizations or withdrew their applications pending the end of the war.

The first attempt at an FM network, the American network, never made it on the air, largely because of difficulties in constructing the needed affiliate stations in sufficient markets. Programs offered on FM were of two types—duplicated AM station signals (the most common type) or recorded music. Because of the duplicated content of existing stations, FM stations had little appeal for advertisers. Another problem was FM audiences. There were some 15,000 FM sets in use at the beginning of 1941 and perhaps 400,000 by the time manufacturing was stopped early in 1942, compared to 30 million AM-equipped households. Most observers expected FM to become an important part of the industry after the war.

Frequency Shift and Decline (1945–57)

The next dozen years—from 1945 through 1957—were both exciting and frustrating as the FM service struggled to become established and successful amidst a broadcasting industry increasingly infatuated with television and still investing considerable sums into the expansion of AM. Initial excitement over FM's potential gave way to a slow decline.

Toward the end of the war, potential operators were already concerned that FM's allocation of 40 channels was not sufficient for expected postwar expansion. To further complicate matters, wartime spectrum and related research suggested that the FM allocation of 42–50 megahertz might be subject to cycles of severe sun spot interference. Concerns about television expansion led to demands by some members of the industry for FM's spectrum space to be reallocated to television.

Extensive FCC hearings in mid-1944 aired some of the technical concerns about the FM band, though wartime security limited what could be discussed. Armstrong and his backers argued to retain (or, better yet, to expand) the existing allocation, in part because stations could easily network by picking up each other's signals and passing them on—something that would be impossible were FM to be moved higher in the spectrum (moving lower was out of the question because of existing services). In January 1945 the FCC proposed moving FM to the 84–102 megahertz band to avoid the expected atmospheric interference and to gain more channels, for a total of 90. Subsequent proceedings continued the industry split over what to do and how. Finally, in June 1945, the FCC made its decision, shifting FM "upstairs" to the 88–108 megahertz band with a total of 100 channels that the service occupies today. Continuing the precedent established in 1941, educational users were assigned to channels reserved for them at 88–92 megahertz. The former FM band would be turned over to television and other services after a three-year transition period.

At first it seemed the shift would only disadvantage those stations actually on the air (46 at the time) and those people with FM sets that could not also receive AM signals (perhaps 30,000 old-band FM-only sets in consumer hands). Generally FM's outlook was good. The FCC issued the first postwar grants for new stations in October 1945, and more applications were piling up. Through 1946 there were always at least 200 applications pending, and although the number of stations actually on the air grew fairly slowly, the number of authorized FM stations exceeded 1,000 by 1948—more than all the AM and FM stations on the air just three years earlier. Most applications were coming in from AM stations hedging their bets on the future. Several government agencies issued optimistic publications encouraging still more FM applicants. Two specialized FM trade magazines began to publish. A number of potential FM networks were in the planning stages, and the first, called the Continental network, began operations with four stations early in 1947.

But all was not well. FM's frequency shift was more damaging in the short term than it had seemed. When stations began to transmit on FM's new frequencies, there were few receivers available to pick up the signals. Manufacturers were trying to meet pent-up wartime demand for new AM sets and had little capacity to devote to FM's needs. Thus FM suffered from the lack of a good-sized audience that might appeal to advertisers. Only token numbers of receivers were available until 1950, and by then demand for television sets was threatening capacity devoted to radio. FM's lack of separate programming (after considerable industry argument both ways, the FCC had allowed co-owned AM and FM stations to simulcast or carry the same material) offered little incentive for consumers to invest in one of the rare and expensive FM receivers. A cheap AM set could tune popular local and network radio programs just as well. FM's better sound quality was not enough of a draw. What independent programming did exist was largely classical music and arts material of interest to a relatively small elite. Advertisers saw no reason to invest in FM, especially when FM time was usually given away with AM advertising purchases. Indeed, AM was thriving—more than doubling the number of stations on the air from 1945 to 1950. And the growing concentration on television by broadcasters, advertisers, and the public made FM seem unnecessary.

As these factors combined and intensified, the results soon became apparent. The number of FM new station applications began to drop off, and then overall FM authorizations declined. By 1948 FM stations already on the air, among them some pioneering operations, began to shut down, returning their licenses to the FCC. FM outlets could not be given away, much less sold. The number of stations on the air declined each year. Faced with the seeming failure of his primary invention, Armstrong took his own life in 1954; with the loss of his financial backing, the Continental network had to close down as well.

Rebound (1958–70)

Then, and at first very slowly, FM began to turn around. Reports in several trade magazines late in 1957 picked up the fact that the number of FCC authorizations for FM stations had increased for the first time in nine years. Slowly the pace of new station construction picked up, first in major markets and then in suburban areas. Several factors underlay this dramatic shift.

First, AM had grown increasingly crowded—there were virtually no vacant channels available in the country's major markets. The number of AM stations had doubled from 1948 to 1958, and about 150 more were going on the air annually. However, an increasing proportion of the new outlets were limited to daytime operation in an FCC attempt to reduce nighttime interference. FM, with no need for daytime-only limitations, was now the only means of entering major markets. In addition, the major spurt of television expansion was over, and this eased up pressure on time, money, and personnel, which could now be applied to FM.

But aside from overcrowding in AM and television, FM itself had more to offer. In 1955 the FCC had approved the use of Subsidiary Communications Authorizations, which allowed stations to multiplex (to send more than one signal from their transmitter) such non-broadcast content as background music for retail outlets ("storecasting"). This provided a needed revenue boost. So did the growing number of listeners interested in good music. These "hi-fi" addicts doted on FM operations, and this interest was evident in the increasing availability and sale of FM receivers. A developing high-end audience led advertisers to begin to pay serious attention to the medium.

Another technical innovation gave FM a further boost: the inception of stereo broadcasting. Beginning as early as 1952, some stations, such as New York's WQXR, offered AM/FM stereo using two stations—AM for one channel of sound and an FM outlet for the other. Occasional network two-station stereo broadcasts began in 1958—the same year commercial stereophonic records first went on sale. By 1960, more than 100 stations were providing the two-station system of stereo. But such simulcasting wasted spectrum (two stations with the same content), and the uneven quality of AM and FM provided poor stereo signals. What was needed was a system to provide stereo signals from a single station, and FM's wide channel seemed to offer the means.

In 1959 the National Stereophonic Radio Committee began industry experiments with several competing multiplexed single-station systems. By October 1960 the committee had recommended that the FCC establish FM stereo technical standards combining parts of systems developed by General Electric and Zenith. The FCC issued the standards in April 1961, and the first FM stereo stations began providing service in June. By 1965, a quarter of all commercial stations were offering stereo; by 1970, 38 percent of FM stations had the capability. Though few saw the future clearly, stereo would be a key factor in FM's ultimate success over long-dominant AM stations.

FM's continued expansion led the FCC to establish three classes of FM station in mid-1962. Lower-powered Class A (up to 3,000 watts of power and a service radius of 15 miles) and B stations (up to 50,000 watts of power and a service radius up to 40 miles) would be granted in the crowded northeastern section of the country as well as in southern California. Higher-powered C stations (up to 100,000 watts of power providing a service radius of 65 miles) could be granted elsewhere. A five-year FCC freeze on most new AM station grants beginning in 1968 helped funnel still more industry expansion into FM as the FCC began to see AM and FM as parts of an integrated radio service.

Of even greater importance to FM's continued growth was a series of landmark FCC decisions from 1964 to 1966 requiring separate programming on co-owned AM and FM stations in the largest markets (those with populations over 100,000). Long concerned about the effect of wasting spectrum space by allowing the same programs to run on both AM and FM, the FCC had been persuaded by industry leaders to allow the practice when FM was weak. Indeed, many FM broadcasters expressed great concern about losing their ability to carry popular AM programming. But FM's growth in numbers and economic strength prompted the move—which further accelerated creation of new FM stations. In just a few years the importance of the FCC decisions (which by the late 1960s had been extended to smaller markets) became apparent as FM audiences increased sharply—bringing, in turn, greater advertiser interest and expenditure to make FM economically viable for the first time in its history. By the early 1980s, when the AM-FM non-duplication requirement was eliminated in a deregulatory move, FM stations were dominant in large part because of their unique programming.

That FM had achieved its own identity was exemplified when one of the big-three networks, the American Broadcasting Companies (ABC), initiated a network of FM stations in 1968. Although relatively short-lived, as the industry increasingly began to think of FM as radio rather than something different, the recognition that such a network gave to FM radio was a tremendous boost in the advertising community. Another indicator was Philadelphia's WDVR, which within four months of first airing in 1963 was the number-one FM station in the city, competing for top spot with long-established AM outlets, an inconceivable development just a few years earlier. Five years later, the same station became the first FM outlet to bill more than $1 million in advertising time. The FM business as a whole reported positive operating income in 1968 for the first time (it happened for the second time in 1973, after which the industry as a whole remained profitable).

The key measure of FM's coming of age, of course, is actual audience use of the service. In 1958, for example, FM was available in about one-third of all homes in such major urban markets as Cleveland, Miami, Philadelphia, and Kansas City. By 1961 the receiver penetration figures for major cities were creeping up to about 40 percent, and national FM penetration was estimated at about 10 percent, showing how few FM listeners lived in smaller markets and rural areas, many of which still lacked FM stations. By the mid-1960s, FM household penetration in major markets was hovering at the two-thirds mark, and national FM penetration stood at about half that level. Although stereo and car FM radios were initially expensive, increasing production dropped prices and helped to further expand FM availability.

Dominance (The 1970s and Since)

After the many FM industry and policy changes of the 1960s, the 1970s saw FM becoming increasingly and rapidly important economically. Where FM attracted 25 percent of the national radio audience in 1972, just two years later survey data showed FM accounted for one-third of all national radio listening—although only 14 percent of all radio revenues. By 1979 FM achieved a long-sought goal when for the first time, total national FM listening surpassed that of AM stations. Every major market had at least four FM stations among the top 10 radio outlets. Indeed, FM would never lose that primacy, slowly expanding its role until by the turn of the century, FM listening accounted for nearly 80 percent of all radio listening.

Getting there had not been easy and had taken far longer than early proponents had expected. In part, FM's own success got in the way. After years of promoting FM's upper-scale (though small) audiences, often prejudicially dubbed eggheads and high-fidelity buffs, it was hard to shift gears and promote FM's large and growing audience as being tuned to simply "radio." (Indeed, the number of commercial FM classical music stations had actually declined by half since 1963, to only 30 by 1973.) At the same time, the number of educational FM stations expanded dramatically after 1965, greatly aided by the creation of National Public Radio and the appeal of its programs as well as by the availability of increased funding for station development and operation.

But with success came pressure to keep up. As news and talk formats increasingly defined AM (where the poorer sound quality did not matter), FM flowered with a full cornucopia of musical formats and styles. By the early 1970s, FM stations in the nation's largest markets were developing formats every bit as tight and narrow as those of their AM forebears. Each station and its advertisers were appealing to a specific segment of the once-mass radio audience in an attempt to build listener loyalty in a marketplace often defined by too many stations in most cities. By the late 1980s, FM's primary target market was that defined by its advertisers: listeners aged 26 to 34, followed by those 35 to 44 years of age. Only a relative handful of stations target teens, and fewer than 30 percent are interested in listeners aged 55 or older. As compared with its earlier days, FM has become positively mainstream.

FM's success is also seen in the usual marketplace measure—the price of FM stations being sold on the open market. Where top-market stations could literally not be given away in the early 1950s, by the late 1960s, the first million-dollar prices were being quoted. Three decades later, FM stand-alone stations in top markets sold for tens of millions of dollars, and some have sold for well over $100 million. On the other hand, many miss the old days of FM programs aimed at a small, elitist, sometimes cranky but usually appreciative audience. A

1999 FCC proposal to create scores of low-power FM outlets was intended to bring back some of that spirit, but was severely curtailed by Congress in 2000.

FM Outside the U.S.

FM or VHF radio developed more slowly outside of the United States. In Europe, for example, postwar radio reconstruction in most countries focused first on established medium wave and long wave services and then on television; few countries had economies strong enough to develop FM services at the same time as these other initiatives. And politics played a part, because Europe hoped in the meantime to find a European technical solution to its substantial problems of interference and static.

Given the total destruction of its broadcasting system, Germany had to start over and thus led Europe in beginning FM broadcasting. The first transmitters were on the air by 1949, and most of West Germany was covered with FM signals by 1951. Sale of FM receivers was brisk (some were exported to the United States), partly because television was not a competitor until 1952. By 1955 there were 100 FM transmitters in operation. With a severe shortage of medium wave frequencies, Italy followed suit, providing its first VHF radio services in the early 1950s.

At about the same time, other European nations, working through the European Broadcasting Union (EBU), began to reconsider FM's potential, because they had largely completed the process of repairing or replacing wartime AM radio losses. FM was seen as the only means of reducing serious medium wave overcrowding and resulting interference problems as well as serving regions largely unreached by existing stations, and FM could do so less expensively than could medium wave facilities. Countries also sought additional program channels. Interestingly, the same debate over whether FM should carry the same or different programs (as existing medium wave services) divided industries and governments in Europe as it had in the United States. By the late 1950s, EBU member nations were working together to build a system integrating existing and new VHF radio stations. And, as in the United States, the new services were increasingly programmed independently.

After experimenting with FM in London as early as 1950, the British Broadcasting Company (BBC) began introducing a chain of VHF radio stations in 1955. By 1960, most of the country was reached with the new transmitters, which largely simulcasted the medium wave station signals, though receiver penetration hovered at only about 15 percent, rising to 30 percent five years later. The planned role of the VHF transmitters was to introduce local programming for specific audiences—something that had been lacking in Britain since the early 1920s. By the early 1960s, VHF radio transmitters outnumbered medium wave facilities by 160 to 57. A decade later, there were 252 VHF transmitters in Britain. Lower FM receiver prices prompted rapid ownership growth.

Even by the mid-1980s, however, only about 20 countries (most of them in Europe) had extensively developed VHF radio. Despite its potential value to tropical countries, which are plagued by static on their AM or medium wave broadcast stations, few Third World nations had embarked on FM service. They lack the funds and even the need, because they have not fully utilized available medium wave channels. South Africa is an exception, having embarked in 1961 on development of VHF radio to cover the nation. Apartheid politics may have played a role here, because the VHF transmitters made it more difficult for Africans to hear foreign broadcasts, none of which were available on FM. Other African nations only experimented with FM in this period.

In the Far East, Japan experimented with FM for a decade before stations opened in major cities in 1969. The Ministry of Posts and Telecommunications sought to have an FM station in every prefecture and at least two in major cities. All of these are advertising-supported local stations. For a time in the 1970s and 1980s, a raft of mini–FM transmitters called "free radio," which covered a radius of only about 3,000 feet, were very popular, playing music and advertising. Few were licensed, however, and many were closed down in the late 1980s. The service came later to Australia, where what would become FM frequencies had been originally allocated to television. Reallocation of that service made initiation of FM service possible there in the 1980s.

Perhaps the most extreme examples of the FM-based "free radio" movement took place in the 1970s in both France and Italy. A number of unlicensed small local Italian FM stations went on the air in late 1974 and into 1975. When an Italian court held that the state broadcasting authority did not have a monopoly on local radio, hundreds more followed in 1976. By mid-1978, some 2,200 were on the air, providing Italians with the most radio per capita of any nation on earth. Stations programmed music and advertising and often expressed strong political viewpoints on both the right and left. France went through something similar in the late 1970s—by the early 1980s there were more than 100 such stations in Paris alone. Most gave way to a 1982 government decision to provide licenses to many of the stations as well as official permission to advertise.

CHRISTOPHER H. STERLING

See also Armstrong, Edwin Howard; Don Lee Network; Educational Radio to 1967; Federal Communications Commission; FM Trade Organizations; Low-Power Radio/Microradio; Radio Corporation of America; Receivers; Sarnoff, David; Shepard, John; Stereo; Subsidiary Communications Authorization; United States; Yankee Network

Further Reading

Armstrong, Edwin H., "Evolution of Frequency Modulation," *Electrical Engineering* 59 (December 1940)
Besen, Stanley M., "AM versus FM: The Battle of the Bands," *Industrial and Corporate Change* 1 (1992)
Boutwell, William Dow, Ronald Redvers Lowdermilk, and Gertrude G. Broderick, *FM for Education,* Washington, D.C.: United States Government Printing Office, 1944; revised edition, 1948
Codding, George A., "Frequency Modulation Broadcasting," in *Broadcasting without Barriers,* Paris: UNESCO, 1959
Cox Looks at FM Radio: Past, Present, and Future, Atlanta, Georgia: Cox Broadcasting, 1976
Eshelman, David, "The Emergence of Educational FM," *NAEB Journal* 26 (March–April 1967)
"Evolution of FM Radio: Extracts from FCC Annual Reports," *Journal of Broadcasting* 5 (Spring 1961 and Fall 1961), 6 (Summer 1962), 7 (Fall 1963)
Inglis, Andrew F., "FM Broadcasting," in *Behind the Tube: A History of Broadcasting Technology and Business,* Boston: Focal Press, 1990
Lewis, Peter M., and Jerry Booth, *The Invisible Medium: Public, Commercial, and Community Radio,* Washington, D.C.: Howard University Press, and London: Macmillan, 1989
Longley, Lawrence D., "The FM Shift in 1945," *Journal of Broadcasting* 12 (Fall 1968)
Sleeper, Milton B., editor, *FM Radio Handbook,* Great Barrington, Massachusetts: FM Company, 1946
Slotten, Hugh Richard, "Rainbow in the Sky: FM Radio, Technical Superiority, and Regulatory Decision-Making," *Technology and Culture* 37 (October 1996)
"Special Report: FM Sniffs Sweet Smell of Success," *Broadcasting* (31 July 1967)
Sterling, Christopher H., "Second Service: Some Keys to the Development of FM Broadcasting," *Journal of Broadcasting* 15 (Fall 1971)
United States Congress, Senate Special Committee to Study Problems of American Small Business, *Small Business Opportunities in FM Broadcasting,* 79th Congress, 2nd Session, April 10, 1946, Washington, D.C.: GPO, 1946
Wedell, E. George, and Philip Crookes, *Radio 2000: The Opportunities for Public and Private Radio Services in Europe,* Geneva: European Broadcasting Union, 1991

FM Trade Associations

Promoting Radio's Second Service

From the inception of commercial FM radio in 1941, a series of five industry trade organizations appeared—and disappeared—in parallel with the medium's struggles and eventual success. Each was different in its outlook and focus.

The Early Struggle

The first FM group, the National Association of FM Broadcasters Incorporated (FMBI), was created in 1940 to promote the technology as much as the industry. Spearheaded by John Shepherd III of the New England–based Yankee Network and by Walter J. Damm of the *Milwaukee Journal* radio stations (one of which was the first FM station west of the Alleghenies), FMBI published thousands of copies of *Broadcasting's Better Mousetrap* to promote FM's better sound and other qualities. Before and during World War II, a mimeographed newsletter edited by Dick Dorrance appeared regularly to record the slow initial development of the business. FMBI had 43 members by 1943—most of those either on the air or building new stations—and 137 by September 1944. Among its campaigns was a successful move to persuade the Federal Communications Commission (FCC) to modify FM station call letters from letter and number combinations denoting the channel of the station (e.g., W55M, which was on 45.5 megahertz) to the more familiar all-letter system used with AM stations. The FCC adopted the plan in 1943. Although FMBI fought the shift of FM frequencies that came in 1945, it worked to put the new spectrum into action.

With the end of the war, FMBI voted in 1946 to merge its activities into the FM Department (later Committee) of the National Association of Broadcasters (NAB), a pattern that would be repeated several times. Initially headed by Robert Bartley, later an FCC commissioner, this arm of the main industry trade association sought a place for FM within an

Frederick, Pauline 1908–1990

U.S. Radio Network Journalist

One of the first female network news correspondents, Pauline Frederick became best known for more than two decades of reporting from the United Nations for National Broadcasting Company (NBC) radio and television. For her first dozen years on network television—until 1960—she remained the only female reporter of lasting duration in the medium.

Frederick grew up in Harrisburg, Pennsylvania, and got her first journalistic experience covering social news for the local paper. She left to earn her bachelor's degree in political science (1929) and a master's degree in international law (1931), both from American University in Washington, D.C. She originally intended to be an attorney but grew more interested in journalism while in school. She worked as a freelance reporter for, among other media, the *Washington Star* on women's issues, and some of her interviews were syndicated by the North American Newspaper Alliance (NANA).

Her first broadcasting work was for NBC Blue, assisting commentator H.R. Baukage by conducting interviews with various newsmakers beginning in 1938 with the wife of the Czech ambassador as Germany occupied that country. She continued at NBC into the 1940s, undertaking scriptwriting and other assignments, few of them on air in an era when female broadcast journalists were almost unheard of. Unable to break that gender barrier, she left the network in 1945 to work full-time for NANA while freelancing occasional "women's news" and other reports for what had become American Broadcasting Company (ABC) radio. This period of intensive international experience included travel to 19 countries in 1945–46; she sent reports from several countries in the Far East, from the Nuremberg trials of Nazi war criminals in Germany, and from Poland.

Frederick returned to broadcasting when, despite being turned down by both the Columbia Broadcasting System (CBS) and NBC (who still felt women's voices were not authoritative enough for news), she obtained a part-time position with fledgling ABC News in 1946, initially focusing (again) on "women's news." Her impressive output—including an exclusive interview with General Eisenhower and later coverage of a foreign ministers' conference—raised her status to full-time employment in 1948. That year she helped anchor ABC's television coverage of the national political conventions, and from August 1948 to March 1949 she hosted the Saturday evening 15-minute *Pauline Frederick's Guest Books* of television news interviews. She was heard on several radio network newscasts, her radio focus becoming international affairs, including the then-new United Nations, on which she quickly became an authority.

In 1954 Frederick returned to NBC and continued covering the United Nations for 21 years until her mandatory retirement (because of her age—she was 65), about which she learned from a story in the *New York Times* in 1974. In this two-decade period she became the voice of the United Nations for many Americans. She was also heard covering political conventions, tensions and wars in the Middle East, the Cuban Missile Crisis and other parts of the Cold War, and the war in Vietnam. At the same time, she continued on NBC radio with *Pauline Frederick Reporting,* a 15-minute daily program.

After retiring from NBC radio and television news, Frederick commented on United Nations affairs for National Public Radio. She also became the first female journalist to moderate a presidential candidates' debate when she presided over one of the 1976 televised forums pitting President Gerald Ford against Governor Jimmy Carter. She retired in 1980.

Frederick received 23 honorary doctorates and was the first woman to win both the Alfred I. DuPont award (for commentary, in 1954 and 1956) and the George Foster Peabody award (1954). She was also the first woman to win the Paul White Award from the Radio-Television News Directors Association (1980). She was a model for many aspiring female journalists and was one of the first to succeed over a long career.

CHRISTOPHER H. STERLING

Pauline Frederick. Born in Gallitzen, Pennsylvania, 13 February 1908. B.A. in political science, 1929 and M.A. in international law, 1931, American University. Performed freelance journalism work, 1920s and 1930s; interviewer, NBC, Washington, D.C., 1938–45; war correspondent, North American Newspaper Alliance, 1945–46; joined ABC radio, 1946–53; returned to NBC, becoming first woman to report serious telelvision news, 1954–74; first woman elected president of the UN Correspondents Association, 1959; international affairs analyst, NPR, 1974–80; first woman to moderate a presidential debate, Jimmy Carter and Gerald Ford, 1976. Recipient: Alfred DuPont Award for meritorious service to the American people, 1953; twenty-three honorary degrees; Headliner Award, Theta Sigma Phi; George Foster Peabody Award, School of Journalism, University of Georgia, 1954; First Pennsylvania Journalism Achievement Award; first woman recipient, Paul White Award, Radio-Television News Directors Association, 1980. Died in Lake Forest, Illinois, 9 May 1990.

Pauline Frederick
Courtesy AP/Wide World Photos

Radio Series

1949–53 *ABC News*
1953–56 *NBC News*
1954–55 *At the UN*

Selected Publications

Ten First Ladies of the World, 1967

Further Reading

Hosley, David H., and Gayle K. Yamada, *Hard News: Women in Broadcast Journalism*, New York: Greenwood Press, 1987

Marzolf, Marion, *Up from the Footnote: A History of Women Journalists*, New York: Hastings House, 1977

Nobile, Philip, "TV News and the Older Woman," *New York Times* (10 August 1981)

Sanders, Marlene, and Marcia Rock, *Waiting for Prime-Time: The Women of Television News*, Urbana: University of Illinois Press, 1988

Talese, Gay, "Perils of Pauline," *The Saturday Evening Post* (26 January 1963)

Freed, Alan 1921–1965

U.S. Disc Jockey

In the 1950s, Alan Freed became the first nationally recognized disc jockey in the U.S. to feature the emerging rhythm and blues and rock musical forms. During a time when many white "platter pilots" were reluctant to play songs by African-American performers, Freed was not; as a result, he helped to advance the careers of a number of artists. He has also been credited with helping to popularize the term *rock and roll.*

Freed, who grew up in Salem, Ohio, began his career in radio as an announcer in New Castle, Pennsylvania, after World War II, and by 1949 he had a popular music request show over WAKR in Akron, Ohio. In 1950 he moved to Cleveland and landed a job at WXEL-TV; however, by June 1951 Freed had returned to radio, hosting a record show over Cleveland's WJMO from 6:00 to 7:00 in the evenings. Less than a month later, he moved over to WJW in Cleveland to host what became a very popular late-evening request show.

Although Freed had started playing rhythm and blues on the air while he was in Akron, the inspiration for his career in rhythm and blues came from Joe Mintz, the owner of a Cleveland record store called the Record Rendezvous. Mintz was convinced that a rhythm and blues show would be popular, because both blacks and whites were buying rhythm and blues records in his store. He convinced Freed to give the music a chance on the air, and the result was *The Moondog Show*, which proved to be a great success.

Freed received wide attention when, in 1952, he took his show on the road. He decided to host a live concert in the Cleveland/Akron area called the Moondog Ball. The show, which featured a variety of acts, drew a crowd of 25,000 people to a 10,000-seat arena, and it was heralded as a successful, though raucous, event. The near riot created by the ticketless crowd outside the theater attracted press attention to the music Freed was playing.

By 1954 WNJR in New Jersey had begun to air taped copies of Freed's programs, and on 1 May he hosted an Eastern Moondog Coronation Ball at the Sussex Avenue Armory in Newark. Years later, the *New York Times* noted, "Going to one of Alan Freed's rock 'n' roll musicales has always been something like having an aisle seat for the San Francisco earthquake."

Freed moved from Cleveland to New York City in July 1954 after signing with WINS radio for the largest annual salary paid to an independent rhythm and blues jockey up to that point—$75,000. However, blind street musician Louis "Moondog" Hardin objected to Freed's use of the Moondog moniker, and in December of 1954, Hardin won a court injunction against Freed's use of the term. Freed changed the name of his WINS program to *Alan Freed's Rock and Roll Party*, and the age of rock began.

Freed moved to WABC in 1958, but he lost both his prestigious radio program and a television show at WNEW-TV in 1959 as a result of the Congressional quiz-show investigations. The legal action was prompted by accusations that TV networks were rigging popular quiz shows of the day, but the inquiries shifted to radio after Burt Lane, a representative for the American Guild of Authors and Publishers, sent a letter to the Congressional Special Subcommittee on Legislative Oversight citing examples of commercial bribery. Freed and a number of popular disk jockeys were accused of accepting money from record companies in return for playing those companies'

Alan Freed at WABC, June 1958
Courtesy AP/Wide World Photos

songs on the air. Freed pleaded guilty to a charge of taking bribes in 1962 and subsequently left New York City. Freed died in 1965 at 43 years of age in Palm Springs, California.

CHARLES F. GANZERT

Alan Freed. Born Albert James Freed, near Johnstown, Pennsylvania, 21 December 1921. Attended Ohio State University, B.S. in mechanical engineering, 1943; served two years in U.S. Army; disc jockey, WKST, New Castle, Pennsylvania, 1945; radio host, WAKR, Akron, Ohio, 1947; hosted television dance show, WXEL, Cleveland, Ohio, 1950; disc jockey, WJW, Cleveland, Ohio, 1951–54; staged "Big Beat," all-African-American-talent rock and roll shows, 1952–58; hosted *Rock 'n' Roll Party,* CBS-TV, 1957; disc jockey, WINS, New York City, 1954–58; disc jockey, WABC, New York City, 1958; hosted dance party, WNEW -TV, New York City, 1959; disc jockey, KDAY, Los Angeles, California, 1960; worked at WQAM, Miami, Florida, 1962; jazz disc jockey, KNOB, Los Angeles, California, 1964. Elected to the Rock and Roll Hall of Fame, 1986. Died in Palm Springs, California, 20 January 1965.

Radio Series

1947 *Request Review*
1951–54 *The Moon Dog Show*
1954–58 *Alan Freed's Rock 'n' Roll Party*

Films

Rock around the Clock, 1956; *Don't Knock the Rock,* 1956; *Rock, Rock, Rock,* 1957; *Go Johnny Go,* 1959

Further Reading

Fornatale, Peter, and Joshua E. Mills, *Radio in the Television Age,* Woodstock, New York: Overlook Press, 1980
Passman, Arnold, *The Deejays,* New York: Macmillan, 1971
Smith, Wes, *The Pied Pipers of Rock 'n' Roll: Radio Deejays of the 50s and 60s,* Marietta, Georgia: Longstreet Press, 1989
Sterling, Christopher H., and John M. Kittross, *Stay Tuned: A History of American Broadcasting,* 3rd edition, Mahwah, New Jersey: Lawrence Erlbaum, 2002
United States Congress, House Committee on Interstate and Foreign Commerce, *Responsibilities of Broadcasting Licensees and Station Personnel,* 2 vols., Washington, D.C.: GPO, 1960
Ward, Ed, Geoffrey Stokes, and Ken Tucker, *Rock of Ages: The Rolling Stone History of Rock and Roll,* New York: Rolling Stone Press, 1986

Freed, Paul 1918–1996

U.S. Religious Broadcaster

Paul E. Freed founded Trans World Radio in 1952 in Tangiers, Morocco. By 1999 Trans World Radio was broadcasting in 150 languages from 12 locations in the world. Its gospel message is broadcast more than 1,400 hours each week and reaches listeners on three continents.

Freed, the son of missionaries, was born in Detroit, Michigan, in 1918. As a young boy, his family moved to the Middle East, where his parents served with the Christian and Missionary Alliance Church. Freed's early education was sporadic; he attended English and German school in Jerusalem, was home schooled, and worked with tutors. When his parents were home on furlough, he attended Wheaton Academy in Illinois. From the Academy, he matriculated at Wheaton College, where he earned his bachelor's degree. After graduating from Wheaton College, Freed attended Nyack Bible College and graduated from its missions program. Shortly after his marriage to Betty Jane Seawell in 1945, Freed left his employment as pastor of a small church in Greenville, South Carolina. Torrey Johnson, founder of the Youth for Christ movement, recruited Freed as the director of the Greensboro, North Carolina, program.

In 1948 Johnson sent Freed to the Youth for Christ conference in Switzerland. Despite protestations from Freed, Johnson insisted it was God's calling for him. While in Switzerland, Freed was convinced there was a need to transmit the message of God to the evangelical Protestant youth of Franco's Spain. Following a trip to Spain, Freed laid the foundation for a radio organization to bring the gospel message to the people of Spain. The Spanish government refused to sanction such work. Frustrated by his experience, Freed returned to the United States to find a way to fulfill his mission of evangelical radio for the people of Spain.

Rev. Paul E. Freed reports to White House after a visit to Spain, 21 December 1951
Courtesy AP/Wide World Photos

Origins of Trans World Radio

In February 1952 Freed founded International Evangelism (later to be known as Trans World Radio) in the international city of Tangiers, Morocco. Freed acquired a small piece of land directly across the narrow Strait of Gibraltar from Spain. By 1954, at 61 years of age, Freed's father, Ralph Freed, accepted a new ministry to become director of the radio station in Tangiers. Dr. Ralph Freed transmitted the first Christian message from Tangiers, Morocco, on a 2,500-watt transmitter. Working together, by 1956 the Freeds built the Voice of Tangiers into an organization that broadcast the gospel message to 40 countries in over 20 languages. In 1959, when Morocco became a politically independent nation, government officials for the new regime ordered all radio stations to become nationalized. Freed was forced to move his station. Freed was worried that the 80 million listeners of the Voice of Tangiers would be lost, because there were no other full-time gospel radio stations on the air. The tiny Riviera country of Monaco welcomed Trans World Radio and its gospel message. Freed and Trans World Radio began broadcasting from a transmitter originally built during World War II by the Germans for propaganda purposes.

By 1960 Freed had completed a dissertation at the New York University School of Education. He combined international relations, mass communication, and religious education in preparation for his goal to expand Trans World Radio's broadcast of the gospel message to other areas of Europe. Freed believed that millions of people around the world would be receptive to the gospel message if they could hear it. European radio was not receptive to gospel programs or preachers. Most radio in Europe was controlled by the government, and the few countries that allowed preaching charged extremely high rates for even their lowest-rated times. Other countries that closed their borders to missionaries and the gospel message were also targeted by Freed for radio broadcasts.

The station in Monte Carlo broadcast with 100,000 watts. Freed targeted Spain, Portugal, the British Isles, Scandinavia, the then the Soviet Union, Central Europe, Southern Europe, the Middle East, and North Africa as the primary areas to be reached from Monte Carlo. The first programs aired by Trans World Radio were broadcast in 24 languages. The staff in Monte Carlo received 18,000 letters the first year offering support for their programming efforts. Trans World Radio continued to grow, setting up branch offices in different areas of Europe. Their ministry spread across Europe and into the Middle East.

In August 1964 Trans World Radio added a transmitting station in Bonaire Island, part of the Netherlands Antilles. Through this facility, Trans World Radio programmed 70 hours a week of gospel messages to the Caribbean and the northern part of South America. By 1980 Trans World Radio had established new transmitting stations in Swaziland to reach sub-Saharan Africa and Pakistan, in Cyprus to reach people in 21 countries in the Middle East and North Africa, and in Guam (broadcasting in 35 languages) to reach listeners in Central Asia, South Asia, Southeast Asia, and the Asian Pacific area. In the 1980s Trans World Radio established operations in Uruguay in partnership with "Radio Rural" to carry its gospel message to listeners in Uruguay and northern Argentina.

As Trans World Radio moved into the 1990s, Freed continued to work to expand listenership. Stations were added in Albania, Russia, Johannesburg, and Poland. By 2000 Trans World Radio broadcast 1,400 hours of gospel programs from 12 locations around the world. These programs were broadcast in 150 languages to an estimated 2 billion people. Each year, Trans World Radio receives over 1.4 million letters from listeners in 160 countries. Trans World Radio initially transmitted on AM at 800 kilohertz and has since added shortwave transmissions to reach more listeners. Shortly after Freed's death on 1 December 1996, he was inducted into the National Religious Broadcaster's Hall of Fame.

MARGARET FINUCANE

See also International Radio; Religion on Radio; Shortwave Radio

Paul Ernest Freed. Born in Detroit, Michigan, 29 August 1918. Son of missionaries Ralph and Mildred Freed; attended Wheaton College, B.A.. 1940; graduate of Missions Program, Nyack Bible College, 1944; attended Columbia University, M.A. 1956; New York University, Ph.D. in Mass Communication, 1960; founded Trans World Radio, 1952; awarded "President's silver Medallion for Service to the Kingdom of God," Toccoa Falls College, 1996; inducted into the National Religious Broadcaster's Hall of Fame, 1997. Died in Cary, North Carolina, 1 December 1996.

Selected Publications

Towers to Eternity, 1968

Let the Earth Hear: The Thrilling Story of How Radio Goes over Barriers to Bring the Gospel of Christ to Unreached Millions, 1980

Further Reading

Melton, J. Gordon, et al., "Paul Freed" and "Trans World Radio," in *Prime-Time Religion: An Encyclopedia of Religious Broadcasting*, Phoenix: Oryx Press, 1997

Free Form Format

During the 1960s, FM was ripe for a new form of radio—radio that burst through established format boundaries, emphasizing wholeness over separation and communal action over atomistic listening. *Free form*—in which imaginative disc jockeys combined many types of recorded and live music, sound effects, poetry, interviews, and calls from listeners—was the aural representation of the counterculture movement. Eschewing the slick professionalism, high-pressure salesmanship, and tight formats of AM radio, free form was—and sometimes still is—distinctly spontaneous, experimental, and challenging. At its best, free form is an exhilarating art form in its own right—a synergistic combination of disparate musical forms and spoken words. At its worst, free form may be pandering and self-indulgent.

Origins

Free form's roots developed in both commercial and noncommercial settings. During the mid-1960s, noncommercial community stations were developing across the country, following the lead established by Pacifica stations in California, New York, and Texas. These stations depended heavily on low-paid (often volunteer) programmers whose anti-establishment agendas rejected the tight structure of most corporate, commercial media. At the same time, commercial FM was still in its infancy, and disc jockeys were encouraged to experiment with longer segments and album cuts. Free form developed amidst these experimental venues, catching on quickly among community stations and some commercial FM stations—albeit late at night and on weekends.

Free form most likely originated at WBAI in New York City around 1963–64, with three different deejays: Bop Fass (Radio Unnameable), Larry Josephson (In the Beginning), and the following year with Steve Post's the Outside. Soon, it spread to other stations, notably Pacifica stations KPFA in Berkeley and KPFK in Los Angeles, and privately owned KMPX in San Francisco. KMPX's general manager, Tom Donahue, is often credited as being the driving force behind the "underground radio" movement. Although he did not invent free form, Donahue nurtured it and allowed it to grow from a program shift to an entire format (although *anti-format* might be a better term).

Style

Free-form programmers featured everything from cutting-edge musicians such as Bob Dylan and the Grateful Dead to comedy routines from W.C. Fields and Jonathan Winters. Indian ragas and classical music were heard back to back. Shows started late and ran overtime. Guests wandered in and out of control rooms, sometimes speaking on air, at other times just being part of the scene. Disc jockeys pontificated on the day's topics, their delivery styles ranging from chats with listeners and studio guests to rambling, witty monologues—often within the same program. Interviews and announcements regarding the counterculture and antiwar protests peppered broadcasts increasingly as the 1960s wore on.

Free form's deliberately anarchistic and undisciplined sound was, in effect, a form of participatory theater and gained a considerable following within the counterculture. Listeners called in to programs and were often heard on the air, rallies were announced (and broadcast), and listeners met at live remotes and events sponsored by stations (such as WBAI's 1967 "fly in" at Kennedy Airport, organized by WBAI free-form host Bob Foss).

Challenges to Free Form

The popularity of free form reached its peak between 1965 and 1970 and ultimately waned for three primary reasons. Ironically, once established through the success of free form, commercial FM became bound by the same tight formats that defined AM. Also, leaders among free-form disc jockeys, notably WBAI's Larry Josephson, grew weary of underground radio and moved on to other pastures. And the counterculture movement that nurtured free form eventually evolved beyond its communal sentiments. As the movement splintered into subgroups focusing upon sexuality, gender, race, and ethnicity, free form gave way to specialty shows on community radio and to the newer, more professional "public" stations affiliated with National Public Radio (NPR).

The 1980s were particularly difficult for free form, as community stations and NPR affiliates began programming more syndicated programming and professionalizing their sound, especially following NPR's near bankruptcy in 1983. Severe internal battles over station control were sometimes waged, with the fate of free form hanging in the balance. Proponents argued that free form was a unique means of expression that the new professionals simply failed to understand. The latter charged that free form's time had passed and that free form appealed to only a tiny fraction of the potential market. Despite such challenges, free form continued to survive at some stations, albeit most often during the late-night hours where it had originally developed.

Contemporary Free Form

Among the community stations and a dwindling number of public stations that still program it, free form has taken on an air of sanctity, hearkening back to the good old days when community radio was central to the underground movement. Yet without a symbiotic cultural context to fuel and inform it, contemporary free form lacks the immediacy and connection with the public that it once held. As such, free form has become a much more personal medium among disc jockeys, and a successful program is one that has smoothly combined a wide variation of sounds reflective of the programmer's moods and inclinations at the moment. Whereas 1960s free form was jarring and often disturbing in its quirky juxtapositions, contemporary free form is more often about flow and seamless segues.

Besides community and public radio stations, most college radio stations also program free-form music to some extent, although the preferred term is "alternative radio." College radio programmers, however, typically lack a historical awareness of free form and have little concept of its cultural implications. Also, college radio's alternative programming is rarely as diverse as free form heard on community and public stations.

Free form's most recent manifestation is on the internet. Community, public, and college stations increasingly broadcast via the web, and some internet-only stations—often the efforts of individuals working from home—advertise themselves as free-form radio. The internet is also an important meeting place for free form enthusiasts, whose web pages and chat groups provide means of sharing information and ideas.

WARREN BAREISS

See also Internet Radio; KPFA; Pacifica Foundation; WBAI

Further Reading

Armstrong, David, *A Trumpet to Arms: Alternative Media in America,* Los Angeles: Tarcher, 1981

Bareiss, Warren, "Space, Identity, and Public-Access Media: A Case Study of Alternative Radio Station KUNM-Albuquerque," Ph.D. diss., Indiana University, 1997

"Digging FM Rock," *Newsweek* (4 March 1968)

Engelman, Ralph, *Public Radio and Television in America: A Political History,* Thousand Oaks, California: Sage, 1996

Keith, Michael C., *Voices in the Purple Haze: Underground Radio and the Sixties,* Westport, Connecticut: Praeger, 1997

Knopper, Steve, "Free-Form Radio Still on the Dial," *Billboard* (14 January 1995)

Krieger, Susan, *Hip Capitalism,* Beverly Hills, California: Sage, 1979

Land, Jeffrey Richard, *Active Radio: Pacifica's "Brash Experiment,"* Minneapolis: University of Minnesota Press, 1999

Lasar, Matthew, *Pacifica Radio: The Rise of an Alternative Network,* Philadelphia, Pennsylvania: Temple University Press, 1999

Mayor, Alfred, "Accent: Radio Free New York," *Atlantic Monthly* (May 1968)

Frequency Allocation

Providing Spectrum for Broadcasting

Governments allocate bands of frequencies, including radio frequencies, for specific uses. Frequency allocation meshes technical limits and options with political and economic realities to create the compromise solutions behind today's broadcast services.

Three definitions are useful. *Allocation* is the broadest division of the electromagnetic spectrum into designated bands for given services (such as AM radio in the medium waves or FM in the very-high-frequency [VHF] spectrum). *Allotments* fall within allocations—they are given channels that are designated for specific places (only FM and television broadcasting have allotments). Finally, *assignments* are allotments that have actual users operating on them (such as a given station using 98.1 megahertz, for example)—they are virtually the same as a license to operate.

Frequency allocation can be examined under three broad rubrics. First, frequencies are allocated to classes of service. All radio signals that travel through the air use frequencies that are part of the electromagnetic spectrum. By international agreement this natural resource is divided into bands in which certain kinds of broadcasting occur. Medium-wave (or AM) radio occurs in one part of the spectrum, VHF (or FM) radio in another, VHF and ultrahigh-frequency (UHF) television in others, cellular telephony in another, satellite communication in another, and so on.

The frequencies used by different services are a function of three circumstances. First is history. Early experimentation with certain kinds of broadcasting resulted in assumptions that final allocations for that service should occur in the bands, or at the frequencies, originally used. This is because radio and television sets are designed to detect and amplify certain frequencies. Therefore, once such devices begin to be sold, changing the frequency of the service they were designed to use would make them obsolete. This can happen—for instance, in the United States the frequencies used to broadcast VHF/FM were changed in 1945—but the presumption is against such changes if they can be avoided.

Second are the technical needs of a particular service compared to the characteristics of certain portions of the spectrum. For instance, lower frequencies, such as those used for medium-wave/AM radio or shortwave radio, travel farther and propagate in ways that make it possible for them to bypass barriers more effectively than higher frequencies, such as those used for VHF/FM. This makes shortwave an effective means to broadcast transcontinentally or across oceans, medium-wave an effective means to provide national radio services (or international services to contiguous countries), and VHF an effective means to provide local radio services. Satellite television signals are at such high frequencies that they are effectively blocked by buildings, trees, or other obstacles. Such frequencies would be relatively useless if they were used by terrestrial (or land-based) transmitters, but because the satellite signals travel essentially vertically (from the sky to the earth), they can be used for this service as long as the dishes for receiving them are clear of obstacles. They can be affected by electrical storms or heavy thunderstorms, however, so some disruptions of service are inevitable.

Third are the political compromises made by the signatory administrations (or countries) that sign the allocation agreements under the auspices of the International Telecommunication Union (ITU). Such agreements, for instance, can result in an altered frequency band assignment for a particular service, despite uses of another portion of the spectrum in some countries. This is usually the result of a recognized need to rationalize frequency allocations so that transmission and reception devices can be designed using a worldwide standard. Otherwise the economies of scale may not achieve maximum impact, and the devices made may not be manufactured or sold as inexpensively as they would be otherwise. International broadcasting would be impossible if there were not an international allocation for such services, because the radio sets used to listen to them could not tune the same frequencies from country to country.

Frequencies are also allocated within classes of service to particular countries. Some frequencies assigned are exclusive, and others are shared. The less powerful a station is, the less distance its signals travel. Consequently, it can share its frequency with other stations located at a sufficient distance to avoid interference. This is easier with VHF/FM than with medium-wave/AM, because FM signals travel only by line of sight, whereas the propagation characteristics of amplitude-modulated (AM) signals change at dusk, traveling farther via night-time sky waves, which bounce off a layer of the ionosphere and return many hundreds of miles from their origination point. Countries contiguous to one another must share the total frequency allocation for a particular type of service within its region.

Frequencies are also assigned within particular allocations to particular users (or broadcasters). Different carrier frequencies (the center point of a channel—the frequency that appears on your receiver when you tune a specific station) are assigned to individual stations. In the United States, frequencies are assigned by the Federal Communications Commission (FCC) by means of broadcasting licenses. These licenses stipulate the channel (or band of frequencies centered on the carrier) that a station is to use to broadcast, the power it can use (according to the class of service it is licensed to provide), and its hours of operation. AM stations, for instance, use the 535-to1705-kilohertz band, and FM stations use the 88.1-to107.9-megahertz band. There are 117 AM carrier frequencies that can be assigned and 100 FM frequencies. Because of the propagation characteristics of these two services and the differences in bandwidth (10-kilohertz bandwidths for AM stations and 200-kilohertz bandwidths for FM), there were in 2000 about 4,900 AM stations and more than 6,700 FM stations in the United States. In January 2000 the FCC also began a new class of FM service, allowing both 100-watt and 10-watt stations, which will add many new low-power FM stations to the American broadcast landscape.

The principal exception to these general rules for frequency allocation is the frequencies used for international broadcasting in the shortwave portion of the spectrum. Here, individual stations are not assigned particular frequencies or broadcast power to use. Because the amplitude-modulated carrier waves of shortwave stations have the same propagation characteristics as AM waves generally, shortwave stations must change their frequencies as the seasons change (because the sunspot cycle moves from inactive to active every 17 years) and often as the time of day changes. This is why such stations register their "demands" with the international Frequency Registration Board (FRB), part of the ITU. By registering, they can discover whether they are attempting to use the same frequency as another broadcaster in the same part of the world. Often stations will also collaborate to ensure that their broadcasts will not interfere with one another's.

When radio was just beginning to be used in the early part of the 20th century, scientists believed that there were a limited number of frequencies suitable for broadcasting. When the first stations began to go on the air in about 1919, they used the same few frequencies, and there was significant interference between stations. Shortwave was given its name because people believed that any wavelengths shorter than those first used for radio would be unusable. This was because the shorter the wavelength (and thus the greater the frequency per second with which a wave crosses a particular plane), the more power it takes to move a wave a given distance. In other words, the longer the wave, the farther it will travel with a given transmitter. People thought that if wavelengths became shorter than those used by shortwave, the power required to make them usable would be prohibitively high.

Transmitters have become more efficient, however, and new forms of broadcasting (such as frequency modulation and digital broadcasting) have developed that continue to open up new frequencies for use. At the 1992 World Administrative Radio Conference (WARC), the participants provided new allocations for broadcast satellite service. For audio (or sound) broadcasting, the frequencies 1452–1492 megahertz, 2310–2360 megahertz, and 2535–2655 megahertz were agreed to, and the FCC subsequently allocated the spectrum 2310–2360 megahertz based on the international allocation adopted for the United States by the 1992 WARC for a Digital Audio Radio Service. The 1992 WARC also adopted an even higher set of frequencies for broadcast satellite service for high-definition television, with 17.3–17.8 gigahertz assigned to region 2 and 21.4–22.0 gigahertz for regions 1 and 3. All these new allocations will become effective 1 April 2007.

As seen in Table 1, the ITU has allocated the bands in the electromagnetic spectrum for various uses.

The pattern in these allocations is easy to see. It is useful to note that there are only 20,000 hertz in band 4, 299,970 hertz in band 5, and 2,999,700 hertz in band 6. As the frequencies used for broadcasting rise, the total amount of spectrum available increases not arithmetically (as, for instance, the band numbers do), but exponentially. What this means in practical terms is that the amount of spectrum now available for services has enabled enormously more service, more competition, and more exclusive service allocations at ever-higher frequencies. For broadcasting, this has also meant the opportunity to expand bandwidth as the frequency allocations have risen, thus allowing for higher-fidelity transmissions. Whereas in the AM band, bandwidths of 10 kilohertz only allow stations to broadcast about half of the frequency response that is within human hearing range, with VHF/FM broadcasting two signals (left and right) can be broadcast using the entire 20-kilohertz range and still leave room for sideband broadcasting, guard bands to prevent cross-channel interference, and a broadcasting envelope to prevent atmospheric or manmade interference. Use of even higher frequencies allows the broadcasting of multiple CD-quality digital signals in the same channel, which digital radio delivered by satellite will deliver.

All frequency allocations are based on the use of hertz (or cycles per second) generated by a broadcast transmitter (hertz are named for Heinrich Hertz, whose experiments led to recognition of cycles generated by sound). Human hearing, for instance, can decipher the frequencies from about 20 hertz to 20,000 hertz (or 20 kilohertz). Any vibrating object creates waves at a particular frequency. Large objects (such as kettle drums or tubas) generate mostly low frequencies, whereas smaller ones (such as flutes or piccolos) generate mostly high frequencies. Tuba sounds travel farther than piccolo sounds do. The same principle applies to broadcast transmitters that generate the carrier waves upon or within which sound is carried to radio or television receivers, with some reservations. Low-frequency signals tend to travel along the ground, and much of their power is absorbed by the earth. As the frequencies increase, more of the signal travels through the air than along the ground, and gradually more of it also becomes a sky wave, which travels up and bounces back to earth. These char-

Table 1. International Telecommunication Union Band Allocations

Band Number	*Frequencies*	*Designation*	*Some Designated Uses*
4	10–30 kHz	Very low frequency	Long distance point-to-point broadcasting
5	30–300 kHz	Low frequency	Medium distance point-to-point broadcasting, radio navigation, aeronautical mobile, low-frequency broadcasting
6	300–3000 kHz	Medium frequency	AM broadcasting, short-range communication, international distress
7	3–30 MHz	High frequency	International radio broadcasting; air-to-ground, ship-to-shore, and international point-to-point broadcasting
8	30–300 MHz	Very high frequency	Line-of-sight communication, VHF television broadcasting, FM broadcasting, aeronautical distress
9	300–3000 MHz	Ultrahigh frequency	UHF television broadcasting, space communication, radar, citizens band radio
10	3–30 GHz	Superhigh frequency	Microwave communication, space communication
11	30–300 GHz	Extremely high frequency	Microwave communication, space communication, radar, radio astronomy

acteristics mean that less power will actually move a wave at a higher frequency farther than a wave at very low frequencies, despite the fact that the wavelengths are lower at the higher frequency (wavelength and frequency are in inverse relationship). Therefore, when shortwave propagation was discovered in 1921, it was possible to reach as far with a 1-kilowatt transmitter as organizations had used 200 kilowatts to do before using the ground wave of low-frequency broadcasting.

Frequency allocations thus have to be made with several interrelated factors in mind: (1) the type of propagation that will occur at a given frequency (ground, direct, or sky); (2) the type of service that is to be accomplished with a particular allocation (local, national, or international, via terrestrial or satellite transmission); (3) the fidelity required for the service to be provided (for instance, voice, music, video, or CD quality) and the bandwidth necessary to provide that service; (4) whether the allocations must be exclusive or can be shared with other services; (5) existing experimental or other uses that a particular set of frequencies have been put to (thus providing what are called "squatter's rights"); and (6) the political realities of allocation among the different administrations that seek to employ the frequencies for particular uses.

ROBERT S. FORTNER

See also AM Radio; Clear Channel Stations; Digital Satellite Radio; Federal Communications Commission; FM Radio; Ground Wave; Hertz, Heinrich; International Telecommunication Union; Licensing; North American Regional Broadcasting Agreement; Portable Radio Stations; Shortwave Radio; Subsidiary Communication Authorization; Ten-Watt Stations

Further Reading

Glatzer, Hal, *Who Owns the Rainbow? Conserving the Radio Spectrum,* Indianapolis, Indiana: Sams, 1984

Gosling, William, *Radio Spectrum Conservation,* Oxford and Boston: Newnes, 2000

Jackson, Charles, "The Frequency Spectrum," *Scientific American* (September 1980)

Joint Technical Advisory Committee, *Radio Spectrum Conservation: A Program of Conservation Based on Present Uses and Future Needs,* New York: McGraw-Hill, 1952; revised and expanded edition, as *Radio Spectrum Utilization: A Program for the Administration of the Radio Spectrum,* New York: Institute of Electrical and Electronics Engineers, 1964

Levin, Harvey Joshua, *The Invisible Resource: Use and Regulation of the Radio Spectrum,* Baltimore, Maryland: Johns Hopkins University Press, 1971

Withers, David J., *Radio Spectrum Management: Management of the Spectrum and Regulation of Radio Services,* London: Peregrinus, 1991; 2nd edition, London: Institution of Electrical Engineers, 1999

Fresh Air

Public Radio Arts and Issues Program

Fresh Air host Terry Gross refers to herself and her production team as "culture scouts," seeking the latest in arts, ideas, and issues. The program is one of the most popular on public radio, drawing a weekly audience of more than 4 million listeners on some 435 stations. *Fresh Air*'s Peabody Award citation in 1994 noted that "unlike the cacophony of voices that sometimes obscure and polarize contemporary debate, Ms. Gross asks thoughtful, unexpected questions, and allows her subjects time to frame their answers." *Fresh Air*'s guests have ranged from former First Lady Nancy Reagan to filmmaker Martin Scorsese, from hostage negotiator Terry Waite to novelist Joyce Carol Oates, singer Tony Bennett, playwright David Mamet, and thousands more.

Over the years, *Fresh Air* has evolved from a live, three-hour local program to a highly produced hour-long program that runs nationally. David Karpoff created the show in 1974 when he was program director at WHYY (then WUHY) in Philadelphia; he modeled it on *This Is Radio,* a program he had worked on at WBFO in Buffalo. Karpoff was the first host, interspersing live interviews with classical music. He was followed as host by Judy Blank, and when she moved on, Karpoff in 1975 hired Terry Gross, who had been co-hosting and producing *This Is Radio* in Buffalo.

Gross drew complaints by changing *Fresh Air*'s music to jazz, blues, and rock and roll, but she won listeners over with an interview style that was thoughtful and direct. Gross, who was 24 at the time, had broken into radio just two years earlier when she helped produce and host a feminist program at WBFO.

By 1978, when Bill Siemering arrived as station manager at WHYY, Gross was carrying on the entire three-hour program by herself, "playing records," Siemering recalls, "that were just long enough to show one guest out and lead another one in." As a former station manager at WBFO, Siemering had created *This Is Radio* out of the turmoil of campus protest and had then gone on to develop *All Things Considered* as a program director at National Public Radio (NPR). When he arrived at WHYY (then WUHY), Siemering says the station was in a rundown building in West Philadelphia, where the ladies' room plumbing leaked onto Gross' desk. Siemering got a Corporation for Public Broadcasting grant to upgrade the station and was able to hire intern Danny Miller as an assistant producer for *Fresh Air.*

Miller, who eventually became the program's co-executive producer, says the fact that *Fresh Air* began as a local show is an important source of its strength: "The show had years to mature before it went national." Part of that maturation involved cutting back the amount of time on the air from three hours to two in 1983, because, as Gross says, "Danny and I often felt that in order to fill the airtime, we were forced to focus more on the quantity than the quality of guests." At the same time, Gross and Miller added a weekend "best of" edition of the show that became the seed for the weekly national edition that was to follow.

Gross sees the development of the show as a step-by-step evolution. The next step came in the spring of 1985, when WHYY premiered a weekly 30-minute version of *Fresh Air,* distributed by NPR. It appeared at a time when public radio stations on the East Coast were pressing for an earlier start time for the popular newsmagazine *All Things Considered,* which would enable them to capture more of the drive-time audience. Robert Siegel, then the news director at NPR, resisted the idea, feeling that the show was already stretched to meet a 5 P.M. deadline. Siegel saw *Fresh Air* as an answer to the demand for a 4 P.M. start, because its sensibility matched that of *All Things Considered* without duplicating its news content. "It was a very good program," he says, "and Terry is the best interviewer in public radio."

Gross says the program was reconceived in 1987 as a daily arts-and-culture companion to *All Things Considered.* To integrate it further into the *All Things Considered* sound, the new format included a drop-in newscast. *Fresh Air*'s shorter interviews, reviews, and other features were put in the second half hour so that its pace would match that of the newsmagazine as listeners went from one program to the next. The new *Fresh Air* also featured a recorded interchange between Gross and the hosts of *All Things Considered* in which they discussed what was coming up.

The national version of *Fresh Air* was a hit, both with audiences and with program directors, who liked the show's predictable format because it gave them the flexibility to drop in local material during drive time. However, Gross says she and co-producer Danny Miller grew to feel imprisoned by the rigidity of the format. They were glad, therefore, when the 1991 Gulf War brought new demands on everyone. Even though *Fresh Air* had concentrated on arts and culture, Gross says, "we had to address the war. Everybody was rightly obsessed with it." The producers sought interviews that could supplement the news, looking for what Danny Miller calls "the great explainers," experts on the culture and the history of the region. "Emergencies require change," Gross says, "and emergencies justify change. If an interview ran more than a half hour, we let it." The war coverage restored some of *Fresh Air's* flexibility and expanded its portfolio to include a full range of contemporary issues.

Terry Gross, host of *Fresh Air*
Courtesy National Public Radio

Regardless of the subject, Gross applies the same demanding preparation for each interview, reading each author's books, viewing the films, and listening to the CDs that she will discuss. Interviews typically last between 20 and 40 minutes, during which Gross gives her guests the opportunity to rethink and rephrase their answers if they feel they can express themselves better. She prides herself on treating guests with respect, but "that doesn't mean I won't challenge you." Gross' critics have complained that she does not ask confrontational questions, but Bill Siemering says her method is much more effective. "If you're on the attack, all you get is their defense. If you're respectful, you get a lot more."

After nearly three decades on the air, *Fresh Air*'s archives are filled with the voices and thoughts of some of the era's most interesting people, what Gross calls "a scattershot history of American culture."

COREY FLINTOFF

See also National Public Radio; Peabody Awards; Public Radio Since 1967; Siemering, William

Hosts
David Karpoff, 1974
Judy Blank, 1974–75
Terry Gross, 1975–

Producer/Creator
David Karpoff

Co-producers
Terry Gross and Danny Miller

Programming History

WHYY	1974–85
National Public Radio	1985–

(Note: WHYY continued to produce the program after NPR commenced distribution in 1985)

Further Reading

Clark, Kenneth R., "A Talent for Conversation: NPR's Terry Gross Provides Talk Radio with a Breath of Fresh Air," *Chicago Tribune* (19 November 1993)

Dart, Bob, "Fresh and Funky: 'Cultural Scout' for Public Radio Homes In on Truth," *Cox News Service* (January 1994)

Scheib, Rebecca, "Media Diet: Terry Gross," *Utne Reader* (March–April 1997)

White, Edmund, "Talk Radio: Terry Gross," *Vogue* (May 1997)

Friendly, Fred 1915–1998

U.S. Broadcast Journalist

Although best known for his work in television journalism and public-affairs programming at the Columbia Broadcasting System (CBS) and at the Public Broadcasting Service (PBS), Fred Friendly's broadcasting roots were in radio. His experience in radio and recording prior to 1950—in Providence, Rhode Island; in the China-Burma-India theater during World War II; and in New York City at the National Broadcasting Company (NBC) and CBS in the late 1940s—provided his entrée into network journalism and influenced his subsequent work in television.

In 1921, Friendly and his father, a jewelry maker, built a radio receiver and listened to the Jack Dempsey–Georges Carpentier championship boxing match broadcast by KDKA in Pittsburgh. Friendly first went on the air in the early 1930s as a high school student on WJAR in Providence, Rhode Island, to perform in a one-act play, *The Valiant,* written by H.E. Porter and R. Middlemass.

Friendly's broadcast career began in earnest in 1936 at WEAN, Rhode Island's first radio station, which was part of the Yankee network and affiliated with the NBC Blue network. At WEAN, Mowry Lowe, a Rhode Island broadcasting pioneer, served as Friendly's mentor. From 1937 to 1941, Friendly produced and broadcast *Footprints in the Sands of Time,* daily five-minute biographies of important historical figures, for which a sponsor paid Friendly $5 per program, in addition to his $35-a-week salary (see Bliss, 1991). Decca Records later purchased the series. Friendly also worked as an announcer at WEAN and appeared on Mowry Lowe's *Sidewalk Backtalk,* a man-in-the-street interview program. During the same period, he operated the Fred Friendly Company, characterized on its letterhead as "A Radio Production Service to Advertising Agencies."

At WEAN, Friendly made occasional news reports. For example, he helped cover the great hurricane of September 1938 in Rhode Island. In July 1941 the Mutual Broadcasting System carried his report on the opening of the naval marine air station in Quonset, Rhode Island. On 1 October 1999, National Public Radio rebroadcast the report on *All Things Considered* as "a piece of radio poetry." The broadcast evoked the verse of Walt Whitman and anticipated the themes of the army wartime film series *Why We Fight.* Friendly celebrated "$30 million worth of cement and steel . . . and mortar and sweat to keep America strong," accomplished by "a melting pot of O'Neils, Murphys, Gustafsons, and Joneses, and Cohens and Marinos."

During World War II, as a result of his radio experience, Friendly served in the Signal Corps and in the Information and Education Section of the army as a master sergeant. In the China-Burma-India theater, he lectured to troops and wrote for the army newspaper, *CBI Roundup.* In addition, Friendly made recordings of bomber runs for the historical archives of the army. He accomplished this by hooking up a wire recorder to a plane's intercom in order to capture the crew's communication, to which he would add his own running account of the mission. He also made recorded combat reports from the front lines of the Eastern Theater for the fledgling Armed Forces Radio Network. In addition, after visiting Europe to cover the Allied victory in the West and the liberation of the Mauthausen concentration camp in Austria, Friendly gave a radio report for Armed Forces Radio from New Delhi.

Following the war, Friendly had an idea for a record album made up of historical figures of the 1930s and 1940s, an extension of the idea behind *Footprints in the Sands of Time,* a project facilitated by the advent of magnetic tape. In 1947, the agent J.P. Gude introduced Friendly to Edward R. Murrow, who provided narration for the 46 voices—from Roosevelt and Churchill to Will Rogers and Lou Gehrig—heard on *I Can Hear It Now,* issued in December 1948 by Columbia Records. Friendly produced five additional records for the *I Can Hear It Now* series. The first album, a spectacular critical and commercial success, initiated the Friendly-Murrow collaboration, which has been called "the most productive, most influential partnership in the whole history of broadcast journalism" (Bliss, 1991).

On 2 July 1948 NBC first aired a radio show, conceived and produced by Friendly, called *Who Said That?* On the show, a panel of journalists and celebrities was given quotations reported during the previous week and tried to identify their sources. Robert Trout served as host, John Cameron Swayze as a regular panelist. In December 1948, Friendly launched a television version of *Who Said That?* on NBC that continued to be carried on radio, one of the first programs to be simulcast. The radio version remained on the air until 1950, the television program until 1955, by which time it had moved to ABC-TV.

In 1950, Friendly produced and directed for NBC *The Quick and the Dead,* a four-part radio dramatization and documentary on the history and future implications of atomic power. Friendly had witnessed the dawn of the nuclear age firsthand, flying in a reconnaissance plane over Hiroshima and Nagasaki several days after their destruction. In the radio documentary, Bob Hope played an average citizen posing questions that were answered by actors playing Albert Einstein and other figures instrumental in the development of the atomic bomb. Friendly collaborated with *New York Times* science

Fred Friendly
Courtesy of Ruth W. Friendly

reporter William Laurence on the program, which included interviews with actual scientists and statesmen as well as with crew members of the plane that dropped the first atomic bomb.

In 1950, as a result of critical acclaim for *The Quick and the Dead,* Sig Mickelson, director of public affairs at CBS, hired Friendly to join the network's radio documentary unit. Friendly immediately teamed up with Murrow to produce *Hear It Now,* a prime-time hour-long radio news magazine carried on 173 stations. The acclaimed program, which combined audiotaped actualities and interviews with newsmakers, prompted imitations at NBC (*Voices and Events*) and at ABC (*Week Around the World*).

Hear It Now lasted only one season before being transformed by the Murrow–Friendly team into the groundbreaking television program *See It Now* (1951–58), with Friendly as its executive producer. Friendly subsequently worked at CBS television as executive director of *CBS Reports* (1959–64) and as president of CBS News (1964–66). After leaving CBS, Friendly served as broadcasting advisor to the Ford Foundation (1966–80), as Edward R. Murrow Professor of Journalism at Columbia University (1968–79), and as originator and director of the Columbia University Seminars on Media and Society (1974–98), which have been broadcast since 1981 on PBS.

A radio sensibility persisted throughout Friendly's television career. Don Hewitt, executive producer of *60 Minutes* and a member of the original *See It Now* team, said that Fred Friendly taught him his most valuable lesson in television: "It is your ear more than your eye that keeps you at the television set. . . . The picture brings you there, and what you hear keeps you there."

RALPH ENGELMAN

See also Hear It Now; Murrow, Edward R.; News; Peabody Awards; World War II and U.S. Radio

Fred W. Friendly. Born Ferdinand Friendly Wachenheimer in New York City, 30 October 1915. Attended Nichols Junior College, 1936; served in U.S. Army, Information and Education Section, Signal Corps instructor, 1941–45; announcer, newscaster, and producer, WEAN-AM, Providence, Rhode Island, 1937–41; correspondent for *CBI Roundup*, U.S. Army, 1941–45; co-creator with Edward R. Murrow, *I Can Hear It Now,* 1948–51; writer and producer of various radio and television shows, 1948–64; president of CBS news, 1964–66; broadcasting advisor, Ford Foundation, 1966–80; Edward R. Murrow Professor of Journalism, Columbia University, 1966–98; director, Columbia University Seminars on Media and Society, 1984–98; director, Michele Clark Program for minority journalists, Columbia University, 1968–75; member, Mayor's Task Force on CATV and Telecommunications, New York City, 1968; visiting professor, Bryn Mawr College, 1981; visiting professor, Yale University, 1984; commissioner, Charter Revision Committee for City of New York, 1986–90; Montgomery fellow, Dartmouth College, 1986. Received Decorated Legion of Merit and four battle stars; Soldier's Medal for heroism; 35 awards for *See It Now,* including the Overseas Press Club Award, Page One Award, New York Newspaper Guild, and National Headliners Club Award, 1954; 40 awards for *CBS Reports;* Theatre Library Association Award, 1977; 10 George Foster Peabody Broadcasting Awards; honorary degrees from various U.S. universities. Died in New York City, 3 March 1998.

Radio Series

1938–40	*Footprints in the Sands of Time*
1948–50	*Who Said That?*
1950	*The Quick and the Dead*
1950–51	*Hear It Now*

Television Series

See It Now, 1948–55; *CBS Reports,* 1959–64; *Columbia University Seminars on Media and Society* (also known as *The Fred Friendly Seminars),* 1984–98

Selected Publications

See It Now (with Edward R. Murrow), 1955
Due to Circumstances beyond Our Control, 1967
The Good Guys, the Bad Guys, and the First Amendment: Free Speech v. Fairness in Broadcasting, 1976
Minnesota Rag: The Dramatic Story of the Landmark Supreme Court Case That Gave New Meaning to Freedom of the Press, 1981
The Constitution: That Delicate Balance (with Martha J. H. Elliot), 1984

Further Reading

Barnouw, Erik, *A History of Broadcasting in the United States,* 3 vols., New York: Oxford University Press, 1966–70
Bliss, Edward, *Now the News: The Story of Broadcast Journalism,* New York: Columbia University Press, 1991
"A Bloody Test of Wills at CBS," *Broadcasting* (21 February 1966)
Gates, Gary Paul, *Air Time: The Inside Story of CBS News,* New York: Harper and Row, 1978
Halberstam, David, *The Powers That Be,* New York: Knopf, 1979
Mickelson, Sig, *The Decade That Shaped Television News: CBS in the 1950s,* Westport, Connecticut: Praeger, 1998
Pace, Eric, "Fred W. Friendly, CBS Executive," *New York Times* (5 March 1998)
Smith, Sally Bedell, *In All His Glory: The Life of William S. Paley,* New York: Simon and Schuster, 1990
Sperber, Ann M., *Murrow, His Life and Times,* New York: Freundlich Books, 1986

video clips of life and love projected across their television screen. For some musicians, it was time for the next step—queer music, in which gays and lesbians would share their experiences about lovers and life partners in their music.

The formation of the National Lesbian and Gay Country Music Association in 1998 was just one example of the trend toward acceptance of queer music in the United States. Other signs toward changing times include the acceptance of country musicians such as Mark Weigle, an independent singer/songwriter who has received regular airplay on Americana shows in Europe. He was nominated (along with Ani DiFranco, Rufus Wainwright, and the B-52's) for two 1999 Gay/Lesbian American Music Awards (GLAMA)—for both Debut Artist and Out Song "If It Wasn't Love." Weigle has been praised by music critics in gay and mainstream music publications such as *Genre, Billboard,* and *Performing Songwriter.* GLAMA is the first and only national music awards program to celebrate and honor the music of queer musicians and songwriters. Its first annual ceremony took place in New York's Webster Hall in October 1996, with about 700 people in attendance. The judges comprised music reviewers from the gay and mainstream media; leaders in radio distribution; executives of major and independent record labels; and those working in performing rights, talent management, and retail. GLAMA's final awards ceremony was held in April 2000; increasing administrative costs were cited as the reason for its demise.

Beyond the Gay Ghetto

In April 1988 Los Angeles radio producer Greg Gordon, a former *imru* producer and host (along with his volunteer staff), created the 30-minute newsmagazine program called *This Way Out,* which began with a weekly distribution to 26 public stations in the United States and Canada. It now airs on more than 125 radio stations—public and commercial—in six countries. The program contains news, author interviews, AIDS updates, humor, poetry, and music recordings by openly gay and lesbian performers rarely heard on commercial radio. The Gay and Lesbian Press Association honored *This Way Out* with its Outstanding Achievement Award in 1988, and the National Federation of Community Broadcasters presented the "Silver Reel" award to its producers in 1991 for their ongoing news and public-affairs commitment to cultural diversity. One affiliate station, Kansas City's KKFI-FM, boasts that more than 500 individuals have presented a myriad of gay, lesbian, and transgender issues across its airwaves. In fact, KKFI is the first and thus far the only station in its market to air a local queer radio program. To some gay media activists, however, the failure to move beyond noncommercial radio and into the mainstream has been a form of "gay ghettoization," or what some queer broadcasters have referred to as "preaching to the choir." For years, noncommercial radio has been the primary vehicle for communicating the queer perspective to the gay community, and increasingly to a straight audience. In 1992 WFNX-FM, a commercial Boston station, debuted *One in Ten,* a three-hour show with a mix of news, entertainment, music, and call-in discussions. In the early 1990s, several other commercial stations experimented, although unsuccessfully, with locally produced programs targeting gay and lesbian audiences.

In 1990 Thomas Davis became the president and general manager of two Amherst commercial stations located in a renowned gay and lesbian community in central Massachusetts. His company's mission was to target listeners outside the gay community, in addition to gays and lesbians themselves. In doing so, the stations would attempt to convey the idea that gays and lesbians function much like any other members of society on a daily basis. As institutional members of the Gay and Lesbian Business Coalition, the stations have been supported on the air by many gay- or lesbian-owned businesses. Programming includes news stories relevant to gays and lesbians, as well as fund-raising efforts for a number of queer community concerns, such as AIDS research and hospice funding.

1992 was the year of the largest commercial venture for gay radio. The KGAY Radio network signed on the air on 28 November 1992 in Denver. KGAY, with its motto "All Gay, All Day," was the inspiration of Clay Henderson and Will Guthrie. Their previous efforts had included a short-lived 30-minute weekly gay commercial radio show and a weekly gay and lesbian news show called the *Lambda Report,* which first aired on public access cable television in 1989. KGAY Radio was the first attempt to market a 24-hour all gay and lesbian format in America. The KGAY founders planned to use a local Denver station as the headquarters for what they hoped would become a national cable FM operation. The decision to broadcast by satellite from Denver seemed easier and less expensive to the owners than purchasing a commercial radio station in a large or medium market. KGAY's programming was uplinked to satellite dishes in North America, Canada, and the Caribbean. The music and news network was promoted as the first daily media vehicle for the gay and lesbian community in North America. Less than a year after it began, however, the KGAY network, with only a few sponsors and a mostly volunteer staff, failed.

In May 1994 another commercial radio venture was born on adult contemporary WCBR-FM in north Chicago. LesBiGay Radio founder Alan Amberg conversed insightfully every weekday afternoon on America's only drive-time gay radio show, as his signal reached into the Chicago neighborhoods where many gays and lesbians resided. By April 2001, his radio enterprise had logged more than 3,000 hours of programming and had a number of prominent national and local sponsors. Indeed, the show was hailed as the most successful commercial queer radio venture of the 1990s in America, but Amberg was forced to end it in April 2001 because of financial difficulties. The internet is also home to a number of gay radio programs that can only be heard via the web, and this appears to be the trend among gay radio broadcasters who wish to seek a larger audience base to justify commercial sponsorship.

PHYLIS JOHNSON

See also Affirmative Action; Pacifica Foundation; Stereotypes on Radio

Further Reading

Alwood, Edward, *Straight News: Gays, Lesbians, and the News Media,* New York: Columbia University Press, 1996
Gross, Larry, and James D. Woods, editors, *The Columbia Reader on Lesbians and Gay Men in Media, Society, and Politics,* New York: Columbia University Press, 1999
Hendriks, Aart, Rob Tielman, and Evert van der Veen, editors, *The Third Pink Book,* Buffalo, New York: Prometheus Books, 1993
Hogan, Steve, and Lee Hudson, *Completely Queer: The Gay and Lesbian Encyclopedia,* New York: Holt, 1998
Johnson, Phylis, Charles Hoy, and Dhyana Ziegler, "A Case Study of KGAY: The Rise and Fall of the First 'Gay and Lesbian' Radio Network," *Journal of Radio Studies* 3 (1995–96)
Johnson, Phylis, and Michael C. Keith, *Queer Airways: The Story of Gay and Lesbian Broadcasting,* Armonk, New York: M.E. Sharpe, 2001
Lasar, Matthew, *Pacifica Radio: The Rise of an Alternative Network,* Philadelphia, Pennsylvania: Temple University Press, 1999
Schulman, Sarah, *My American History: Lesbian and Gay Life during the Reagan/Bush Years,* New York: Routledge, 1994
"Trends in the Making," *Advertising* (11 August 1998)

General Electric

Manufacturer of Consumer Electronics

General Electric (GE) was instrumental in the shaping of early American radio broadcasting through its creation of the Radio Corporation of America (RCA) in 1919 and its subsequent operation of several pioneering radio stations. GE is a diversified company with holdings in consumer services, technology, and manufacturing. GE operates in more than 100 countries and employs nearly 340,000 people worldwide, including 197,000 in the United States.

Origins

General Electric traces its history to the Edison Electric Light Company, established in 1878 by Thomas Edison, and to the Thomson-Houston Electric Company, established by Elihu Thomson and Edwin Houston in the early 1880s. Both companies grew by the 1890s into leaders in their field and battled over adoption of electrical current standards for the United States. Thomson-Houston promoted alternating current, whereas Edison championed use of direct current. Alternating current was adopted in the United States, and in 1892 the two merged to form GE. GE is the only company that has been listed continuously on the Dow Jones Industrial Index since its inception in 1896.

After the merger, the new company could boast of having some of the best minds in the country. Thomson's financial genius Charles Coffin became GE's first president, and Edison became a director. Thomson helped establish a program of scientific research that led to the creation of GE's Research Laboratory in 1900 under the direction of Dr. Willis R. Whitney. In a career spanning five decades, Thomson was awarded 696 U.S. patents for devices as varied as arc lights, generators, X-ray tubes, and electric welding machines. His successful "recording wattmeter" was a practical method of measuring the amount of electricity used by a home or business. In 1893 a young German, Charles Steinmentz, joined GE, and he designed new methods of designing machinery using alternating current. After the turn of the century, GE expanded its power-generation business by developing the first steam turbine-generator large enough to power cities. In 1903 GE purchased the Stanley Electrical Manufacturing Company of Pittsfield, Massachusetts. William Stanley, the head of that company, joined GE and pioneered electrical line transmission equipment. He is credited with inventing the transformer, which became the heart of the electrical distribution system. In 1910 GE developed ductile tungsten for light bulb filaments; ductile tungsten is still used in virtually every incandescent lamp.

GE and Radio

By the turn of the 20th century, GE began developmental work in wireless radio, and in 1906, E.F.W. Alexanderson developed a practical alternator to produce the high frequencies needed for reliable long-distance transmission. Later, Dr. Irving Langmuir of GE's Research Laboratory designed an amplifier for Alexanderson's alternator, completing the components of a transoceanic radio-transmitting system. This system was the most powerful generator of radio waves then known, and it became the pivotal point for negotiations with American Marconi, the U.S. subsidiary of British Marconi's worldwide wireless enterprise; these negotiations eventually led to the formation of RCA.

During World War I, the navy had operated most radio stations in the United States, including those using GE's Alexanderson alternator. Naval radio experts became convinced that this equipment was vital to U.S. interests, and after the war Admiral W.H.G. Bullard and Captain Stanford Hooper convinced GE executive Owen D. Young not to sell improved vacuum tubes and the exclusive rights to the Alexanderson alternator to the British subsidiary. In a meeting in May 1919, Young told E.J. Nally of American Marconi that GE would not sell the equipment because the U.S. government did not want control of this equipment to pass into foreign hands. Over the summer the two companies' officers negotiated the sale of American Marconi's assets to GE, and the deal they struck resulted in RCA's incorporation on 17 October 1919.

Shortly after its incorporation, RCA and GE entered into patent cross-licensing agreements with American Telephone and Telegraph (AT&T) and Westinghouse. Under these agreements, all signatories shared their patents. With this patent pooling, RCA quickly became one of the leading companies manufacturing and selling wireless radio equipment. By 1922 the company had moved into wireless' newest application—broadcasting. Radio station WGY began broadcasting from Schenectady, New York, with one of the first U.S. radio dramas, *The Wolf*. As this new phenomenon caught on with the American public, GE, with RCA and Westinghouse, became engaged in a series of intercompany battles with AT&T and its manufacturing arm, Western Electric, over who held what rights in broadcasting. To resolve this issue, AT&T sold its radio interests, and RCA created the National Broadcasting Corporation (NBC) in 1926 to oversee network endeavors in broadcasting.

In the early 1930s the Justice Department filed an antitrust suit against General Electric, Westinghouse, and RCA, alleging restraint of trade. In 1932 the companies signed a consent decree, and GE was forced to divest itself of any interest in RCA.

During the Depression years, GE made important improvements and contributions to X rays, electric ship turbines, and the efficiency of electrical light and appliances. In 1940 GE expanded its business by relaying television broadcasts from New York City and by starting FM broadcasts. During World War II, these operations were suspended, as GE turned its efforts to helping win the war. The company supplied much help to the war effort, from aircraft gun turrets and jet engines, to radar and radio equipment, to electrically heated flying suits. GE made propulsion units for nearly 75 percent of the Navy's ships, and from GE laboratories came new systems for the detection of enemy aircraft and ships.

After the war, GE carried out an extensive program of expansion and decentralization. As a result, autonomous product departments developed. Over the years, GE also expanded its research and development divisions as well as its international efforts. Its many manufacturing lines ranged from consumer products such as light bulbs and consumer electronics, to major appliances such as refrigerators and television sets, to industrial and military equipment such as locomotives, aircraft engines, nuclear reactors, and ICBM guidance systems. In addition, GE moved into services such as insurance, consumer credit, and data processing. In 1957 GE opened the world's first licensed nuclear power plant and entered the mainframe computer business, which it sold in 1970 to Honeywell. Two years later GE developed the TIROS 1 weather satellite. During the 1960s and 1970s, the company continued expanding its lighting, aircraft engine, and electrical equipment businesses both domestically and abroad.

Modern GE

In late 1986 GE bought RCA (and NBC) in a $6.28 billion deal that some say ironically came full circle to the 1930s divestiture. The following July, GE sold off the NBC Radio Network to Westwood One for $50 million so that GE/RCA could concentrate on television and newer consumer media. At the same time, the company began selling the NBC-owned-and-operated radio stations. Over the next ten years, GE and RCA combined their consumer electronics businesses and subsequently sold them to Thomson in exchange for its medical equipment business and $800 million in cash. The companies' combined defense business was also sold, to Martin Marietta for $3 billion. GE/RCA retained ownership of the NBC television operations, and during the 1990s these NBC operations became exceedingly profitable for GE.

In 1989, NBC launched the business financial cable television network CNBC, and GE formed a mobile communications joint venture with Ericsson of Sweden. In 1991 NBC acquired the Financial News network (FNN) and sold its interest in the RCA Columbia Home Video joint venture. In 1994 GE created one of the first major industrial websites, \\www.ge.com\\, and two years later NBC and Microsoft joined forces to launch MSNBC, a 24-hour television and internet service. In 1999 GE began e-Business as a key growth initiative. NBC launched NBC Internet (NBCi), a publicly traded internet company that merged the network's interactive properties with XOOM.com and the internet portal Snap.com to form the seventh-largest internet site and the first publicly traded internet company integrated with a major broadcaster. NBCi will use Snap.com as an umbrella consumer brand, integrating broadcast, portal, and e-commerce services. NBC also maintains equity interests in cable channels Arts and Entertainment (A&E) and the History Channel. NBC also has an equity stake in Rainbow Programming Holdings, a leading media company with a wide array of entertainment and sports cable channels. It also holds interests in CNET, Talk City, iVillage, Telescan, Hoover's, and 24/7Media. In partnership with National Geographic and Fox/BskyB, NBC owns and operates the National Geographic Channel in Europe and Asia.

LOUISE BENJAMIN

See also Alexanderson, E.F.W.; National Broadcasting Company; Radio Corporation of America; Westinghouse

Further Reading

Alexander, Charles, "A Reunion of Technological Titans," *Time* (23 December 1985)

Barnum III, Frederick O., *His Master's Voice In America: Ninety Years of Communications Pioneering and Progress: Victor Talking Machine Company, Radio Corporation of America, General Electric Company,* Camden, New Jersey: General Electric Company, 1991

Case, Josephine Young, and Everett Needham Case, *Owen D. Young and American Enterprise: A Biography,* Boston: Godine, 1982

A Century of Progress: The General Electric Story, 1876–1978, Schenectady, New York: GE Hall of History, 1981

GE website, <www.ge.com>

Hammond, John Winthrop, *Men and Volts: The Story of General Electric,* Philadelphia: Lippincott, 1941

Lewis, Tom, *Empire of the Air: The Men Who Made Radio,* New York: Burlingame Books, 1991

Loomis, Carol J., "Ten Years After," *Fortune* (17 February 1997)

Nye, David E., *Image Worlds: Corporate Identities at General Electric, 1890–1930,* Cambridge, Massachusetts: MIT Press, 1985

Reich, Leonard S., *The Making of American Industrial Research: Science and Business and GE and Bell, 1876–1926,* New York: Cambridge University Press, 1985

The George Burns and Gracie Allen Show

Comedy Series

For 17 years, Burns and Allen provided one of radio's most enduring comedy series based on the vaudeville tradition.

Vaudeville Origins

George Burns, born Nathan Birnbaum in New York City on 20 January 1896, left school after fourth grade to sing professionally with the PeeWee Quartet. That move led to his career in vaudeville as a singer, dancer, and monologuist (1910–31). Grace Ethel Cecile Rosalie Allen was born 26 July 1906 in San Francisco. She met George after a 1922 New Jersey vaudeville performance he had done with his partner, Billy Lorraine. Gracie, who began in vaudeville at 14 with her three sisters, wanted to work with Lorraine. She teamed with George instead, and they were married 7 January 1926. They performed together in vaudeville, film, radio, and TV from 1922 until her retirement in 1958.

Gracie played "straight man" until George discovered she was funnier being nice yet dim-witted. George never considered changing Gracie, whom he loved because of her befuddlement and inverted logic. Gracie was a nervous performer, but audiences thought it was an act, part of her giddy, scatterbrained persona. She was uninterested in business, but George enjoyed other responsibilities as script supervisor and manager.

In 1929 Burns and Allen were invited to perform their Vaudeville act in London. After 21 weeks there performing to appreciative audiences at various nightclubs, the duo was asked to go on BBC radio—a successful stint that lasted 26 weeks. The team believed they could experience similar success on American radio.

George Burns and Gracie Allen
Courtesy Radio Hall of Fame

Network Success

In 1930 the National Broadcasting Company (NBC) rejected them because of Gracie's squeaky, high-pitched voice, but she performed on Eddie Cantor's show in 1931 to rave reviews. A week later Burns and Allen played Rudy Vallee's show, which led to their 15 February 1932 debut on the Columbia Broadcasting System's (CBS) *The Robert Burns Panatella Program.* In 1933 their popularity earned them their own show, which lasted 17 years. *The George Burns and Gracie Allen Show*—also known as *Maxwell House Coffee Time, Burns and Allen,* and *The Adventures of Gracie*—presented sketches, music, and vaudeville routines. Among the show's characters was Mel Blanc's Happy Postman, who spoke pleasant and cheerful thoughts but always sounded depressed and near crying.

Originally based in New York, the show made road broadcasts from other cities and military installations throughout the United States. By 1934, the couple began work in film and relocated to Beverly Hills with two adopted children, Sandra Jean and Ronald John. In 1942 ratings dropped temporarily and the sponsor in turn dropped the program. The premise of Burns and Allen as boyfriend and girlfriend no longer worked. George explained: "Everybody knew we were married and had growing children . . . you have to have truth in a joke just the way you do in anything else to make it any good. If it's basically dishonest, it isn't funny." George finally realized the problem was that he and Gracie were too old to do the boyfriend/girlfriend premise. The jokes were stale, so the format was changed to reflect their status as a married couple and the program experienced a renewal of popularity, attracting 45 million listeners per week.

Many promotional strategies featuring Gracie were employed for the program throughout its run. She went on a show-to-show search for her mythical brother. She ran for president in 1940, receiving several hundred votes. In 1942 bandleader Paul Whiteman wrote *Gracie Allen's Concerto for Index Finger,* which Gracie mentioned constantly on the show. Ultimately, she performed the number at Carnegie Hall and with major orchestras, including the Boston Pops.

George and Gracie were not social commentators, prevailing instead with timeless humor and talented performances. In 1950 they successfully moved to TV, and their 239 episodes continue to air in reruns decades later. George explained: "We talked in vaudeville, we talked in radio, we talked in television. It wasn't hard to go from one medium to another." Gracie retired after the final TV episode on 4 June 1958, to be Mrs. George Burns, a mother and a grandmother. Only after her 27 August 1964 death did the public discover she had retired because of heart problems.

In his book *Gracie: A Love Story,* George's first line read: "For forty years my act consisted of one joke. And then she died." George continued in TV as an actor, developer, and producer of *Wendy and Me, No Time for Sergeants, Mona McCluskey,* and *Mr. Ed.* In 1975 he returned to film, replacing his late friend Jack Benny in *The Sunshine Boys.* He also starred in the *Oh, God!* trilogy and a handful of other films. George continued TV guest appearances before being slowed by a 1994 fall. He performed in Las Vegas until his death on 9 March 1996.

W.A. KELLY HUFF

See also Comedy; Vaudeville

Cast

George Burns	Himself
Gracie Allen	Herself
The Happy Postman	Mel Blanc
Tootsie Stagwell	Elvia Allman
Mrs. Billingsley	Margaret Brayton
Muriel	Sara Berner
Waldo	Dick Crenna
Herman, the duck	Clarence Nash
Also featured	Gale Gordon, Hans Conried, Henry Blair
Vocalists	Milton Watson, Tony Martin, Jimmy Cash, Dick Foran
Bandleaders	Jacques Renard, Ray Noble, Paul Whiteman, Meredith Willson
Announcers	Ted Husing, Harry Von Zell, Jimmy Wallington, Bill Goodwin, Toby Reed

Directors
Ralph Levy, Al Kaye, Ed Gardner

Writers
Paul Henning, Keith Fowler, Harmon J. Alexander, Henry Garson, Aaron J. Ruben, Helen Gould Harvey, Hal Block, John P. Medbury

Producer/Creator
George Burns

Programming History
CBS and NBC 15 February 1932–17 May 1950

Further Reading
Blythe, Cheryl, and Susan Sackett, *Say Goodnight, Gracie! The Story of Burns and Allen,* New York: Dutton, 1986

Burns, George, *Gracie: A Love Story,* New York: Putnam, 1988

Burns, George, *All My Best Friends,* New York: Putnam, 1989

Campbell, Robert, *The Golden Years of Broadcasting: A Celebration of the First 50 Years of Radio and TV on NBC,* New York: Scribner, 1976

Finkelstein, Norman H., *Sounds in the Air: The Golden Age of Radio,* New York: Scribner, 1993
Harmon, Jim, *The Great Radio Comedians,* Garden City, New York: Doubleday, 1970
O'Connell, Mary C., *Connections: Reflections on Sixty Years of Broadcasting,* N.p.: National Broadcasting Company, 1986
Wertheim, Arthur Frank, "Scatterbrains," in *Radio Comedy,* New York: Oxford University Press, 1979

German Wireless Pioneers

Arco, Braun, Slaby, and Telefunken

In addition to the seminal figure of Heinrich Hertz, the first experimenter to prove that James Clerk Maxwell's theories concerning the potential of electromagnetic radiation for communication were correct, several later German innovators helped to pioneer the practical introduction of wireless telegraphy at the turn of the 20th century. This entry briefly outlines the lives and contributions of three of them and the early development of the Telefunken company, which in 1903 merged their efforts.

Origins

Born early in 1849, Adolph Slaby spent much of his life as an academic researcher at a succession of prestigious German institutions. Into the early 1890s, he focused his efforts on motors and gas engines, thermodynamics, and some aspects of electricity, and in 1893 he had become the science and technology advisor to Kaiser Wilhelm II, a testimonial to his position in the German academic world. Drawing on the findings of Hertz, Slaby began working in wireless in the mid-1890s, but he was unable to transmit more than about 300 feet (in other words only within the technical high school building where he taught), no matter what changes in his apparatus he tried.

Born in 1850, Ferdinand Braun also became an academic. After earning his Ph.D. at the University of Berlin in 1872, he served as a faculty member at Marburg, the University of Strasbourg, Karlsruhe, and finally again at Strasbourg, becoming the director of the Physical Institute there. Braun applied for his first wireless patent in 1898, for a detector circuit to determine frequency. He would later become best known as the inventor of the cathode-ray oscilloscope, precursor of the modern television tube.

George von Arco, a member of the German nobility, was 20 years Slaby's junior. He would join Slaby as a research assistant and would become an important figure in the early development of Telefunken.

Early Wireless

Thanks to diplomatic channels through William Preece, chief electrician of the British Post Office, Slaby was invited to be present for the important May 1897 Marconi tests across the Bristol Channel from England to Wales that achieved transmission over eight miles. Marconi was unhappy over aiding a possible competitor (as turned out to be the case), but had no choice as the Post Office controlled the experiment.

Slaby returned home impressed with what he had witnessed and determined to improve on the system Marconi had demonstrated. Just two months later he conducted the first German wireless transmissions between suburbs of Berlin where, among others, the Kaiser was in attendance. With the Kaiser's full support in terms of funds and personnel, Slaby made rapid progress, and in the fall of 1897 he initiated wireless experiments up to a distance of about a dozen miles, using aerials held up by captive balloons.

Count von Arco joined Slaby as an assistant in 1898 in the Charlottenburg Technical High School's department of wireless telegraphy. The developing Slaby-Arco system made many improvements. One was to move the signal detector (a coherer) away from the base of the antenna, providing far better results. Soon Germany naval and merchant vessels were experimenting with Slaby-Arco devices. Transmitting stations were established in Austria, Sweden, Norway, Portugal, Russia, Denmark, and Chile in addition to Germany. In 1899, the German General Electric Company (AEG) purchased Slaby-Arco patent rights and provided funding for further experimentation.

By 1897–98, Ferdinand Braun was actively experimenting with wireless systems as well. He focused on four elements, including higher transmitter power and use of an oscillating circuit plus a crude early form of tuning to develop a system, for which he would share the 1909 Nobel Prize in Physics with Marconi. In 1898 he received his first wireless patent for his conduction system that worked for limited distances through water or earth (and was thus sufficiently different from Marconi's work to survive any patent appeal). A short time later, Braun realized what was limiting the range of all wireless experimenters when he added two types of coil to the transmitting circuit and quickly began to achieve greater range, efficiency, and safety in operation.

By 1899 Braun's patents were supported by a commercial syndicate known as Telebraun, which was, in turn, taken over by the Siemens-Halske firm in 1901 with the intention of commercializing the system.

Despite all this innovation, all was not well. Early in 1902, when a brother of the German Kaiser returned from a visit to America, Marconi-equipped ships and shore stations refused to communicate with the Slaby-Arco equipped German passenger liner on which he sailed. Or at least that was the German position—Marconi interests claimed the problem was poor performance and lack of distance achieved by the German radio equipment. At the same time this international rivalry was sharpening, fierce patent litigation developed in German courts between Slaby-Arco (whose patents were held by AEG) and Braun (backed by the Siemens-Halske firm). Concerned that Germany might lose out in the world competition to develop viable wireless systems (and thus would have to rely on Marconi or American wireless equipment), Kaiser Wilhelm II stepped in.

Telefunken

At the Kaiser's insistence, on 27 May 1903 some 30 Slaby-Arco (AEG) and Braun (Siemens-Halske) patents (and as many personnel) were combined to create a new company, Gesellschaft für drahtlose Telegraphie, soon better known as Telefunken. Thanks to substantial government support, the new company soon operated more than 500 stations. Telefunken sold rather than leased (as Marconi did) equipment that initially used coherers as signal detectors but soon transferred to electrolytic detectors and headphones. The quality German equipment and growing number of transmission stations soon created strong competition with Marconi, aided by cooperation among German electrical corporations under the umbrella of government purchases and investment.

In part as a result of the 1902 incident of lack of maritime cooperation, the German government required that all German facilities make use of Slaby-Arco equipment. Imperial Germany also hosted the first international wireless conventions in Berlin in 1903 and 1906 to encourage inter-system communication. These meetings bore testament to the growing importance of German radio efforts in the face of Marconi's attempted maritime radio monopoly.

By 1906 von Arco had conducted wireless-telephony (voice) experiments over distances greater than 20 miles, and Telefunken had developed "singing spark" and "quenched spark" transmitters, which use shorter aerials and were more efficient than earlier devices. The German army was setting up portable Slaby-Arco wireless telegraphy units for use in the field. The U.S. Navy had relied heavily on Slaby-Arco equipment beginning in 1903 for lack of a reliable American source of wireless equipment and dislike of Marconi policies. The Russian Navy did likewise, though its loss in the Russo-Japanese War of 1904–05 did not help the image of German wireless, especially as the winning Japanese fleet had made better use of its Marconi equipment.

Telefunken began to build a giant wireless transmission station at Nauen, outside of Berlin, to enable faster communication with distant German colonies. By 1910 it had developed into the largest transmitting station in the world, using a von Arco–developed alternator for transmission with antennas covering more than a square mile. By 1914, German wireless stations were communicating regularly with North America and Africa, and stations were being completed to allow regular communication to the Pacific as well. German wireless was certainly equal to anything created by Marconi in Britain or by various competing American companies.

Decline

Compared to the Marconi or American inventors' stories, however, there is comparatively little evident in English-language sources today about these early German innovators and companies.

Slaby died in 1913, and Braun, who had traveled to the U.S. to deal with patent infringement problems, died as an enemy alien in New York five years later. While von Arco continued with Telefunken and lived into the early part of World War II, outside of Germany only Braun is known today, primarily for his pre-television work.

But the end of Germany's growing radio success was due largely to the impact of World War I. Inception of the war in mid-1914 cut off Germany from her own colonies in Africa and elsewhere and isolated her business and industry. Most overseas German radio facilities were destroyed or captured by 1915. When the United States entered the war in April 1917, all German-owned wireless facilities there were taken over as enemy contraband. Losing the war in 1918 left Germany destitute and her industries, including wireless, in disarray. In just four years, Germany had lost her extensive network of international wireless stations and markets while many British and American concerns thrived as radio expanded into the 1920s.

CHRISTOPHER H. STERLING

See also Early Wireless; Hertz, Heinrich; Lodge, Oliver J.; Marconi, Guglielmo; Popov, Alexander

George Wilhelm Alexander Hans (Count) von Arco. Born in Grossgorschuetz, Germany, 30 August 1869. Joined Slaby as a research assistant in 1898. Appointed manager of German telegraphs in 1903. Eventually chief engineer of Telefunken. Died in Berlin, 7 May 1940.

Karl Ferdinand Braun. Born in Fulda, Germany, 6 June 1850, sixth of seven children of Johann Conrad Braun, a government official, and Franziska Göhring. Ph.D., University of Berlin, 1872. Associate Professor, Marburg, 1877–79; Associate Professor, University of Strasbourg, 1880–82; Professor, Karlsruhe, 1883–85; Professor of physics, later Director of Physical Institute, Strasbourg, 1895–1918. First wireless patent applied for, 1898. Shared 1909 Nobel Prize in Physics with Marconi. Died in Brooklyn, New York, 20 April 1918.

Adolph Karl Heinrich Slaby. Born in Berlin, 18 April 1849. Studied mechanical engineering and mathematics, Royal Trade School, Potsdam. Appointed to Berlin Trade Academy in 1876. First chair of electrical engineering (1882) and director (1884), Electrotechnical Laboratory of Technical High School at Charlottenburg. Honorary Professor, University of Berlin, 1892. Technical advisor to the German Kaiser, 1893–1912. Died in Charlottenburg, Germany, 6 April 1913.

Further Reading

Blaine, Robert Gordon, "German Systems," in *Ætheric or Wireless Telegraphy,* by Blaine, London: Biggs, 1905

Fleming, J.A., "The Work of Slaby and Von Arco [and] of Professor F. Braun . . . the Braun-Siemens Practical System . . . the Telefunken System," in *The Principles of Electric Wave Telegraphy and Telephony,* by Fleming, London and New York: Longmans Green, 1910; 4th edition, 1919; reprint, 1978

Headrick, Daniel R., *The Invisible Weapon: Telecommunications and International Politics, 1851–1945,* London and New York: Oxford University Press, 1991

Kurylo, Friedrich, *Ferdinand Braun: Leben und Wirken des Erfinders der Braunschen Röhre, Nobelpreisträger, 1909,* Munich: Moos, 1965; revised edition, as *Ferdinand Braun: A Life of the Nobel Prizewinner and Inventor of the Cathode-Ray Oscilloscope,* translated and adapted by Charles Susskind, Cambridge, Massachusetts: MIT Press, 1981

Sewall, Charles Henry, "Methods of Dr. Slaby," in *Wireless Telegraphy: Its Origins, Development, Inventions, and Apparatus,* by Sewall, New York: Van Nostrand Reinhold, 1903; reprint, Waterbury, Connecticut: Brohan Press, 2000

Slaby, Adolph Karl Heinrich, *Die Funkentelegraphie,* Berlin: Simion, 1897

Slaby, Adolph Karl Heinrich, "The New Telegraphy," *Century Magazine* (April 1898)

Germany

Despite increasing competition from other electronic media, radio continues to play an important role in Germany. The country's radio system is divided into the long-established network of public service stations—the Association of German Broadcasters (ARD)—and (since 1984) their commercial counterparts. Most stations are FM outlets and have regional or local reach. Although music and entertainment programs are important for German listeners, there are also many niche programs that offer anything from science magazines and minority programs to the renowned *Hörspiel* (radio drama) experimentations. Moreover, radio still is an information medium; even music stations are expected to offer at least news on the hour and headlines in between.

Over the last two decades, the industry has experienced tremendous change stimulated by deregulation. This has led to an exponential increase of stations and intensified competition resulting in many new radio formats and audience fragmentation. On average, Germans listen to radio for almost three hours per day, which is as much time as they spend watching television.

Origins

On 29 October 1923 the first German radio program was aired from Berlin. As was common in many European countries, the first German radio stations were not privately owned enterprises; instead, they were directly or indirectly controlled by national or regional governments. After World War I, when a nonmilitary radio network became a possibility, several wire services, newspaper publishers, the electrical industry, and other private investors tried to negotiate with the Weimar government to establish commercial radio stations. But ultimately the government prevailed, keeping broadcasts under their control.

A national broadcasting organization and nine regional publicly traded radio organizations were launched. These were controlled by the postal ministry but allowed limited private investments. Two programming organizations produced most of the content. Drahtloser Dienst AG (DRADAG; Wireless Service, Inc.) controlled by the department of internal affairs, produced political information (newscasts). *Deutsche Stunde* (German Hour), controlled by the postal and foreign ministries, was responsible for providing music and cultural programs. The system was financed through a license fee that was added to the price of new radio sets. In the last chaotic years of the Weimar Republic, broadcasting became progressively more centralized; by 1932, the national government took full power over the new medium.

Radio in the Third Reich

After the Nazis seized power in 1933, Joseph Goebbels quickly took control of all cultural and media institutions, restructuring them as part of his propaganda machine. As Minister for Public Enlightenment and Propaganda and head of the Chamber of Culture, he controlled all cultural productions including print, radio, film, theater, literature, and music.

An inexpensively produced radio set called *Volksempfänger* (people's receiver) became one of the Nazis' most successful propaganda instruments because it made tuning other non-German stations almost impossible. Because of these new sets, radio soon reached nearly 100 percent of the population and ensured the almost complete penetration of Third Reich propaganda. Both programming and staffing were monitored by the Chamber of Culture's department for broadcasting. From 1940 on, all German radio stations had to air the same government-created program.

As the war expanded, Allied countries made increasingly effective use of radio in their efforts to counter German propaganda broadcasts. The British Broadcasting Corporation (BBC) and others aired programs in German and other languages of occupied territories. Despite being threatened by a death penalty, an increasing number of listeners turned to these programs as the war continued and Goebbels' propaganda lies became more and more obvious. As the war drew to a close, German radio programs became more limited. Many stations were destroyed until in May 1945 the last intact Nazi-controlled radio station at Flensburg announced Germany's unconditional surrender.

A Divided Postwar System

West Germany

At the end of World War II, the occupying powers (the United States, Britain, and France) established a new broadcasting system. In the beginning, all three Western allies tried to establish their respective media systems in their occupied zones. Yet a commercial system such as that in the United States was not viable because war-torn and destroyed Germany could not generate enough advertising support to finance it. On the other hand, the centralized French system resembled too closely the former Weimar system that had allowed an effortless takeover by the Nazis. Therefore, all Western allies at last accepted the British public service system as the basis for the new German radio operation.

It was important to all of the Allies to keep broadcasting away from the influence of the German national government and to decentralize the powerful medium. At first the Allied representatives were directly in charge of programming and station management, but they soon hired German nationals for increasingly responsible positions. The first regional broadcasting organizations were launched right after the war and by the mid-1950s, the now sovereign West Germany had nine regional public broadcasting corporations. They were organized into the Association of German Broadcasters (ARD). Each of the nine independent regional broadcasting corporations covered either one large or several smaller federal states, and each subsequently increased its stations to as many as five that aired different niche programs.

Reflecting Germany's decentralized political structure, each state organized its own media systems. The ARD broadcast corporations were monitored by boards composed of representatives of political parties, unions and industry, youth organizations, religious groups, and other groups that are considered relevant for society. These boards set general programming guidelines and voted on major administrational and editorial staff. Public stations were required to offer a variety of programs that catered to all groups of the public they served. There were also boards that controlled financial matters.

Critics of this system argue that the political parties have too much influence on broadcasting, denying the constitutional freedom of the media. The system is financed by a monthly broadcast fee levied on every household with a radio and by the sales of some limited commercial time. Most ARD organizations had one general station with a broad pop music selection, in addition to news and short features. The other stations were educational and youth talk-based stations, classical and jazz stations (many ARD stations own symphony orchestras and sometimes big band and jazz orchestras), and stations playing traditional music catering to older audiences. More recently, some have added stations with all-news formats. The indirect financing through listener fees allowed the stations to produce programs other than just those with mass appeal, which has made radio an important cultural form. From the 1950s through the 1970s, niche departments (not infrequently headed by Ph.D.s) flourished, producing genres such as *Hörspiel* (radio drama) and other experimental radio work.

Reacting to new technological developments that ended channel scarcity, in the mid-1980s the West German broadcasting system, including radio, was deregulated and allowed commercial competition.

East Germany

The East German media system was organized according to the Soviet-totalitarian model. Following the Marxist-Leninist-Stalinist concepts of media, broadcasting outlets were state-owned and centrally organized and their main purpose was to educate people about communism (usually and rather confusingly termed "socialism" in practice).

The Soviets developed stations in Berlin and other major East German cities—all under a central administration. But by 1952 Soviet forces relinquished control and censorship rights to a state broadcasting committee that was directly monitored by the East German government. East Germany abolished the states as political entities and centralized the media system. After 1955 there were the *Berliner Rundfunk* (Berlin Broadcasting) for East (and potentially West) Berlin; the *Radio DDR* for all East Germans; *Deutschlandsender* (Germany's Station; later *Stimme der DDR* [Voice of German Democratic Republic]) broadcast for West German listeners; and *Radio Berlin International*, which produced programs in German and ten other foreign languages for international listeners. During the height of the Cold War, AM propaganda channels were added: *Freiheitssender 904* (Freedom Station 904) aimed at communists outside the country after 1956; and *Soldatensender 935* (Soldiers' Station 935) geared toward Western soldiers after 1960. They were on the air until a phase of détente in the early 1970s eased the Cold War rhetoric.

A reform in 1987 resulted in a new channel for young people—the soon popular *Jugendradio DT64* (Youth Radio DT64). In addition, the established stations changed to more distinctive formats, under different names. *Radio DDR 1* aired information and music; *Radio DDR 2* aired regional programming in the morning and cultural or educational programming during the day. The *Berliner Rundfunk* focused on Berlin current events. The broadcasting system in East Germany was mainly financed by viewer's fees and state subsidies. All radio and TV sets had to be officially registered.

A special problem for the East German government was that an estimated 80 percent of its people could tune to broadcast signals from the West. Besides the public TV stations, people especially listened to radio stations close to the German border. Although at first the political elite tried to enforce a strict ban on tuning in foreign broadcasts, it later gave in to their popularity. During the first few years of East Germany's existence, members of the state-controlled youth organization were sent out to destroy antennae that were positioned to receive Western broadcasts. Sometimes the government tried to jam (electronically interfere with) transmissions from the West, but that always led to protest from audiences. Under the rule of state council chairman Erich Honecker in the early 1970s, the East German government officially accepted West German broadcasting. Only the most rigid party members then refrained from listening and viewing.

In fact, West German broadcasting became a part of the political agenda. Often East German politicians countered Western news that had not been aired on eastern channels. It has been argued that these years of Western broadcasting had a major impact on destabilizing East Germany, which finally led to the fall of the Berlin Wall in 1989. The constant exposure to unreachable Western consumerism definitely led to a serious level of dissatisfaction among the population. After a phase of several short-lived media reorganizations during the upheaval, with the final German reunification in 1990, the East and West German systems were merged, following the Western model.

Today: A Dual System

In 1984 the West German media system changed to a so-called dual system allowing commercial as well as state governmental radio and television stations. After the reunification, the whole country adopted this system. The new political situation let to the launch of two new ARD organizations in former East Germany and the merger of two Western ones; thus there are now ten regional public service networks under the ARD umbrella. Further deregulation has let to a tremendous growth in the number of commercial radio stations in East and West. Reflecting the decentralized political system, each state has a different setup for commercial radio broadcasting, but most states permit a combination of local and statewide commercial radio stations. Some states only have five or six statewide private stations. Sometimes they are even guaranteed a commercial broadcasting monopoly in their area to guarantee enough advertising support; more competitive systems such as Bavaria have 50 or more private stations. There are some ownership regulations, but they have not stopped many newspaper, magazine, and book publishers from acquiring stations. A number of local newspaper publishers have taken advantage of their existing editorial base to launch radio stations.

The dual system has been redefined repeatedly based on several decisions by the highest constitutional court (*Bundesverfassungsgericht*), state legislation, and joint accords of all states. The latest legislation tries to guarantee survival of public as well as private broadcast entities. Although the German constitution guarantees freedom of the press and independence from government interference, there are several content regulations that apply to public as well as commercial stations. ARD stations are controlled by their boards, but even commercial broadcasters must follow certain rules similar to the former U.S. Fairness Doctrine. These concern the protection of children and young people and balanced coverage of topics of public importance. Each state sets up differently structured boards that license the stations and monitor commercial stations' compliance with these requirements.

Two additional radio organizations do not fit into either the public or the commercial category. *Deutschlandradio* is another result of the reunification. It was launched in 1994 by combining the former national public service AM station *Deutschlandfunk* (Germany Radio) with RIAS (Radio in the American Sector, a station of the U.S. Information Service in Berlin) and a cultural station of former East Germany. Jointly operated by ARD and the German TV network ZDF, it airs two programs from Cologne and Berlin. It is commercial free and governed by its mission to support the reunification of the two Germanys. Its main focus is on political and cultural programs that are broadcast on FM, AM, and shortwave. The second organization, *Deutsche Welle* (*DW*; German Wave), broadcasts programs worldwide via satellite and shortwave radio in 34 languages. Based in Cologne and Berlin, it also broadcasts television programs via satellite. The reunification brought about a new legal situation for *DW*, which was formerly a West German government-controlled station. It is now legally monitored by the ministry of interior affairs and financed by the government, commercials, and sponsoring. In accordance with constitutional broadcasting freedom, however, no government department can directly control its programming.

As of early 2003, every household owning at least one radio paid Euro 5.32 (US$5.60); households with both television and radio paid Euro 16.50 (US$17.30). Twenty-eight percent of the overall collected fees were used to finance radio stations. A committee of representatives from all states decides on changes in fee structure or levels. With commercial competition it has become more difficult for public service broadcasters to defend fee increases. Public service radio stations also air commercials that are by law restricted to Monday-through-Saturday broadcasts at certain times of the day. But commercial local and regional radio broadcasters, which are fully supported by advertising, face almost no advertising regulations.

In 1987 there were only about 44 stations in West Germany, but the 1980s deregulation and German reunification increased that number to about 246 in 1999 in the united Germany. Of those, 59 are public service stations. This number has been fairly stable since the mid-1990s. After a strong growth phase in the first decade after deregulation, Germany seems to have reached a saturation point for its local/regional markets. Recently, new transmission technology has triggered some new national private stations and there are also several internet-only stations in Germany.

Audiences

According to media research in 1999, radio stations reach an average of 82 percent of the German people over age 14, on any given day some time between 5:00 A.M. and 12:00 A.M. Germans listen to the radio an average of 179 minutes a day. These numbers have slowly but steadily been increasing in recent years. Radio has remained popular despite increased competition from other traditional and new media. German listeners seem to be fairly loyal to their stations. When asked how many stations they heard the day before, they name only 1.4, on average, and they name only 2.8 stations heard when asked about their listening habits over the last two-week period. Overall, the public stations in the ARD network are losing listeners to commercial stations; the market share of ARD was 53 percent in 1999, compared to 45 percent for commercial radio. Only seven years earlier, in 1992, the ARD network had a market share of 70 percent, compared to 29 percent for commercial stations.

When asked what they consider to be the most important aspect of radio, 69 percent of women and 65 percent of men name news, 63 percent of women/60 percent of men name music, and 55 percent of women/54 percent of men name services such as weather and traffic reports and time announcements. Other popular genres are general background information on political events and political reporting, as well as regional programs and consumer news. As in the United States, radio listening peaks on Mondays through Fridays during the 7:00 A.M. to 8:00 A.M. "drive time" or commuting period, when almost 40 percent of women and 33 percent of men tune in. Listening slowly declines to another light peak around lunchtime with 22 percent of women and 16 percent of men listening. The afternoon sees another decline, but listening picks up again from 4:00 P.M. to 6:00 P.M., with 21 percent of men and 18 percent of women listening.

Many ARD and commercial stations stream their programs via the internet. There are currently pilot projects in several states testing new digital radio transmission technology (DAB). It is expected that new technology will lead to new stations and might shift radio from local and regional to more national format based stations. Other important factors in the future are the tough competition between public service and commercial broadcasters and the increasing audience fragmentation. German media researchers have discovered a distinctive audience split. Younger, less-educated, less-affluent audiences seem to favor commercial stations, whereas older, better-educated, more-affluent people tend to listen to public radio. The maturation of the "dual" system and new technology will provide more challenges for public and private stations in finding and keeping their audiences.

ELFRIEDE FÜRSICH

See also German Wireless Pioneers; Propaganda by Radio; Radio in the American Sector (Berlin)

Further Reading

Bergmeier, Horst J.P., and Rainer E. Lotz, *Hitler's Airwaves: The Inside Story of Nazi Radio Broadcasting and Propaganda Swing,* New Haven, Connecticut: Yale University Press, 1997

Browne, Donald R., *Electronic Media and Industrialized Nations: A Comparative Study,* Ames: Iowa State University Press, 1999

Hoffmann-Riem, Wolfgang, *Regulating Media: The Licensing and Supervision of Broadcasting in Six Countries,* New York: Guilford Press, 1996

Kuhn, Raymond, editor, *Broadcasting and Politics in Western Europe,* Totowa, New Jersey, and London: Cass, 1985

Noelle-Neumann, Elisabeth, Winfried Schultz, and Jürgen Wilke, editors, *Publizistik, Massenkommunikation,* Frankfurt: Fischer Taschenbuch Verlag, 1989; 6th edition, 2000

Short, Kenneth R.M., editor, *Western Broadcasting over the Iron Curtain,* New York: St. Martin's Press, and London: Croom Helm, 1986

Gillard, Frank 1908–1998

British Correspondent and Executive

Frank Gillard's radio career may be divided into three important phases. He was described as "one of the BBC's most distinguished war correspondents" during World War II. Then during the 1950s and 1960s he was a key management executive determining reforms in British national and local radio. Finally, during his "retirement," he was responsible for acquiring archive interviews and writing the history of the British Broadcasting Corporation (BBC). He was also an adviser and consultant in North America and may have had some influence on the development of U.S. public radio. After his death on 20 October 1998, the then Director-General Sir John Birt described him as having made "an unrivalled contribution to the BBC."

War Correspondent

Following his death, a number of newspaper obituaries praised Gillard as a World War II war correspondent. For example, the London *Times* said, "Gillard's was one of the best known voices on British war time radio . . . his incisive reports kept listeners at home abreast of events from the raid on Dieppe to the Allied entry into Berlin and the final German surrender."

In fact, toward the end of his life Gillard had expressed "shame and disgrace" about the distorted and misleading nature of his BBC report on the ill-fated raid at Dieppe in 1942. This is hardly surprising given the fact that he had broadcast the extraordinary line: "As a combined operation this raid was an all-time model." It can be argued that no amount of censorship required him to express this degree of enthusiasm for an event regarded as one of the greatest disasters of modern military history. He also regretted that he had followed the army's optimistic line too often in his reports and wished he had been more independent. (He was the only BBC correspondent with the Dieppe force because he had been the regional correspondent closest to Portsmouth.)

Throughout the war, Gillard had to be "on side" and operate in the context of a "white propaganda" role in war reporting. There was strict censorship, and journalistic presence depended on the approval and goodwill of commanding generals. Gillard had been a schoolteacher, not a journalist, before working as a part-time broadcaster and talks assistant for the BBC in western England. His war reports do not, therefore, indicate the instinct of the critical reporter seeking to "tell it as it is" but instead fall into the category of a performer operating as a public relations officer. He replaced Richard Dimbleby in North Africa because Dimbleby had been too closely associated with the failures of General Auchinleck's command. Gillard developed a rapport with Auchinleck's replacement General Bernard Montgomery. But after the Allied victory in North Africa, Gillard fell foul of Supreme Allied Commander General Alexander, who had mistakenly accused him of evading official censorship. But an examination of Gillard's war reports again reveals his role as a broadcast performer rather than a reporter detailing the truth of Allied slaughter in Bocage country between the invasion and break out.

An example of the wartime bias of his journalism is illustrated by this extract from a report broadcast 25 June 1944: "Almost the first words a German says when he is brought in to a British or Canadian headquarters on this front are 'Please don't shoot.' Thousands have been amazed to find that we've no intention or thought of killing them." The fact was that there had been several occasions when Canadian troops did not take prisoners, and instances of prisoners on either side being shot dead or having their throats cut were being covered up by Allied reporting. Gillard and his colleagues failed to report, for example, that German tanks and anti-tank weapons were vastly superior to Allied equipment; that the Nazis had more automatic guns than the Allies; that British and Canadian army tactics in the Bocage were inadequate and contributed to an acceleration of casualties that eventually led to the British Army running out of reserves; that a number of Allied troops were killed by friendly fire; and that the bombing of Caen had been futile, resulting in the deaths of many more French civilians than German combatants. In fact, the Scottish playwright William Douglas-Home was court-martialed for refusing orders in relation to the Caen operation. An attempt to break out with British and Canadian armor in "Operation Goodwood" on 18 July 1944 was an unmitigated disaster and significant German military victory, but this was not even hinted at by any of Gillard's reports.

Management

The second phase of Gillard's career in broadcasting was much more significant. He demonstrated a cunning political ability to lobby for and to effect significant change in structures of BBC institutions and broadcasting. Beginning in 1956, he built up the resources, power, and importance of the BBC West Region and was appointed director of sound broadcasting in 1963.

Gillard went on a tour of the United States in 1954 and there realized that, despite the growing ascendancy of television, he did not meet a single person who thought that radio was dying. He noted that the changes were affecting "big battalion broadcasting" (networks) and "an increasing range of local stations." In particular, Gillard was impressed by the small station WVPO Stroudsburg in the Pocono mountains of Pennsylvania, which operated in daylight hours only and served a community of 15,000 people with a staff of 13. Gillard wrote a special report for the BBC, "Radio in the USA: A Visitor's View," which emphasized that WVPO "spoke to its listeners as a familiar friend and neighbor and the whole operation was conducted with the utmost informality."

His successes included the establishment of a network of BBC local radio stations; the redefining of BBC national networks to incorporate the popular musical format channel *Radio 1;* the introduction of more live, unscripted programming to reflect the changing nature of British popular culture; and the preservation of the BBC monopoly in radio broadcasting long after it had been challenged in television. He could be criticized for axing the radio features department without establishing a context whereby its management could continue with the same force and creativity. He could also be condemned for canceling the *Children's Hour* without finding an alternative method of attracting and developing a new generation of young radio listeners. And his ruthless campaigning against a more democratic distribution of taxation for U.K. public radio and the suppression of commercial radio licensing may have had a negative impact on development and choice in British radio.

Historian

Gillard officially retired from the BBC in 1969, but he remained central to its establishment voice. In the 1970s he became the BBC's archival historian, and according to the *Guardian* journalist Maggie Brown, he was "the custodian of its inner secrets." He originated and managed a project of recording on audiocassette and film the memoirs of BBC personnel, which could only be used "on the BBC's centenary in 2022" when "everyone was safely dead or otherwise." As it has been argued that cultural memory and power are often determined by those who write history, Gillard's influence on historical portrayals/evaluations of the BBC is likely to endure. This is probably why the *Guardian* called him "a BBC Mandarin."

As Leonard Miall noted in his *Independent* newspaper obituary, Gillard, "would have been horrified that BBC News managed to get his age wrong in their announcement of his death." The BBC celebrated his memory by inaugurating awards for local radio broadcasting in his name.

TIM CROOK

See also British Broadcasting Corporation; British Radio Journalists

Frank Gillard. Born in Tiverton, Devon, England, 1 December 1908. Educated at Wellington School, Somerset, and St. Luke's College, Exeter; attended London University, 1929–31: B.S.; schoolmaster and freelance broadcaster, BBC West Region, 1936–40; talks assistant, BBC Bristol, 1941; war correspondent with Southern Command, 1942; accredited to operation on Dieppe, 19 August 1942; war correspondent BBC, accredited to North Africa, Italy, Normandy, Berlin, 1942–45; head of BBC West Regional Programmes, 1945–55; controller, West Region, 1956–63; director and later Managing Director, BBC Radio, 1963–69; consultant to U.S. Corporation of Public Broadcasting and Australian Broadcasting Commission, 1969–70. Officer of the Order of the British Empire, 1946; Commander of the Order of the British Empire, 1961; distinguished fellow of the U.S. Corporation for Public

Broadcasting, 1960–66; honorary LL.D., Exeter, 1987. Died in London, 20 October 1998.

Further Reading

Briggs, Asa, *The History of Broadcasting in the United Kingdom,* 5 vols., Oxford and New York: Oxford University Press, 1961–95 (see especially vols. 3–5)

Crook, Tim, *International Radio Journalism: History, Theory, and Practice,* New York and London: Routledge, 1998

Donovan, Paul, *The Radio Companion,* London: HarperCollins, 1991

Hawkins, Desmond, editor, *War Report, D-Day to VE-Day: Radio Reports from the Western Front, 1944–45,* London: BBC Books, 1946

Hickman, Tom, *What Did You Do in the War, Auntie?* London: BBC Books, 1995

Miall, Leonard, *Inside the BBC: British Broadcasting Characters,* London: Weidenfeld and Nicolson, 1994

Miall, Leonard, "Frank Gillard Obituary" *The Independent* (23 October 1998)

Purser, Philip, "Frank Gillard, Obituary" *The Guardian* (23 October 1998)

Godfrey, Arthur 1903–1983

U.S. Radio Personality

During the years following World War II, Arthur Godfrey was one of the great stars of radio, keeping alive a tradition of variety programming through his simulcast *Talent Scouts* amateur program and also doing a popular morning talk and variety program with a stable of stars. These were Godfrey's radio venues, but he did do other radio programs—particularly during the early 1950s—as Columbia Broadcasting System (CBS) executives sought to take full advantage of his growing appeal. No one challenged the popularity of his other principle venue through the 1950s—a regular Wednesday night television variety program, *Arthur Godfrey and His Friends.*

Radio Origins

The red-haired ukulele player, whose gift for gab was a favorite of radio and television audiences through the post–World War II era, started his radio career in 1930 at WFBR-AM in Baltimore. Later, in the 1930s, he moved to Washington, D.C., first to WMAL-AM and then to CBS affiliate WJSV-AM (later WTOP-AM) for an early-morning program that would begin Godfrey's life-long association with the CBS network. Godfrey remained in Washington, D.C., until 1941, when he moved to New York and CBS's WABC-AM. During this decade and a half, Godfrey perfected what would be his fabled ability to communicate on an almost personal level with his listeners.

Based on his narration for the CBS broadcast of the funeral procession down Pennsylvania Avenue for Franklin Delano Roosevelt in April 1945, CBS executives moved Godfrey to New York to begin a radio daytime series, *Arthur Godfrey Time.* This program would remain his staple for 27 years, frequently changing starting times but always running about an hour, and always ending before the noon CBS news.

In 1946 CBS added *Arthur Godfrey's Talent Scouts,* a copy of the long-running Major Bowes and later Ted Mack programs. *Talent Scouts* was Godfrey's first venture into prime time, and it became a big hit. Indeed, his move into television came in 1948 when he agreed to simulcast *Talent Scouts* on television. This would prove to be his launching pad to national fame.

Although others abandoned radio for television as soon as their popularity made it possible, Godfrey never gave up his morning radio chat fest. He did the morning program and then turned around on Monday nights—with the help of a capable production staff led by Janette Davis—and did *Talent Scouts.* Godfrey was smart enough to know where his comparative advantage lay, and he never gave up the morning radio show. (Indeed, once his popularity inevitably waned, he returned exclusively to radio, continuing his radio career until 1972.)

A Talent Scout

While his morning show touched an audience of housewives and baby boomer children, his fame, fortune, and pioneering activities in prime time brought him to the cover of *Time* magazine. Variety show *Arthur Godfrey's Talent Scouts* proved to be a far greater hit than Major Bowes' or Ted Mack's shows ever were. *Arthur Godfrey's Talent Scouts* ran on both radio and television on Monday nights for a decade and was clearly Godfrey's best prime-time showcase. The show reached number one in the TV ratings in the 1951–52 season but fell behind *I Love Lucy* the next season and thereafter finished behind, but not too far behind, Hollywood-produced shows. It remained a radio staple even as other variety shows switched to TV and abandoned radio altogether.

The formula for *Arthur Godfrey's Talent Scouts* was simple enough, as old as the amateur hour idea from vaudeville, which had been popularized by Major Bowes on radio. For Godfrey's version, "scouts" brought on their "discoveries" to perform live before a national radio and television audience. Most of these "discoveries" were in fact struggling professionals looking for a break, and so the quality of the talent proved quite high, as thousands auditioned with the hope that exposure on Godfrey's show would provide that needed national boost. The winner, chosen by a fabled audience applause meter, often joined Godfrey during the following week on his morning radio show and less frequently as part of his televised *Arthur Godfrey and His Friends*. Some even joined his regular talent pool.

Godfrey and Janette Davis did well with pop and country singers. Through the late 1940s and 1950s, Godfrey significantly assisted the careers of Pat Boone, Tony Bennett, Eddie Fisher, Connie Francis, Leslie Uggams, Lenny Bruce, Steve Lawrence, Connie Francis, Roy Clark, George Hamilton IV, and Patsy Cline. But both Elvis Presley and Buddy Holly never made it past the audition. This Monday night 8:30 P.M. talent search was no place for early rockers, nor for African-Americans. That Godfrey lived in Leesburg, Virginia, and Janette Davis was from Arkansas seemed to enable them to find country talent on the fringes, but they always worked best when mining aspiring stars in the Tin Pan Alley tradition.

The "discovery" of Patsy Cline on 21 January 1957 was typical. Her scout (actually her mother), Hilda Hensley, presented Patsy, who sang a recent recording "Walkin' After Midnight." Though this was a country song, recorded in Nashville, and Cline wanted to wear one of her mother's hand-crafted cowgirl outfits that she wore while appearing on local television in Washington, D.C., Davis forced Cline to wear a cocktail dress. Still, the audience's ovations stopped the meter at its apex, and for a couple of months thereafter Cline appeared regularly on Godfrey's radio program. But despite the fact that Cline had been performing for nearly a decade, had been recording in Nashville since 1954, and had been a regular on local Washington, D.C., radio for more than a year, it is often stated that Godfrey, because of the great ratings and fame of his *Talent Scouts,* discovered Patsy Cline.

For *Arthur Godfrey Time,* Godfrey hosted a straight variety show, employing a resident cast of singers that at times included Julius La Rosa, Frank Parker, Lu Ann Simms, Pat Boone, and the Cordettes. Tony Marvin was both the announcer and Godfrey's "second banana," as he was on *Arthur Godfrey's Talent Scouts*. Although the fame of *Arthur Godfrey Time* never matched that of *Talent Scouts,* the assembled company of singers, all clean-cut young people, remained a 1950s fixture in homes as Godfrey played host and impresario.

Fame and Fortune

By the early 1950s, Godfrey seemed unable to do anything wrong, despite a press that could find little reason for his popularity. During the summer of 1950, Godfrey even appeared twice a week on CBS, in addition to his television and daily radio shows, offering lessons on the ukulele to his television audience on Tuesday and Friday nights. This proved his peak of popularity.

His fall from grace began in October 1953 when he publicly fired the popular Julius La Rosa on the air. Through the mid- and late 1950s Godfrey feuded with newspaper critics who complained of his insensitivity to La Rosa and other employees. Although Godfrey's bout with cancer drew international sympathy, falling TV ratings led to his retreat back to radio.

But he did not give up on television. He flopped on *Candid Camera.* But CBS kept trying, and Godfrey continued to do specials: *Arthur Godfrey in Hollywood* on 11 October 1963, *Arthur Godfrey Loves Animals* on 18 March 1963, and so on once or twice a year. His final television special came on 28 March 1973: *Arthur Godfrey's Portable Electric Medicine Show,* filmed and shown on National Broadcasting Company (NBC) TV, was ironically his only major effort for that network. Radio was where Godfrey started and ended his career—and it was always his best medium.

Despite his critics, many of whom argued that Godfrey had no talent, he was one of the important on-air stars of radio and television for the 15 years after the end of World War II. Indeed, one can credibly argue that through the 1950s, there was no bigger star than this freckle-faced, ukulele-playing variety show host. Through most of the decade of the 1950s, Godfrey appeared in his two top ten prime-time television shows, as well as doing a daily radio program, all for CBS. Despite his fame as a discoverer of talent and as a host, in the end it was something about Godfrey's infectious chuckle, his offbeat sense of humor, and his ability to connect to middle-class Americans that made his fans tune in not one, but two, three, or four times per week, in the morning and during prime time.

To industry insiders, Godfrey's shows were simply vehicles for television's first great pitchman. Godfrey blended a Southern folksiness with enough sophistication to charm a national audience measured in the millions through the 1950s. For CBS in particular, Godfrey was one of the network's most valuable stars, generating millions of dollars in advertising billings each

Arthur Godfrey
Courtesy Radio Hall of Fame

year, with no real talent save being the most congenial of hosts.

On radio and television, Godfrey frequently kidded his sponsors, but he always "sold from the heart," only hawking products he had actually tried or that he regularly used. No listener or viewer during the 1950s doubted that Godfrey really did love what he sold, for Godfrey's rich, warm, resonant timbre made him sound like he was confiding to each audience member. Godfrey delighted in tossing aside prepared scripts and telling his audience: "Aw, who wrote this stuff? Everybody knows Lipton's is the best tea you can buy. So why get fancy about it? Getcha some Lipton's, hot the pot with plain hot water for a few minutes, then put fresh hot water on the tea and let it just sit there." So, despite all his irreverent kidding and the uneven quality of his assembled talent, advertisers loved Godfrey and made him and his various shows a key part of broadcasting during the 1950s.

He did the same magic for Pillsbury, Frigidaire, and Toni (hair) products. Here was a friend recommending the product, no snake-oil salesman hawking useless or overpriced merchandise. Godfrey drove CBS network efficiency experts crazy because he refused to simply read his advertising copy in the allocated 60 seconds. Instead, Godfrey talked for as long as he felt was necessary to convince his viewers, frequently running over. CBS chairman William S. Paley detested Godfrey but bowed to his incredible popularity. CBS president (and chief number cruncher) Frank Stanton loved Godfrey because his shows were so cheap to produce yet drew consistently high ratings.

By 1950 Godfrey was making well in excess of a million dollars per year and was among the highest-paid persons in the United States at the time. So popular was Godfrey in his heyday that in 1959, when he underwent one of the first successful operations for the removal of a cancerous lung, it was front-page news across the nation. At one point, in the mid-1950s, he had an estimated audience of 40 million and had more than 80 sponsors for his daily morning show. He received 60,000 letters a week. Because of his homey approach and sometimes sly, off-color wit, radio great Fred Allen dubbed Godfrey "the Huck Finn of radio."

The late 1950s and early 1960s were not kind to Godfrey. He battled cancer and saw his million-dollar yearly salary plunge. Godfrey finally retired from radio in 1972, by then a symbol of another era.

DOUGLAS GOMERY

See also Variety Shows

Arthur Godfrey. Born in New York City, 31 August 1903. Attended Naval Radio School, 1921; Naval Radio Material School, 1929; served in U.S. Navy as radio operator, 1920–24; served in United States Coast Guard, 1927–30; radio announcer and entertainer, WFBR-AM, Baltimore, Maryland, 1930; staff announcer, WRC-AM, Washington, D.C., 1933; worked at other local stations, particularly WJSV-AM (later WTOP-AM), a CBS affiliate in Washington, D.C., 1934–45; joined CBS network, 1945; host of several radio and television programs, 1940s and 1950s; returned exclusively to radio morning show,1960; retired, 1972. Died in New York City, 16 March 1983.

Radio Series

1945–72 *Arthur Godfrey Time*
1946–56 *Arthur Godfrey's Talent Scouts* (on television until 1958)
1949–59 *Arthur Godfrey and His Friends*

Television Series

Arthur Godfrey and His Friends, 1949–59

Further Reading

"Arthur Godfrey: Oceans of Empathy," *Time* (27 February 1950)

Crosby, John, "Arthur Godfrey: It Doesn't Seem Like Old Times," *Collier's Magazine* (30 September 1955)

Curtis, Joan, editor, *The Arthur Godfrey Nobody Knows* (Famous People Series), New York: Fawcett, 1950

Godfrey, Arthur, "This Is My Story" (as told to Pete Martin), *Saturday Evening Post* (8 parts; 5 November–24 December 1955)

Godfrey, Jean, and Kathy Godfrey, *Genius in the Family,* New York: Putnam, 1962

MacDonald, J. Fred, *Don't Touch That Dial! Radio Programming in American Life, 1920–1960,* Chicago: Nelson-Hall, 1979

Marder, Irving, "The Great Godfrey," *Sponsor* (5 June 1950)

O'Brian, Jack, *Godfrey the Great: The Life Story of Arthur Godfrey,* New York: Cross, 1951

Rhoads, B. Eric, *Blast from the Past: A Pictorial History of Radio's First 75 Years,* West Palm Beach, Florida: Streamline, 1996

Singer, Arthur J., *Arthur Godfrey: The Adventures of an American Broadcaster,* Jefferson, North Carolina: McFarland, 2000

Slocum, William J., "The Arthur Godfrey You Never See," *Coronet* (June 1953)

The Goldbergs

Comedy Serial Program

As early as 1937, just 8 years into its 16-year run on radio, *The Goldbergs* was selected by the industry magazine *Radio Daily* as one of the "Programs that Have Made History." Indeed, *The Goldbergs* was a groundbreaking show that influenced both the form and content of later radio and television programming. Amid early radio fare, such as music variety shows and public talks, *The Goldbergs* (along with *Real Folks* and *Amos 'n' Andy)* was one of the first dramatic serials on network radio and one of the first serials to concentrate on family life. *The Goldbergs* demonstrated the power of serials to attract a loyal audience. Its immediate success in 1929 prompted interest in radio programs that regularly featured familiar domestic situations, recurring characters, and continuing story lines.

The Goldbergs was a hybrid program, part comedy, part drama, and part serial; with its continuing story line and domestic focus, the program was the prototype for both later situation comedies and daytime soap operas. Yet *The Goldbergs* is most fondly remembered for its ethnic content. Among the first urban, ethnic comedies in broadcasting, the program spoke eloquently about the experience of immigrants during the Depression and their struggles to assimilate in their adopted country. *The Goldbergs* remains one of the relatively few programs in the history of radio and television to offer a sustained ethnic perspective on American life.

The stories of the Goldberg family—at 1038 East Tremont Avenue, Apartment 3B, in the Bronx—emerged solely from the creator and writer of the series, Gertrude Berg. As Michele Hilmes (1997) writes, "no other daily serial drama reflected so explicitly its creator's own ethnic background." Gertrude Edelstein Berg (1899–1966) was born in Harlem to Russian Jewish parents. As a teenager at her father's resort in the Poconos, Berg began writing plays to entertain the guests. This hobby continued even after Berg married and gave birth to two children. She soon developed a popular skit featuring a wife and mother named Maltke Talnitsky—modeled after her grandmother, her mother, and herself—her no-good husband, and her children. These characters were the earliest forms of the Goldberg family—Molly Goldberg, a Jewish immigrant mother; Jake Goldberg, a tailor and her sympathetic husband; and her two children, Sammy and Rosalie. Gertrude Berg sent a sample script through a family acquaintance to a New York radio station. Berg was offered jobs writing continuity and translating commercials and recipes into Yiddish on radio.

Although Berg's first network offering, about two working-class salesgirls, *Effie and Laurie,* was picked up by the Columbia Broadcasting System (CBS) in 1929, it was canceled after just one broadcast. After this failure, Berg began to shop around her idea for a family comedy based on her earlier skits. Berg claimed that initially "radio studio big wigs" believed audiences would reject a program about Yiddish life. However, the National Broadcasting Company (NBC) saw the promise of radio's first Jewish comedy and aired *The Rise of the Goldbergs* as a sustaining weekly evening series starting on 20 November 1929. Paid $75 a week to write the series and produce the program, Berg controlled all aspects of the show's development, from scripting the program in longhand to paying the performers. Berg, who voiced the character of Molly Goldberg, assembled a cast of New York stage actors to bring the rest of the Goldberg family to life: James Waters as Jake Goldberg, Alfred Ryder as Sammy, and Roslyn Silber as Rosalie. Berg's importance to the series was recognized by fans; nearly 37,000 letters poured in when Berg became ill and was off the air for a week. NBC acknowledged Berg's role, as well: Berg soon earned more than $7,500 a week for the program. By fans and the industry, Berg was considered one of the most important personalities in broadcasting and one of the greatest women in radio.

By the 1931–32 season, the series, retitled *The Goldbergs,* aired six times a week and had become one of the highest-rated programs on radio. On 13 July 1931, the show was picked up by Pepsodent, who sponsored it for the next three years. Berg ended the serial briefly in 1934 to take the cast and the series on a nationwide promotional tour. When the networks cleared serial dramas from the night-time air in 1936, *The Goldbergs* moved to daytime until 1945. In January 1938 Berg was signed to a five-year, million-dollar contract to write and star on *The Goldbergs,* making her one of the highest-paid writers on radio. Oxydol and Procter and Gamble picked up sponsorship of the show until the end of its run.

Gertrude Berg of *The Goldbergs*
Courtesy CBS Photo Archive

The Goldbergs was a serial that spoke about the economic and social tensions of the 1930s and 1940s, an assimilationist drama about an immigrant Jewish family living on New York's Lower East Side. Many programs focused on typical domestic situations—report cards, dinner guests, schoolyard loves, and Molly's worries about her family. In the early years of *The Goldbergs,* Berg described life in an urban tenement and the attempts of this immigrant family to achieve economic security during the Depression. But, most important, the serial vividly depicted the clash between old and new, yesterday's traditions and today's values, Old World parents and American-born children.

At the heart of this serial drama was the struggle of an immigrant family to assimilate culturally while still retaining their ethnic identity. Early episodes were marked by generational conflict over how much the family should adapt to life in the United States. The parents, Molly and Jake Goldberg, were ethnic immigrants with "Old World" values. Molly's voice revealed her immigrant background: she spoke with a heavy Yiddish accent and was famous for her "Mollyprops," Yiddish malapropisms that twisted common phrases ("If it's nobody, I'll call back" or "I'm putting on my bathrobe and condescending the stairs"). Their American-born children, Sammy and Rosalie, spoke with relatively little accent and often challenged the traditions of their parents. However, by the end of the serial, the Americanization of Molly's family was nearly complete; like so many other immigrants, the upwardly mobile Goldbergs eventually moved from their New York apartment to the suburbs.

The immigrant experience recounted in *The Goldbergs* clearly resonated in an era characterized by both massive immigration and calls for greater national unity. Although Gertrude Berg moved in more assimilated, upper-middle-class circles, she worked diligently to maintain the "realism" of the ethnic immigrant experience detailed in the series. In 1936, for example, Berg took Dan Wheeler, a writer from the *Radio Mirror,* to the Lower East Side to witness her research. In an article entitled "How the Ghetto Guides *The Goldbergs,*" Wheeler recounted Berg's conversations with street vendors and immigrant women and her anonymous participation in a Lower East Side charitable club. Her efforts to represent the ethnic experience were appreciated by contemporary audiences. The program was cited at the time by groups such as the National Conference of Christians and Jews for promoting religious and ethnic tolerance.

The continued popularity of the radio program spawned the 1948 Broadway play *Molly and Me,* a comic strip, several vaudeville skits, and a 1950 film. After the program's demise, *The Goldbergs* was briefly revived in 1949–50 for CBS as a 30-minute weekly radio series, but it endured in American culture as a television situation comedy from 1949 to 1955. *The Goldbergs* became one of the most popular comedies of early television, earning Berg the first Emmy Award for Best Actress in 1950. In 1961–62, Molly Goldberg inspired yet another television series, *Mrs. Goldberg Goes to College* (or *The Gertrude Berg Show*). Because of its acuity in representing a common immigrant experience, the magazine *TV Show* appropriately labeled *The Goldbergs* an "American institution" (Merritt, 1951).

JENNIFER HYLAND WANG

See also Comedy; Jewish Radio; Situation Comedy; Stereotypes on Radio

Cast

Molly Goldberg	Gertrude Berg
Jake Goldberg	James Waters, Phillip Loeb (1949–50)
Sammy Goldberg	Alfred Ryder, Everett Sloane (late 1930s), Larry Robinson (1949–50)
Rosalie Goldberg	Roslyn Silber, Arlene McQuade (1949–50)
Uncle David	Menasha Skulnik, Eli Mintz (1949–50)
Joyce	Anne Teeman
Edna	Helene Dumas
Solly	Sidney Slon
Jane	Joan Tetzel
Seymour Fingerhood	Arnold Stang, Eddie Firestone, Jr.
Sylvia	Zina Provendie
Mr. Fowler	Bruno Wick
Mr. Schneider	Artie Auerback
Esther	Joan Vitez
Mickey Bloom	Howard Merrill
Martha Wilberforce	Carrie Weller
Libby	Jeanette Chinley
Uncle Carlo	Tito Vuolo
Mrs. Bloom	Minerva Pious
Announcer	Clayton "Bud" Collyer, Alan Kent, Art Millet

Creator/Writer

Gertrude Berg

Programming History

NBC 1929–34; 1937 (briefly); 1941 (briefly)
CBS 1935–50

Further Reading

Berg, Gertrude, *The Rise of the Goldbergs,* New York: Barse, 1931

Berg, Gertrude, and Cherney Berg, *Molly and Me,* New York: McGraw Hill, 1961

Hilmes, Michele, *Radio Voices: American Broadcasting, 1922–1952,* Minneapolis: University of Minnesota Press, 1997

Kutner, Nanette, "The Nine Greatest Women in Radio," *Radio Stars* (December 1934)

Merritt, Joan, "Calling: Molly Goldberg," *TV Show* 1, no. 1 (May 1951)

O'Dell, Cary, *Women Pioneers in Television: Biographies of Fifteen Industry Leaders,* Jefferson, North Carolina: McFarland, 1996

"Programs That Have Made History," *Radio Daily* (3 November 1937)

Seldes, Gilbert, "The Great Gertrude," *Saturday Review* (2 June 1956)

Wheeler, Dan, "How the Ghetto Guides the Goldbergs," *Radio Mirror* 6, no. 2 (June 1936)

Goldsmith, Alfred Norton 1888–1974

U.S. Radio Engineer and Inventor

Alfred Norton Goldsmith—teacher, engineer, and inventor—won the admiration of two generations of engineers for his contributions to the field of communication technology.

Goldsmith taught electrical engineering at City College of New York while working as director of research for the American Marconi Company. After World War I, he became the first director of research in the newly organized Radio Corporation of America (RCA), and he ultimately became vice president and general engineer of that company. Though RCA was organized to consolidate the patent rights held by different radio interests, under Goldsmith RCA also began to develop research interests in a number of key wireless technologies. His broad knowledge, both theoretical and practical, served RCA well as the company rapidly expanded.

Goldsmith was a prolific inventor with 134 American patents. He was a consulting engineer after 1930, and he served as editor (1912–54) of the highly influential *Proceedings of the IRE (Institute of Radio Engineers).*

During a career that spanned more than 50 years, Goldsmith inspired those around him to apply rigorous and thoughtful science to the day-to-day engineering problems encountered by the infant broadcasting industry. The outgrowth of that research significantly influenced modern communications technology. He was responsible for patents related to many fields of communication technology, including radio, television, phonographs, aircraft guidance, and motion pictures. Included among Goldsmith's inventions were the shadow-mask cathode-ray tube used in modern color televisions; interlace scanning, which solved early television flicker problems; the radio diversity-reception system, which reduced signal fading; ultrasonic remote control systems; color facsimile transmission; an instrument landing system for aircraft; and the combination radio-phonograph.

Goldsmith was widely regarded as one of radio's outstanding early engineers. Friends, students, and colleagues described him as brilliant, citing his breadth of knowledge and quick insights, both of which won him the respect of those who worked with him and of the administrators who relied on his expertise. Archer (1939) claims that Goldsmith was responsible for christening RCA's early radios "Radiola," a trade name that became synonymous with RCA products. Later, Goldsmith became a freelance consultant and worked for RCA, Kodak, Radio-Keith-Orpheum (RKO), and other companies, but he remained a friend and trusted adviser to RCA's David Sarnoff.

Goldsmith was also an important voice in the formulation of broadcast policy, serving as a member of the Radio Advisory Committee for the National Bureau of Standards at the four Radio Conferences held from 1922–1925. Later, he led the Board of Consulting Engineers, which helped the Federal Radio Commission develop a policy for international relay broadcasting and television. Goldsmith was able to popularize theoretical problems about early radio problems in language that was easily understood, and he published more than 100 papers and books. He edited several anthologies of technical papers by RCA engineers on facsimile, radio at ultrahigh frequencies, and electron tubes.

Although Goldsmith's engineering capabilities and inventions won him fame, perhaps his greatest contribution to the art of communication was his service to the engineering profession. He served as president of the Institute of Radio Engineers (IRE), an early professional society that he cofounded with Robert Marriot, and he edited the institute's journal, *Proceedings of the IRE,* for 42 years. Goldsmith also served as president of the Society of Motion Picture and Television Engineers. Throughout his career, he continued an active interest in these professional organizations, publishing dozens of technical papers on a remarkably wide range of different communication technologies.

FRITZ MESSERE

Alfred Norton Goldsmith. Born in New York City, 15 September 1888. Attended City College of New York, B.S. 1907; Columbia University, Ph.D., 1911; electrical engineering teacher, City College of New York, 1907–19; consultant, General Electric, 1915–17; Director of Research, American Marconi Wireless Telegraph Company, 1917–19; joined Radio Corporation of America (RCA) as Director of Research; promoted to Vice-President of Research and General Engineer, RCA, 1919–31; became independent consulting engineer for NBC, RCA, and Eastman Kodak, 1931; co-founded Institute of Radio Engineers (IRE; predecessor of IEEE), 1912; edited IRE journal *Proceedings,* 1912–54; served as president of the Society of Motion Picture and Television Engineers, 1932–34, and on Radio Advisory Committee for Bureau of Standards, National Television System Committee (NTSC); held 200 patents related to radio and television, 1919–72, as well as patents in motions pictures, communication technology, and air conditioning. Received Medal of Honor and the Founders Award of the IRE; Modern Pioneer Award, National Association of Manufacturers; Townsend Harris Medal, City College of New York; first recipient of the Haraden Pratt Award. Died in St. Petersburg, Florida, 2 July 1974.

Selected Publications

Radio Telephony, 1918

This Thing Called Broadcasting: A Simple Tale of an Idea, an Experiment, a Mighty Industry, a Daily Habit, and a Basic Influence in Our Modern Civilization (with Austin C. Lescarboura), 1930

Further Reading

"Alfred N. Goldsmith, Biographical Notes," *Journal of the SMPTE* 81 (November 1972)

Archer, Gleason Leonard, *Big Business and Radio,* New York: American Historical Society, 1939; reprint, New York: Arno Press, 1971

Dreher, Carl, "His Colleagues Remember 'The Doctor'" *IEEE Spectrum* (August 1974)

Dunlap, Orrin Elmer, Jr., *Radio's 100 Men of Science: Biographical Narratives of Pathfinders in Electronics and Television,* New York: Harper, 1944

Kraeuter, David W., "Alfred N. Goldsmith," in *Radio and Television Pioneers: A Patent Bibliography,* edited by Kraeuter, Metuchen, New Jersey: Scarecrow Press, 1992

The Goon Show

BBC Comedy Program

British radio comedy reached its peak in the 1950s, before television took over as the dominant medium around 1960. Of all the varied series, *The Goon Show* was the most outrageous, the most inventive, and the most demanding in its use of the medium. Rooted in off-the-wall humor such as Lewis Carroll, the Marx Brothers, and Hollywood cartoons, it became a cult item—loved by schoolboys and jazz musicians and hated by schoolmasters and the British Broadcasting Corporation (BBC) establishment.

It was the creation of four young performers, all of whom had been in the Armed Forces during World War II and had been involved with troop entertainments to some degree—Michael Bentine, Spike Milligan, Harry Secombe, and Peter Sellers. In the postwar period they performed in theatrical variety and made occasional radio appearances; Milligan made contributions to radio comedy scripts through his friendship with Jimmy Grafton, who ran a public house, Grafton's, in London's Victoria area and who wrote scripts for the conventional *Derek Roy Show.*

From meetings and experimental performances in Grafton's, the four put together a comedy show that they suggested to the BBC. After a pilot recording and some false starts, the first season began on 28 May 1951. The four had wanted to call it *The Goon Show,* borrowing from the Goons in the Popeye comic strip, but the BBC, not understanding the name (one official asked, "What is this *Go On Show*?") insisted that it be called *Crazy People.*

There were 17 shows in the first season and 25 in the second, which started on 22 January 1952. The format of the program included short sketches, very cartoonish in nature and gradually building up the use of idiot logic and outlandish sound effects. After the second season, Michael Bentine left to follow his own slightly different comedy style, and the three others formed the basic cast for the rest of the show's life.

The third season, now at last called *The Goon Show,* offered 25 shows starting on 11 November 1952. It continued in the sketch format, with two straight musical items from mouth organist Max Geldray and Ray Ellingon's jazz-style quartet. With the fourth season (30 shows from 2 October 1953), the shows took on the format that would become the standard, telling a single story that was interspersed with the two musical items. Plots included sending the Albert Memorial to the moon and chasing the Kippered Herring Gang.

From the beginning of the fifth season (26 shows) on 28 September 1954, the series gained a worldwide audience through recordings sold to overseas radio stations by the BBC's Transcription Services; the show's overseas popularity outlasted the series itself by many years through repeats. The style had by now settled down, through the use not only of strong, daft plots but also of strongly drawn characters, created by scriptwriter Milligan both alone and in collaboration with Eric Sykes and Larry Stephens.

The main character, around whom all the plots revolved, was Neddie Seagoon (Secombe)—short, fat, stupid, noisy, patriotic, and well-meaning but gullible. In most cases, he was deceived into carrying out some ridiculous task—such as ferrying snow to the Sahara in cardboard boxes—for apparently patriotic reasons but in fact at the behest of the two master criminals, Hercules Grytpype-Thynne (Sellers), a smooth George Sanders type, and Count Jim Moriarty (Milligan), who degenerated over the series into a cringing, starving wreck who dragged Grytpype-Thynne down with him ("Keep still, Moriarty—do you want us both out of this suit?").

In the course of his adventures Seagoon would meet up with other eccentric couples. The senile Minnie Bannister (Milligan) ("We'll all be murdered in our beds!") and Henry Crun (Sellers) ("You can't get the wood, you know") were based, Milligan later claimed, on his parents: their dithering and bickering marked every appearance.

Another couple, usually cast as Seagoon's helpers, were Eccles (Milligan) and Bluebottle (Sellers). Eccles was a combination of Walt Disney's Goofy and the village idiot, and Bluebottle was the eternally hopeful Boy Scout, who almost always got blown up.

The main remaining character, Major Dennis Bloodok (Sellers), was a loose-boweled, lecherous, and greedy army officer of dubious morals. There were also many minor characters, including the baby Little Jim (Milligan), whose only line, whenever appropriate, was "He's fallen in the water!" and the excitable Indians Lalkaka and Bannerjee.

In most radio series, the actors played characters who took part in the plot, but in *The Goon Show* it was the characters themselves who performed the plot, often making asides that showed they were indeed "only acting," which gave an odd extra dimension to the comedy.

Apart from an occasional special show, the remaining seasons were the sixth (10 shows from 20 September 1955), the seventh (25 from 4 October 1956), the eighth (26 from 30 September 1957), a special season of reworked repeats for Transcription Services (14, recorded in 1957–58), the ninth (17 from 3 November 1958), and the final short season of 6 episodes beginning 24 December 1959. By this time the strain of writing and performing the shows had caused Milligan a nervous breakdown and a good deal of subsequent strain—which was showing in the writing—and the series came to an end on 28 January 1960.

Subsequently there were a few special appearances by the trio and a strange television series, *The Telegoons,* reworking old scripts with new recordings of the performers (without an audience) and puppets for the visuals: it was not successful.

The final appearance of the Goons was in the special show *The Last Goon Show of All,* recorded on 30 April 1972 for radio and also televised. The subsequent death of Sellers in 1980 made further reunions impossible.

The Goon Show had a lasting effect on British comedy. It took radio to the boundaries of what was possible through the use of detailed production and complex sound effects, and its irreverent and wildly illogical look at British life influenced many subsequent shows and performers—most famously television's *Monty Python* team. The shows have continued to be extremely popular for many years through repeats and issues on records, tapes, and CDs.

ROGER WILMUT

See also British Broadcasting Corporation; British Radio Formats

Cast

Neddie Seagoon (in various guises)	Harry Secombe
Eccles	Spike Milligan
Miss Minnie Bannister	Spike Milligan
Count Jim Moriarty	Spike Milligan
Bluebottle	Peter Sellers
Mr. Henry Crun	Peter Sellers
Major Denis Bloodnok	Peter Sellers
Hercules Grytpype-Thynne	Peter Sellers

Creators

Spike Milligan, Jimmy Grafton, Peter Sellers, Harry Secombe, Michael Bentine, Eric Sykes, Larry Stephens

Programming History

BBC 1951–60

Further Reading

Milligan, Spike, *The Goon Show Scripts,* London: Woburn Press, and New York: St. Martin's Press, 1972

Milligan, Spike, *More Goon Show Scripts,* London: Woburn Press, 1973; New York: St. Martin's Press, 1974

Milligan, Spike, *The Book of the Goons,* London: Robson, and New York: St. Martin's Press, 1974

Took, Barry, *Laughter in the Air: An Informal History of British Radio Comedy,* London: Robson, 1976; 2nd edition, 1981

Wilmut, Roger, *The Goon Show Companion: A History and Goonography,* London: Robson, 1976

Gordon, Gale 1905–1995

U.S. Radio, Film, and Television Actor

Gale Gordon, a hard-working radio (and, later, television) actor, may never have been a household name, but his voice and face were known to millions. Born into a theatrical family, he spent the first five years of his life in England. His father, Charles Aldrich, was a quick-change artist and his mother was actress Gloria Gordon, best known for her portrayal of Mrs. O'Reilly on radio's *My Friend Irma.*

Born with a cleft palate, Gordon endured two painful operations to correct the problem, and his parents pushed him into show business to help him perfect his speech. By the time he was 17, Gordon's voice was so richly developed that it became his trademark. "His voice was colorful and powerful so he could bend and shape the dialogue he was given," according to Lucie Arnaz. "He also used his whole body for effect, not just his voice," Arnaz added.

While still a teenager, Gordon made his Broadway debut with Richard Bennett in *The Dancers* in the 1920s. He then moved over to try the new medium of radio. Radio offered something new for actors: steady work. Gordon soon became one of the busiest actors in radio, sometimes appearing in 20 to 30 shows per week. In 1929, he moved to the West Coast for the role of Judas in *The Pilgrimage Play.* After a year of touring, he returned to Hollywood where he worked as a freelance actor and announcer. No role daunted him—he supplied the voices of villains, Tarzan, and Flash Gordon, in addition to playing straight dramatic roles in syndicated shows such as *The Adventures of Fu Manchu* and *English Coronets.*

Hy Averback, noted television and movie producer, described Gordon as a combination of Laurence Olivier and Charley Chaplin. Called the master of the slow burn and stack blowing, Gordon found his true forte in comedy. He became everyone's idea of the perfect stuffed shirt and comic foil. His bellowing voice proved to be indispensable and assured gainful employment in dozens of character roles on radio, including his well-known long-running radio roles as Mayor LaTrivia on *Fibber McGee and Molly,* the harried sponsor on *The Phil Harris and Alice Faye Show,* and the apoplectic banker Attenbury on Lucille Ball's *My Favorite Husband.*

Gordon was Lucille Ball's first choice for next-door neighbor and landlord Fred Mertz when *I Love Lucy* was brought to television in 1951. "My mother could always depend on Gale and trusted his acting choices on the radio show," Lucie Arnaz said. "And my mother wasn't fond of change; just for change's sake she would have gladly given Gale the part of Fred Mertz. But Gale was under exclusive contract with CBS Radio so William Frawley got the part."

Gordon worked with many notable actors in radio, including Jimmy Durante, Doris Singleton, Eve Arden, George Burns and Gracie Allen, Richard Crenna, Claire Trevor, Edward G. Robinson, Bea Benaderet, Jack Webb, Mary Pickford, Jack Haley, and Dennis Day. The only actor with whom he seemed to conflict was Mel Blanc, creator of the voice of Bugs Bunny and many other Warner Brothers cartoon characters; the two were constantly in competition for the same parts.

In October 1941, Gordon began playing the role of Mayor LaTrivia on *Fibber McGee and Molly.* (Mayor LaTrivia would arrive at the McGee house, start an argument, become tongue-tied, and blow his stack.) Gordon's only break from the show during 15 seasons was his service in the U.S. Coast Guard, where he served as chief gunner's mate during World War II.

In 1948 Gordon was given the role of Osgood Conklin in *Our Miss Brooks,* a situation comedy about a high-school English teacher forced to endure the supervision of this stuffy, by-the-rules, pain-in-the-rear school principal. Gordon's Conklin was dry, cynical, blustery, and explosive, all at once. It was to be his last role in radio; the show moved to television in 1952 and ran until 1956, with most of the original radio cast, including Gale Gordon, making a successful transition to the new visual medium.

Gordon met his wife, Virginia Curley, while working on the *Death Valley Days* radio program. They married in 1937 and later played together as Mr. and Mrs. Osgood Conklin in both the radio and television versions of *Our Miss Brooks;* they remained together until their deaths a few weeks apart in 1995.

ANNE SANDERLIN

See also Fibber McGee and Molly; Our Miss Brooks

Gale Gordon. Born Charles T. Aldrich in New York City, 2 February 1906. Made Broadway debut in "The Dancers," 1920s; radio debut, *The Church Mouse,* 1934; featured on *Michael and Mary, Flash Gordon,* and many other shows; served in the U.S. Coast Guard during World War II; returned to radio, 1945. Died in Escondido, California, 30 June 1995.

Radio Series

1934	*The Church Mouse*
1934	*Michael and Mary*
1938–40	*Irene Rich*
1940	*Those Who Love; Crossroads*
1942	*Maxwell House Coffee Time*
1941–42, 1945–56	*Fibber McGee and Molly*
1946–47	*The Fabulous Dr. Tweedy*

1947	*The Irene Show*
1947–50	*The Phil Harris and Alice Faye Show*
1948–51	*The Judy Canova Show*
1948–51	*My Favorite Husband*
1948–57	*Our Miss Brooks*
1950	*The Penny Singleton Show*
1950–53	*Fibber McGee and Molly*

Films

Here We Go Again, 1942; *A Woman of Distinction,* 1950; *Here Come the Nelsons,* 1952; *Francis Covers the Big Town,* 1953; *Our Miss Brooks,* 1956; *Rally 'Round the Flag, Boys!* 1959; *The 30 Foot Bride of Candy Rock,* 1959; *Visit to a Small Planet,* 1960; *Give Up the Ship,* 1961; *All Hands on Deck,* 1961; *Dondi,* 1961; *All in a Nights Work,* 1961; *Sergeant Deadhead,* 1965; *Speedway,* 1968; *The Burbs,* 1989

Television

Our Miss Brooks, 1952–56; *The Brothers,* 1956–57; *Sally Bascomb Bleacher,* 1958; *Pete and Gladys,* 1960–62; *Dennis the Menace,* 1962–63; *The Lucy Show,* 1963–68; *Here's Lucy,* 1968–74; *Life with Lucy,* 1986

Further Reading

Brooks, Tim, *The Complete Directory of Primetime Television Stars, 1946–Present,* New York: Ballantine, 1987

DeLong, Thomas A., *Radio Stars: An Illustrated Biographical Dictionary of 953 Performers, 1920 through 1960,* Jefferson, North Carolina: McFarland, 1996

Fidelman, Geoffrey Mark, *The Lucy Book: A Complete Guide to Her Five Decades on Television,* Los Angeles: Renaissance Books, 1999

Gospel Music Format

The gospel music format is a popular genre of radio programming that features generally upbeat music with a Christian message.

Gospel music has long been a staple of radio, particularly in the Bible Belt, and it was featured on several early radio stations, including WFOR in Hattiesburg, Mississippi; KWKH in Shreveport, Louisiana; WSM in Nashville, Tennessee; WVOK in Birmingham, Alabama; and WKOZ in Kosciusko, Mississippi.

Today's gospel music format originated with and grew out of special programming. As local radio stations began to dot the landscape of the United States, many of them featured gospel music on Sundays, or perhaps 30 minutes to an hour each day. Typically, this special programming featured either white Southern quartets, groups, and soloists (including the Statesmen Quartet, the Blackwood Brothers, the Chuck Wagon Gang, and Tennessee Ernie Ford), or black groups and soloists (such as the Jackson Southernaires, Edwin Hawkins Singers, or Mahalia Jackson).

During the 1980s, some radio stations began to program gospel music exclusively, billing themselves as "all gospel all the time." With deregulation of the broadcasting industry in the 1980s, the number of radio stations significantly increased, and program directors created specialized music formats to target specific audiences. The gospel music format began to burgeon; a number of stations chose it as the only type music they featured, complete with the "clock hour," which delineates and specifies every element of programming during each hour. The clock hour, also known as the "format wheel" or "programming wheel," could be compared to a pie cut into approximately 25 parts. For example, the top of the hour on the format wheel might include five minutes of national news from a network (not a few gospel stations are affiliated with the *USA Radio Network* because its content and style of reporting correlate with issues with which some Christians are concerned). At five minutes past the hour, a number-one song from yesteryear might be featured, followed by a top-ten gospel hit, followed by a totally new selection that might prompt the disc jockey to remark, "And you heard it first right here." In this way, the gospel music format is similar to the structure of a Top-40 station, with on-air personalities using their names, throwing in some pleasantries, and striving to create an image appealing to the target audience—in the case of gospel, a bright, happy, encouraging sound.

Top radio groups such as Jacor, Clear Channel, and Infinity see gospel as a viable format. Capitalizing on audience loyalty to the music, a number of stations have improved their ratings after switching to the gospel format, increasing their average quarter hour listening shares and paving the way for advertising acceptance. The gospel music format is typically appealing to advertisers, although gospel music stations decline to advertise certain products or services, such as alcoholic beverages or night clubs. If a gospel music station is airing a sporting event, typically the operator at the station has been instructed to block any network advertisements which pertain to alcohol or tobacco products.

Radio stations have increasingly subscribed to gospel music programming via satellite services. Listeners to such services may well feel that the syndicated announcers are actually present at the local station playing the music and making the comments, even though these DJs are merely part of the download. One of the most popular formats is called *The Light*, which features a mix of black urban gospel artists and includes cross-over collaborations by mainstream artists such as Cheryl "Salt" James of *Salt-n-Pepa* and Stevie Wonder.

Some stations have recently begun to feature specific gospel subgenres, such as southern, African American, country, jazz, contemporary, bluegrass, and even rap.

DON RODNEY VAUGHAN

See also Christian Contemporary Music Format; Evangelists/Evangelical Radio; National Religious Broadcasters; Religion on Radio

Further Reading

Collins, Lisa, "Urban Radio Sees the Light," *Billboard Magazine* (2 August 1997)
Schwirtz, Mira, "No Longer up in the Air," *Media Week* (11 May 1998)

Goulding, Ray. *See* Bob and Ray

Grand Ole Opry

Country Music Variety Program

As the 20th century ended, the *Grand Ole Opry* was the most famous and longest-running live radio broadcast still on the air. A traditional radio barn dance, originating on WSM radio from Nashville, Tennessee, the *Opry* has reached homes across the eastern half of the United States. Although it started as a local show and later reverted to that status, through the 1940s and most of the 1950s the *Grand Ole Opry* was a staple on Saturday nights on the National Broadcasting Company (NBC). To most Americans, the *Opry* defined what a radio barn dance was and is. Because of its triumph over all major rivals, exemplified by the centralization of the country music industry in Nashville, the *Grand Ole Opry* occupies an important place in both radio and recording industry history.

The *Grand Ole Opry* made its debut on 28 November 1925 on WSM-AM's Studio B to an audience of 200 people. C.A. Craig, one of the founders of the National Accident Insurance Company of Nashville, Tennessee, owned a radio station during the early 1920s (later called WSM for "We Shield Millions"); in 1925 he hired George D. Hay away from WLS-AM in Chicago to develop a barn dance show for WSM, as Hay had done for WLS with the *National Barn Dance*. Hay began in November 1925, and within a month the new show was a two-hour-long Saturday night staple.

By 1927, as an NBC affiliate, this two-hour local country hoedown followed the network broadcast of the National Symphony Orchestra, which aired Saturday nights from 8:00 to 9:00 P.M. One night, probably 8 December 1928, Hay reportedly stated, "For the past hour you have heard music largely taken from grand opera; now we will present the *Grand Ole Opry*." True or not, the title is now world famous.

Regional success can be measured by the show's need within a few years to move to a new studio, Studio C, which held 500 persons; later, after a series of temporary moves, the show made its permanent home in the 2,000-seat Ryman Auditorium at Fourth and Broadway in downtown Nashville in 1943. The show remained at that location throughout its network radio days and then, in 1974, moved to a new auditorium as part of the opening of the Opryland theme park in suburban Nashville.

The show had started informally as what scholar Charles Wolfe calls "a good natured riot." But although the program seems informal, getting on was always a struggle for the new

Marty Robbins on stage at the Grand Ole Opry at the Ryman Auditorium
Courtesy Radio Hall of Fame

artist, and many argued that it became more and more commercialized. By the 1930s "hillbilly" stars had been developed, and some dead singers were immortalized. The music was spread thanks to the diffusion and growth of the population, and even greater stars emerged during the post–World War II era. The Opry management, particularly James Denny, took advantage of this interest, and the network (and many clear channel stations) carrying *Grand Ole Opry* enabled the broadcast to become one of the most popular radio programs in the country. Denny and his colleagues also worked with leading record labels to make Nashville the center of the "country and western" universe.

The *National Barn Dance,* from Chicago's WLS, was already an NBC fixture when a half-hour segment of the *Grand Ole Opry* was added to a number of NBC's regional broadcasts, including 26 stations in the Tennessee area. In 1939 NBC began to carry the *Opry* regularly on a regional basis. Two years later the *Opry* went out all across the NBC network.

The coming of TV and format country radio signaled the end of the barn dance radio show. Yet WSM-AM stuck with the *Grand Ole Opry,* and Nashville became not just one center for the making of country music but the leading one. Indeed, many of the early Nashville recordings were done in WSM studios, until Owen Bradley and others began to fashion "Music Row" several miles west of the Ryman Auditorium. And by the time that Bradley at Decca and Chet Atkins at the Radio Corporation of America (RCA) began to remake "hillbilly" music into crossover country music with stars such as Hank Williams and Patsy Cline, the "Athens of the South" had become "Music City."

DOUGLAS GOMERY

See also Country Music Format; National Barn Dance; WSM

Cast

Announcers
George Dewey Hay, Grant Turner

Comedienne
Cousin Minnie Pearl

Singers (partial listing)
Roy Acuff, Hank Williams, Bill Monroe, Patsy Cline, Kitty Wells, Red Foley, George Morgan, Ernest Tubb, Grandpa Jones, DeFord Bailey, Uncle Dave Macon, Eddy Arnold, Loretta Lynn, Hank Snow, Little Jimmy Dickens, Lorrie Morgan, Trisha Yearwood, Vince Gill, Garth Brooks, Emmylou Harris, Ricky Skaggs

Programming History

WSM (and other local Tennessee stations at various times)	1925–present
NBC	1939–57

Further Reading

Eiland, William U., *Nashville's Mother Church: The History of the Ryman Auditorium,* Nashville, Tennessee: Opryland USA, 1992

Hagan, Chet, *Grand Ole Opry,* New York: Holt, 1989

Hickerson, Jay, *The Ultimate History of Network Radio Programming and Guide to All Circulating Shows,* Hamden, Connecticut: Hickerson, 1992; 3rd edition, as *The New, Revised, Ultimate History of Network Radio Programming and Guide to All Circulating Shows,* 1996

Kingsbury, Paul, *The Grand Ole Opry History of Country Music: 70 Years of the Songs, the Stars, and the Stories,* New York: Villard Books, 1995

Malone, Bill C., *Country Music U.S.A.,* Austin: University of Texas Press, 1968; revised edition, 1985

Wolfe, Charles K., *A Good-Natured Riot: The Birth of the Grand Ole Opry,* Nashville, Tennessee: Country Music Foundation Press and Vanderbilt University Press, 1999

The Great Gildersleeve

Situation Comedy

"Great" is the perfect epithet for the character Throckmorton P. Gildersleeve. First a foil on *Fibber McGee and Molly,* then star of his own program, he was a large man in the tradition of Shakespeare's pleasure-loving Falstaff. Loud but never mean, he began sparring with Fibber, the archetypal windbag, in 1937. His character was honed as the bumptious, explosive Gildersleeve, who typically ended a duel with his exasperated phrase, "You're a harrrd man, McGee!" His very name, coined by script writer Don Quinn, combined dignity (Basil Gildersleeve was a famous Victorian classicist) and inside joke (the actor playing Gildersleeve lived on Throckmorton Place). The character left *McGee,* taking the train from the National Broadcasting Company's (NBC) Wistful Vista to Summerfield on 31 August 1941, thus becoming radio's first successful spinoff. The program would last 16 years, until March 1957. Gildersleeve, or "Gildy" to some, was played by actor Hal Peary to 1950 and by Willard Waterman from 1950 to 1957.

In his new town, Gildy's abrasive personality mellowed as he embraced home, work, and social life. Each contact in Summerville deflated his grandiosity and humanized him. Once he planned to attend a costume ball dressed as an ancestor. He daydreamed about possible relatives—a romantic castaway on a tropical island, a dashing pirate, a Gilded Age tycoon—only to learn that he was descended from Goldslob the Pennsylvania butcher (24 March 1948). His appetites kept housekeeper Birdie Lee Coggins busy. More feisty than most of radio's black domestics, Birdie moderated his pomposity by repeating herself ("You know what I said? That's right! That's what I said"). Birdie often mirrored her employer. Both belonged to fraternal groups (she to "The Mysterious and Bewildering Order of the Daughters of Cleopatra," he to "The Jolly Boys"); both sang well. Her talents and industriousness silently rebuked Gildy's natural sloth. Her chocolate cake won a prize, and when she went on vacation, no one could prepare a suitable dinner. She also provided a mother surrogate for Gildy's wards, niece Marjorie Forrester and nephew Leroy.

In an era when single-parent families usually implied a widow with children, Gildy was unusual. Like other unmarried guardians (Donald Duck, Sky King), he coped with the younger generation by combining bossiness, wheedling, and exasperation. Marjorie usually abided by his rules, but she began dating a series of boys who fell short of Gildy's expectations. After wedding Bronco Thompson on 10 May 1950 and bearing twins (21 February 1951), she set up her own household next door. Although she dutifully catered to Gildy's whims, double dating with him before the marriage and asking him to babysit after the kids arrived, her in-laws tried his patience. Used to dominating her husband, Mrs. Thompson openly defied Gildy until they bonded on a picnic. But everyone else had become accustomed to their bickering, so they obligingly pretended to spat (22 March 1950).

Leroy gave little promise of accepting maturity: he reacted to his "Unk's" apparently foolish directions with an exasperated "Oh, for corn sakes." Gildy tried patience but often resorted to the ultimate threat, a menacingly drawn-out "Leeeroy." Certainly Leroy needed direction. His academic work would have embarrassed anyone, but it particularly discomfited Gildy, who courted the school principal, Eve Goodwin. Leroy's troubles with bullies, jobs, attractive girls, stolen lumber, and toothaches often defied logical advice, yet they somehow solved themselves.

Gildy seemed to be an unlikely source of practical wisdom. He loitered through his job as water commissioner, sometimes aided by his simple secretary Bessie. His campaign for mayor in 1944 floundered when he lost his temper on a political broadcast. When he discovered that no one in the city knew him or his job, he hatched a publicity stunt: to dive into the reservoir (23 April 1952). The bungled descent temporarily dampened his quest for recognition. Romance, at any rate, interested him more. Various women with descriptive names like Eve Goodwin and Adeline Fairchild prompted him to buy perfumes and candy from the crusty druggist, Peavey. Gildy should have imitated Peavey's famous tag line, "Well, now, I wouldn't say that," because his amorous crusades never led to the altar. His closest approach, with Leila Ransom, a flirtatious Southern widow, ended when her supposedly dead spouse Beauregard turned out to be alive (27 June 1948).

Gildy's male friends provided enough excitement to compensate for these losses. Judge Horace Hooker, the "old goat" who monitored Gildy's care of Marjorie and Leroy, diminished his ego by staying when Gildy wished to court a lady friend or by demanding vegetarian food when he came for dinner. So did Rumson Bullard, Gildy's wealthy and insulting neighbor, who drove a big car and disdained to invite him to a neighborhood party. The "Jolly Boys"—Peavey, Hooker, Floyd Munson the barber, and police chief Gates—met to gossip and sing. The bonding sometimes frayed, inspiring Chief Gates to plead, "Aw, fellows; let's be Jolly Boys!"

Network politics might have caused disaster, because the original Gildy, Hal Peary, launched his own short-lived Columbia Broadcasting System (CBS) show, *Honest Harold,* in 1950. Luckily, Willard Waterman, a friend of Peary's who often teamed with him on other shows, sounded like him and took over the lead until the show ended in 1957. Both men had fine singing voices and incorporated easy listening songs into the plot.

The two prolonged Gildy's life in movies and television. Peary appeared briefly in four amusing films (*Comin' Round the Mountain,* 1940; *Look Who's Laughing,* 1941; *Country Fair,* 1941; *Unusual Occupations,* 1944) and starred in four others (*The Great Gildersleeve,* 1942; *Gildersleeve's Bad Day,* 1943; *Gildersleeve on Broadway,* 1943; *Gildersleeve's Ghost,* 1944). Waterman was featured in 39 TV episodes (September 1955–September 1956). Radio writers Paul West, John Elliotte, and Andy White followed, sometimes recycling story lines (Gildy's aforementioned dive into the reservoir; his attraction to Bullard's sister from 19 September 1951). However, they overemphasized Gildy's womanizing tendencies for the first 26 programs. Other shows toned down Leroy's mischief, substituted new actors (only three originals remained), and lost a major sponsor. Still, both films and TV communicated some of the great man's foibles and successes familiar to radio fans.

JAMES A. FREEMAN

See also Comedy; Fibber McGee and Molly

Cast

Throckmorton P. Gildersleeve	Hal Peary (1941–50), Willard Waterman (1950–57)
Leroy Forrester	Walter Tetley
Marjorie Forrester	Lorene Tuttle (1941–44), Louise Erickson (mid 1940s), Mary Lee Robb (mid 1940s–56)
Judge Horace Hooker	Earle Ross
Birdie Lee Coggins	Lillian Randolph
Mr. Peavey	Richard Legrand, Forrest Lewis
Floyd Munson	Arthur Q. Bryan
Police Chief Gates	Ken Christy
Leila Ransom	Shirley Mitchell
Adeline Fairchild	Una Merkel
Eve Goodwin	Bea Benaderet (1944)
Nurse Kathryn Milford	Cathy Lewis (1950s)
Bashful Ben	Ben Alexander (mid 1940s)
Bronco Thompson	Richard Crenna
Rumson Bullard	Gale Gordon, Jim Backus (1952)
Craig Bullard	Tommy Bernard
Bessie	Pauline Drake, Gloria Holliday
Announcer	Jim Bannon (1941–42), Ken Carpenter (1942–45), John Laing (1945–47), John Wald (1947–49), Jay Stewart (1949–50), Jim Doyle, John Hiestand

Producers/Directors

Cecil Underwood, Frank Pittman, Fran Van Hartesveldt, Virgil Reimer, and Karl Gruener

Programming History

NBC 1941–57

Further Reading

Harmon, Jim, *The Great Radio Comedians,* Garden City, New York: Doubleday, 1970

"Helpful Hints to Husbands," *Tune In* 4 (July 1946)

Salomonson, Terry, *The Great Gildersleeve: A Radio Broadcast Log of the Comedy Program,* Howell, Michigan: Salomonson, 1997

Stumpf, Charles, and Ben Ohmart, *The Great Gildersleeve,* Boalsburg, Pennsylvania: BearManor Media, 2002

Thomsen, Elizabeth B., The Great Gildersleeve, <www.ethomsen.com/gildy>

"Throckmorton P," *Newsweek* (13 December 1943)

Greece

Radio began in Greece later than in most other European nations, and the country was one of the last to develop an official government station. In a radio industry characterized by its state of constant flux, private stations and others controlled by the armed forces have existed alongside government outlets. Advertising has been accepted on some outlets, not on others.

Origins

Experimental radio stations were operated in Greece in the 1920s, but the first station offering a regular program schedule was established in Thessaloniki in 1928 as a private operation. Greek radio listeners relied primarily on stations in other nations, and considerable confusion and change was evident as the ministry in charge of radio kept switching. Despite several attempts, no other regularly operated stations existed in the country until the 1930s. In 1936 the government decided to develop a national broadcasting system that could educate Greek society. In 1938, Greece became one of the last European nations to develop a state-run radio station. The Ethnikon Idryma Radiophonias (EIR), or National Radio Foundation, provided programs consisting largely of music by the station's orchestra, choirs, and news.

At the beginning of World War II, the station helped boost the morale of the soldiers and the public. Eventually, the occupying German forces sealed radio receivers in Athens to receive only the national station, which they controlled. In the countryside people were ordered to turn in their radios or face severe punishment.

In 1945, EIR closed down the private station after an unsuccessful attempt to force it to broadcast the national signal. In the late 1940s, additional stations were built by EIR, as well as by private interests and the Armed Forces. Armed Forces stations were established to "enlighten" the people of Northern Greece about the dangers of communism. The United States, which had taken a major role in the Greek Civil War, built two radio stations that transmitted Voice of America programs part of the day and state programming the rest of the day.

The Armed Forces stations were financed through army funds and advertising. In 1952, as Armed Forces stations gained popularity, EIR established a Second Program. This service carried commercials and more popular music, unlike the original station's (First Program) more serious orientation. That service carried news, information, and fine arts programming, but no commercials. In 1954, EIR established the Third Program that primarily provided classical music broadcasts. All stations from 1946 to 1953 were under strict government censorship, while some form of censorship continued until 1975.

A new Greek constitution drafted after the fall of the military junta (1967–1974) placed radio and television "under the immediate control of the state." A new state organization, Hellenic Radio-Television (ERT), heavily controlled by the government, was created to operate the public stations.

Structure

According to a 1975 law, and subsequent broadcast laws, the purpose of ERT is to provide "information, education, and recreation for the Greek people (through) the organization, operation and development of radio and television." In addition, the act states that "ERT programs must be imbued with democratic spirit, awareness of cultural responsibility, humanitarianism and objectivity, and must take into account the local situation." Finally, this law stated: "The transmission of sound or pictures of any kind by radio or television by any natural person or legal entity other than ERT and the Armed Forces Information Service (YENED) shall be prohibited" ("Greek Radio-TV Law," 1976). This brought an end to any legal private broadcasting in Greece.

In 1982, the socialist governing party, PASOK, placed all YENED stations under civilian government control. Nevertheless, there was a great reluctance to open the airwaves to private citizens. No political party was willing to part with control of broadcasting, which they could use for self-promotion when in power.

In 1987, a new legal structure for broadcasting created one company, Hellenic Radio-Television S.A. (ERT), now a corporation owned by the state, to control all public service broadcasting in Greece. One of its entities is Hellenic Radio (ERA), made up of the four domestic radio services (ERA-1, ERA-2, ERA-3, ERA-4) and the Voice of Greece shortwave radio service (ERA-5).

Nevertheless, the major importance of the 1987 law was that, despite initial opposition from the government, it provided for the establishment of private stations. The impetus for this change came from political pressure brought on by opposition political parties that wanted a piece of the airwaves. Candidates running for local elections pledged to build municipal radio stations. Eventually, opposition candidates won the mayoral races in Greece's three largest cities and the breakdown of the state broadcasting monopoly was imminent.

The new mayor of Athens insisted that if a legal structure for "free" radio was not created by the end of March 1987, he would build a municipal radio station anyway. On May 31, 1987, municipal station "Athens 98.4 FM" went on the air without a license.

A major problem with the 1987 law, and relevant decrees that followed, was that it did not deal with the issue of the number of frequencies available. Nevertheless, the first non-state radio station licenses were approved in May 1988, most going to municipalities and publishing companies (Roumeliotis, 1991). However, the Licensing Commission never announced the criteria used in allocating these licenses and did not tie them to specific frequencies.

Following the initial allocation of licenses, which had to be renewed in two years, the government failed to implement legislation regulating private broadcasting. Related laws and decrees were passed, but since no action ever took place, all radio station licenses expired at the end of the initial two-year period. Thus, technically some 1000 radio stations in Greece were operating illegally.

Current Scene

It was not until May 2001 that the first "permanent" radio station licenses were allocated for the Athens area. This license approval process is to continue for the rest of the country. However, given its history, Greek radio likely will not stabilize for some time to come. The radio environment in Greece, despite some maturation that forced many stations to go out of business, remains in a state of anarchy. In addition to hundreds of illegal stations, there are stations that never even requested licenses and others that hold more than one frequency, while most do not pay the relevant taxes as required by the broadcast laws (Zaharopoulos, 1993).

State radio broadcasting has dramatically diminished in importance as its audiences have dwindled. Nevertheless, ERA still has 19 local and regional stations on AM, two AM relay stations, and 40 FM transmitters throughout the country. The ERA networks have also undergone certain changes. In 2002 the First Program was renamed NET Radio—NET (New Hellenic Television) being the name of Greek television's second channel, which has a serious orientation. The Second Program (ERA-2) carries more popular music and a few magazine and public affairs programs. One of the Second Program's frequencies in Athens was used to create ERA's fifth domestic service called "Kosmos" radio in 2002, carrying primarily world music. ERA-4 became ERA Sport, carrying mostly sports programming.

Municipal stations have also diminished in importance. Many have closed down, as their audiences were won over by commercial stations, and as cities could no longer afford to subsidize municipal radio.

Private stations built by former pirates have generally not succeeded either. Many of them either went out of business, were sold to larger corporations, or still broadcast as "mom and pop" operations. The real winners in the Greek radio scene have been the large media corporations or other industrial companies with their own stations. Most Greek industrialists who went into broadcasting did so in order to use their media voices to gain government contracts for their other businesses. Thus, even if most of these stations today lose money, their owners still view them as valuable sources of revenue.

Despite the trend toward program specialization resulting from a great number of stations, the two most popular stations in Athens (Sky, Antenna) feature variety formats with emphasis on news, political talk, and music entertainment. Another successful station is Flash, which is a news and information station. Generally, on the Greek airwaves today one can find any radio format imaginable. At the same time, public service broadcasting struggles to survive, while municipal public service broadcasting, with a few exceptions, is nearly dead. The next wave in Greek radio is expected to see foreign conglomerates entering the Greek market by buying existing stations.

THIMIOS ZAHAROPOULOS

Further Reading

Emery, Walter B., "Greece: Anxiety and Instability," in *National and International Systems of Broadcasting: Their History, Operation, and Control*, by Emery, East Lansing: Michigan State University Press, 1969

Keshishoglou, John E., "The Development of Broadcasting in Greece," Master's thesis, University of Iowa, 1962
Roumeliotis, A., *Imaste ston aera* [We're on the Air], Thessaloniki: Paratiritis, 1991
Zaharopoulos, Thimios, and Manny E. Paraschos, *Mass Media in Greece: Power, Politics, and Privatization*, Westport, Connecticut: Praeger, 1993

The Green Hornet

Juvenile Drama Series

Joining a number of American fictional superheroes already entertaining the large radio audience from the late 1930s through the war and into the 1950s, *The Green Hornet* debuted over Detroit station WXYZ on 31 January 1936. The series began simply as *The Hornet* (the descriptive color was added later in order to copyright the title, according to radio historian Jim Harmon). Another brainchild of Detroit station operator George W. Trendle, who also created *The Lone Ranger,* the two half-hour action dramas shared more than classical music themes (for which no copyright fees needed to be paid) and the same creator. In this case, the famous sound of a buzzing hornet was made by a musical instrument called the theremin, while the music was Rimsky-Korsakov's "The Flight of the Bumblebee."

Russo (2001) reports that the first 260 episodes of the series lacked individual titles (they were simply numbered), but those broadcast after 9 August 1938 carried episode names as well. Most if not all of the early scripts (perhaps the first five years) were written by Fran Striker (who also authored *The Lone Ranger*), but they increasingly became a WXYZ team effort for the remainder of the 16-year run. At different times the half-hour drama appeared weekly or twice-weekly.

The protagonist was Britt Reid (also the Green Hornet), who served as a wealthy young newspaper publisher of *The Daily Sentinel* during the day and transformed into evil's arch enemy after sunset. The Green Hornet's mission, according to the opening narration, was to protect us (the law-abiding American citizen) from those "who sought to destroy our way of life." If one listened carefully, one of the series' conceits was made clear—the familial connection of the Green Hornet with his great uncle, the Lone Ranger. References to the earlier legendary figure were abundant. Young Reid was seen as carrying on the family tradition of fighting for justice and the American way. Against the backdrop of an uncertain world, the Green Hornet reassured listeners that the forces of good would always triumph over the forces of darkness. Only three characters knew that Reid was also the Hornet—his father (who appeared rarely), his secretary (who never lets on until late in the series), and Kato.

Kato served as the Green Hornet's faithful valet and partner in crime fighting. Kato also drove the Hornet's famous high-speed car, Black Beauty, during countless breathtaking chases in pursuit of the bad guys. A famous radio legend has it that Kato, who had been described for five years as Japanese, became a Filipino overnight after the 1941 Pearl Harbor attack. Harmon and other sources say, however, that Kato had been described as a Filipino of Japanese ancestry well before the war began. The role partially reflected a continuing American fascination with things oriental.

Unlike most crime fighters, the Green Hornet did not use lethal weaponry; his gun fired a knockout gas instead of bullets. And in contrast with most other radio superheroes, the Green Hornet often assumed a bad guy persona in order to capture criminals, and this frequently confused law enforcement officials, richly adding to the plot line. He and Kato would always escape the crime scene just before law enforcement officers (and reporters from Reid's own paper) arrived. The final scene would usually feature a newspaper boy hawking the latest headlines of the Hornet's ventures as featured in *The Daily Sentinel*, noting that the Hornet was "still at large" and being sought by police. They never did catch him.

Like other superhero programs of the day, *The Green Hornet* adventure series had its genesis in the pulp detective novels of the 1920s, and the characters also appeared in comic books during and well after the radio broadcasts. The series was resurrected for one season on ABC television in the mid-1960s, riding on the coattails of the tremendously popular *Batman* series.

MICHAEL C. KEITH AND CHRISTOPHER H. STERLING

See also Lone Ranger; Striker, Fran; WXYZ

Cast

Britt Reid (The Green Hornet)	Al Hodge (to 1943); Donovan Faust (1943); Bob Hall (1944–51); Jack McCarthy (1951–52)
Kato	Tokutaro Hayashi, Rollon Parker, Michael Tolan
Lenore Casey Case	Leonore Allman
Michael Axford	Jim Irwin (to 1938), Gil Shea
Ed Lowery	Jack Petruzzi
Dan Reid	John Todd
Newsboy	Rollon Parker

Announcers
Charles Woods, Mike Wallace, Fielden Farrington, Bob Hite, Hal Neal

Director
James Jewell

Writer
Fran Striker and several others

Programming History

WXYZ, Detroit	January 1936–April 1938
Mutual	April 1938–November 1939
Blue Network/ABC	November 1939–December 1952
ABC (Television)	September 1966–July 1967

Further Reading

Bickel, Mary, *George W. Trendle,* New York: Exposition, 1973

Harmon, Jim, "From the Studios of WXYZ-III (*The Green Hornet*)," in *The Great Radio Heroes,* Garden City, New York: Doubleday, 1967; revised edition, Jefferson, North Carolina: McFarland, 2001

Osgood, Dick, *WYXIE Wonderland: An Unauthorized 50-Year Diary of WXYZ Detroit,* Bowling Green, Ohio: Bowling Green University Popular Press, 1981

Russo, Alexander, "A Dark(ened) Figure on the Airwaves: Race, Nation, and *The Green Hornet,*" in *Radio Reader: Essays in the Cultural History of Radio,* edited by Michele Hilmes and Jason Loviglio, New York: Routledge, 2001

Striker, Fran, Jr., *His Typewriter Grew Spurs: A Biography of Fran Striker—Writer,* Lansdale, Pennsylvania: Questco, 1983

Van Hise, James, *The Green Hornet Book,* Las Vegas, Nevada: Pioneer, 1989

Greenwald, James L. 1927–

U.S. Radio Marketing Executive

James L. Greenwald was one of the most important leaders of the station representative business. He was also an early proponent of the commercial value of FM radio.

Early Career

Greenwald began his career as a songwriter. In 1955 he determined that it might be more profitable to sell "air." Thus began a career in the national media sales business that spanned 41 years at the Katz Agency, Katz Communications, and Katz Media Corporation, where he served as chairman and chief executive officer until 1995. When Greenwald joined Katz Radio as a salesman, the firm represented only 25 stations. When he retired, the firm represented over 2,500 stations.

He was named assistant manager of the radio division in 1963 and president in 1970. With a company that by then had 65 clients, he began building an organization that launched many innovative sales and marketing concepts. Greenwald was among the first national radio sales executives to foresee the emergence of FM radio. In a speech before the West Virginia Broadcasters Association, he said: "The days of FM stations being looked upon as supplements to AM stations are over. FM is radio."

In 1972 one of his first major steps as president of Katz Radio was to begin selling FM radio audience. Until then, most FM stations, if they were sold to national advertisers at all, were coupled with sister AM stations. Nearly all FM stations, except those that programmed classical music, simulcast programming with their AM counterparts, and Katz Radio was particularly steeped in the history of selling large AM stations only.

Greenwald visited with the owners of the major Katz AM stations that also had FM stations and first convinced them to sell their fledging FM stations in combination with their AM stations. In many instances the additional audience, which was essentially sold for the same price as the AM-only audience, resulted in higher rates and larger shares of budgets for the AM station. The Katz clients responded favorably. When the Federal Communications Commission (FCC) in 1965 passed the rule limiting simulcasting to 50 percent of the program day, Greenwald formed an in-house programming consulting unit within the Radio Division that urged the owners of FM stations to aggressively program their FM properties independently. Greenwald foresaw a national sales market rapidly developing that was willing to spend large sums to reach the emerging FM audience. By 1976, national sales on Katz-represented FM stations had grown to represent over 20 percent of the company's total volume. By 1980 it had eclipsed 35 percent, and by 1990, 70 percent.

Greenwald was an early believer in packaging audiences for advertisers. He recognized that with the represented FM stations emerging with strong audience, there was an opportunity to package stations for a larger share of market dollars. He created a concept called the Katz AID Plan, which provided an advertiser a substantial discount based on the share of budget placed exclusively on the Katz-represented stations.

During the early 1970s, the only network selling of national radio was being done by the traditional wired networks. Greenwald, having expanded the list of Katz Radio–represented stations to over 200, developed the non-wired radio network and formally organized the Katz Radio Network. This sales unit combined the Katz-represented stations on a customized basis to fit a particular advertiser's audience and demographic needs. Furthermore, the advertiser was offered the convenience of one invoice and could thus avoid dealing with hundreds of stations individually. It was part of Greenwald's marketing plan to make radio advertising easier to plan and buy.

Radio audience research was another Katz Radio innovation fostered by Greenwald. He strongly believed that radio should not sell against newspapers and other media but rather with them. Under his leadership, the Katz Radio Probe Research System was developed in 1975 using early computer technology; the system demonstrated how a radio schedule could enhance the reach of a newspaper buy by utilizing a relatively small portion of the newspaper budget. Later, the same concept was enhanced to include the combination of radio with television and outdoor, and the name Katz Probe Media Mix was created.

By the end of the 1970s, Greenwald had been promoted to executive vice president of the corporate entity, Katz Communications. His responsibilities were expanded to include the Katz Television representation. However, he never removed himself from close contact with and oversight of the radio division.

In 1976 Katz Radio, with Greenwald's support, took the bold step of representing more than one station in a single market when it simultaneously served WRNG and WGST in Atlanta and later WAIT-AM and WLAK-FM in Chicago. Greenwald reasoned that two weak stations had a better chance of getting larger budgets if they combined their sales pitch. The concept of dual representation quickly spread within Katz markets, because even though exclusive representation was considered desirable by many owners, the huge advantages of the research, network sales operations, and the expansion of sales offices into regional territories was compelling to the stations. They understood that having a representative who could partner them with compatible stations would allow them to gain larger shares of budgets. This concept was the beginning of the consolidation of the representation industry and later of the radio industry in general.

Greenwald was instrumental in establishing an employee stock ownership plan at Katz that made the purchase of the firm from the Katz family possible, and Katz became the first employee-owned station representative firm in the industry.

The next giant step for Katz under Greenwald's leadership came in 1984, when Katz Communications purchased Christal Radio Sales from Robert Duffy. Katz-owned representation companies now competed directly against one another in nearly all of the top 100 radio markets. Katz also purchased the Jack Masla Company, Eastman Radio, Blair Radio, RKO Radio Sales, and Metro Radio Sales. These purchases set in motion the total consolidation of the radio representation industry. By 1992 there were only three major national representatives, Katz Radio Group, Interep, and Columbia Broadcasting System (CBS) Radio Spot Sales. CBS Radio Spot Sales was sold to Interep in 1995, leaving just two.

From his position as executive vice president of Katz Communications, Greenwald went on to become chairman, president, and chief executive officer of the newly named Katz Media Corporation. During Greenwald's tenure as the leader of the company, Katz entered broadcast ownership for the first time in its 100-year history when it purchased Park City Communications from Richard Ferguson in 1982. Katz sold the radio group in 1986. During his leadership of Katz Media Corporation, Greenwald expanded Katz Television into dual-ownership representation within a single market; purchased Seltel, a competing television representation firm; and instituted many of the innovative selling and research strategies created during his tenure at Katz Radio. Greenwald retired in 1995 and remains chairman emeritus of Katz Media Corporation.

GORDON H. HASTINGS

See also FM Radio; Station Rep Firms

James L. Greenwald. Born in New York City, 2 April 1927. Began career as songwriter; salesman, Katz Communications and Katz Media Corporation, 1955, assistant manager, radio division, 1963; formally organized Katz Radio Network;

executive vice president of corporate entity and assumed responsibility for Katz Television representation, 1970–75; president, 1975–82; chairman of the board of directors and chief executive officer, 1975–95; sold radio group of Katz Media Corporation, 1986; director, Granite Broadcasting Company, Paxson Communications Corporation and the Young Adult Institute; honorary trustee, Foundation of American Women in Radio and Television; past president, International Radio and Television Foundation; past president, Station Representatives Association; chairman emeritus, Katz Media Corporation, 1995–; director, Source Media, 1996–.

Ground Wave

A ground wave is a radio signal that propagates along the surface of the earth. It is one of two basic types of AM signal propagation, the other being the sky wave, which travels skyward from the transmitting antenna and then may be refracted back toward earth by the atmosphere. The behavioral characteristics of both types of wave are important both to frequency allocation and to the nature of various radio services.

The term *ground wave* includes three different types of waves: surface waves, direct waves, and ground-reflected waves. Surface waves travel directly along the surface of the earth, following terrain features such as hills and valleys. Direct waves follow a "line-of-sight" path directly from the transmitting antenna to the receiving antenna, and ground-reflected waves actually bounce off the surface of the earth.

Both ground-wave and sky-wave signals can be used to provide radio communication. The distance each type of signal can travel is determined by a number of factors, among them frequency, power, atmospheric conditions, time, and—in the case of ground waves—terrain and soil conductivity.

The principal determinant of which type of signal provides the communication is transmitting frequency. At very low frequencies (below 300 kilohertz), signal propagation takes place mostly by surface ground waves, which at these frequencies may provide a reliable signal for several thousand miles. At medium frequencies (300 kilohertz to 3 megahertz), surface ground waves may propagate hundreds of miles, and sky waves may travel thousands of miles. At high frequencies (3 to 30 megahertz), sky waves provide the principal means of signal propagation, and they may provide usable signals for many thousands of miles. At very high frequencies (30 megahertz and above), propagation is largely by ground-reflected and direct ground waves, although at these frequencies the waves generally travel less than 100 miles.

The standard broadcast (AM) band (535–1705 kilohertz) is a medium-frequency band and is thus characterized by both ground-wave and sky-wave signals. During the day, AM propagation takes place mainly by ground-wave signals; sky-wave signals generally travel through the atmosphere and into space. However, during night-time hours, changes in a portion of the ionosphere known as the Kennelly–Heavyside layer cause the sky waves to be reflected back toward the earth's surface. These refracted sky waves can then provide usable service over many hundreds—or even thousands—of miles, although sky waves are generally more susceptible to interference and fading than are ground waves. A certain amount of AM sky-wave propagation also takes place in the hours immediately before sunset and immediately after sunrise. In contrast, propagation in the FM band (88–108 megahertz) takes place by line-of-sight or near–line-of-sight direct and ground-reflected ground waves only.

The complexities of signal propagation in the AM band have presented significant challenges for the allocation of frequencies since the inception of broadcasting in the 1920s. Primary among these, of course, is the presence of both ground-wave and sky-wave signals at various times of the day. The Federal Communications Commission (FCC) has established three service area categories for AM stations: (1) primary service area, in which the ground-wave signal is not subject to objectionable interference or fading; (2) intermittent service area, in which the ground-wave signal may be subject to some interference or fading; and (3) secondary service area, in which the sky-wave signal is not subject to objectionable interference.

Another significant factor is that AM signals—both ground-wave and sky-wave signals—can cause objectionable interference over a much wider area than that for which they can provide usable service. For example, although a given station may not be able to provide a listenable signal more than 50 miles from its transmitter, that station's signal can still create objectionable interference to other stations on the same frequency over a much wider area. Signals in the FM band do not create this type of wide-area interference, and thus FM stations can be placed geographically closer together on the same frequency.

Perhaps the best illustration of the problems of allocation in the AM band is the dispute over clear channel stations, which began in the 1930s and was not completely resolved until 1980. Clear channel stations were originally created to provide wide-area service to rural audiences through their vast secondary service areas; other stations that were assigned to clear channel frequencies had to sign off at sunset in order to avoid interfering with the dominant stations' sky-wave signals. Clear channel stations sought to maintain and enhance their status by seeking power increases and the maintenance of their clear night-time frequencies. Other classes of stations called for the "breakdown" of clear channels by adding more stations to clear frequencies and by allowing daytime-only stations to broadcast full-time. At the heart of this dispute was an engineering argument over the best way to provide radio service to isolated areas: Clear channel stations maintained that the only way to provide effective rural service was by increasing the power of clear channel stations so that their secondary service areas would expand. On the other hand, other classes of stations called for more stations, located in close geographic proximity to the isolated rural areas, to provide ground-wave service to those areas.

Ultimately, the FCC decided to assign additional full-time stations to use clear channel frequencies, but the FCC protected a substantial portion of the clear channel stations' existing secondary service areas (a roughly 700- to 750-mile radius). Only clear channel stations (now called "Class A" stations) receive protection from interference in their secondary service areas.

The characteristics of ground waves and sky waves are in many cases the determining factors in the purposes for which radio services at various frequencies are used. AM broadcast service can provide reliable ground-wave communication at all times of the day and somewhat less reliable sky-wave communication at night. FM broadcast service can provide reliable line-of-sight service over shorter distances, with less blanketing of interference. Broadcasters in high-frequency bands (shortwave) can provide international sky-wave service.

JAMES C. FOUST

See also AM Radio; Antenna; Clear Channel Stations; DXers/DXing; Federal Communications Commission; FM Radio; Frequency Allocation; Shortwave Radio

Further Reading

Federal Communications Commission, *Radio Broadcast Services,* part 73, 47 CFR 73 (2001)

Foust, James C., *Big Voices of the Air: The Battle over Clear Channel Radio,* Ames: Iowa State University Press, 2000

Jordan, Edward C., editor, *Reference Data for Engineers: Radio, Electronics, Computer, and Communications,* Indianapolis, Indiana: Sams, 1985; 7th edition, 1999

Orr, William Ittner, *Radio Handbook,* Indianapolis, Indiana: Sams, 1972; 23rd edition, 1992

Smith, Albert A., Jr., *Radio Frequency Principles and Applications: The Generation, Propagation, and Reception of Signals and Noise,* New York: IEEE Press, 1998

Group W

Westinghouse Radio Stations

Westinghouse Broadcasting (Group W after 1963) remained active in radio broadcasting longer than any other company—beginning with the initial airing of Pittsburgh's KDKA in November 1920 and lasting into the late 1990s. Thanks to constant retelling of the KDKA story, the earliest years of Westinghouse Broadcasting are well known, but throughout the history of broadcasting the company was an important owner of both radio and later television stations, eventually merging into the once-independent Columbia Broadcasting System (CBS) network.

Getting into Radio (to 1931)

A Westinghouse engineer, Frank Conrad, had been experimenting with wireless for a number of years, and in 1919–20 he operated amateur station 8XK, playing recorded music one or two nights a week. A September 1920 newspaper advertisement by a local department store seeking to sell receivers to people who wanted to hear Conrad's broadcasts caught the eye of Harry Phillips Davis, a Westinghouse vice president in charge of radio work. Davis perceived that making receivers for a possible new radio service could be the answer to Westinghouse's predicament. He urged Conrad to develop his hobby station into something bigger, and the inauguration of station KDKA on 2 November 1920 was the result.

The success of that initial operation led Davis to the development of a second station, WJZ, at the company meter plant in Newark, New Jersey, in September 1921 (the station was sold to the Radio Corporation of America [RCA] a year later). In the same month, station WBZ took to the air in Springfield, Massachusetts, followed by station KYW in Chicago in December and by WBZA in Boston in 1924. These pioneering outlets made Westinghouse an important early station operator that pioneered many types of program service.

Westinghouse stations were initially located at the factories, which meant that early performers had to learn to entertain in a room filled with electronic equipment, with only an engineer as their audience. Ultimately, the studios were moved to more aesthetically pleasing locations, such as hotels or office buildings.

Experimenting with the potential of shortwave technology, Westinghouse placed KFKX on the air in Hastings, Nebraska, in 1923 to make KDKA's signal more widely available. Another experimental shortwave station, W8XK, was established in Pittsburgh and was soon broadcasting 18 hours per day including a "far north" service to the Arctic. It was joined in 1930 by yet a third station, W1XAZ, in East Springfield, Massachusetts. By the late 1920s, Westinghouse was also pioneering in television research.

Evans Years (1931–55)

On Davis' death in 1931, Walter Evans became the next Westinghouse radio chief. He had joined the company as chief of operations in 1929 and would serve for more than two decades. Evans took a different approach to managing the company's stations and in 1933 signed a contract with the National Broadcasting Company (NBC) network to manage them all, including provision of all local and national programs selling advertising time. The agreement lasted until 1940, when it ended as part of the Federal Communications Commission's (FCC) investigation of national networks, and Westinghouse took over day-to-day operations itself.

Westinghouse operations expanded in the prewar years. Station KYW was moved from Chicago to Philadelphia in late 1934 as part of a deal with the FCC to provide service in underserved areas. Two years later, Westinghouse purchased its first station (it had built its previous operations), WOWO in Fort Wayne, Indiana, along with WGL in the same city, both of which were licensed to a new entity, Westinghouse Radio Stations (WRS). In 1940, WRS took control of all Westinghouse stations, separating the broadcast operations from other company functions.

Westinghouse shortwave (international) stations consolidated operations in Pittsburgh (8XS, which became WPIT in 1939 but closed a year later) and Boston (WIXT, which became WBOS). Westinghouse's international broadcasting was a multilingual operation that by late 1941 was providing 12 hours of programming a day: 5 hours to Europe and 7 to Latin America. Government programs expanded that total to 16 hours just a few months later. Early in 1942, the Boston station was taken over by the government's Office of War Information.

Westinghouse was an early player in FM radio: by 1943 the company owned five FM stations in cities where it also operated AM outlets. Original programming was provided on those FM facilities, but they had all reverted to simulcasting by the end of 1948. That same year, Westinghouse placed its first television station, WBZ-TV in Boston, on the air. As the operation continued to grow, the broadcast subsidiary's headquarters moved several times, finally ending up in New York in 1953, when it became known as Westinghouse Broadcasting Company (WBC), in part because of the addition of television.

McGannon Years (1955–81)

The man who served longest as head of the Westinghouse stations was Donald H. McGannon, who ran the operation from 1955 (after a few interim leaders) until 1981. McGannon soon earned a reputation as a man concerned about public service and program quality as well as profit. He brought Westinghouse back to Chicago with the 1956 purchase of WIND for $5.3 million—at that point the highest price paid for a station. He also began a Washington news bureau to serve his stations in 1957. That same year WBC initiated an arts and classical music format from 4 P.M. to midnight on the four FM stations it still owned. But FM was then a weak service, and by 1970 Westinghouse was down to just two FM outlets, one in Boston and the other in Philadelphia, both programming classical music and suffering from a lack of promotion or advertising.

McGannon faced three serious policy crises early in his tenure. The first concerned Philadelphia's KYW AM and TV. Under at least an implied threat of losing NBC network affiliations for its Philadelphia and Boston television stations, Westinghouse agreed to "trade" its Philadelphia radio and television stations (KYW) for NBC outlets in smaller Cleveland in 1955. Over the next decade various business, FCC, and Congressional investigations brought to light the network threats that had created the deal, and it was undone in 1965, with the KYW stations returning to Philadelphia.

Two other problems briefly threatened Westinghouse licenses. Its Cleveland and Boston radio stations were implicated in the national payola scandal of the late 1950s, and several disc jockeys were fired. Their activities figured in widely covered Congressional investigations. In 1961 antitrust price-fixing convictions against Westinghouse threatened the company's ownership of broadcast stations. Because of the independence of WBC from the parent manufacturing company, its licenses were renewed after several months of threatened FCC hearings. In mid-1963, WBC was renamed Group W.

Because the Westinghouse stations were situated in major cities, they produced sufficient revenue to allow for further acquisitions. In 1962, for example, Westinghouse shut down WBZA in Springfield (which had mainly been simulcasting the much more successful WBZ in Boston for years), making it possible to buy another station, KFWB in Los Angeles and by 1965 was offering an all-news format on WINS in New York. By the early 1960s, Westinghouse had begun using the term "Group W" to make its owned radio and TV stations more memorable.

A contemporary move was the successful implementation of all-news operations at three major-market radio stations. The conversion began with station WINS in New York, purchased in 1962, which suffered from a weak rock music format. Likewise, KYW had returned to Philadelphia to follow a weak decade of NBC station operation in its place. Westinghouse stations became known for a commitment to news and public affairs (Group W had operated its own news bureau in Washington, D.C., since 1957). All-news operations began KYW and WINS in 1965. KFWB in Los Angeles was purchased in 1966, and two years later it was also converted from rock music to an all-news format. Although not first with the format, Group W was the first to make it a lasting success in major markets.

Only toward the end of McGannon's tenure as Group W chief did the company begin to reconsider FM radio, which by 1980 was dominating national radio listening for the first time. That year Group W purchased two major-market Texas FM stations, KOAX (FM) in Dallas, which soon was renamed KQZY (FM), and KODA-FM in Houston. But although there were adjustments in station lineup, Group W's overall size and contribution to the Westinghouse bottom line (roughly 15 to 20 percent of annual revenues) remained remarkably stable.

Final Decades (1981–2000)

On McGannon's retirement, Daniel L. Ritchie became Group W's leader and served into the late 1980s, to be succeeded in turn by Burton B. Staniar. The expansion into FM continued with the purchase of KJQY (FM) in San Diego and KOSI (FM) in Denver in 1981. A San Antonio FM station, KQXT, was purchased in 1984; KMEO AM and FM in Phoenix were bought a year later; and WNEW-FM in New York was added to the Group W stable in 1989. Although some outlets were spun off, the overall effect was to slowly grow the company—and to increase the proportion of FM to AM stations.

Group W switched AM outlets in Chicago as well. In the aftermath of Martin Luther King, Jr.'s 1968 assassination, WIND suspended popular on-air figure Howard Miller, whose statements about race and the police were becoming more strident. A breach of contract suit was settled out of court, but the station lost audience steadily for years thereafter and was sold in 1985 to a Spanish language broadcaster. Three years later, Group W purchased WMAQ, the one-time NBC outlet, thus resuming a role in Chicago radio.

In the late 1990s, Group W underwent a series of mergers and acquisitions that changed the face of the company and eventually caused it to disappear into other entities. The process began with the 1995 purchase by Westinghouse Electric of the weakened CBS network from Lawrence Tisch for $5.4 billion dollars. The deal needed and received several cross-ownership waivers from the FCC, as the radio and television stations of Group W and the network were located in many of the same cities. The new entity controlled 39 radio stations, worth $1.4–1.7 billion, and became the largest group owner in terms of revenues. In buying CBS, Westinghouse purchased a radio division that provided two services Westinghouse had been paying other companies to provide—network news and national sales representation.

In late 1996 CBS/Westinghouse merged with Infinity Broadcasting, combining under the CBS Radio Group name 83 stations, for a time the largest single ownership block in the industry. Over the next couple of years, Westinghouse sold off its traditional manufacturing base (power systems, which had been losing money, and electronic and environmental systems) and its original name to concentrate on the development of its radio and television holdings under the CBS name. The Group W trademark was briefly retained to identify technical support for television distribution and sports marketing, and as owner of record of six AM radio stations, the original KDKA (Pittsburgh), WBZ (Boston), and KYW (Philadelphia), as well as the later-acquired outlets WNEW (New York), WMAQ (Chicago), and KTWV (Houston).

By the turn of the 21st century, however, the radio group was operating under the Infinity name and the one-time Westinghouse (or Group W) stations were merely one integrated part of what had become the country's third largest group owner of radio stations.

MARY E. BEADLE, DONNA L. HALPER, AND CHRISTOPHER H. STERLING

See also Columbia Broadcasting System; Conrad, Frank; FM radio; KDKA; KYW; McGannon, Don; National Broadcasting Company; Network Monopoly Probe; Radio Corporation of America; Westinghouse; WBZ; WINS; WMAQ; WNEW

Further Reading

Broadcasting Yearbook (1935)

Davis, H.P., "The Early History of Broadcasting in the United States," in *The Radio Industry: The Story of Its Development,* Chicago and New York: Shaw, and London: Shaw Limited, 1928; reprint, New York: Arno Press, 1974

Douglas, George H., *The Early Days of Radio Broadcasting,* Jefferson, North Carolina: McFarland, 1987

Westinghouse Electric Corporation, *The History of Radio Broadcasting and KDKA,* East Pittsburgh, Pennsylvania: Westinghouse Electric Corporation, 1940

Gunsmoke

Western Series

Gunsmoke, a western on the CBS Radio Network, was introduced at a time when most radio drama was disappearing. It not only lasted nearly a decade but also spawned television's longest-running drama series. *Gunsmoke*'s devoted fans, who have praised its historical accuracy and realism, would likely attribute the program's longevity to its brilliant writing and acting. Others would say, "it's just a good story."

The stories on *Gunsmoke* provided the groundwork for the so-called adult western, a dominant TV genre for nearly 20 years. There were many radio westerns before *Gunsmoke,* particularly as daytime serials or evening programs, but most of them, such as the even longer-running *Lone Ranger,* were aimed primarily at a young audience.

Origins

Gunsmoke was the result of a collaboration between several writers and producers (all "urban oriented," according to William N. Robeson), including John Meston and producer Norman Macdonnell, who worked together at CBS from 1947 on *Escape* and other radio dramas. Robeson, who created *Escape,* admitted that that program was "pretty darned close to *Suspense.*" Macdonnell was an assistant director with Robeson, and William Conrad was the announcer on *Escape.* With writer John Meston, several experimental western stories were tried between 1947 and 1950. In 1949 the team also produced two pilots of what they conceived of as an "adult western" with a hero named Mark Dillon.

The first *Gunsmoke* program came about when another program was abruptly canceled. Norman Macdonnell and writer Walter Brown Newman used elements from several of Macdonnell's earlier western stories to create "Billy the Kid," the first episode, which ran on 26 April 1952. Although unforeseen at the time, after "Billy the Kid" there would be 412 more episodes of *Gunsmoke;* the final show aired 18 June 1961.

Raymond Burr and Robert Stack (both of whom later became famous TV actors) were considered for the lead role, but at the last moment the job went to William Conrad. While he was the announcer on *Escape* and a veteran radio actor (he would also later star in several TV series), Conrad had just finished the movie *The Killers* and was considered a "heavy." In the pilot, the hero had to narrate much of the story in voice-over. Chester, his assistant, appeared in the first episode, but the part quickly grew larger, and the important characters of Doc and Miss Kitty evolved. From the first show to the last, *Gunsmoke* kept its cast of William Conrad as Marshal Matt Dillon, Georgia Ellis as Kitty Russell, Howard McNear as Doctor Charles Adams, and Parley Baer as Chester Wesley Proudfoot. While the relations between the four characters held the program together, it was the deep, booming voice of Conrad that provided the program's unmistakable signature. In addition to these regulars, veteran radio actors appeared in episode after episode.

Among fans of radio drama, *Gunsmoke* is considered the best western ever made. The series was marked by high-caliber writing and used only a score of authors during its entire run. Meston wrote 183 stories, and there were three years in which he wrote more than 45 episodes per year. His scripts often concerned the difficulties of frontier life, particularly for women.

After a light Christmastime show in 1952, Meston wrote "The Cabin," broadcast on 27 December, to assure the audience that *Gunsmoke* had not "gone soft." The episode concerned a young woman named Belle who had been raped. By the end of the program the marshal has killed the men who raped her (though this is not acted out explicitly) and is asked by the woman if he is married. "Too chancy," he replies. Matt tells Belle, "Don't let all this make you bitter, there are lot of good men in the world," and she replies, "So they say." As he heads back to Dodge, Matt ends by noting that the blizzard

1960s began to emerge, the media reinstated some elements of free speech, including several news and public-affairs shows that dealt with controversial issues, and more talk shows. It was not until after full-service radio networks disappeared in the early 1950s and specialized limited-time networks consisting mostly of music took their place—along with community-targeted narrowcasting, also mostly music, on local stations—that talk show hosts with set opinions emerged in force.

The far right seemed to understand more fully than the middle or the left the power of talk radio and quickly deluged stations with calls and opinions and stimulated a demand for, as some put it, loud-mouthed right-wing talk show hosts. One such host was Joe Pyne, who became one of the most popular talk personalities in the country with glib, biting, and unabashedly opinionated comments. A number of hosts, such as Bob Grant, later became famous for using Pyne's caustic approach. Grant often referred to African-Americans as "savages" and used expletives about other targets freely. Grant attributed his reputation to show business techniques, not bigotry. Ira Blue, who hosted a talk show in San Francisco in the 1960s, openly admitted during his on-air reign that the radio talker succeeds most when he is brazenly opinionated.

As the 1950s and 1960s progressed, many right-wing talk show programs and hosts became more subtle, using twisted logic rather than blatant vituperation to persuade their audiences. Meanwhile, right-wing rancor on talk shows went in two distinct directions during the decades that followed. As the number of stations increased, more on-air opportunities existed for fringe advocators. Ranters and ravers, some affiliated with organizations dedicated to violence, found microphones available to them. At the same time, soft-spoken intellectuals dispensing the same bottom line also had their access.

Late Millennium Waves of Rancor

The 1990s saw the greatest rise in the use of radio by far-right extremist groups, among them white supremacists, armed militias, survivalists, conspiracy theorists, and neo-Nazis. Many of these groups had effectively used the shortwave radio medium to promote their dark agendas in the 1980s, and they sought to go more mainstream with their messages by utilizing the AM and FM bands in the 1990s. Dozens of broadcast stations around the country gave airtime to organizations and individuals intent on denigrating people of color as well as those with non-Christian religious orientations. The bulk of these stations were smaller AM outlets, many of which were battling for their economic survival in the face of vastly declining audiences and shrinking revenues. Far-right programs were a source of income.

Today, far-right hate groups still promote their ideologies over radio stations, but not to the degree that they did prior to the Oklahoma City federal building bombing in 1995. This tragic extremist deed prompted the president to issue an antiterrorism bill making it clear that anyone employing the airwaves to promote violence and hatred would be hunted down and prosecuted. Technology, however, provided right-wing radicals with yet another way to propagate their racist and antigovernment views. The internet soon became the new home and the preferred medium for hate groups, which relished the freedom and lack of censorship that cyberspace afforded them.

In 1996 a white supremacist organization calling itself Stormfront launched what is considered the first extremist website. A former grand dragon of the Ku Klux Klan, Don Black, operates the internet site, which also features a link for children—"Stormfront for Kids"—run by his 11-year-old son. Within four years, a nearly incalculable number of radical-right websites, replete with sophisticated graphics and chat rooms, were in full operation, and many claimed thousands of hits each week. Several watchdog groups, among them the Simon Weisenthal Center, the Anti-Defamation League, Political Research Associates, and the Southern Poverty Law Center, reported that many of the same organizations and individuals, once so dependent on the airwaves to get their messages out to the public, now download their proclamations of hatred to thousands of websites, thus relegating radio to a secondary medium for their egregious purposes. One cannot help but note an ironic analogy in this migration of radio users to another medium. However, one suspects that the impact of this conversion will be far less traumatic for radio than the one brought on by the rise of television.

MICHAEL C. KEITH AND ROBERT L. HILLIARD

See also Controversial Issues; Coughlin, Father Charles; Pyne, Joe; Winchell, Walter

Further Reading

Berlet, Chip, editor, *Eyes Right! Challenging the Right Wing Backlash,* Boston: South End Press, 1995

Hilliard, Robert L., and Michael C. Keith, *Waves of Rancor: Tuning In the Radical Right,* Armonk, New York: Sharpe, 1999

Kurtz, Howard, *Hot Air,* New York: Basic Books, 1996

Laufer, Peter, *Inside Talk Radio: America's Voice or Just Hot Air?* Secaucus, New Jersey: Carol Group, 1995

Hear It Now

CBS Documentary Program

Hosted by esteemed newsman Edward R. Murrow, *Hear It Now* was more important in broadcasting's history that its mere six-month run on the Columbia Broadcasting System (CBS) radio network would suggest. The program developed from a series of successful documentary record albums and helped to pave the way for the even more important television documentary series *See It Now,* which presented some of Murrow's finest work.

The Recordings

The idea of making record albums featuring the actual sounds of historical events originated with Fred Friendly, a World War II veteran working as a producer for the National Broadcasting Company (NBC). Friendly saw that the relatively new medium of magnetic tape would make editing sounds recorded during historical events far easier. When Friendly realized that he needed a narrator for his recordings, CBS producer Jap Gude introduced him to newsman Edward R. Murrow and a team was born. Sometime around 1947, Friendly and Murrow approached Decca Records, but that firm was not interested in "talking" records, which were usually money losers. On the other hand, Columbia Records (a subsidiary of CBS) had available capacity as well as interest: a "scrapbook for the ear" they called it.

The initial recording, *I Can Hear It Now, 1933–1945,* was released in the winter of 1948 as a boxed set of five 78-rpm records (10 sides, about 45 minutes total). Murrow provided the historical context and narrated the many sound bites. There was no music or sound other than those of the actual events. To the surprise of virtually everyone involved in the project, the set sold 250,000 copies in the first year, highly unusual for talking records. It was said to be the first financially successful non-musical album.

A second album covering the postwar years (1945–48), with sound bites drawn largely from the extensive archives of the British Broadcasting Corporation (BBC), came out a year later, and a third, *I Can Hear It Now: 1919–32* (which did use actors for some events when recordings were not available), appeared in 1950. Although the reasons are no longer clear, plans for a forth album were abandoned. All were issued as 78-rpm albums originally, but some were later reissued in long-playing (LP) format. The title was resurrected in the 1990s for *I Can Hear It Now: The Sixties*, a two-compact-disc set narrated by Walter Cronkite.

The Broadcasts

The idea of broadcasting recorded sounds of historical or present-day events was anything but new. It had been done for *The March of Time* series beginning in 1931, and was often used during World War II. But there had been limited use of such material in part due to the networks' long-standing ban on use of recordings on the air. Something of a pilot program existed in CBS files—a 1948 proposed "Sunday with Murrow" documentary that had never aired for lack of advertiser support. Murrow and Gude now proposed a new program, called simply *Hear It Now* to stress its current-events emphasis, for a half-hour time slot. CBS Chairman William S. Paley so liked the idea, however, that he urged them to make it an hour-long program. Part of the difference in acceptance from the 1948 attempt to 1950 was a change in world events: the Korean War had begun and people once again were interested in world events.

When CBS began to look for a producer for the series, Gude suggested Friendly and CBS hired him away from NBC while the second *I Can Hear It Now* record album was being made. It was an easy choice, given Friendly's role as producer for the recordings and with several radio documentaries for the senior network. The series also provided a vehicle for Murrow upon his return from reporting from the Korean War battlefront.

As it aired, *Hear It Now* included "columnists" covering different subjects: CBS correspondent Don Hollenbeck discussed the media, Abe Burrows dealt with entertainment, and sportscaster Red Barber covered professional teams in several sports. The program had an original musical score by U.S. composer Virgil Thomson. Murrow decided which topics would be included and the order in which they were presented, but he and Friendly wrote the program together. Friendly was the key editor of essential sound bites. The result was a "magazine of sorts, covering the news events of the previous six days in the voices of the newsmakers themselves, by transcription and hot live microphones" in an era before recordings of actual events were common in radio news (quoted in Dunning, 1998). *Show Business Weekly* said the program, which won a Peabody Award, was "almost breath-taking in scope and concept." It was carried on 173 CBS affiliates. Other networks quickly caught on to the idea and imitated it. NBC's *Voices and Events* and ABC's *Week Around the World* provided essentially the same sort of content, but without Murrow and Friendly.

Fred Friendly (left) and Edward R. Murrow on *Hear It Now,* 1950
Courtesy CBS Photo Archive

In many ways, the program was a radio vehicle for Murrow, who was then little interested in (and indeed, uneasy about) television. But as the realization became clear that pictures would add considerably to the *Hear It Now* idea, the audio version left the air while video preparations began. *See it Now* premiered on 18 November 1951 and ran until 1955.

CHRISTOPHER H. STERLING

See also Columbia Broadcasting System; Documentary Programs; Friendly, Fred; March of Time; Murrow, Edward R.; News

Commentators

Edward R. Murrow, Red Barber, Abe Burrows, Don Hollenbeck

Writer/Producer

Fred Friendly

Programming History

CBS 15 December 1950–15 June 1951

Further Reading

Bliss, Edward, Jr., *Now the News: The Story of Broadcast Journalism,* New York: Columbia University Press, 1991

Dunning, John, *On the Air: The Encyclopedia of Old-Time Radio,* New York: Oxford University Press, 1998

Kendrick, Alexander, *Prime Time: The Life of Edward R. Murrow,* Boston: Little Brown, 1969

Sperber, Ann M., *Murrow: His Life and Times,* New York: Freundlich Books, 1986

Heavy Metal/Active Rock Format

The "Heavy Metal/Active Rock" format encompasses a musical genre that has played a marginal role in commercial radio programming, while the medium of radio has at times played crucial roles in the acceptance, rejection, and content of the music itself. The moniker is the radio industry's term for a category that has gone in and out of style while maintaining a core fan subculture since heavy metal's emergence in the early 1970s. The music in this format is characterized by a distorted guitar sound, a heavy bass-and-drums rhythm section, and a vocal approach that eschews traditional melodic conventions in favor of an aggressive, emotionally raw sound. The name "heavy metal" was at first uncomfortably accepted by first-generation rock bands such as Led Zeppelin, Deep Purple, and Black Sabbath and later happily adopted by second-generation acts such as AC/DC, Kiss, Blue Öyster Cult, Motörhead, and others.

Initially, few heavy metal bands found a place on commercial radio. Commercial rock radio itself was only then taking shape on the FM dial, as progressive/free-form stations (such as WOR in New York and KMPX in San Francisco) gave way to the new album-oriented rock (AOR) format, at stations such as WNEW in New York and WMMS in Cleveland, designed to reach a larger listening and buying public. Thus, heavy metal was not welcome in the earlier progressive rock format that grew out of the late-1960s counterculture, whose ideology did not complement the nihilism of groups like Black Sabbath. A few bands, however, did find their place on the FM dial in the early 1970s. One band, Led Zeppelin, in fact became central to AOR playlists through much of the decade.

If radio largely ignored heavy metal during the 1970s, in the 1980s the format would find new popularity. By the late 1970s, music on the radio was still functioning much as it had earlier in the decade, due to the continuing influence of AOR, which increasingly programmed the most benign rock music to appeal to the largest audience possible. The heavy metal music that did find its way onto the airwaves was limited to a few songs played endlessly in a station's rotation. The effect was that a few songs on AOR radio came to stand in for heavy metal as a musical genre and in the process became emblematic of the genre's perceived creative bankruptcy by the end of the decade.

In the late 1970s, competition from new genres such as disco and punk had some influence on American radio formats, but both musical styles returned to the level of subculture within a few years. Meanwhile, heavy metal was being reinvigorated, first by British bands such as Def Leppard, Iron Maiden, and Judas Priest, and then by U.S. bands such as Quiet Riot and Guns N' Roses. By the mid-1980s, heavy metal from bands like Scorpions and Mötley Crüe could regularly be heard on American radio. Still, some saw a certain sacrifice in the newfound popularity of the genre on the airwaves. The anticommercial heavy metal of the 1970s gave way, slowly but surely, to a new style sometimes derisively called "lite metal," which meant less emphasis on long instrumental breaks featuring virtuosic guitar solos and greater emphasis on radio-friendly melodies and more traditional pop song structures.

In recent years, the heavy metal format has struggled amidst the relative fragmentation of radio into new formats that have eroded a once-loyal listenership. The format has become one choice in a sea of others, and stations carrying the format increasingly find themselves fighting for audiences in an ever-smaller market share. The emergence of the classic rock format in the mid-1980s and the alternative rock format in the 1990s has meant the loss of both older listeners alienated by newer bands and younger listeners with little allegiance to older heavy metal. Interestingly, the classic rock format has largely not designated earlier heavy metal music as classic. Heavy metal radio was slow to incorporate the music of alternative rock formats after the "grunge" explosion of the 1990s, headed by bands such as Nirvana, Pearl Jam, and Soundgarden. While grunge had stylistic links to the heavy metal bands of the 1970s, it also embraced the punk aesthetic.

The music played on heavy metal radio in the late 1980s and early 1990s, in comparison, largely catered to the mainstream music industry. Alternative rock formats encroached on both heavy metal and classic rock listeners with the success of stations such as KNDD in Seattle and WHTZ in New York, both of which made significant gains in their respective radio markets. Later in the decade, newer bands stylistically associated with heavy metal, such as Limp Bizkit and Rage against the Machine, developed. These groups owed much to the emergence of hip-hop music as a predominant popular style in the 1990s. The heavy metal/active rock format has embraced these groups in order to garner younger listeners, while the format expands and absorbs influences in a confusing radio market environment.

As the radio industry continues to change and programmers again return to once-outmoded formats, it will be interesting to see whether the heavy metal/active rock format will thrive in the new fragmented format environment. Other new venues for radio, including both internet and low-power radio initiatives, may serve heavy metal music fans in ways that commercial radio cannot.

KYLE S. BARNETT

See also Album-Oriented Rock Format; Alternative Rock Format; Classic Rock Format; Contemporary Hit Radio Format/Top 40; Music; Progressive Rock Format

Further Reading

Keith, Michael C., *Radio Programming: Consultancy and Formatics,* Boston and London: Focal Press, 1987

O'Donnell, Lewis B., Philip Benoit, and Carl Hausman, *Modern Radio Production,* Belmont, California: Wadsworth, 1986

Ross, Sean, "Music Radio—The Fickleness of Fragmentation," in *Radio: The Forgotten Medium,* edited by Edward C. Pease and Everette E. Dennis, London and New Brunswick, New Jersey: Transaction, 1995

Walser, Robert, *Running with the Devil: Power, Gender, and Madness in Heavy Metal Music,* Hanover, New Hampshire: University Press of New England, 1993

Weinstein, Deena, *Heavy Metal: A Cultural Sociology,* New York: Lexington Books, and Toronto: Macmillan Canada, 1991

Herrold, Charles D. 1875–1948

U.S. Broadcasting Pioneer

Charles Herrold is a relatively little-known broadcasting pioneer whose most significant work took place between 1912 and 1917. Although today most historians believe Herrold's claim that he was the first to broadcast radio entertainment and information for an audience on a regularly scheduled, preannounced basis, he is dismissed as a minor figure because he failed to have a long-lasting impact on the radio industry. Nevertheless, his early broadcasts show innovation and originality and are of interest because Herrold is symbolic of many of the early broadcast pioneers: unknown, underfinanced, and overshadowed by the major corporations that would control broadcasting beginning in 1920.

Herrold was born in 1875 in the Midwest and grew up in San Jose, California. In high school he was recognized by his teachers and classmates for his superior grasp of mechanical and scientific subjects. After graduation, he attended Stanford University in nearby Palo Alto, but he dropped out for health reasons. In 1900 Herrold set up an electrical manufacturing company in San Francisco, but when the Great Earthquake of 1906 destroyed his residence, he moved to Stockton, California, to teach at a technical college. Building on his work as an inventor and his experience with students, Herrold returned in 1909 to San Jose and opened a vocational school. The Herrold College of Wireless was a way to provide an income and at the same time to allow him access to the laboratory environment necessary to continue his research.

Like Lee de Forest, Reginald Fessenden, and others, Herrold was most interested in inventing a radiotelephone system that would make him rich and famous. His contributions to the technology of the radiotelephone were lacking in scientific originality, although his device did allow him to broadcast. As the inventor of an arc transmitting system, he spent years attempting to differentiate his system from that of Danish inventor Valdemar Poulsen. Herrold received six U.S. patents for his devices, patents that certainly would have been challenged by the Poulsen people were it not for the fact that by 1917 the perfection of the vacuum tube as the basis of all future radiotelephones made such arc-based devices obsolete.

Early notice of Herrold's use of the radiotelephone to "broadcast" to an audience is a notarized statement by Herrold, published in an ad for wireless equipment in the 1910 catalog of the Electro-Importing Company: "We have been giving wireless phonograph concerts to amateur men in the Santa Clara Valley," a statement prophetic of what broadcasting was to become. And although his 1910 listeners were amateurs and hobbyists, he did broadcast to public audiences daily during the 1915 San Francisco World's Fair. But Herrold's most significant contribution was that between 1912 and 1917 he operated a radio station providing programs of information and entertainment for an audience on a regular schedule, many of them announced previously in the newspapers. That he accidentally stumbled onto what was to become radio broadcasting may have been a function of his role as the headmaster of a wireless trade school. Because Herrold had the responsibility of providing daily technical activities for hundreds of eager young boys, it is likely that the broadcasting of the popular music of the day by his students to an audience of friends, families, and possible future students was the cauldron from which broadcasting emerged, Charles Herrold style.

Interviews with former students indicate that Herrold was broadcasting to a sizable audience: "It was a religion for 'Prof' Herrold to have his equipment ready every Wednesday night at nine o'clock. He would have his records ready, all laid out, and what he wanted to say. And the public or listeners, it became a habit for them to wait for it," according to Herrold's assistant, Ray Newby. Recalled Newby in an interview with Gordon

Greb, "We even had a San Jose music store that supplied us records, of course free of charge, and I think we played them all. We would take the *Mercury-Herald* in San Jose, and we would read headlines and discuss them a little bit, just something to yak about and make it interesting at the same time, to develop an audience, I would say." A local news story of a typical Herrold broadcast illustrates the method:

> For more than two hours they conducted a concert in Mr. Herrold's office, which was heard for many miles around. The music was played on a phonograph furnished by the Wiley B. Allen Music company. Immediately after the first record was played numerous amateurs from various points in the valley notified (the announcer) that they had heard the music distinctly. He gave the names of the records he had on hand and asked those listening to signify their choice (San Jose *Mercury-Herald*, 22 July 1912).

The question remains: if Herrold's was the first broadcasting station, why do not more people know about it? Some of the misunderstanding surrounding the "who was first" broadcasting claims can be traced to an early historian, George Clark. In 1921 Clark, RCA's in-house historian, dismissed the claims of all who broadcast before KDKA, because, as he wrote, "ordinary citizens" could not buy radios until KDKA, and therefore men like Herrold and de Forest were not really broadcasters, because their audiences were engineers or amateurs, not "citizens." Still, Herrold was the first to use radio to broadcast entertainment programs to an audience on a regular basis. He was not the first to broadcast pre-announced to an audience—that was Fessenden in 1906; he was not the first to broadcast election returns—that was de Forest in 1916; he was not the first to get a broadcast license—that was Conrad in 1920. Herrold returned to the air in 1921 licensed as KQW, ran the station until 1925, and later specialized in radio advertising. During World War II, he worked as a janitor at the Oakland shipyards. He died in 1948. Until 1958, when his story was uncovered by San Jose journalism professor Gordon Greb, almost no one outside of northern California had ever heard of Charles Herrold.

MICHAEL H. ADAMS

See also De Forest, Lee; Fessenden, Reginald; KCBS/KQW

Charles D. Herrold. Born in Fulton, Illinois, 16 November 1875. Studied astronomy, Stanford University (no degree), 1895; head of technical department, Heald's College, Stockton, California, 1906–08; started Herrold College of Wireless and Engineering, San Jose, California, 1909; began a regularly scheduled, pre-announced broadcast operation, 1912–17; invented a wireless radiotelephone and developed other new products in various fields; broadcast operation licensed as KQW, 1921; sold airtime for several Bay area radio station, 1925–30; worked at Oakland Public Schools as media technician, 1932; worked as janitor at Oakland shipyards during World War II. Died in Hayward, California, 1 July 1948.

Further Reading

Adams, Michael H., "The Race for Radiotelephone: 1900–1920," *AWA Review* 10 (1996)

Adams, Michael H., et al., *Broadcasting's Forgotten Father: The Charles Herrold Story* (videorecording), Los Altos, California: Electronics Museum of the Perham Foundation, 1994

Barnouw, Erik, *A History of Broadcasting in the United States*, 3 vols., New York: Oxford University Press, 1966–70; see especially vol. 1, *A Tower in Babel: To 1933*, 1966

Greb, Gordon, "The Golden Anniversary of Broadcasting," *Journal of Broadcasting* 3, no. 1 (Winter 1958–59)

The Charles Herrold Historic Site, <www.charlesherrold.org>

Hertz, Heinrich 1857–1894

German Physicist

In a series of laboratory experiments in 1887–88, Heinrich Hertz verified James Clerk Maxwell's theory that electromagnetic waves, or wireless transmissions, existed, and that these invisible forms of radiant energy traveled at the speed of light. Hertz paved the way for the development of wireless radio communications by notables such as Edouard Branly, Sir Oliver Lodge, Guglielmo Marconi, Reginald Fessenden, Sir John Ambrose Fleming, E.F.W. Alexanderson, and Lee de Forest.

Hertz's research placed the field of electrodynamics on a firm footing, sparked enormous activity among scientists, and laid the foundations for the development of wireless telegraphy. He showed that electricity could be propagated or transmitted as electromagnetic waves through space (without wire connections) and could be detected at a point distant from the transmitting source. The discoveries that he described in his May 1888 paper "Electromagnetic Waves in Air and Their Reflection" would later move the scientific community to call these waves "Hertzian" in his honor.

Hertz began wireless research in 1879 just as mathematical physics began to be recognized as a separate subdiscipline in Germany. Hermann von Helmholtz was a pre-eminent figure in the 19th-century scientific renaissance in Germany and was Hertz's lifelong mentor in Munich. Helmholtz encouraged Hertz to participate in a competition designed to solve a problem posed by James Clerk Maxwell's theories. Hertz's active wireless experimentation began in 1886 while he was a faculty member at the Technische Hochschule (Technical High School) in Karlsruhe. His discoveries began as he was conducting a class demonstration in electricity.

Hertz experimented with gaps in a wire coil connected to a Leyden jar to generate waves at various frequencies. The inner and outer foils of the Leyden jar became the two arms of a dipole or two-part transmitting antenna. Hertz tuned two flat coils of wire or metal strips to the same frequency so that the waves generated and received were identical. The Leyden jar stored the electrical charges while an induction coil magnified them as the spark gap and metal plate radiated the charges into the ether (or across the metal-laden laboratory). Thus, Hertz demonstrated the transfer of energy that Maxwell had predicted.

Hertz conducted his experiments in less-than-ideal laboratory conditions: steel and lead in the lab affected his measurements. His scientific apparatus (use of a spark gap) was crude. He used simple laboratory devices and often built his own devices to confirm Maxwell's prediction that electromagnetic waves existed, both as light and radio waves. The physically confined space Hertz used for his experiments required him to work with fast laboratory oscillations and short wavelengths. Eventually, Hertz showed that, like light waves, electromagnetic waves were reflected and refracted and, most important, that they traveled at the speed of light but had a much longer wavelength.

European scientists were reluctant to accept the results of Hertz's experiments. Researchers were slow to see the significance of Hertzian waves and to accept his conclusion that such waves traveled at the speed of light. They tended to cling to older, more familiar concepts, such as the corpuscular theory of light, which implied that light traveled in the form of material particles. However, Hertz's mentor, Helmholtz, never wavered in his support for his protégé.

Hertz did influence several scientific studies and research efforts around the world. He was generally regarded not as an innovator but as an uncommonly critical and lucid intelligence who addressed the conceptual problems of physics. Hertz helped connect the traditional study of mechanics and the evolving study of electrodynamics.

Other scientists credited Hertz for their achievements. Albert Einstein, in an 1899 letter to his fiancée, indicated that Hertz's *Electric Waves* stimulated his interest in the electrodynamics of moving bodies; this interest would ultimately lead to his special theory of relativity. Hertz's work on the photoelectric effect (although the existence of electrons was unknown at the time) helped to verify Einstein's quantum equation for the photoelectric effect and identified certain phenomena as being clearly associated with the ultraviolet portion of the spectrum. Hertz also influenced Max Planck's development of quantum physics by providing equations for the emission and absorption of energy by oscillators. Wilhelm Conrad Roentgen credits his discovery of X rays in late 1895 to Hertz and Lenard's cathode-ray experiments.

Hertz did not extend his work to include potential applications of his discoveries. He was not interested in going beyond the theoretical stage of study. His experimental work on electromagnetic waves, however, led directly to radio telegraphy and to later innovations. The material he read about Hertzian waves influenced Guglielmo Marconi significantly. Marconi became the practical inventor and took Hertz's studies further.

Hertz helped move wireless theories and experiments to practical reality and refinement. He helped formulate suggestions on the proper approach theoretical physicists should take toward the physical universe. He suggested future research to build on his discoveries. His fusion of theory and experiment with a creative interest in philosophical and logical foundations is unique among scientists. Hertz succeeded in delineating central concepts and pointing the way to fruitful future research initiatives, and his achievements were lauded in science publications, the popular press, and in public speeches. He was a model for many future generations of physicists.

Leading physicists of the day had profound respect for Hertz both as a physicist and as a man. John Ambrose Fleming, who would use Hertz's discoveries for his own experiments, noted that Hertz's work marked a fresh epoch in electrical discovery. Hertz's name is used internationally to indicate a unit of frequency (what had been called a cycle, as in kilocycle, is now hertz or kilohertz) in his honor.

Hertz's final years were devoted almost entirely to exploring the theoretical implications of Maxwell's electrodynamics for the rest of physics. While at the peak of his productivity, Hertz died tragically at age 37 of chronic blood poisoning. At his funeral in Hamburg, Hertz was praised for his noble simplicity and genuine modesty.

Heinrich Hertz was one of the last classical physicists. Had he lived longer, he would doubtless have been a major participant in the development of modern physics.

PETER E. MAYEUX

Heinrich Rudolf Hertz. Born in Hamburg, Germany, 22 February 1857. Attended private school of Richard Lange, 1863 and 1872–74; admitted to upper class of *Johanneum Gymnasium,* Hamburg, 1874; began engineering studies at Dresden Polytechnic, 1876; switched from engineering to physics at University of Munich, 1877–78; attended Friedrich-Wilhelm University, Berlin, doctoral degree *magna cum laude,* 1880; Master builder intern, Frankfurt Public Works Department, 1875–76; military service with railway regiment, Berlin, 1876–77; assistantship with Hermann von Helmholtz, Berlin Physics Institute, 1880–83; lecturer, mathematical physics, Kiel, 1883; appointed professor of physics, Technische Hochschule in Karlsruhe, 1885; appointed professor of physics, Friedrich-Wilhelm University, Bonn, 1888. Received Philosophical Faculty prize, 1879 (leading to the discovery of electromagnetic waves); Matteucci Medal, Italian Scientific Society, 1888; Baumgartner Prize, Vienna Academy of Sciences, 1889; La Caze Prize, Paris Academy of Sciences, 1889; Rumford Medal, British Royal Society, 1890; Bressa Prize, Turin Royal Academy, 1891; elected corresponding member of several major scientific societies, including the Berlin Academy of Sciences, the Manchester Literary and Philosophical Society, the Cambridge Philosophical Society, and the Accademia dei Lincei; official adoption by the International Electrotechnical Commission of the name *hertz* as a unit of frequency, 1933. Died in Bonn, Germany, 1 January 1894.

Selected Publications

Erinnerungen Briefe Tagebucher, 1927; *Memoirs, Letters, Diaries,* 2nd (German-English) enlarged edition, edited by Mathilde Hertz and Charles Susskind, 1977

Collected Works, edited by Philipp Werke, 1896–99

Further Reading

Aitken, Hugh G.J., *Syntony and Spark: The Origins of Radio,* New York: Wiley, 1976

Baird, Davis, R.I.G. Hughes, and Alfred Nordmann, editors, *Heinrich Hertz: Classical Physicist, Modern Philosopher,* Boston: Kluwer Academic, 1998

Bryant, John H., *Heinrich Hertz, The Beginning of Microwaves,* New York: Institute of Electrical and Electronics Engineers, 1988

Cichon, D.J., and W. Wiesbeck, "Heinrich Hertz, Wireless Experiments at Karlsruhe in the View of Modern Communication," in *International Conference on 100 Years of Radio,* London: Institution of Electrical Engineers, 1995

Kniestedt, J., "Heinrich Hertz: The Discovery of Electromagnetic Waves 100 Years Ago," *Archiv für das Post- und Fernmeldewesen* 41 (February 1989)

Lodge, Oliver, *Signalling across Space without Wires: Being a Description of the Work of Hertz and His Successors,* London: The Electrician, 1894; 4th edition, 1911

Mulligan, Joseph F., "Heinrich Hertz and the Development of Physics," *Physics Today* 42 (March 1989)

O'Hara, James.G., and Willibald Pricha, *Hertz and the Maxwellians: A Study and Documentation of Electromagnetic Wave Radiation, 1873–1894,* London: Peregrinus, 1987

Plank, M., "Gedächtnisrede auf Heinrich Hertz," *Verhandlungen der Physikalischen Gesellschaft zu Berlin* 13 (1894)

Hicks, Tom 1946–

U.S. Broadcast Executive

Thomas O. "Tom" Hicks is the president and chief executive officer (CEO) of Hicks, Muse, Tate, and Furst, a Dallas-based leveraged-buyout firm. At one point during the late 1990s, Hicks and his investment firm held ownership interests in more radio stations in the United States than any other company or individual. Hicks, Muse, Tate, and Furst came into existence in 1989 when Hicks and three partners formed the private investment firm after several successful industry buyouts. Today, the portfolio of companies controlled by the firm includes real estate, consumer products, movie theaters, sports franchises, and broadcast stations.

Hicks became a common name in the radio industry during the 1990s, when he served as CEO for AMFM Incorporated, formerly Chancellor Media Company. Under Hicks, Chancellor invested heavily in radio prior to the passage of the 1996 Telecommunications Act. With the elimination of ownership limits under the new legislation, Chancellor was well positioned to take advantage of further consolidation in radio.

Within a few months, Chancellor became one of the largest radio operators in the United States, at one point owning more than 400 radio stations.

Thomas O. Hicks was born in 1946 in Houston, Texas. One of four sons of a Dallas media representative, Hicks grew up with a strong understanding of and appreciation for the radio business. His father became an owner of a few radio stations in a number of small Texas markets, including Beaumont, Bryan, Port Arthur, Laredo, and Big Springs. As a teenager, Hicks worked as a radio announcer at his father's station in Port Arthur. Following high school, Hicks graduated from the University of Texas with a business degree in 1968. He then moved west, completing a master's of business administration degree at the University of Southern California.

Hicks and his partner, Bobby Haas, established a strong reputation for putting together profitable deals for investors, acquiring companies such as Dr. Pepper, Sybron Corporation, 7-Up, Thermadyne Industries, and Spectradyne. In 1989 Hicks formed Hicks, Muse, Tate, and Furst, as a venture capital firm with a buy-and-build philosophy.

Hicks began acquiring radio stations in earnest in 1993, with Chancellor Broadcasting and CapStar Broadcasting forming the cornerstones of the radio group as part of Hicks, Muse, Tate, and Furst's overall investment strategy. Acquisitions continued in 1996 and 1997 as Chancellor was renamed Chancellor Media. The company acquired several existing radio group holdings, including Evergreen, SFX, and Viacom.

The firm also ventured into television with the purchase of stations owned by LIN Television, as well as several outdoor advertising companies. In July 1999 shareholders approved the merger of the former Chancellor Media and CapStar into a new company known as AMFM Incorporated to reflect the emphasis on radio as well as to match the name of the company's national radio network. Early in 1999, AMFM owned 460 radio stations.

Aside from being one of the largest radio owners in the country, Hicks established a new type of entrepreneurial spirit in the radio industry. By clustering stations in geographical areas and appealing to different target audiences, Hicks capitalized on the changing economics of radio and the resultant cash flow that would come with streamlined operations. In an interview published in *Broadcasting and Cable* in 1997, Hicks called radio "one of the all-time great businesses for pre-cash flows." Hicks understood that by clustering operations, fixed costs could be lowered, while profit margins would increase.

Hicks' business philosophy toward radio spurred other groups to consolidation in order to maintain a national presence in the radio industry. Hicks demonstrated to other investors that radio was still a profitable investment. Although the consolidation movement in radio was not without controversy, there is no doubt that the radio industry experienced renewed interest among the investment community and higher valuation as an industry group with Hicks as one of its leading advocates.

Throughout 1999, the stock value of AMFM remained flat amid investor concerns that the company was carrying too much debt. The company surprised the radio industry by disclosing in March 1999 that it would consider a possible sale or merger. In October 1999, a dramatic $23.5 billion merger was announced between Clear Channel Communications and AMFM, creating the world's largest radio company. Once the merger was finalized in 2000, Clear Channel owned a total of 830 radio stations, capable of reaching an audience of more than 100 million listeners.

With the merger, Tom Hicks moved into a new role as the vice chairman of Clear Channel, working closely with Chairman L. Lowry Mays. Hicks remains one of the company's largest stockholders. Although he maintains less of a public role in his new position, he will continue to influence the radio industry with his presence on Clear Channel's board of directors.

ALAN B. ALBARRAN

See also Clear Channel Communications; Ownership, Mergers and Acquisitions; Telecommunications Act of 1996

Thomas O. Hicks. Born in Houston, Texas, 7 February 1946. Attended University of Texas, B.A. in business administration, 1968; University of Southern California, M.B.A. 1970; became disc jockey at father's radio station at age 15; worked at Continental Illinois; investment officer, Morgan Guaranty Trust Company; president of venture capital affiliate, National Bank of Dallas, First Dallas Capital Corporation; first leveraged buyout with Louis Marx, Jr., 1977; co-managing partner, Summit Partners, LBO; co-chairman and CEO, Hicks and Haas, 1980s; formed buyout firm Hicks, Muse, Tate and Furst, 1989; owns Southwest Sports Group; chairman, University of Texas Investment Management Company.

Further Reading

Albarran, Alan B., and Gregory G. Pitts, *The Radio Broadcasting Industry,* Boston: Allyn and Bacon, 2000

Hurt, Harry, "Texas-Tall Buyouts: Dallas-Based Hicks, Muse Outguns Wall Street Firms," *U.S. News and World Report* (26 January 1998)

Michaels, James, "Tuning in with Tom Hicks," *Forbes* (1 June 1998)

Rathburn, Elizabeth, "Count 'em: 830," *Broadcasting and Cable* (11 October 1999)

"Thomas Hicks Follows the Radio-TV Muse," *Broadcasting and Cable* (23 June 1997)

Whitley, Glenna, "The Player," *Texas Monthly* 19 (March 1994)

High Fidelity

High fidelity is a term used to mean the highly accurate reproduction of sounds within the spectrum of human hearing, usually considered to be between 20 hertz and 20,000 hertz. English engineer Harold Hartley first applied the term in 1926. Much of today's understanding of what constitutes high fidelity reproduction stems from pioneering research into the way humans hear and interpret sound done by Harry Olsen for the Radio Corporation of America (RCA) and Harvey Fletcher at the American Telephone and Telegraph Company's (AT&T) Bell Laboratories.

Origins

Early sound reproduction devices such as Edison's cylindrical phonograph (1877) and Emile Berliner's Gramophone disk developed a decade later demonstrated the feasibility of recording, but they produced tinny sound with significant distortion and limited reproduction of voice and music. The near-simultaneous developments of radio broadcasting and sound motion pictures led engineers to search for ways to improve sound quality.

E.C. Wente's invention of the condenser microphone (1916) and improvements in loudspeaker technology by Rice and Kellogg at General Electric, Peter Jensen and others (1925) greatly improved the ability to record and reproduce audio. In the 1920s Edwin Armstrong's development of the heterodyne circuit improved the sensitivity and selectivity of radio receivers, and Harold Black's discovery of negative feedback provided improved audio reproduction, but several obstacles still prevented accurate reproduction of sound. The surface noise associated with records, coupled with their limited audio range and short playing time, sharply curtailed improvements in mechanical sound reproduction. AM radio transmissions were subject to significant noise and static interference. Engineers thought that reducing the audio bandwidth would reduce annoying whistles and associated distortions.

Simultaneous research into improved audio occurred in Britain, Germany, and the United States, but it was AT&T that spearheaded high-quality audio development. AT&T's Bell Laboratories undertook long-term development of sound reproduction in conjunction with high-quality long-distance telephone service. With the 1922 construction of WEAF, AT&T's flagship New York City radio station, the telephone company carried out research to improve broadcasting microphones, consoles, and transmitters. Bell Labs also developed the transcription turntable using a slower speed (33 1/3 revolutions per minute) to increase playing time to 30 minutes to meet the needs of broadcasters and motion picture engineers. By 1929 the introduction of the matched-impedance recorder, coupled with development of gold master records, increased the attainable frequency response to 10,000 Hz and greatly reduced surface noise for records.

By 1930 both RCA and Bell Labs were experimenting with various means of improving audio quality for records. One year later Leopold Stokowski, the famed conductor of the Philadelphia Orchestra, enlisted Bell Labs' help in setting up an audio test room at the Academy of Music. The first disk recordings capable of accurate sonic reproduction were cut with Stokowski's help, and Bell Labs made more than 125 high-quality recordings of the 1931–32 Philadelphia musical season. During this time Stokowski recorded the first binaural recording using AT&T's new two-styli cutter, developed by Arthur C. Keller, and in 1933 the first U.S. stereophonic transmission over telephone lines occurred when Bell Labs demonstrated a three-channel audio system in Constitution Hall in Washington, D.C. In 1938 Keller received a patent for a single-groove stereophonic disk record system.

Improving Radio

Although various advancements in the technology allowed AM radio to improve substantially, the narrow channel bandwidth adopted by the Federal Radio Commission and static interference problems created technical limitations to full high fidelity transmission. By 1935 radio stations that specialized in quality music were eager to adopt improved technology. The Federal Communications Commission (FCC) licensed four stations on three channels at the high end of the AM frequencies (in the 1500–1600 kHz region) to experiment with high fidelity broadcasting using a wider channel bandwidth. WHAM, a clear channel station that originated Rochester Philharmonic Orchestra broadcasts on the National Broadcasting Company's Blue Network, and WQXR in New York were among the early pioneers of high fidelity AM broadcasting. These stations used new Western Electric transmitters boasting better frequency response with a wider dynamic range. Improved radio receivers capable of better fidelity were manufactured by WHAM's parent company, Stromberg-Carlson, E.H. Scott, and others.

Regularly scheduled high fidelity FM broadcasts began on 18 July 1939 as Edwin Armstrong's station retransmitted classical music programs from New York's WQXR via special telephone lines. That same year, the Yankee radio network began high fidelity FM broadcasting, soon followed by General Electric and others. In 1944 Britain's Decca records introduced full fidelity recordings capable of reproducing most of the audio spectrum.

authorities campaigned to ban Spanish from the airwaves. Although many stations continued to program Spanish language blocks, others wishing to reach Mexican-Americans moved their operations to the Mexican side of the border out of the reach of U.S. authorities. The tension created by the contradictory responses of the Anglo establishment to the Mexican community—commercially welcoming, but politically and culturally rejecting—would continue to shape the development of a Hispanic audience.

The Early Transnational Hispanic Audience

Emilio Azcárraga, patriarch of the Mexican entertainment conglomerate today known as *Televisa,* began his broadcasting empire with radio stations in the 1930s. Shortly thereafter, he began transmitting music from his Mexico City station XEW, *La Voz de América Latina* (The Voice of Latin America), to a radio station in Los Angeles, which then relayed it to other U.S. stations. In addition, Azcárraga owned five radio stations along the United States–Mexico border that transmitted directly into the United States. For Azcárraga and his fledgling broadcasting empire, the border that separates the United States from Mexico was little more than a bureaucratic nuisance. Mexicans who listened to radio lived on both sides of the official separation of the two countries.

By the 1940s U.S. broadcasters were discovering that the emotional impact of an advertising message delivered in a listener's first language and suggestively enfolded in a program of music or drama, evoking the most nostalgic memories of a listener's far-away birthplace, was infinitely greater than the same message in English. These Spanish language radio programs were broadcast weekly, not daily, in four states: New York, Arizona, Texas, and California, most of them in the off hours.

The early Spanish language radio audience in the United States was defined by its "otherness," particularly its continuing close ties to Mexico. When the commercial establishment began to imagine Spanish speakers as members of *their* marketplace, they began to mold Spanish language radio for an imagined audience more commensurate with that of the dominant, majority society. Immigrant program hosts were urged to shorten their commentary and pick up their pace, so as to better match the quick tempo of the new advertisements they were reading. The length of the music selections was also shortened to make room for more advertising breaks.

Changing the Immigrant Paradigm

In the postwar period outside the Southwest, Spanish language radio shared off-hour time slots with other foreign language radio. By the 1950s, German, Polish, Scandinavian, and other foreign language radio broadcasting began a steady decline. This was largely attributable to the assimilation of European immigrants into the dominant culture. As these peoples were recognized as predominantly English monolingual, the commercial appeal of foreign language radio programs declined; these consumers could be reached with general radio programming and advertising. As such, foreign language broadcasting was not as attractive to advertisers and thus not as appealing to radio station owners.

During this period, the number of weekly hours of U.S. Spanish language radio doubled. Two-thirds originated in the Southwest, the region most heavily populated with Spanish speakers. By 1960 Spanish language radio accounted for more than 60 percent of all U.S. foreign language radio. Spanish was the only foreign language to command entire stations and entire broadcast days. Because of continuing immigration, U.S. Spanish language radio was growing at a time when other foreign language broadcasting was dying.

Radio station owners and their advertisers were among the first to notice (in commercial terms) that the European paradigm of immigration to the United States was not identical to that of Latin American immigration. Most European immigrants, within a generation or two of their arrival, were socially and economically integrated into the majority culture, losing their European "mother tongue" in favor of English monolingualism. In addition, European immigration to the United States was discontinuous, disrupted by two world wars and the vastness of the Atlantic Ocean. Once new German immigrants, for example, stopped arriving, a generation or so later all but a few reduced their use of German or stopped speaking German completely. Consequently, the market for German language radio dropped off precipitously.

In contrast, immigrants from Latin American countries, primarily Mexico, have arrived in a steady stream (of varying size) to the United States for most of this century. Monolingual Spanish speakers settling in the United States renew the life of the language and provide a core audience for Spanish language radio programming. Today, the Spanish-speaking audience is in many ways the ideal specialized audience. Language, race, and continued close association with Latin America made it an easily identifiable audience. Between 1960 and 1974, spurred by immigration from Cuba and Puerto Rico, the number of radio stations carrying Spanish language programming doubled.

In the next quarter century, that number doubled again as immigration from Mexico and Central America increased and, in equal measure, the United States born Latino population grew. At the same time, the Hispanic audience was "discovered" by Madison Avenue and the narrowcasting broadcasting industry to be "targetable," that is, definable in market terms, and therefore a potentially profitable "niche market."

Today there are more than 400 Spanish language radio stations in the United States. Like general market U.S. radio, it is shaped by ownership chains, with the Spanish Broadcasting System being the largest. Spanish language radio formats vary from news/talk to different kinds of music—*salsa*, *norteña*, and *rock en español*. In Los Angeles and Miami, the two U.S. cities with the highest concentrations of Latinos, Spanish language radio has been consistently rated first by audience measurement firms such as Arbitron throughout the 1990s.

Reflecting the permanence and diversity of Latino communities, radio programmers have begun experimenting with bilingual radio. Youth-oriented music dominates this format, with the disc jockeys' patter and the advertisements in Spanish *and* English, as well as in Spanglish.

AMÉRICA RODRÍGUEZ

See also Border Radio; Mexico; South America

Further Reading

Fowler, Gene, and Bill Crawford, *Border Radio: Quacks, Yodelers, Pitchmen, Psychics, and Other Amazing Broadcasters of the American Airwaves*, Austin: Texas Monthly Press, 1987

Gutiérrez, Félix, and Jorge Reina Schement, *Spanish Language Radio in the Southwestern United States*, Austin: Center for Mexican-American Studies, University of Texas, 1979

Rodriguez, América, "Objectivity and Ethnicity in the Production of the *Noticiero Univisión*," *Critical Studies in Mass Communication* 13, no. 1 (1996)

Rodriguez, América, *Making Latino News: Race, Language, Class*, Thousand Oaks, California: Sage, 1999

Rodriguez, América, "Creating an Audience and Remapping a Nation: A Brief History of U.S. Spanish-Language Broadcasting, 1930–1980," *Quarterly Review of Film and Video* 16, nos. 3–4 (2000)

Hoaxes

Pranks, Policies, and FCC Rulings

Until the early 1990s the Federal Communications Commission (FCC) had taken a relatively laissez-faire attitude toward radio hoaxes, admonishing offenders but avoiding several penalties. From 1975 to 1985 the Commission threw out a number of programming content policies initiated a decade earlier. In 1985 the FCC, under Chairman Mark Fowler's leadership, voted to eliminate its policy of restricting "scare" announcements as part of its deregulation initiatives during the Reagan administration. When the public became subject to a number of hoax abuses, however, in May 1992 the FCC issued a ruling prohibiting pranks that cause immediate public harm or divert resources from law enforcement.

War of the Worlds

Orson Welles perpetrated the first hoax in radio history in his 1938 radio play, "War of the Worlds." The national public panic over a well-crafted imaginary Martian invasion of the east coast of the United States was the ultimate demonstration of radio's impact on an audience. Subsequently, after the Welles broadcast, the FCC warned broadcasters not to use the words *bulletin* or news *flash* in entertainment programs and to provide adequate cautionary language in the airing of dramatizations.

Beginning in the 1950s legendary deejays such as Dick Biondi and Wolfman Jack attempted to shock their audiences with crazy stunts and wild antics. Radio stations across the nation undertook many pranks and trickery, such as turkeys thrown out of airplanes and a scavenger hunt for a $1,000 bill hidden in a public library, in the name of fun and higher market ratings. In the 1960s the FCC began a new era of regulation, in part because of what it perceived to be the public's vulnerability to deceptive programming and promotions. In 1960 it issued a policy statement that addressed intentional distortion or falsification of programming (i.e., news staging). In 1966 the FCC issued a stronger policy, which warned against airing "scare" announcements. The 1966 policy was a reaction to specific complaints about radio contests that disrupted traffic, caused property damage, diverted law enforcement, alarmed listeners with imaginary dangers, and threatened life. The FCC stand slowed down the occurrence of hoaxes over the next several years.

Then, in 1974, Rhode Island's WPRO-FM recreated "War of the Worlds." The program had been promoted as a spoof throughout the day. During the actual broadcast, however, 45 minutes elapsed before the station aired a public disclaimer. One hundred and forty listeners called the radio station. While station personnel had warned the local police department of its intent to air the program, the FCC admonished WPRO on the basis of its 1966 statement concerning broadcast of scare announcements.

In another instance, that same year a Tucson, Arizona radio personality, with the help of the news director, faked his own kidnapping. The commission failed to renew the license of KIKX-AM, specifically based on its violation of FCC policies related to the "fake kidnapping" (i.e., news staging, false newscasts, and licensee failure to exercise adequate control over station operations) and to a lesser extent on its technical violations on several station program logs and its Equal Employment Opportunity record. The commission affirmed its position in 1980, and two years later the District of Columbia Circuit Court upheld the decision.

Serious Radio Hoaxes

After the FCC eliminated its scare announcement policy, between 1989 and 1991 a number of serious radio hoaxes popped up across the United States; at least five are documented in the commission's ruling "Regarding Broadcast Hoaxes." Four of these incidents resulted in admonishment by the FCC, although only one resulted in a $25,000 fine. On 2 October 1989 the FCC admonished KSLX-FM in Scottsdale, Arizona, for a stunt that faked the station being taken hostage by terrorist activity. In July 1990 the commission admonished WCCC-AM/FM in Hartford, Connecticut, for reporting a nearby volcanic eruption.

In 1991 three other serious hoaxes were perpetrated by St. Louis' KSHE, Los Angeles' KROQ, and Rhode Island's WALE. On 29 January 1991 KSHE morning personality John Ulett staged a mock nuclear alert during the morning drive time, complete with a simulated Emergency Broadcast System (EBS) tone and an authentic-sounding civil defense warning that announced that the nation was under nuclear attack. There was no disclaimer until two hours after the broadcast. Four hundred listeners called the station. The FCC fined KSHE $25,000 based on the false use of EBS during the hoax. The KROQ morning team staged a false confession from an anonymous caller who claimed to have brutally murdered his girlfriend. Police spent nearly 150 hours investigating the case and the incident was featured twice on the syndicated TV program *Unsolved Mysteries.* On 9 July 1991 the WALE news director in Rhode Island announced that the overnight on-air personality had been shot in the head. Police and media rushed to investigate the incident. Upon hearing the broadcast, the program director called the station and told the producer to cease the hoax. When the producer failed to do so, the program director shut off the transmitter. The station went back on the air one minute later, with a disclaimer that aired every 30 minutes for the following 30 hours. Although the program director terminated the news director, the talk show host, and the producer, the FCC admonished WALE for broadcasting false and misleading information and stated that the licensee was not excused by subsequent remedial action.

Anti-Hoax Ruling

The Commission's 1992 anti-hoax rule (Section 73.1217) did not discourage the morning crew at WNOR-FM in Norfolk, Virginia, from staging a series of news reports that the city park built over a landfill was about to explode. Local police, overwhelmed with concerned calls from listeners, filed complaints with the FCC. A month after the WNOR-FM incident, the FCC issued its anti-hoax ruling in an effort to target those incidents involving a false report of a crime or catastrophe. The FCC was eager to clear its docket of what appeared to be a stream of hoax violations and to enact a middle range of enforcement. The commission said that its ruling would provide enforcement flexibility by allowing fines that could range up to $25,000 a day. In the ruling the commission states:

> No licensee or permittee of any broadcast station shall broadcast false information concerning a crime or catastrophe if (a) the licensee knows this information is false, (b) it is foreseeable that broadcasting the information will cause substantial public harm. Any programming accompanied by a disclaimer will be presumed not to pose foreseeable harm if the disclaimer clearly characterizes the program as fiction and is presented in a way that is reasonable under the circumstances (amendment to Part 73 Regarding Broadcast Hoaxes, Communications Act, Report and Order, 7FCCRcd4106 [1992]).

This ruling clearly demonstrated the FCC's desire to manage promotional content abuses on the airwaves by assigning specific monetary punitive actions for serious hoaxes that posed a substantial threat to the public safety and welfare.

PHYLIS JOHNSON

See also War of the Worlds

Further Reading

Cantril, Hadley, *The Invasion of Mars: A Study in the Psychology of Panic,* Princeton, New Jersey: Princeton University Press, 1940

Cobo, Lucia, "False Radio Broadcast Evokes FCC Investigation," *Broadcasting* (4 February 1991)

"FCC Admonishes Station for Airing Hoaxes," *The News, Media, and the Law* 16, no. 1 (Winter 1992)

"FCC Moves to Revamp EBS," *Radio and Records* (25 September 1992)

Fedler, Fred, *Media Hoaxes,* Ames: Iowa State University Press, 1989

Jessell, Harry A., "FCC Picks up Pace on Indecency Enforcement," *Broadcasting* (31 August 1992)
Johnson, Phylis, and Joe S. Foote, "Pranks and Policy: Martians, Nuclear Bombs, and the 1992 Ruling on Broadcast Hoaxes," *Journal of Radio Studies* 2 (1993–94)
"KROQ Team Suspended after Murder Hoaxes," *Radio and Records* (19 April 1991)
"KSHE Radio Jolts Listeners with Fake Nuclear Bomb News," *St. Louis Post-Dispatch* (30 January 1991)
Ray, William B., *FCC: The Ups and Downs of Radio-TV Regulation,* Ames: Iowa State University Press, 1990
Tunstall, Jeremy, *Communications Deregulation: The Unleashing of America's Communications Industry,* New York and Oxford: Blackwell, 1986

Hogan, John V.L. "Jack" 1890–1960

U.S. Radio Inventor and Engineer

Jack Hogan was a key figure in early American wireless and radio broadcast development, combining invention with practical innovation. He developed the single-dial radio tuner, the first high-fidelity radio station, and later turned to mechanical television and then facsimile.

Origins

Hogan was born in Bayonne, New Jersey, in 1890. He built his first amateur radio station in 1902 at age 12, using a coherer as a detector. (A coherer was a device in which iron fillings cohered in the presence of an electrical signal, making it a useful "detector" of those signals.) Just four years later he was working as a laboratory assistant with Lee de Forest. Hogan attended the Sheffield Scientific School at Yale University, as had de Forest, specializing in physics, mathematics, and electric waves. In 1910 he went to work for Reginald Fessenden's National Electric Signaling Company (NESCO) at its Brant Rock, Massachusetts, station, as a telegraph operator. Fessenden was so impressed with Hogan that he assigned him to supervise erection of a terminal station in Brooklyn, New York. There he developed perhaps the first ink tape siphon (a method for permanently recording on a paper tape) for recording transatlantic radio signals, using an Audion amplifier. Hogan remained with Fessenden until 1914.

Radio

Hogan was a cofounder of the Institute of Radio Engineers in 1912 (he served as IRE president in 1920). In 1913 he directed acceptance tests of the U.S. Navy's first high-power station at Arlington and served as the Navy's chief research engineer until 1917, focusing on high-speed recorders for long-distance wireless. In 1917 he became commercial manager of the International Radio Signal Company. He was placed in charge of operations and manufacturing with emphasis on radio for what were then called submarine chasers (destroyers) and aircraft. In 1918 he was made manager of the International Telegraph Company.

Hogan established his own consulting practice in 1921 (founding Radio Inventions, Inc. in 1929), where he specialized in broadcast apparatus and radio regulations. During this period he also wrote *The Outline of Radio* (1923), a guide for the general public about how radio worked; the book was so well received that several subsequent editions appeared. Always interested in tonal quality, Hogan built the first high-fidelity experimental radio station, licensed as W2XR, in 1929. Hogan and Elliott Sanger converted it to become commercial classical music station WQXR in New York City in 1936.

Graphic Communication

In the 1930s Hogan began to work on television technology, and then on facsimile systems. Since his television efforts had concentrated on what turned out to be outmoded mechanical scanning systems, Hogan soon dropped television to concentrate on the more promising field of facsimile transmission, which he worked on well into the 1940s. His system was said to be both faster and to provide more fidelity of reproduction than the crude systems that preceded his. He demonstrated his system to industry observers in New York and Milwaukee in the late 1930s, using experimental radio transmitters licensed by the Federal Communications Commission (FCC). The Radio Corporation of America (RCA) nearly adopted the Hogan system, but World War II intervened.

Hogan wrote a number of articles in the 1940s, for both specialist and lay readers, on what facsimile systems could do. During World War II, he acted as advisor to several government agencies on radar, guidance systems, and missiles. He also chaired Panel 7 of the radio industry's Radio Technical Planning Board during World War II, which concerned itself with spectrum needs and technical standards for facsimile and made recommendations about these topics to the FCC.

After the war, Hogan returned to his facsimile work (resigning from WQXR in 1949 to focus his efforts further), but the systems then being touted did not develop commercially and are unrelated to facsimile as we use it today. Today Hogan is perhaps most remembered for his invention of single-dial tuning and as one of three cofounders of the Institute for Radio Engineers, the predecessor of today's Institute of Electrical and Electronic Engineers (IEEE).

CHRISTOPHER H. STERLING

See also De Forest, Lee; Fessenden, Reginald; Technical Unions; WQXR

John Vincent Lawless ("Jack") Hogan. Born in Bayonne, New Jersey, 14 February 1890. Son of John Lawless Hogan, salesperson, and Louise Eleanor Shimer, writer and musician; worked with Lee de Forest, 1906; attended Sheffield School, Yale University, 1908–10; telegraph operator with Fessenden's NESCO, 1910–14; patented single-dial radio tuning system, 1912; co-founder, Institute of Radio Engineers, 1912; chief research engineer, U.S. Navy, 1914–17; commercial manager, International Signal Company, 1917; manager, International Radio Telegraph Company, 1918–21; consultant, 1921; formed Radio Inventions, 1929; founded station W2XR (later WQXR) in New York, 1929; worked with government agencies during World War II, including Office of Scientific Research and Development; chaired Panel 7 of Radio Technical Planning Board, 1943–45; resigned as president of WQXR, 1949. Died in Forest Hills, Queens, New York, 29 December 1960.

Selected Publications

The Outline of Radio, 1923

Further Reading

Dunlap, Orrin Elmer, Jr., *Radio's 100 Men of Science: Biographical Narratives*, New York and London: Harper, 1944

Kraeuter, David W., *Radio and Television Pioneers: A Patent Bibliography*, Metuchen, New Jersey: Scarecrow Press, 1992

Hollywood and Radio

In this era of studio-owned television networks, it is difficult to remember that only a few years ago, accounts of the history of television and broadcasting cast their relationship in terms of a bitter bicoastal rivalry. Hollywood hated and resisted television as it had radio, these historians said: they turned their back on it and refused to let their stars appear on it, and one studio even forbade television sets from appearing in its films.

If we take a closer look at history, however, nothing could be further from the truth. From the earliest years of radio, Hollywood studios regarded the upstart sound-only medium with a great deal of interest—despite the fact that in those days movies had no voice at all. After a period of experimentation with movie/radio cross-promotion, a few studios attempted to enter the network business. Thwarted by both economics and regulation, the film industry turned to steady and profitable production for radio, to the point that in the mid-1930s both major networks, the National Broadcasting Company (NBC) and the Columbia Broadcasting System (CBS), constructed major studios of their own in the heart of filmland. This productive relationship continued through the early years of television, and although radio lost its importance in the Hollywood scheme of things as television quickly took over the production of dramatic programs, film companies still maintained a presence in radio station ownership and also in the production of recorded music, so vital to radio's new format mode. The merger mania of the 1980s and 1990s consolidated these cross-ownership positions, as radio, television, film, music, and new media became interlocking parts of the same communications conglomerates.

Origins

One of the earliest instances of film/radio cooperation took place not in Hollywood but on the stage of the Capitol Theater in New York City, part of the Loews/Metro Goldwyn Mayer (MGM) chain. In 1923 theater manager Samuel L. Rothafel entered into an agreement with American Telephone and Telegraph (AT&T) to broadcast his prefilm stage show over their new station, WEAF. The results were so positive that the show quickly became a regular feature, called *Roxy and His Gang,* one of the earliest hits of radio broadcasting. Soon other theaters jumped on the bandwagon.

Movies might not have been able to talk, but that didn't mean there wasn't a lot of musical entertainment in the theaters. Many big-city theaters featured elaborate stage shows and enormous theater organs, whose musical accompaniments animated their film showings. Concerts by theater organists were broadcast over WMAC, WGN, and KWY in Chicago and in many other cities starting in 1925. That year, Harry Warner of Warner Brothers Studios proposed that the film industry as a whole should start a radio network to publicize their pictures. He began by opening a Warner Brothers radio station, KFWB in Los Angeles, and in 1926 a second one, WBPI in New York City. Other studios took note. Pathe, producers of newsreels, announced that they would begin distributing a script version of their news films for delivery over local stations. By 1927 Universal chief Carl Laemmle inaugurated the *Carl Laemmle Hour* over WOR-New York, presenting vaudeville and film stars and giving previews of upcoming pictures. MGM experimented with the world's first "telemovie": a dramatic, blow-by-blow narration of MGM's new release, *Love,* starring Greta Garbo and John Gilbert, delivered on the air by WPAP's announcer Ted Husing (usually known for his sports coverage) as it unreeled before his eyes in the Embassy Theater in New York.

That same year, MGM announced an ambitious project with the Loews theater chain: a planned network based on movie materials and promotion that would link over 60 stations in more than 40 cities. This proposal followed a more detailed one announced the previous spring by Paramount Pictures Corporation. Paramount, in conjunction with the Postal Telegraph Company, planned to start up the Keystone network "for dramatizing and advertising first-run motion pictures." Because AT&T had a lock on the land lines vitally needed to link stations together into a network, and because AT&T had an exclusive contract with the existing radio networks, Paramount needed Postal Telegraph to provide its lines. Despite much excitement in the industry, neither the Keystone Chain nor the MGM/Loews network reached fruition. A combination of regulatory discouragement, exhibitor opposition, and competition from other sources diverted studios' radio ideas in other directions. Paramount shortly thereafter purchased a 49 percent interest in the CBS network, still struggling to compete with its deep-pocketed competitor. Meanwhile, NBC's parent company, the Radio Corporation of America (RCA), acquired its own film studio, Radio-Keith-Orpheum (RKO), in 1929. Though this was intended more as a way to capitalize on RCA's new sound-on-film system than as a radio venture, the era of "talking pictures" would facilitate a renewed interest in the potential of film/radio cooperation.

Depression Years

By 1932 America had been hard hit by the Depression. Film industry profits suffered, as theaters went out of business and box office receipts slowed to a trickle. Radio, however, continued to thrive. As advertising agencies began to take the broadcast medium seriously as an outlet for their customers' campaigns, a new and influential partnership was about to emerge. Dissatisfied with CBS's and NBC's staid approach to programming, several aggressive advertising firms turned their attention to Hollywood's untapped potential for radio-based product promotion. One of the most influential in this Hollywood/agency alliance was John U. Reber of the J. Walter Thompson Company (JWT), whose plan for radio advertising envisioned big-budget, star-studded productions sponsored by JWT clients over the major radio networks. He determined to form a working relationship with the proven entertainment producers in Hollywood, and by the mid-1930s JWT was producing at least five shows out of each year's top ten, most of them featuring Hollywood talent. Other major agencies included Young and Rubicam, Blackett-Sample-Hummert, and Dancer Fitzgerald. When in 1936 AT&T, as a result of an investigation by the Federal Communications Commission, reduced their land line rates to the West Coast, a "rush to Hollywood" resulted, and most major agencies, along with the two national networks, opened up studios in Los Angeles. Radio had gone Hollywood.

This productive and profitable association would have great impact on both the radio and film industries. A variety of radio programs developed that centered on movie industry stars, properties, and Hollywood celebrities. The most prestigious was the movie adaptation format pioneered by JWT's *Lux Radio Theater.* Hosted by celebrity director Cecil B. DeMille, *Lux* presented hour-long radio adaptations of recent Hollywood film releases, introduced and narrated by DeMille and featuring well-known film stars. It started on NBC in 1934 but jumped in 1935 to CBS, where it ran until 1954. From 1936 on, the program was produced in Hollywood. Others in this format, often referred to at the time as "prestige drama," included *Screen Guild Theater, Hollywood Premiere, Academy Award Theater, Dreft Star Playhouse, Hollywood Startime,* and *Screen Directors' Playhouse.* A popular feature of these programs was the intimate, casual interviews with famous stars; DeMille, for instance, would chat at the end of each show with that night's leading actors, often casually working in a mention of the sponsor's product.

The second major venue for Hollywood stars and film promotion was radio's leading genre, the big-name variety show. Starting with the *Rudy Vallee Show* in 1929, almost all of the top-rated programs on the major networks in the 1930s belonged to this genre: the *Kate Smith Hour, Maxwell House Showboat, Shell Chateau* (Al Jolson), the *Chase and Sanborn*

Hour (Bergen and McCarthy), the *Jack Benny Program, Kraft Music Hall* (Bing Crosby), *Texaco Star Theater,* the *Eddie Cantor Show, Burns and Allen, Town Hall Tonight* (Fred Allen), and many more. All featured regular guest appearances from Hollywood's best and brightest, often promoting their latest pictures or acting out skits related to film properties. Many stars eventually began hosting such programs themselves, especially in the late 1930s and early 1940s. Adolph Menjou and John Barrymore served as hosts for *Texaco Star Theater;* Al Jolson appeared on radio almost exclusively after 1935; and William Powell and Herbert Marshall hosted *Hollywood Hotel* at various times. Some directors also got into the act: Orson Welles was a frequent variety show guest and often guest-hosted for Fred Allen, and Alfred Hitchcock established a reputation on radio before becoming a television personality. Furthermore, a whole set of Hollywood's secondary ladies became more famous via radio performances than their film careers had permitted: Lucille Ball, Dinah Shore, Joan Davis, Hattie McDaniel, Ann Sothern, and many others began as frequent guest stars, then headlined their own continuing programs on radio and later television.

Dramatic series programs also featured Hollywood talent. Most were the anthology-style programs that would also become early television's most prestigious fare. *First Nighter, Cavalcade of America, Hollywood Playhouse, Grand Central Station, Four Star Playhouse, Ford Theater, Everyman's Theater,* and many others brought film stars to radio in a wide range of stand-alone drama and comedy pieces. During the war years, Hollywood generously donated its talent to morale-boosting programs, sometimes on the regular networks and sometimes for the Armed Forces Radio Service only, such as *Command Performance, Free World Theater, Everything for the Boys, The Doctor Fights,* and many more. Hollywood stars moved freely between film and radio, and they would host and perform just as frequently on television's early dramas. Only in the mid- to late 1940s, however, did film stars begin turning up as leading actors in series comedies and dramas. The situation comedy form, pioneered by radio programs such as *Amos 'n' Andy, The Goldbergs, Fibber McGee and Molly,* and *Vic and Sade,* would be given a new gloss and prestige as Hollywood luminaries, particularly the comediennes mentioned above, moved into regular series production in shows such as *Joan Davis Time, My Favorite Husband, My Friend Irma, Maisie, Our Miss Brooks,* and *Beulah.*

Finally, mention should be made of the ever-popular genre of Hollywood gossip and talk. Many leading figures built their reputations on film industry chitchat, including the print divas Louella Parsons and Hedda Hopper. Walter Winchell also started in print but achieved full status on radio, combining gossip with news-related material. Ed Sullivan, Earl Wilson, and Jimmy Fidler all trafficked in celebrity news and views. A late-developing genre, the so-called breakfast program, presaged the television morning show *Today* with a combination of host chatter, celebrity guest interviews, and light news. Journalist Mary Margaret McBride pioneered the talk show format on radio in her long-running program of the same name. Another writer, Pegeen Fitzgerald, tried out McBride's formula in an early-morning show called *Pegeen Prefers;* she and her husband Ed would develop the first of the big-time breakfast shows, *The Fitzgeralds.* Others in this genre were *Tex and Jinx* (Tex McCrary and Jinx Falkeberg) and *Breakfast with Dorothy and Dick* (Dorothy Kilgallen and Richard Kollmar).

The film industry came increasingly to rely on the star-producing capabilities of radio as well. Radio personalities starred in many popular Hollywood films, from Freeman Gosden and Charles Correll ("Amos" and "Andy") in *Check and Double Check* in 1929, to special "radio" movies such as *The Big Broadcast of 1936* (and *1937* and *1938*), to the Bing Crosby/Bob Hope/Dorothy Lamour "Road" movies in the 1940s (*Road to Morocco, Road to Zanzibar, Road to Rio,* etc.). Rudy Vallee, Eddie Cantor, and Jack Benny all met with box-office success. Orson Welles' flamboyant production of *War of the Worlds* for the CBS *Mercury Theater of the Air* won him the contract to make *Citizen Kane* in Hollywood.

Radio and Television

As television loomed on the horizon after World War II, movie studios stood in a strong position to move into television production. A combination of network economics, the emphasis on "live" programming during television's early days, and royalty disputes within the film industry would delay the Hollywood/television alliance until the late 1950s. Though the nature of radio changed dramatically once television came onto the scene, some studios did maintain a persistent presence in radio ownership and production. Warner Brothers, Paramount, RKO, and MGM all owned radio stations, and they were to get in on television station ownership early on as well. MGM went into syndicated radio program production and distribution in the late 1940s with such programs as *MGM Theater of the Air* and *Maisie,* starring Ann Sothern. As attention and dollars shifted to television in the late 1940s and early 1950s, and as radio became once again primarily a musical medium, Hollywood stars and on-air production would migrate to the newer medium as well. Soon film studios would dominate prime-time television programming, though this would not translate into network power until passage of the FCC's financial interest and syndication rules broke up the networks' tight vertical integration in the 1970s. However, just as film companies diversified into television, they also began to acquire interests in the music industry, the new backbone of radio, with frequent cross-promotion between music and film.

Merger Mania

As the 1980s wave of mergers and acquisitions continued into the 1990s, the film majors of yore became part of diversified media conglomerates. Warner became part of the Time/Warner/Turner empire, with more than 50 labels under its imprint, including Warner Music International, Atlantic, Elektra, Rhino, Sire, and Warner Brothers. The conglomerate also has interests in music publishing, record clubs, recording technology, and music distribution. Time Warner accounted for 21 percent of U.S. music sales in 1997. Columbia Pictures was acquired by the Sony Corporation, owner of Columbia Records (acquired from CBS) and associated labels, the Columbia House music club, and other manufacturing and distribution arms, all of which accounted for 15 percent of U.S. music revenues. Universal became a part of the Music Corporation of America (MCA), which was later acquired by Seagram. MCA has long been a major presence in the music industry, with 11 percent of the U.S. market. Its labels include A&M, Decca, Def Jam, Deutsche Grammophon, Interscope, Geffen, MCA, MCA Nashville, Motown, Island, Phillips, Polydor, Universal, and Verve.

In 1995 Paramount was acquired by Viacom, owner of MTV and related cable music channels (M2, VH1), a considerable power in the music business. MTV produces radio programming as well, including radio versions of *MTV Unplugged, MTV News,* and *Weekend Revolution.* In 1995 Viacom announced a partnership with radio's largest program syndicator and station groups, Westwood One, to launch a new MTV Radio network featuring music-related material. The Disney Corporation also holds extensive interests in music recording, and with its merger with American Broadcasting Company (ABC) in 1995, it now owns radio stations that reach 24 percent of U.S. households. Twentieth Century Fox was purchased by Rupert Murdoch's News Corporation in the 1980s and is now linked with satellite music channels worldwide. News Corporation also owns the Australian Mushroom and Festival record labels. And in this age of synergy, the tie between movies and music has become tighter than ever before, with movie sound tracks used to promote artists and recordings, and sound track releases often achieving billions of dollars in sales.

In the era of new media, where the lines between film, radio, television, music, recordings, and the internet seem to be growing more blurry every day, the integrated entertainment corporations formerly designated by the term *Hollywood* have fingers in nearly every form of media that reaches into the home—or that reaches the viewer anywhere she or he might be. Now internet radio technology gives companies the ability to go on-line with their own "radio" services. DisneyRadio.com already provides a schedule of music and features from its films and artists, oriented toward children. Television shows on studio-owned networks promote recordings distributed by the company's record arm, which become hits on pop radio. Recording stars launch film careers; even radio personalities such as Howard Stern might receive a moment of celluloid fame. Though in the United States the days of radio drama and comedy faded, transferring their stars and audiences to television, the film industry continues to play a vital behind-the-scenes role linking radio to a host of other media. Without Hollywood, American radio could never have risen to the heights of creativity and popularity it achieved in its heyday. That the older medium bequeathed this tradition to a newer medium might be radio's loss, but it was television's gain.

MICHELE HILMES

See also Film Depictions of Radio; Television Depictions of Radio

Further Reading

Anderson, Christopher, *Hollywood TV: The Studio System in the Fifties,* Austin: University of Texas Press, 1994

Balio, Tino, editor, *Hollywood in the Age of Television,* Boston: Unwin Hyman, 1990

Hilmes, Michele, *Hollywood and Broadcasting: From Radio to Cable,* Urbana: University of Illinois Press, 1990

Jewell, Richard B., "Hollywood and Radio: Competition and Partnership in the 1930s," *Historical Journal of Film, Radio, and Television* 4, no. 2 (1984)

Lucich, Bernard, "The Lux Radio Theatre," in *American Broadcasting: A Source Book on the History of Radio and Television,* compiled by Lawrence Wilson Lichty and Malachi C. Topping, New York: Hastings House, 1975

Rothafel, Samuel Lionel, *Broadcasting, Its New Day,* New York and London: Century, 1925

Smoodin, Eric, "Motion Pictures and Television, 1930–1945," *Journal of the University Film and Video Association* 34, no. 3 (Summer 1982)

Watt, Kenneth, "One Minute to Go," *Saturday Evening Post* (2 April 1938 and 9 April 1938)

Hooperatings

Radio Ratings Service

Hooperatings was radio's best known and most widely quoted rating service in radio broadcasting during its heyday from 1934 to 1950. C.E. Hooper (1898–1955) pioneered a technique, the coincidental telephone call, that became an industry standard. Hooper sold subscriptions to his ratings information, making his service the first commercial venture in the field of radio ratings. C.E. Hooper, known as "Hoop," was imbued with a mission, and through his salesmanship he made his ratings service famous not just to the broadcasting industry but also to the public.

Origins

Hooper began his business career by selling aluminum utensils from door to door. He went on to earn an MBA (1923) from Harvard Graduate School of Business Administration. He took a job in Yakima, Washington, as assistant manager of the Liberty Savings and Loan Company. Between 1924 and 1926, he was advertising manager at the *Harvard Business Review*. He then took a similar job at *Scribner's Magazine*. He switched from selling space to buying space in 1929 as an account executive for Doremus and Company. After two years, at age 33, he entered the market research field as a member of the Daniel Starch organization.

Daniel Starch had taught business psychology at Harvard when Hooper was a student there. Starch conducted pioneering radio audience research for the new NBC network in 1928 and 1930, and in 1931 established the first continuous service for measuring the readerships of magazine and newspaper advertisements.

Since March 1930, Archibald Crossley had been "rating" broadcasts to estimate audience size for advertisers and for agencies that supported the Cooperative Analysis of Broadcasting (CAB). Crossley used a telephone recall method to ask listeners about their previous day's listening. He limited his surveys to areas of equal network opportunity, the 32 cities where all four networks (National Broadcasting Company [NBC] Red and Blue, Columbia Broadcasting System [CBS], and, after 1934, Mutual) could be heard with equal ease.

Hooperatings Begin

In 1934 Hooper left Starch to go into business as president of Clark–Hooper, a service that measured magazine and newspaper effectiveness. He also entered the field of radio audience measurement that same year using telephone *coincidental* calling when the audience was still listening, a method suggested to him by George Gallup. The team of Clark–Hooper, Inc. was encouraged by a group of magazine publishers who wanted to set up a more valid measure of radio's advertising effectiveness. These publishers were convinced that Crossley's ratings overstated the actual number of radio homes. More popular programs under Crossley's rating system would achieve ratings as high as 40 to 60 percent of the radio audience. To make matters worse, despite the fact that Crossley merely provided a rating index, many broadcasters persisted in projecting CAB ratings to total radio homes, resulting in an astronomical number of radio homes. The reason for this rating inflation was that Crossley initially used only the "identified listening audience," or what is now called the "share" (proportion of people tuned to a given channel based on all those using radio receivers at that time), as the base for his ratings. All of these factors hurt magazines and were factors that Clark–Hooper, Inc. undertook to correct.

Hooper's first important publicity came when, in collaboration with CBS, he estimated the number of adult listeners to President Roosevelt's fireside chat of 10 June 1936 in time for the next day's newspapers.

By 1938 the team of Clark–Hooper, Inc. had disbanded, and Hooper continued alone in the field of radio measurement. Although magazine publishers encouraged Hooper's service, they did not underwrite it. Hooper's method allowed him to innovate such features as the available audience base, resulting in ratings half the size of Crossley's. (The available audience included those not listening as well as those listening, whereas Crossley used only those listening.) Hooper also supplied an average audience rating, rather than give the total program listeners as CAB had done. An average audience rating was a program's total audience divided by the time intervals. Crossley's method, by comparison, presented only the total listeners to a given program in a sample or only the program's total audience. Furthermore, the coincidental technique eliminated what Hooper considered another major flaw with the recall method: the memory factor.

Hooper managed to make his name a household word. CAB reports had been primarily available to the buyers of advertising time, and consequently reports were guarded. Hooper, on the other hand, openly courted the press, making himself and his ratings newsworthy. His name began to appear in a vast assortment of trade magazines. In addition, he was written up in daily newspapers and even garnered a feature article in the *Saturday Evening Post* in 1947. Publicity surrounding Hooperatings rode such a crest during this period that Crossley was later to remark wryly that his defeat could

active in the world of international affairs. In addition to having served as the spokesperson for the U.S. ambassador to the United Nations, he returned to radio in 1993 as moderator of *America and the World,* an interview show sponsored by the Council on Foreign Relations that aired on National Public Radio (NPR). In 1995 Hottelet served as the NPR representative on a panel program concerning President Franklin D. Roosevelt and radio, which aired on C-SPAN. His expertise in analyzing foreign affairs found several media outlets well into the end of the century; during the 1990s he narrated a series of audiobooks on political hot spots around the world and continued to write on foreign affairs for the *Christian Science Monitor* into the early 2000s.

ERIKA ENGSTROM

See also Murrow, Edward R.; National Public Radio; News; Office of War Information; World War II and U.S. Radio

Richard Curt Hottelet. Born in New York City, 22 September 1917. Attended Brooklyn College, B.A. in philosophy, 1937. Started as stringer and then became United Press correspondent, Berlin, Germany, 1938–41; only U.S. reporter arrested by the Gestapo, 1941; worked in UP Washington bureau, 1942; U.S. Office of War Information, London, England, 1941–43; radio and television correspondent, originally hired by Edward R. Murrow, CBS TV and radio, 1944–85; host, *America and the World,* National Public Radio, 1993; foreign affairs columnist, *Christian Science Monitor,* late1990s and as of late 2002.

Selected Publications

The United Nations, vol. 5 in *The Dynamics of World Power* series, general editor Arthur M. Schlesinger, Jr., 1973

World's Political Hot Spots (audiobook series), 1994

Kosovo, Serbia, Bosnia: All You Want to Know: The History Behind the Conflict in Central Europe (audio cassette), 1999

Further Reading

Auster, Albert, "Richard C. Hottelet," in *Encyclopedia of Television News,* edited by Michael D. Murray, Phoenix, Arizona: Oryx Press, 1999

Bliss, Edward, Jr., *Now the News: The Story of Broadcast Journalism,* New York: Columbia University Press, 1991

Cloud, Stanley, and Lynne Olson, *The Murrow Boys,* Boston: Houghton Mifflin, 1996

Mickelson, Sig, *The Decade That Shaped Television News: CBS in the 1950s,* New York: Praeger, 1998

Howe, Quincy 1900–1977

U.S. Radio Commentator

In 1939 Quincy Howe began broadcasting news and commentary on radio station WQXR in New York. In 1942 he moved to the Columbia Broadcasting System (CBS), where he helped make analysis an accepted part of news reporting. He was one of the first radio journalists to bring the news of World War II into American homes.

Howe was born on 17 August 1900 in Boston, Massachusetts, to Fanny Howe and Mark Howe, who worked as an editor and a writer at the Atlantic Monthly Company. Howe attended St. George's School in Newport, Rhode Island, and then matriculated to Harvard University, from which he graduated *magna cum laude* in 1921. Howe then studied for a year at Christ's Church, Cambridge University, in England. When he returned to Boston in 1922, he became an editor for *Living Age,* a magazine published by the Atlantic Monthly Company. *Living Age* was sold in 1928, however, and Howe soon moved to New York. Archibald Watson, who had purchased the magazine, hired Howe as editor in chief in 1929. Howe selected articles that were topical and also contributed a regular column about world affairs.

Howe, a Boston liberal, married Mary L. Post, with whom he had two children. He helped get food to striking miners in Harlan County, Kentucky, in 1932. Later that year, he became the director of the American Civil Liberties Union (ACLU), which, under his guidance, opposed censorship. Howe served as director of the ACLU until 1940.

In 1934 Howe published *World Diary: 1929–1934,* an arresting study about the causes of the Great Depression as well as a prediction about the growth of nationalism. A year later he became editor in chief of Simon and Schuster. Under Howe's leadership, the company published more topical nonfiction. In 1937 he published his controversial book *England Expects Every American to Do His Duty,* in which he proposed isolationism for the United States. Howe argued that the

Quincy Howe
Courtesy CBS Photo Archive

United States should not get involved in another war merely to protect Britain's empire. Politicians on both sides of the Atlantic discussed his ideas. Two years later, he published *Blood Is Cheaper than Water: The Prudent American's Guide to Peace and War,* which examined the differences between isolationists and interventionists.

Howe gained experience on radio in 1938, when he provided analysis of the Munich Agreement for the Mutual Broadcasting System. Then, beginning in 1939, he provided three 15-minute news commentaries a week for WQXR, a radio station in New York. Howe's New England twang and educated opinions were well suited for radio, and his voice became one of the most recognizable for listeners almost immediately. Howe maintained his isolationist views until the United States entered World War II in December 1941.

Howe moved to CBS in 1942. He was hired to do news commentary, and he helped to make commentary an important ingredient of broadcast journalism. Like other notable commentators at the time, Howe wrote his own scripts, each containing about 1,500 words. These scripts contained informed opinion about newsworthy events, especially about the war in Europe. Howe was able to accomplish the task because he was a professional writer as well as an excellent speaker. He usually opened his analysis with several maxims about world affairs to make his commentaries interesting and coherent. He became known for his insightful analysis. Others who worked in radio viewed him as one of the most authoritative news analysts around, primarily because of his vast knowledge about world affairs. Howe did more than read the news: he informed his listeners as to how national and international events would affect their lives.

Under H.V. Kaltenborn's leadership, Howe and 30 other commentators based in New York helped organize the Association of Radio News Analysts in 1942. The guild advocated that commentators be permitted to comment when presenting news. In 1943 Howe wrote about the power of those who advertised on radio in "Policing the Commentator: A News Analysis," published in the November *Atlantic Monthly.* He warned that advertisers could cause news and commentary to become slanted.

When World War II ended, Howe attempted to enter television. He worked as a commentator on the CBS evening news until the network was urged by an advertiser to let him go because he was too bombastic and too liberal. Howe was dropped from the evening news, but he continued to work as a reporter and narrator of documentaries for the network. In 1948, for instance, he covered the Republican and Democratic national conventions.

In 1949 he published the first volume of his three-volume history of the 20th century, *A World History of Our Times* (the last volume appeared in 1972). Later that year he left CBS and taught journalism at the University of Illinois until 1954; he left when the American Broadcasting Companies (ABC) hired him. At ABC Howe covered world affairs. In addition, he moderated the last presidential debate between John F. Kennedy and Richard M. Nixon in 1960 as well as the trial of Adolf Eichmann in Israel a year later. He received several awards, including the George Foster Peabody Award and the Overseas Press Club Award, for his work in broadcast journalism.

In addition to his work for ABC, Howe contributed articles to the *Saturday Review of Literature* and other magazines. In 1961 he became the editor of *Atlas: The Magazine of the World Press,* a monthly that featured articles from the foreign press. He served as editor until 1965. Howe died in 1977.

EDD APPLEGATE

See also Commentators; News; Peabody Awards

Quincy Howe. Born in Boston, Massachusetts, 17 August 1900. Served in U.S. Marine Corps, SATC, Harvard Unit, 1918; attended Harvard University, A.B. degree (magna cum laude), 1921; studied at Christ's College, Cambridge University, England, 1921–22; staff member, *Atlantic Monthly*, 1922–29; editor, *Living Age*, 1929–35; editor-in-chief, Simon and Schuster, 1935–42; news commentator, WQXR, New York, 1939–42; news commentator, Columbia Broadcasting System radio and television, 1942–50; associate professor of journalism and news analyst for WILL, University of Illinois, Urbana, 1950–54; news analyst, ABC, 1954–63; editor, *Atlas: The Magazine of the World Press*, New York, 1961–65; news analyst, Radio New York Worldwide, 1966–70; commentator, WTFM, New York, 1973–74; contributing editor, *Atlas World Press Review*, New York, 1974–77. Received George Foster Peabody award, 1955; Overseas Press Club award, 1959; Columbia-Catherwood award, 1962. Died in New York City, 17 February 1977.

Selected Publications

World Diary: 1929–1934, 1934
England Expects Every American to Do His Duty, 1937
Blood Is Cheaper than Water: The Prudent American's Guide to Peace and War, 1939
The News and How to Understand It, in Spite of the Newspapers, in Spite of the Magazines and in Spite of the Radio, 1940, 1968
A World History of Our Times, 3 vols., 1949–72

Further Reading

DeLong, Thomas A., *Radio Stars: An Illustrated Biographical Dictionary of 953 Performers, 1920 through 1960*, Jefferson, North Carolina: McFarland, 1996
Fang, Irving E., *Those Radio Commentators!* Ames: Iowa State University Press, 1977
Howe, Quincy, "Policing the Commentator: A News Analysis," *Atlantic Monthly* 172, no. 5 (November 1943)
Sperber, Ann M., *Murrow: His Life and Times*, New York: Freundlich Books, 1986
Tebbel, John William, *Between Covers: The Rise and Transformation of Book Publishing in America*, New York: Oxford University Press, 1987

Hulbert, Maurice "Hot Rod" Jr. 1916–1996

U.S. Disc Jockey

Maurice "Hot Rod" Hulbert Jr. was one of the most popular black disc jockeys in radio during the 1950s and 1960s. For a period during his heyday, he could be heard hosting different programs on radio stations in Baltimore, Philadelphia, and New York. His fluid and at times nonsensical on-air patter and high profile in the black broadcasting world helped him inspire a generation of black disc jockeys who came of age in the late 1950s and the 1960s.

Hulbert was also one of America's pioneering black disc jockeys, joining WDIA in Memphis, Tennessee, as it was evolving into the nation's first all-black-oriented radio station. Before arriving at WDIA in 1949, Hulbert had worked as a dancer, comedian, bandleader, and emcee in various mid-South nightclubs and traveling tent shows. Hulbert also helped produce musicals with black students in the Memphis school system, and it was while working on one such production that WDIA's general manager, Bert Ferguson, approached him about working at the station. The meeting led to a job, and soon Hulbert was hosting three shows for WDIA: *The Sepia Swing Club*, an afternoon blues and jazz show; *The Delta Melodies*, an early-morning program of spiritual music; and *Moods by Maurice*, a midmorning program tailored to housewives.

At WDIA Hulbert became increasingly adept at switching personas as each show required. On his *Sepia Swing Club*, he became "Hot Rod," describing for listeners his rocket ship on which they would be flown through a solar system of hot music. Listeners to the *Sepia Swing Club* responded enthusiastically, as did those who where charmed by the suave "Maurice the Mood Man" on the *Moods by Maurice* program. By 1951 Hot Rod prepared to board his rocket ship for a galaxy in a larger market called Baltimore.

The management of WITH in Baltimore had cast its net for a popular established black disc jockey to pull in black listeners, and after a national search, they had located Hulbert, whose high ratings in Memphis confirmed his popularity. WITH hired him in 1951, making him the first full-time black disc jockey in Baltimore history. The Hot Rod–rocket man persona followed Hulbert to Baltimore, and almost immediately the rhythm and blues music shows he hosted became popular among young blacks and whites. Listeners enjoyed Hot Rod's outer-space persona as well as the tongue twisters and slick shtick that he rained on Baltimore. They had never heard such on-air wildness before. With good-natured bravado, he might greet the city with, "Not the flower, not the root, but the seed, sometimes called the herb, not the imitator but the originator, the innovator, the true living legend—The Rod!" In addition, he peppered his on-air patter with extra, nonsensical sounds, for example, *ee-us,* as in "This is Hee-us-ot Ree-us-od." This jive complemented the music he played and tickled the ears of his listeners. He ended many programs and pronouncements by proclaiming "VOSA," which meant the "Voice of Sound Advice."

By the late 1950s, Hot Rod moved on to WHAT in Philadelphia where, in his morning time slot, he repeated his Baltimore success. It was in Philadelphia that he became a three-market personality. WWRL in New York, hoping to grab some of Hulbert's luster, paid him to commute to Manhattan for an afternoon show, and then WWIN in Baltimore asked him to tape a program for broadcast there. Few, if any, black disc jockeys enjoyed such exposure in the 1950s and 1960s.

Hot Rod was back in Baltimore exclusively by the late 1960s, hosting a popular live show, first on his old home station WITH and then on WWIN. Hulbert's popularity, particularly among Baltimore's black audiences, was never more evident than during the riots that plagued the city in the wake of Martin Luther King Jr.'s 1968 assassination. Government officials asked the popular local figure to go on television to help calm the violence and tension. "I could get through," he told author Gilbert Williams in 1991, "and I talked to the people, trying to sober them up because people had gone mad. . . . I think many jocks did that all over the country." Hulbert frequently spoke out in advocacy of civil rights for blacks, marching in the streets and discussing related issues on the air.

In the 1970s, Hulbert switched to the sales and management side of radio, working for various Baltimore radio stations. By the time of his retirement in 1993 he was the general manager of WBGR/WEBB. He died in 1996 after a battle with throat cancer.

MICHAEL STREISSGUTH

See also Black-Oriented Radio; Disk Jockeys; WDIA

Maurice Hulbert, Jr. Born in Helena, Arkansas, 30 July 1916. Senior disc jockey on several radio stations, including WDIA, Memphis, Tennessee, 1949–51; became first full-time African-American disc jockey, WITH, Baltimore, Maryland, 1951–late 50s; simultaneous disc jockey positions, WWIN (Baltimore, Maryland), WHAT (Philadelphia, Pennsylvania), WWRL (New York City), 1950s–60s; disc jockey, WITH, then WWIN, Baltimore, Maryland, late 1960s; worked in sales and management for various Baltimore radio stations, including WKTK, WWIN, and WBGR\WEBB, 1974–93; became general manager of WGBR/WEBB, Baltimore, Maryland, 1993. Died in Towson, Maryland, 24 December 1996.

Further Reading

Cantor, Louis J., *Wheelin' on Beale: How WDIA-Memphis Became the Nation's First All-Black Radio Station and Created the Sound That Changed America,* New York: Pharos Books, 1992

Dates, Jannette Lake, and William Barlow, editors, *Split Image: African Americans in the Mass Media,* Washington, D.C.: Howard University Press, 1990; 2nd edition, 1993

Newman, Mark, *Entrepreneurs of Profit and Pride: From Black-Appeal to Radio Soul,* New York: Praeger, 1988; London: Praeger, 1989

Streissguth, Michael James, "WDIA and the Rise of Blues and Rhythm-and-Blues Music," Master's thesis, Purdue University, 1990

Williams, Gilbert Anthony, *Legendary Pioneers of Black Radio,* Westport, Connecticut: Praeger, 1998

Hummert, Anne 1905–1996
Hummert, Frank 1885–1966

U.S. Writers and Producers, Creators of Soap Operas

Anne and Frank Hummert nearly monopolized the creation of radio daytime serials—soap operas—in the 1930s and 1940s. Working within an advertising agency, they took advantage of the advertiser's interest, the new medium, the available audience, and the era. The Hummerts were not the creators of the first soap opera, a form that evolved over a few years, but they were responsible for molding the genre and moving soap operas from evening to daytime, convincing advertisers that women could listen to radio while they were doing their housework. The Hummerts provided programming that included advertising and propaganda (the latter including messages supporting government policies during World War II) plus entertainment for as much as half of daytime network; they also originated sponsored children's programming and many other programs.

Despite speaking to so many people for decades, the Hummerts maintained a very private, secluded life, leaving few biographical details. Frank is reputed to have been a Texas Ranger, a reporter on the *St. Louis Dispatch,* a manager of a writing school, and then a highly regarded copywriter in New York City. He is credited with coming up with the idea of writing advertising as feature news, having written the slogan "Bonds or Bondage" during World War I and Camay's slogan, "For the skin you love to touch." Blackett and Sample lured Frank Hummert away from the Lord and Thomas agency in New York to Chicago with the opportunity to set up his own radio production unit in 1927. Although Hummert was not a partner in the agency, his name was put on the masthead and he retained ownership of all the programs he produced. Hill Blackett and John Glen Sample were interested in attracting clients by providing programming for women during the day, and to get their desired audience's perspective they encouraged Frank Hummert to hire a woman. Anne S. Ashenhurst had just come from a newspaper job with the Paris *Herald,* had a young son, and was looking for a job. Frank reluctantly hired her in 1927. Anne was a very deft writer and was able to communicate with Frank. They married after seven years, and were partners in producing serials for three decades.

After Pepsodent's success with the daily 15-minute program *Amos 'n' Andy,* other writers attempted to produce popular evening serials. As advertisers, Frank and Anne Hummert had program ideas for their clients and hired their writers. It was one way both to control the content and also to produce many hours. In 1931 Anne Ashenhurst and Frank Hummert hired Charles Andrews to write *The Stolen Husband,* a very simple program that was then adapted to become *Betty and Bob.* Their first success was *Just Plain Bill* in 1932, also written by Andrews, and Anne convinced the sponsors to move it to daytime in 1933, arguing that women would be able to listen while doing their chores. Within a few years the Hummerts had as many as 18 daytime serial dramas on the air at the same time, bringing in as much as one-half of network daytime revenues.

The Hummerts used a technique that many have called a soap opera factory to produce the programs. By 1935 the Hummerts had moved their agency to New York City and had a staff of about 20 writers, 6 editors, and 60 clerical workers. Anne and Frank worked out of their house in Connecticut, first creating the title and the rough summary for each show. The staff in the office expanded this, and then the Hummerts created the story line, a sketch of the action for five to six episodes. The Hummerts then sent the theme and story line to five writers, called dialoguers, who would produce sample scripts, with the best script winning the writing job. The writer would then have to stay at least three weeks ahead at all times. The Hummerts would relay any ideas they had for character and plot development to the writer through one of the six editors—all in writing. By 1938 more than 5 million words were being written annually for the Hummert serials alone, the equivalent of 50 full-length novels.

Their control is documented by noting the copy on the first page of all scripts, written in lavender ink: the title followed by the notation that this is a Hummert Radio Feature. The Hummerts gave themselves the credits for the title, original story line, the general supervision of script and production, and ownership. They had a set of rules that covered every eventuality. Nothing could be added in production that was not in the script: no extra sound effects, lest the dialogue not be heard; no overlapping of speeches; and the actors had to have the clearest enunciation—a trait Anne Hummert had herself.

Most of the 36 daytime serials produced by the Hummerts were melodramas about domestic life that moved very slowly and included lengthy advertisements. The Hummerts described them as "successful dramas about unsuccessful people, people who were not wealthy but had successful family lives and were able to help others have good relationships." The themes were often based on relationships of people from different backgrounds, worried about their future happiness. As Anne later said, "Worry, for women, is entertainment. . . . Nobody can understand the phenomenal success of the soaps without knowing when they were born. It was during the Depression. The housewife was at home worrying about everything. Would her husband lose his job? Where was the family's next meal coming from? They found escape in the lives of the people on the soaps."

The format for most of their daytime serials started with a lead-in delivered by the announcer; for example, for *Our Gal Sunday,* the lead-in was, "Can a girl from a little mining town in the West find happiness as the wife of England's richest, most handsome lord?" with "Red River Valley" as theme music. Then the announcer read a chatty commercial for a minute and a half and gave background for this particular program, followed by nine minutes of dialogue, a few leading questions for the next day's show, and a closing commercial. The commercials were written in the problem-solution formula—for example, try new improved Oxydol to make washing easier. Premium promotions were also written in as a means of proving there were large audiences for the advertisers. The offer of a Love Bird Pin just like Helen Trent's with "real simulated-gold flashing," a lavaliere designed by Mary Noble, or a can opener said to have been invented by Lorenzo Jones were written into the scripts months ahead. The audience was asked to send in a dime with a box top or some other proof of purchase of one of the advertised products to receive

Anne and Frank Hummert
Courtesy CBS Photo Archive

the premium. During World War II, at the request of the Office of War Information, the Hummerts wrote their dramas to help overcome the white soldier's fear of the black soldier and to help the war effort in general. After the war, the Hummerts were two of the few writers who never flinched during the McCarthy Era, continuing their work without any concern about the blacklist.

Some say that the reason the Hummerts led such a secretive life was that they were aware of the disdain held by much of the public for their work. Anne did admit that her son disapproved of her work and told a reporter, "As a matter of fact I sit behind my desk with two black eyes." But they did feel they were writing for audiences all over the country. The Hummerts also produced musical programs and crime dramas and were the first to convince advertisers to sponsor children's programming.

When the networks cut back on programming and the soaps were taken off the air in the early 1950s, the Hummerts retired to travel rather than taking their soaps to television.

MARGOT HARDENBERGH

See also Ma Perkins; Premiums; Soap Opera

Anne Hummert. Born in 1905. Graduated from Goucher College, Baltimore, Maryland, 1925; married John Ashenhurst, 1926–27; writer, Paris *Herald*, 1926; writer, Blackett, Sample and Hummert advertising agency, 1927; married Frank Hummert, 1934; co-producer of 36 radio serials. Died in New York City, 5 July 1996.

Frank Hummert. Born circa 1885. Reporter, *St. Louis Dispatch;* copywriter, Chicago, 1920s; worked in New York

City for Lord and Thomas agency; writer for Blackett and Sample when he hired Anne as his assistant; firm became Blackett-Sample-Hummert, although Hummert had no interest in the firm; married Anne Ashenhurst, 1934; started Air Features, a radio production company, where he and Anne produced more than 35 different radio series. Died in New York City, 1966.

Radio Series

1931–51 *American Album of Familiar Music*
1932 *The Stolen Husband*
1932–35 *Skippy*
1932–40 *Betty and Bob*
1932–49 *Manhattan Merry-Go-Round*
1932–55 *Just Plain Bill; Judy and Jane*
1933–48 *Waltz Time*
1933–60 *Ma Perkins; The Romance of Helen Trent*
1934–36 *Lavender and Old Lace*
1935–38 *Mrs. Wiggs of the Cabbage Patch*
1935–59 *Backstage Wife*
1936–37 *Rich Man's Darling*
1936–41 *John's Other Wife*
1936–51 *David Harum*
1937–42 *Arnold Grimm's Daughter*
1937–46 *Second Husband*
1937–55 *Lorenzo Jones; Mr. Keen, Tracer of Lost Persons; Stella Dallas*
1937–59 *Our Gal Sunday*
1938–39 *Alias Jimmy Valentine; Central City; Those Happy Gilmans*
1938–56 *Young Widder Brown*
1939–42 *Orphans of Divorce*
1940–46 *Amanda of Honey Moon Hill*
1941–44 *Helpmate*
1941–48 *American Melody Hour*
1941–54 *Front Page Farrell*
1943–50 *Lora Lawton*
1944–48 *The Strange Romance of Evelyn Winters*
1948–51 *Molle Mystery Theater* (also produced as *Mystery Theater,* with spin offs known as *Mark Sabre* and *Hearthstone of the Death Squad)*
1950–51 *Nona from Nowhere*

Further Reading

Cox, Jim, *Frank and Anne Hummert's Radio Factory: The Programs and Personalities of Broadcasting's Most Prolific Producers,* Jefferson, North Carolina: McFarland, 2003
Edmondson, Madeleine, and David Rounds, *From Mary Noble to Mary Hartman: The Complete Soap Opera Book,* New York: Stein and Day, 1976
Meyers, Cynthia B., "Frank and Anne Hummert's Soap Opera Empire: 'Reason-Why' Advertising Strategies in Early Radio Programming," *Quarterly Review of Film and Video* 16, no. 2 (1997)
Nachman, Gerald, *Raised on Radio: In Quest of the Lone Ranger, Jack Benny . . . ,* New York: Pantheon, 1998
Stedman, Raymond William, *The Serials: Suspense and Drama by Installment,* Norman: University of Oklahoma Press, 1971; 2nd edition, 1977
Thurber, James, "Soapland," in *The Beast in Me and Other Animals,* by Thurber, New York: Harcourt Brace, 1948; London: Hamish Hamilton, 1949

I

I Love a Mystery

Adventure/Mystery Thriller Series

Though relatively short-lived (five years in its original run with a three-year revival based on the original scripts), *I Love a Mystery* (*ILAM*) continued its hold on radio aficionados for several decades after it aired. This was due in part to its creator and writer—Carleton E. Morse—but also to the wide-ranging nature of the adventures of the three key characters.

The Radio Serial

A serial with dozens of continuing stories that were usually presented in three-week units for a total of 1,784 episodes, the program varied from 15-minute to half-hour segments depending on the network carrying it. The program was more of an adventure/thriller than a classic detective story, despite its detective agency basis. *ILAM* originated in Hollywood for its original five-year run, moving to New York when the Mutual Broadcast System reused all but five of the original scripts (and added one new script) with a new cast. Adding to later collector confusion, the Mutual series often used different story titles. Though audition tapes were made in 1954 for a revival on CBS, that series never materialized.

In the program, Jack Packard, a one-time medical student, is head of the Triple A-1 Detective Agency, located "just off Hollywood Boulevard and one flight up," whose motto is "no job too tough, no mystery too baffling." At 37, he is older than the other staff members and is clearly the most cool-headed and clear thinking under pressure. His fellow-adventurers include the Texas-born roughneck "Doc" Long, who loves women and adventure in about that order. The third member of the original trio is Britisher Reggie Yorke, who is refined but also serves as the group's muscle. These original protagonists met in China while fighting the Japanese, and they took over an abandoned detective agency on returning to the Untied States.

Yorke was written out of the series in 1942 when the actor portraying him took his own life. His character was replaced with distaff interest in the form of handsome secretary Jerry Booker. When she joins the WACs during World War II, her secretarial role is taken on by Mary Kay Brown.

Carlton E. Morse, writer-producer of *I Love A Mystery,* working with soundperson
Courtesy CBS Photo Archive

The programs concerned exotic adventures, and while they sometimes had far-fetched aspects, the resolution of the stories was always rational and realistic. Each segment ended with a cliff-hanger situation designed to bring listeners back regularly. Unlike many serials, a given ILAM story ended before another began.

ILAM in Other Media

Three movies (only the first with a script by Morse) were developed from the series, and a 1967 television pilot film, *I Love a Mystery,* was made though not shown until 1973. No series resulted. An earlier (1956) attempt to develop a television series had also failed.

Don Sherwood created a short-lived 1960s comic strip based on the stories and characters of the radio series. *ILAM* creator Carleton E. Morse wrote one related novel and published it before his death; others were planned but did not appear.

CHRISTOPHER H. STERLING

See also One Man's Family

Programming History

NBC West Coast network	January 1939–September 1939
NBC	1939–40
Blue Network	1940–42
CBS	1943–44
Mutual	1949–52

Cast

Jack Packard	Michael Raffetto (1939–44), Russell Thorson (1949–52), Robert Dryden (1952)
Doc Long	Barton Yarborough (1939–44), Jim Bowles (1949–52)
Reggie Yorke	Walter Paterson (to 1942), Tony Randall (1949–52)
Jerry Booker	Gloria Blondell (after 1942), Athena Lord (1949–52)
Mary Kay Brown	Athena Lord (1949–52)

Creator-Writer-Producer-Director

Carleton E. Morse

Films Based on the Series

I Love a Mystery (1945)
The Devil's Mask (1946)
The Unknown (1946)

Further Reading

Grams, Martin, Jr., "Episode Listing of *I Love a Mystery,*" in *Radio Drama: A Comprehensive Chronicle of American Network Programs, 1932–1962,* by Grams, Jefferson, North Carolina: McFarland, 2000

Harmon, Jim, "Jack, Doc, and Reggie," in *The Great Radio Heroes,* by Harmon, Garden City, New York: Doubleday, 1967

Morse, Carleton E., *Stuff the Lady's Hatbox,* Woodside, California: Seven Stones Press, 1988

Packard, Jack, Doc Long, and Reggie York, *I Love a Mystery: The Further Adventures of Jack, Doc, and Reggie,* <www.lofcom.com/nostalgia/ilam>

Unofficial I Love a Mystery Page, <www.angelfire.com/on/ilam>

Imus, Don 1940–

U.S. Radio Disc Jockey and Host

Don Imus moved to WNBC New York in 1971 after just three years in radio. After working at stations in California (Palmdale, Stockton, Sacramento), as well as in Cleveland, Ohio, Imus came to the largest market in the U.S. and to the station with the best combination of a low frequency and high power—arguably the biggest station in the United States. Thirty years later, Imus is still at WNBC.

Origins and Early Radio Years

John Donald Imus, Jr., was born 23 July 1940 in Riverside, California, and lived with his family in the nearby town of Perris. While in high school, he lived briefly in Scottsdale and then Prescott, Arizona. Imus wanted to be a popular singer and participated in theater, but he joined the Marine Corps before his last year of high school. After two years in the military and the failure of a rock band formed with his brother Fred, he worked as a miner in Arizona and for a railroad in California. Using money from an injury lawsuit (and/or the GI bill), he attended the Don Martin School of Broadcasting in Hollywood—then a well-established trade school known for turning out a number of announcers and disc jockeys.

Before graduation, and apparently still owing the school tuition money—which he says he has never paid—Imus got a job at a Palmdale station in the high desert north of Los Angeles. Like most morning DJs, he got attention with stunts like saying he was running for congress in 1968 and holding a press conference—"Put Don Imus on the gravy train." He quickly moved to KJOI in Stockton, where he apparently completed one course in political science at San Joaquin Delta College. Then he moved to KXOA in Sacramento, began calling his show *Imus in the Morning,* and created outrageous characters such as "Judge Hanging," "The Reverend Billy Sol Hargis," and "Crazy Bob," who presented his own suggestive versions of fairy tales. Imus often used phone calls in his stunts—such as calling a local McDonald's one morning, identifying himself as a sergeant in the Air National Guard and ordering 1,200 hamburgers to go. Then he confused the manager by specifying, "on 300 hold the mustard but put on plenty of mayonnaise and lettuce."

Imus began broadcasting in Cleveland in September 1970 and was named by *Billboard* as "the major market DJ of the year." For many of his more memorable bits, he called people on the phone: asking to buy silver bullets as the Lone Ranger, trying to order a rental car to race in the Indianapolis 500, and saying that he had left his clothes in a hotel phone booth after changing from Clark Kent to Superman and asking that they be returned. Often he talked with women whom he told to "get naked."

One character, The Reverend Billy Sol Hargis, who sounded much like Billy Graham, was from "The First Church of the Gooey Death and Discount House of Worship, right here in Del Rio, Texas," and sang "I don't care if it rain or freezes, long as I've got my plastic Jesus riding on the dash board of my car, I can go a 100 miles an hour as long as I got the almighty dollar glued up there by my pair of fuzzy dice."

Imitating President Lyndon Johnson, Imus explained "Why shouldn't the Viet Nam war costs 150 billion dollars?" with: "Let Judge Hanging remind you, my fellow Americans, that you pay for what you get. You don't run down to Sak's Fifth Avenue and pick up some slick suit for $29.95 and you don't wage war for 15 cents. It cost money to dress well and it cost money to kill people."

An Imus character called "Tricky Dick" sold used cars from 1600 Pennsylvania Avenue. Like many other radio comedians, Imus crafted characters that sounded a bit like real life radio

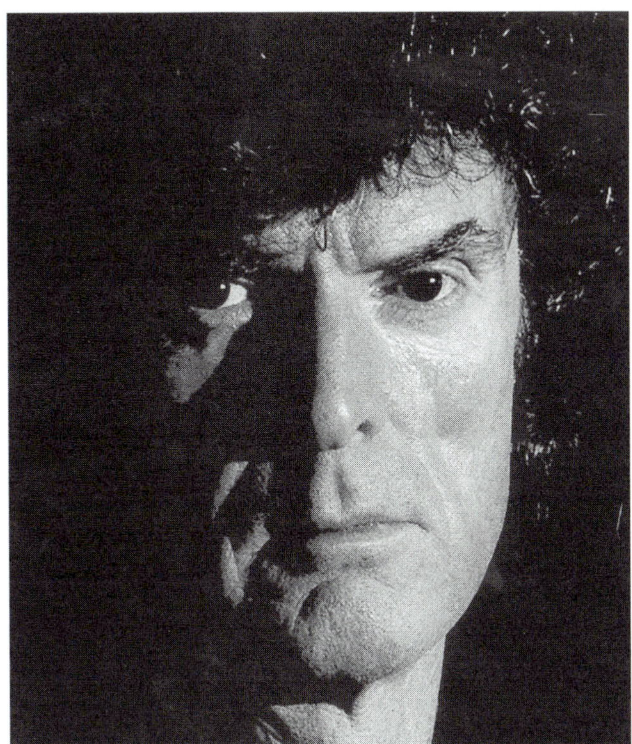

Don Imus
Courtesy of Don Imus

personalities of his youth. Certainly anyone hearing Rev. Hargis was instantly reminded of various paid religious programs from stations across the Mexican border that Imus had heard as a young man in Arizona or while in the Marines at Camp Pendelton.

Cleveland newspaper reviews called Imus "gross, tasteless, vulgar, and nauseating." But in just a few months the station ranked number one in ratings. In December 1971, Imus was hired by WNBC in New York, which then was an adult contemporary to middle-of-the-road music station.

New York Radio

The transition was not easy. Imus missed many days of work—an average of two days a week in 1973—and later admitted to problems with alcohol and cocaine. Later he said: "I was drunk and coked up for 20 years, it's a wonder I'm even alive." In a 1976 interview Imus explained, "I have an aversion to authority, some kind of immaturity . . . it's just the way I am." His appeal, he assumed, was to those many people who rose each morning thinking that they would like to "take their job and shove it," as a song of the time put it.

While insisting that he had 19 lawyers to protect him against his employers, Imus said that he thought he could find a job in radio anywhere. But he was fired from WNBC at the end of August 1977 and remained out of work for some time, eventually ending up back in Cleveland. The shooting star seemed to have fizzled.

Imus did return to WNBC in 1979, but there was more drinking and drugs. In 1985 he was one of the first video jockeys on the cable program service VH1. In 1988 WNBC, with the new call letters of WFAN, adopted a sports talk format, and Imus continued to produce about one half of the station's revenues during his morning shift. He adopted a more mature tone, which better fit with the station's new format and with the fact that fewer AM stations were playing music, but his trend to more serious topics, he says, was accelerated by the Gulf War controversy beginning in the fall of 1990. Now more and more of his phone conversations were with journalists and politicians, the latter particularly during campaigns.

Beginning in July 1993 the program was also syndicated, and by 2001 it was carried on about 55 stations. Since 3 September 1996, it has also been carried on MSNBC and was said at times to be the highest-rated program on the cable channel. While MSNBC uses multiple cameras and adds many graphics and tape clips, Imus insists that he is still doing a radio show and usually ignores or disparages the video coverage.

In 1981 Imus and Charles McCord, the newsman on his program, published a novel, *God's Other Son,* but it did not sell well. Reissued in 1994, it became a best seller. In 1997 Imus also published, with his brother Fred, a book of photographs of the American Southwest.

Yet another controversy involved remarks he made 21 March 1996, at the Radio and Television Correspondents dinner in Washington. Beforehand he had apparently told Bill Clinton it would be a little "rough" and the President said that would be "fine." The President also said that appearing on the Imus show might have helped him get elected. However, as Imus spoke of Clinton's legal problems, his family, and other matters, it was clear from the telecast and from later news reports that the President was angry.

In 1997 *Time* magazine named Imus one of its "most influential" people, and *Newsweek* did a cover story on him in December 1998. Since 1997, he has also operated a ranch in New Mexico for children with cancer. His program is frequently broadcast from the ranch, especially during the summer when the children visit, and fund raising for the ranch is often discussed.

Since the beginning of his radio career in 1968, Imus' strong opinions and remarks about people in public life have been the main attraction for his listeners. Here is a recent sampling:

> Bill Clinton: "is a dirt bag and a low rent weasel."
> Al Gore: "phoniest person on the planet, disgraceful human being."
> George W. Bush: "we're going to be stuck with that moron George Bush [and] Cheney and a bunch of old people will run the country."
> Chris Matthews (who is also on MSNBC): "a blow hard and the most annoying person on television."
> XFL football (carried on channels owned by Viacom which also owns WFAN): "The fans are neanderthal morons. It is lame. It really sucks."

"I'm trying to be entertaining, I'm trying to show a different side of these people. We are not trying to hurt people. We are trying to make people laugh."

LAWRENCE W. LICHTY

John Donald Imus, Jr. Born in Riverside, California, 23 July 1940. Quit high school, but later earned GED. U.S. Marine Corps 1957–1959. Worked at radio stations in California before moving to Cleveland, Ohio, 1970; on-air host at WNBC (later WFAN), New York, 1971–77, 1979–present.

Radio Series

1971–77; 1979–present *Imus in the Morning*

Selected Publications

God's Other Son: The Life and Times of the Rev. Billy Sol Hargus, 1981

Two Guys, Four Corners: Great Photographs, Great Times, and a Million Laughs (with Fred Imus), 1997

Further Reading

Tracy, Kathleen, *Imus: America's Cowboy,* New York: Carroll and Graf, 1999

Indecency. *See* Obscenity/Indecency on Radio

India. *See* All India Radio

Infinity Broadcasting Corporation

Infinity Broadcasting Corporation, a subsidiary of media giant Viacom, is one of the largest radio broadcasting companies in the U.S. Infinity is focused on the "out-of-home" media business, which includes operations in radio broadcasting through Infinity Radio and outdoor advertising through Viacom Outdoor. Infinity's self characterization of being an "out-of-home" media business comes from the fact that the majority of radio listening and practically all viewing of outdoor advertising occurs outside the consumer's home, from places such as automobiles and public transportation systems. The majority of Infinity's revenue, therefore, is generated from the sale of advertising. Infinity Radio consists of more than 180 radio stations serving over 40 markets. Approximately 94 percent of Infinity's radio stations are located in the 50 largest U.S. radio markets. Infinity also manages and holds an equity position in Westwood One, Inc.

The original Infinity Broadcasting Corporation is not the same as the Infinity currently in existence. The original Infinity was formed by two former Metromedia Communications Corporation executives, Gerald Carrus and Michael A. Weiner, in 1972 and acquired its first radio station in May 1973. Carrus and Weiner planned to emulate Metromedia president John W. Kluge's strategy of acquiring unsuccessful radio stations in the country's largest media markets, where the greatest amount of radio advertising dollars are spent, and developing them.

Seeking someone to run the original Infinity, Carrus and Weiner turned to Mel Karmazin in 1981. Karmazin had spent the previous 11 years working for Metromedia, where he managed the company's AM and FM outlets and gained a reputation for paying substantial amounts of money for on-air talent while exercising the tightfistedness he had learned from Kluge to keep operating costs down. When Karmazin requested the opportunity to manage one of Metromedia's TV properties, Kluge turned him down, and Karmazin began to consider other options. By offering him a lucrative salary and equity in the original Infinity, Carrus and Weiner were able to lure Karmazin to the company.

The original Infinity, under Karmazin's leadership, substantially increased its acquisitions by paying record prices for top radio stations in large cities. The success of the original Infinity was also based on its ability to acquire the radio broadcast rights to a number of professional sports teams and to seek out high-profile radio personalities for its stations.

One of those high-profile personalities was "shock jock" Howard Stern, who signed on with the original Infinity in 1985 after being fired from WNBC. The original Infinity provided Stern with a national platform. As a result of Stern's bold activities on the airwaves, the company received numerous warnings from the Federal Communications Commission (FCC) that the *Howard Stern Show* was dangerously close to violating indecency standards. The warnings brought substantial publicity to the original Infinity, and the show's ratings soared, to the dismay of the many national and community watchdog groups working to have the show taken off the air. In 1995, as the fines from the FCC escalated, the original Infinity agreed to put the controversy to rest by paying $1.7 million in exchange for the FCC's dismissal of all pending complaints against the company's stations.

Despite the controversy, the original Infinity became popular not only with the listening audience, but also with Wall Street. The original Infinity went public in 1986 and was then bought back in a leveraged buyout in 1988. The original Infinity was again taken public in 1992. Shares issued in 1992 for $17.50 each were worth $170 when the company was eventually purchased in 1996.

In November 1995 Westinghouse Electric Corporation acquired the Columbia Broadcasting System (CBS), creating the nation's largest TV and radio station group. After Congress passed the Telecommunications Act in February 1996, which permitted the expansion of TV and radio station holdings, Westinghouse began considering its options for growth and, recognizing the potential of the radio industry, purchased the original Infinity in December 1996 for $4.7 billion. After acquiring the original Infinity, Westinghouse decided to sell its industrial businesses and reinvent itself solely as a media company. In December 1997 the new media company was launched as the CBS Corporation, the largest radio and television entity in history, with Karmazin as its president and chief operating officer.

In September 1998 CBS formed the currently existing Infinity Broadcasting Corporation in the largest initial public offering in media history. The "new" Infinity was created as a wholly owned subsidiary to own and operate CBS's radio and outdoor-advertising business. In addition to his duties at CBS, Karmazin was named president and chief executive officer of Infinity. The name of the original Infinity had been changed to Infinity Media Corporation in anticipation of the creation of the new corporation with the same name.

In May 2000 CBS was merged with and into Viacom, Inc. with Farid Suleman becoming Infinity's Chief Executive Officer. In March 2002 John Sykes replaced Suleman as Chief Executive Officer. The company is headquartered in New York City. Among its stations are WFAN-AM in New York (sports format), which is the nation's top-billing station in terms of advertising dollars, and KDKA-AM in Pittsburgh (news/talk format), the oldest regular radio broadcaster in America. In addition to Stern, notable personalities at Infinity include Don Imus, Casey Kasem, Larry King, Charles Osgood, and Tom Snyder.

KARL SCHMID

See also Columbia Broadcasting System; Karmazin, Mel; Westinghouse

Further Reading

Douglas, Susan Jeanne, *Listening In: Radio and the American Imagination: From Amos 'n' Andy and Edward R. Murrow to Wolfman Jack and Howard Stern,* New York: Times Books, 1999

Quaal, Ward L., and Leo A. Martin, *Broadcast Management: Radio, Television,* New York: Hastings House, 1968; revised 2nd edition, by Quaal and James Anthony Brown, 1976; 3rd edition, as *Radio-Television-Cable Management,* by Quaal and Brown, New York: McGraw Hill, 1998

Smith, F. Leslie, *Perspectives on Radio and Television: An Introduction to Broadcasting in the United States,* New York: Harper and Row, 1979; 4th edition, as *Perspectives on Radio and Television: Telecommunication in the United States,* by Smith, John W. Wright II, and David H. Ostroff, Mahwah, New Jersey: Erlbaum, 1998

Inner Sanctum Mysteries

Horror Series

> *Squeeeeeaakkkk!!!!!!* "Good Evening, friends of the inner sanctum. This is your host, Raymond, to welcome you through the squeaking door. Been shopping around for a nice case of murder? Of course you have. And you have come to the right place because the characters on this program simply *kill* themselves to keep you amused. Why only the other day we were accused of making murder our business. But we wouldn't do that friends, oh no, because that would be mixing business with pleasure, and we consider it a *pleasure* to give some stiff the *business* . . . heh heh heh."

So began one of the most famous openings in radio history. The squeaking door and host Raymond's gallows humor marked *Inner Sanctum Mysteries* as a distinctively campy horror series that reveled in its grisly subject matter. It was one of the first and most successful of radio thriller dramas, a genre that peaked in popularity during the 1940s. *Inner Sanctum* was created, produced, and directed throughout its entire run from 1941 to 1952 by Himan Brown, one of radio's most prolific showmen. Brown balanced the program's macabre humor with carefully chosen organ sounds, blood-curdling screams, and other effects, creating some of the most unsettling soundscapes ever heard on radio. Brown used the organ to heighten the listener's fear, incorporating sharp stings to spark terror and creating suspense by using what John Dunning has called "doom chords" to signal approaching trouble. Murders were conveyed in the most disturbing manner possible through sound effects: Jim Harmon notes that when Brown wanted to produce the sound of a head being bashed in, he "devised a special bludgeon with which he would strike a small melon" (1967).

Inner Sanctum took its name from a line of Simon and Schuster mystery novels, but its scripts were generally original (although Edgar Allen Poe's work was a favorite of Brown's and was frequently adapted). Like other programs of the genre, *Inner Sanctum* relied on realism to heighten the listener's fear that "this could happen to me!" Ghostly behavior was commonly explained by the presence of a mad relative or an actual dead body that refused to stay quiet. As critics and even the show's own writers have noted, however, *Inner Sanctum*'s plots were driven by contrivances and coincidences that were highly implausible. The nurse hired by the judge's wife happens to be the girlfriend of the murderer the judge just sent to the gallows (and she's not happy with him!). The wailing of a man's dead wife that haunts him for 40 years is actually caused by a hole in the wall in which he entombed her body (and which he, so terrified, had never thought to investigate earlier). Frequently, the program employed the device of an insane narrator to throw listeners off track and increase their horror at identifying with a murderer. The violence and gore of the program occasionally got Brown into trouble with parents and with the Federal Communications Commission, who were particularly concerned that youth, especially, might be unduly traumatized and might even pick up a thing or two about how to carry out a murder. Brown himself was proud of the fact that "[s]hrinks said [the program] was scaring people out of their wits."

Like other programs in the genre, *Inner Sanctum* stories were a counterpoint (some might even say an antidote) to the suburban ideal of the postwar period. Husbands and wives did not get along well in *Inner Sanctum* stories, which were replete with film noir-type characters (including a healthy number of femme fatales) who murdered each other at terrific rates. Titles such as "Til Death Do Us Part," "Til the Day I Kill You," "Last Time I Killed Her," and "Honeymoon with Death" give some sense of the program's portrayal of marriage. Host Raymond took great glee in the violent disintegration of the postwar family and the impossibility of happy coupling; his closing puns or rhymes commented approvingly on the evening's grim outcome: "He hid her body in a bell, and that's where he made his mistake because she *tolled* on him." "Never tangle with a girl with red hair," he would chuckle, "A man is safer in the electric chair" (which, of course, is where this particular man ended up). The trademark tongue-in-rotting-cheek humor of the program is perhaps best conveyed by some of its more amusing titles, including "Hell Is Where You Find it," "The Dead Want Company," "Death Has a Vacancy," "The Meek Die Slowly," "The Girl and the Gallows," "Death Is a Double-Crosser," "The Long Wait Is Over," "The Man on the Slab," "Ring Around the Morgue," "Corpse on the Town," "Corpse without a Conscience," "The Corpse Who Came to Dinner," "Blood Relative," "One Coffin Too Many," and "The Corpse Nobody Loved."

Screen horror great Boris Karloff was the program's regular star for much of its first season, appearing in the Poe classics "The Tell-Tale Heart" and "The Fall of the House of Usher." According to radio program historian John Dunning, Karloff wanted more gore than the networks would allow, and he appeared much less frequently thereafter. *Inner Sanctum* developed its own stable of stars, which included Larry Haines, Mason Adams, Alice Rhinehart, Everett Sloane, Santos Ortega, Lawson Zerbe, and Elspeth Eric. In addition, up-and-coming film stars such as Mercedes McCambridge and Richard Widmark made frequent guest appearances. Although the programs emphasized plot over character, the alternately haunted and psychotic characters gave the actors a chance to stretch their range; women especially got the rare opportunity to narrate stories and play some very unladylike people.

Paul McGrath replaced Raymond Edward Johnson as host in 1945; he set a lighter tone than his predecessor, but the substance of the programs remained the same. Himan Brown attempted to revive the program in other forms in 1959 (as the *NBC Radio Theatre*) and 1974 (as the *CBS Radio Mystery Theatre*), but neither version proved as successful as the original. Of the more than 500 programs that were produced of *Inner Sanctum Mysteries,* more than 100 are available on tape, providing a unique and still entertaining radio legacy.

ALLISON MCCRACKEN

See also Brown, Himan; Horror Programs

Hosts

Raymond Edward Johnson (1941–45), Paul McGrath (1945–52)

Producer/Creator/Director

Himan Brown

Programming History

January 1941–October 1952 (528 episodes)

Further Reading

DeLong, Thomas A., *Radio Stars: An Illustrated Biographical Dictionary of 953 Performers, 1920 through 1960,* Jefferson, North Carolina: McFarland, 1996

Dunning, John, *Tune in Yesterday: The Ultimate Encyclopedia of Old-Time Radio, 1925–1976,* Englewood Cliffs, New Jersey: Prentice-Hall, 1976; revised edition, as *On the Air: The Encyclopedia of Old-Time Radio,* New York: Oxford University Press, 1998

Harmon, Jim, "And Here Is Your Host," chapter 4 of *The Great Radio Heroes,* New York: Ace, 1967; revised and expanded edition, Jefferson, North Carolina: McFarland, 2000

Intercollegiate Broadcasting System

The Intercollegiate Broadcasting System (IBS) was founded in 1940 by the originators of AM carrier-current college campus radio. Initially, college radio's primary interest concerned exchanging technical information among colleges via this new avenue of transmission. As more college stations were established, the interest evolved to include station management, programming, funding, recruiting, and industry training. Today, IBS is a nonprofit association of student-staffed radio stations located at schools and colleges throughout the United States. Approximately 600 IBS stations operate various types of radio facilities, including closed-circuit, AM carrier-current, cable radio, and Federal Communications Commission (FCC)–licensed FM and AM stations.

The majority of the early college radio stations in the United States were operated under the auspices of campus academic departments of electrical engineering; the primary objectives of these stations focused on the technical aspects of radio broadcasting rather than the public service potential. In 1925, 171 such stations were on the air, but by 1937, only 38 remained in operation. The decline in stations is credited to a general loss of campus interest or funding after the novelty of radio wore off. The few stations that sought to continue as AM broadcasters lost their licenses to commercial interests through comparative hearings before the FCC.

Lobbying in favor of college-based stations led to the FCC's 1938 decision to preserve such stations and to its 1941 and 1945 decisions to reserve FM channels designated for educational use. From the 1960s into the 1980s, the FM stations licensed to colleges and universities in the United States continued to provide leadership for the nation's public radio movement. By the mid-1990s, the majority of the 1,800 noncommercial so-called public radio licenses were granted to colleges and universities.

The formation of IBS was crucial to the preservation of college radio. IBS actively campaigned for reserved FM channels for college radio use. The result was the 1945 continuation of a reserved band of FM frequencies (this time at 88.1 to 91.9 MHz) where most noncommercial stations are now located. IBS was also active in convincing the FCC to establish the category of Class D (10-watt) noncommercial FM stations as an entry-level training ground for college radio. The Class D decision permitted hundreds of fledgling stations to get started; most of these gained momentum and graduated to the increased power of a Class A facility, 100 watts.

Increasingly throughout the years, IBS has taken on the fight for the protection of college radio. In 1978, when copyright laws changed to allow performing rights associations to collect fees for noncommercial broadcast performances, IBS presented testimony that resulted in lower rates being applied to college radio than to other classes of broadcast stations. IBS also filed objections against FCC on a proposal governing underwriting announcements. The IBS favored changes, which were adopted, and gave stations unprecedented latitude in the frequency and content of broadcast announcements, thus encouraging new interest from potential underwriters. Additionally, IBS was the first industry organization to file an FCC Petition for Reconsideration, which resulted in the FCC ruling exempting noncommercial operators from the $35 application permit fee.

IBS is a centralized information source by which college radio remains informed about industry politics, problems, and solutions. IBS lobbies for educational radio through an aggressive campaign of printed materials, e-mail, ground mail, telephone and fax communication, and regional and national seminars and workshops. Beyond addressing the needs of individual member stations, IBS acts as college radio's primary representation before the FCC and other governmental and industry agencies. IBS directors comprise a cross section of professionals representing a broad range of industry-experienced people who contribute their expertise on a voluntary basis.

The volunteer efforts of IBS personnel make sponsorship of new stations possible. IBS assists in launching new stations through a plan of action that includes advisory tips on conducting a frequency search; purchasing an existing station; and implementation of legal alternatives, such as utilizing on-campus carrier-current AM or cable FM piped into existing cable systems. Additionally, IBS offers basic advice regarding the complicated paperwork involved in filing for FCC permits.

IBS also provides helpful tips to member stations on increasing a station's coverage. IBS advises conducting frequency research to see if expansion is possible. They will assist in discussing the pros and cons of increased power versus increased height. For example, maintaining the same power but increasing the antenna height could give the increased coverage desired.

ELIZABETH COX

See also College Radio; Educational Radio to 1967; Low-Power Radio/Microradio; Public Radio Since 1967; Ten-Watt Stations; WHA and Wisconsin Public Radio

Further Reading

Bloch, Louis M., Jr., *The Gas Pipe Networks: A History of College Radio 1936–45*, Cleveland: Bloch and Company, 1980

Engleman, Ralph, *Public Radio and Television in America: A Political History*, Thousand Oaks, California: Sage, 1996

Frost, S.E., Jr., *Education's Own Stations,* Chicago: University of Chicago Press, 1937

Intercollegiate Broadcasting System (IBS) <www.ibsradio.org>

McCluskey, James, J., *Starting a Student Non-Commercial Radio Station,* Boston: Pearson Custom, 1998

International Radio Broadcasting

International radio broadcasting is usually associated with national governments, which certainly do make great use of it to communicate their viewpoints to listeners in other nations. Yet that is only one of many uses by one of many agencies: religious groups, commercial firms, and numerous others have also employed it. By the end of the 20th century, more than half of the world's sovereign nations were or had at one time been hosts to such services, which continued to attract many listeners despite the ending of the Cold War and the rising worldwide popularity of television.

Origins

Although radio amateurs often communicated across national borders, it was not until 1926 that any nation made even occasional use of radio to reach listeners in other nations. The Soviet Union broadcast to the then-Romanian (but formerly Russian) province of Bessarabia in an attempt to intimidate Romania into relinquishing its control of the province. The broadcasts lasted for several days and had no immediate effect. The Soviets also broadcast to miners in Great Britain for a few days during the General Strike of 1926, encouraging their dissatisfaction with the government, but again without visible effect. Both ventures were early examples of short-term tactical uses of the medium, but in 1927 the Netherlands launched the first long-term international broadcasting service when the Phillips electronics company's shortwave radio station PCJ began to broadcast to Dutch citizens living overseas. Over the next five years, Great Britain, France, and Germany launched similar services for their present and former citizens living abroad; the Soviet Union joined their ranks, but mainly for the purpose of reaching communists and gaining converts to communism from around the world.

It is doubtful that any of those services attracted large numbers of listeners since they broadcast almost exclusively over shortwave transmitters, and the shortwave radio sets needed to receive the signals were expensive. France and Great Britain ruled large colonial empires and used the newly created international radio services to keep overseas citizens in touch with their homelands; in those cases, the investment in receivers probably seemed little enough to pay in order to have a touch of "home away from home." However, as fascism began to spread through Europe in the 1930s, Germany, Italy, and Spain began to use mediumwave transmissions to reach nearby nations (in the case of Spain, reaching the opposition during the Spanish Civil War was at least as important as reaching foreign listeners), intimidating some with threats, reassuring others of their good intentions, and even attempting to persuade a few to join them. Because most of Europe was in the midst of an economic depression at the time, the relative economic strength of Germany and Italy provided their international radio services with potentially attractive success stories, and workers in other European nations were a frequent target of those services. Whether the messages were persuasive is an open question, because survey research was still in its infancy.

The increasing level of activity in international radio wasn't limited to politically motivated services. Radio Vaticana had come on the air in 1931, thanks in part to a generous financial donation by Italy's "father of radio," Guglielmo Marconi. Radio Vaticana brought a Catholic message to much of the world, in many languages. It also had company before the end of the year: a Protestant service to Latin America, HCJB, operated from Quito, Ecuador, but received financial and administrative support largely from the United States. HCJB was far more interested in converting its listeners to its version of Protestantism, whereas Radio Vaticana hoped to sustain its listeners in their Catholic faith. A commercially oriented service also appeared during the early 1930s: Radio Luxembourg, taking advantage of its location amid several European nations with little or no commercial broadcasting of their own, began to provide these nations with services in their own languages, heavily laden with popular music and ads.

But it was the politically motivated services that dominated, a domination that only increased as World War II drew nearer. The Soviet service, Radio Moscow, and the German *Weltrundfunk-sender* had been exchanging condemnations of each other's governments since the early 1930s. Italy's Radio Bari began to foment discord in British-ruled Palestine starting in 1934; four years later, the British Broadcasting Corporation (BBC) began its first foreign language service with its Arabic language broadcasts to listeners in the Middle East, largely to defend Great Britain against the continuing attacks of Radio Bari. Japan's Radio Tokyo became active in broadcasting intimidating messages to China in the late 1930s, as the Japanese army moved to occupy parts of the country. Some broadcasters began to make use of a more personality-oriented approach, with conversational styles breaking the pattern of highly formal speech that had predominated on most stations. Some also made use of clandestine (concealed or false identity) stations in hopes of deceiving listeners into believing that broadcasts came from within the listeners' own nations.

World War II

Once World War II had begun, international radio moved into high gear, with most of the major combatants—Great Britain, the United States, Germany, Italy, and Japan—rapidly adding language services, increasing their broadcast hours, and strengthening their transmitter power. The Nazis also attempted to ensure that German citizens would not have access to German language broadcasts from the Western allies by using jamming (electronic interference), by exacting severe penalties for such listening, and by making available only inexpensive radio receivers incapable of picking up the more distant incoming signals. Nevertheless, the Allied services developed an array of specialized programs designed to reach German officers, frontline soldiers, U-boat (submarine) crews, and others; these programs featured fake Germans in roles such as "Der Chef," a German "officer" who spread rumors about misconduct and luxurious lifestyles among high-level German civilians in his broadcasts over Great Britain's Soldatensender—a clandestine radio service for German military personnel. However, the success of any of the broadcasts was difficult to determine. Survey research was not well developed at the time, and conducting surveys in areas governed by the enemy was not an option.

The Cold War

When the war ended in August 1945, there was considerable sentiment in Great Britain and the United States favoring sharp reductions and even elimination of the international radio services. The Voice of America (VOA) was nearly disbanded in 1945–46, and BBC external services were considerably reduced. Radio Moscow at first reduced its services slightly but then expanded them, even as it worked with the Central and Eastern European nations now under Soviet influence to create miniature Radio Moscows in Poland, Hungary, and elsewhere. Influential legislators returning from trips to those nations reported hearing strong anti-Western messages over Radio Moscow, Radio Warsaw, and other stations in the region. This fact, coupled with the virtual disappearance of cooperation between the Western allies and the Soviet Union, helped lead to dramatic reversals of fortune for VOA and BBC, which soon had sizable Czech, Romanian, Russian, and other language services appropriate for the situation.

The United States went even further. The British had operated more clandestine stations during the war, but now the United States took the lead on a grand scale with the creation of Radio Free Europe (RFE) and Radio Liberation from Bolshevism (later Radio Liberty [RL]). The U.S. authorities considered these to be "the sorts of services the captive peoples of eastern Europe and the Soviet Union would want if they had a free choice." Although their broadcast messages did not call for the overthrow of communist governments, RFE and RL at times suggested work slowdowns and other actions that might help weaken those governments and eventually lead to liberation from communism. The stations spent far more time pointing out the rapid economic growth of the West, comparing that growth with the allegedly poor economic progress of the East. The communist international stations reported on strikes and other signs of the "inevitable decline of capitalism," contrasting that decline with what they claimed was the generally robust performance of the communist economies. Increases in broadcasts from the West were met by increases in jamming in the East. However, surveys taken among refugees from Eastern Europe indicated that some of the signals were getting through and that they seemed particularly effective in causing those who heard them to be more and more skeptical of communist media claims that life under communism was far superior to life under capitalism and that the gap between the two was steadily growing.

When China joined the ranks of communist nations in 1949, it soon added its own international broadcast voice, causing the United States to create an Asian equivalent of RFE/RL in the form of Radio Free Asia (RFA, 1951–55; a new Radio Free Asia came on-air in 1996). The small numbers of radio receivers in North Korea and in China, the difficulty of recruiting suitable Asian language speaking talent, and heavy jamming of RFA caused the United States to drop the service, even as China's newly founded Radio Peking (later Beijing) grew to become one of the largest of all international broadcast services, with North Korea's Radio Pyongyang not far behind. In each case, the Soviet Union assisted both materially and ideologically; the Soviet Union played a similar role with respect to Radio Havana Cuba starting in the early 1960s.

The Third World Speaks and Listens

The Soviet Union had become increasingly active in broadcasting to the Third World (industrially developing nations) during the late 1950s, as African, Asian, and Latin American nations emerged from their colonial status. The Chinese, North Korean, and Cuban international services added their voices to Radio Moscow's in denouncing colonialism where it still existed while praising the efforts of the newly independent nations to stake out their own ideological positions in the Cold War. Two of those nations—Egypt and India—were already operating international radio services by the mid-1950s, with Africa and the Islamic nations as prime targets of the former and South and Southeast Asia the targets of the latter. Although they too denounced colonialism, and therefore Western nations such as France and Great Britain, they were not necessarily sympathetic to communism, either. Thus, they constituted yet another international radio voice with yet another cultural and political perspective. As more newly independent nations emerged, some of them also developed international services; Radio Ghana was broadcasting in several West African and European languages by the end of the 1950s. But no single nation among them ever set out to coordinate their efforts into a unified anticolonial voice, as the Soviet Union managed to do to some degree with the other communist international services during the 1950s.

That coordination began to crumble as China increased in strength and in prestige. By the late 1950s, China had begun to part ways with the Soviet Union over the correct interpretation of Marxism–Leninism. The disagreement gave rise to the broadcast of many verbal duels between the two communist powers, generally of little or no interest to anyone but committed Marxist–Leninists. The United States and Great Britain were beginning to see that, as the old colonial system broke apart, they faced both an opportunity and a challenge in reaching African, Asian, and Latin American audiences. It took a few years for each to realize that a new message would have to speak to a new relationship—one that respected the independence of what were now sovereign nations. BBC's World Services and VOA were sufficiently alert to recognize, as Radio Moscow often had not (although it made an effort to do so through Radio Peace and Progress, founded in 1964), that programs dealing with African, Asian, and Latin American events, cultures, and personalities were likely to prove far more attractive than had the one-worldwide-size-fits-all approach generally taken by Moscow. (Radio Peking was generally more sensitive to Third World cultural aspirations and often portrayed the People's Republic of China as a Third World nation.)

By the late 1960s, both VOA and BBC had developed African, Asian, and Latin American services with a variety of cultural and informational programs tailored for those regions. VOA also had introduced a limited-vocabulary/slow-rate "Special English" newscast in 1959; because it was well suited to English language learners, of which there were many in the Third World, the newscast was able to attract listeners. Surveys taken by or for VOA and BBC during the 1960s showed increasing numbers of Third World listeners for both and minuscule numbers for the communist services, with the notable exception of Radio Havana in Latin America. Cuban President Fidel Castro enjoyed considerable popularity among Latinos for his willingness to stand up to the United States, and Radio Havana blended programs featuring Cuban and other Latin American cultures with informational programs that generally avoided the heavy ideological jargon of many such programs broadcast over Radios Moscow, Peking, Pyongyang, and other communist stations.

Religious Voices

There was yet another major entrant in the post–World War II international radio lists, although it grew more slowly than had the communist or Western sectors. Religious stations, few in number in the 1930s, began to multiply in the late 1940s and early 1950s, in part as a reaction to the spread of "godless communism," in part in recognition of the beneficial effects for fund-raising of broadcasting international religious programs for non-Christians and for religion-deprived listeners in communist nations around the world. Most of the new services were financed by U.S. religious groups, most were evangelical, and virtually all were Protestant. They established stations and relay transmitters in Africa (Morocco, Liberia, Ethiopia; in the 1960s, Burundi and Swaziland); Asia (Sri Lanka, the Philippines, Korea); Latin America (Netherlands Antilles); and also in the United States and Europe. Many of the English language programs were rebroadcasts of U.S. evangelical preachers, many of whom had heavy southern accents and made references to people and places that would be unfamiliar to foreign listeners, so it was not surprising that religious stations on the whole did poorly in listener surveys. The exceptions were HCJB in Quito, Ecuador, and FEBC in the Philippines, both of which provided a widely varied assortment of information and entertainment and were careful to show respect to other religious denominations and faiths.

Clandestine Services

Clandestine stations generally flourished in the unstable atmosphere of the Cold War. Not only were they prominent in the conflict between East and West, but they also played roles in the many regional conflicts taking place in Asia, Africa, and the Middle East. Their programs generally were political and often featured exceptionally strong language, including calls for the assassination of political leaders. They also served as channels through which politicians in exile could reach their former homelands with messages criticizing government officials and even encouraging uprisings.

The Cold War Ends

With the collapse of communism in Eastern Europe and in the Soviet Union by the early 1990s, the Cold War came to an end. That also brought radical changes for international radio, which had been so heavily involved in the struggle between communism and capitalism. No longer did many of the governments that had financed the stations see the need for so many language services, so many hours of transmission, and so many foreign transmitter bases, and annual financial appropriations began to decline. Most of the communist countries either cut their international services to the bone or dropped them altogether. The clandestine stations also felt the pinch, since many of them had been partially or wholly financed by the East and the West so that they could play roles in Cold War politics in addition to serving the more narrowly focused ends of some of the groups operating them. Their places were taken to a limited extent by Islamic fundamentalist and paramilitary group operations, and political exile groups continued to use them in Africa and Asia.

Still, there were some bright spots for the international stations during the 1990s. The spread of the internet made it possible for them to begin to provide interference-free service through websites, and by the end of the decade virtually all of the major stations were doing so. There was also the prospect of direct transmission via satellite to individual receivers (the use of satellites to relay signals to foreign transmitter bases had been around since the early 1980s), although the cost of such receivers was discouragingly high. A U.S.-based commercial firm, WorldSpace, launched a satellite-delivered multichannel radio service to Africa in October 1999, but economic data suggested that few African listeners could afford the specialized receivers needed to bring in its transmissions.

As the new millennium dawned, there was every indication that international radio would continue to be viable, even if on a reduced scale. Some of the international commercial radio services from France—Radio Monte Carlo Middle East, Radio Mediterranée Internationale, Afrique Numéro Une (Africa No. 1; for sub-Saharan Africa)—still enjoyed financial success, largely because they provided their audiences with more "worldly" entertainment (largely Western or Western-flavored pop music) and broader and less-biased perspectives on regional and world events than did domestic stations in those areas. Instability in parts of southeast Asia, Africa, the Balkans, and the former Soviet Union helped to ensure that the comprehensive and generally reliable informational broadcasts of the BBC World Service, the Voice of America and Radio Free Europe/Radio Liberty, Germany's Deutsche Welle, and smaller-scale services such as Radio Netherlands, Radio Canada International, Radio Australia, and Radio Japan would have audiences, especially in the areas of conflict themselves.

The Voice of Russia (formerly Radio Moscow), Radio Beijing, Radio Pyongyang, and certainly Radio Havana Cuba also continued to provide their versions of current events to listeners who appreciated Russian, Chinese, North Korean, and Cuban perspectives, even if those perspectives might have become less meaningful with the ending of the Cold War. Religious stations seemed no more or less popular than they had been in earlier decades, but the chief reason for the existence of many of them—to help stimulate contributions from those anxious to bring the gospel message to nonbelievers—remained viable. Increases in the efficiency of distribution of television broadcasts on an international basis, especially when coupled with the invention of modestly priced television sets capable of receiving signals directly from satellites, almost certainly would reduce the attraction of international radio. So would the provision by domestic broadcast services in presently autocratic nations of more balanced and detailed coverage of events at home and abroad. Neither of those changes seems an immediate prospect.

DONALD R. BROWNE

See also Africa No. 1; Axis Sally; BBC World Service; Clandestine Radio; Cold War Radio; Developing Nations; Far East Broadcasting Company; Jamming; Lord Haw-Haw; Propaganda by Radio; Radio Free Asia; Radio Free Europe/Radio Liberty; Radio Luxembourg; Radio Martí; Radio Moscow; Radio Sawa/Middle East Radio Network; Religion on Radio; Tokyo Rose; Vatican Radio; Voice of America; World War II and U.S. Radio

Further Reading

Benhalla, Fouad, *La guerre radiophonique,* Paris: Collection de la Revue Politique et Parlementaire, 1983

Boelcke, Willi A., *Die Macht des Radios: Weltpolitik und Auslandsrundfunk, 1924–1976,* Frankfurt: Ullstein, 1977

Bookmiller, Kirsten Nakjavani, "The War of Words without the War: Radio Moscow, the British Broadcasting Corporation World Service, and the Voice of America in the Old and New International Order," Ph.D. diss., University of Virginia, 1992

Browne, Donald R., *International Radio Broadcasting: The Limits of the Limitless Medium,* New York: Praeger, 1982

Fortner, Robert, *Public Diplomacy and International Politics: The Symbolic Constructs of Summits and International Radio News,* Westport, Connecticut: Praeger, 1994

Hale, Julian, *Radio Power: Propaganda and International Broadcasting,* Philadelphia, Pennsylvania: Temple University Press, and London: Elek, 1975

Krugler, David F., *The Voice of America and the Domestic Propaganda Battles 1945–53,* Columbia: University of Missouri Press, 2000

Mytton, Graham, *Global Audiences: Research for Worldwide Broadcasting, 1993,* London: Libbey, 1993

Nelson, Michael, *War of the Black Heavens: The Battles of Western Broadcasting in the Cold War,* Syracuse, New York: Syracuse University Press, 1997

Rawnsley, Gary, *Radio Diplomacy and Propaganda: The BBC and VOA in International Politics, 1956–64,* New York: St. Martin's Press, and London: Macmillan, 1996

Sémelin, Jacques, *La liberté au bout des ondes: Du coup de Prague à la chute du mur de Berlin*, Paris: Belfond, 1997
Soley, Lawrence C., and John Spicer Nichols, *Clandestine Radio Broadcasting: A Study of Revolutionary and Counterrevolutionary Electronic Communication*, New York: Praeger, 1987
Soley, Lawrence C., *Radio Warfare: OSS and CIA Subversive Propaganda*, New York: Praeger, 1989
Tyson, James L., *U.S. International Broadcasting and National Security*, New York: Ramapo Press, 1983
Wettig, Gerhard, *Broadcasting and Detente: Eastern Policies and Their Implication for East-West Relations*, London: Hurst, and New York: St. Martin's Press, 1977
Wood, James, *History of International Broadcasting*, vol. 1, London: Peregrinus, 1992; vol. 2, London: Institute of Electrical Engineers, 2000

International Telecommunication Union

The International Telecommunication Union (ITU) is a specialized agency operating under the auspices of the United Nations. It attempts to assist in the development of broadcasting and point-to-point communication by providing fora for the discussion of, and adoption of agreements on, issues common to all countries' communications activities, including tariffs, technical standards for interconnection across frontiers, the sharing of broadcasting frequencies, the allocation of geosynchronous satellite locations, and the permissible uses for international communication. It also registers countries' use of shortwave frequencies to provide the basis for frequency coordination among the different users of this broadcasting medium.

Origins

Originally called the International Telegraph Union, the ITU was formed on 17 May 1865 by 20 European countries that desired to facilitate international telegraph communication. Until the convention that established the ITU was signed, telegraph activities occurred entirely within individual countries: messages had to be transcribed at each border, translated into the language of the neighboring country, and then carried across the frontier where they would be rekeyed for further transmission. Each country also used its own telegraph code to safeguard its military and political messages. This made international telegraphy cumbersome at best and prevented the new technology from having a significant positive impact on the relations of the suspicious European powers.

Linking up the different domestic telegraph systems required that three issues be resolved. First, the electrical transmission systems used had to be standardized. Different wire gauges, signal voltages, and wire connection systems had to be standardized so that there would be no technical breaks in lines traveling across frontiers. Otherwise, communication would be impossible. Second, the allocation of revenues had to be agreed on, so that if a message traveled, say, between London and Berlin, the British, French or Belgian, and German telegraph companies (usually state-owned) would each receive an equitable portion of the tariff. Otherwise, there would be no financial inducement to connect the lines. Messages often traveled through a country's system without actually being delivered to anyone within it. (As in the above example, in which the message must travel across France to Germany for delivery.) So a system was devised that compensated different telegraph authorities according to the miles of line used for a message to reach its destination, regardless of origination or destination point. Third, a common language had to be used for messages so that their meaning could be accurately transcribed regardless of the language of the telegraph operator. Morse code, which represented all the letters of the alphabet with a combination of short and long pulses, was adopted as this common language. Also, agreements forbade private codes that shortened messages (such as FYI for "for your information") to assure that those who used the telegraph system were paying their fair share of its costs.

The original Telegraph Convention (or treaty) established the basic rules that the ITU was to administer on behalf of the signatory states and provided the foundation for subsequent agreements as new technologies developed. The ITU was not granted, and still does not have, any enforcement powers to use against states (called administrations) that break the conventions, so it functions as technical advisor, facilitator of new conventions as required, and administrator of agreements between signatory nations.

Since that first convention adopted to coordinate telegraphy, new technologies have called for new agreements and expansion of the ITU's activities. The submarine cable, which connected countries under the sea, began to be widely used in the 1850s, and the first transatlantic cable was successfully laid in 1866. This new form of telegraphy, as well as wired telephony, invented in 1876, both required essentially the same type of agreements between states that the overland telegraph had.

The ITU and Radio

Wireless telegraphy, however, required a different set of agreements. Wireless crossed international frontiers whether a country objected or not. And although wireless operators did not require the use of facilities within every country whose territory they crossed, signals were subject to interference from other transmitters using the same frequency. Furthermore, wireless signals were important as a means to communicate with ships that were not on any country's actual territory. Two issues emerged beyond the earlier concerns. First, should all ships, regardless of country of registry, be required to have both wireless apparatus and trained operators on duty 24 hours per day? Second, were the private monopolies that were being pursued by private companies (such as British Marconi) to take precedence over safety at sea? Marconi, for instance, forbade its operators from communicating with ships using wireless apparatus manufactured by its competitors. This policy was not successfully broken until the *Titanic* disaster (1912) demonstrated the necessity of doing so. The development of wireless telegraphy resulted in the convening of a preliminary radiocommunication conference at Berlin in 1903 and a Radiotelegraph Conference in 1906 that resulted in the first International Radiotelegraph Convention. Three principles emerged from these radiotelegraph meetings. These were (1) that frequencies should be reserved for specific services; (2) that all administrations should take the steps necessary to avoid interference with other users; and (3) that all use of frequencies should be registered.

During the 1920s, three consultative committees were established under the auspices of the ITU to draw up international standards for the telegraph, telephone, and radio. The International Telephone Consultative Committee was set up in 1924, the International Telegraphy Consultative Committee in 1925, and the International Radio Consultative Committee (CCIR) in 1927. These committees all coordinated the technical studies undertaken to develop new standards, developed means to conduct tests and measurements, and made recommendations to conferences convened to adopt new regulations. The telephone and telegraph committees were combined in 1956 into the International Telephone and Telegraph Committee (CCITT).

In 1927 the ITU allocated frequencies to the different radio services that were in operation at that time. These included fixed services (point-to-point wireless radiotelegraphy), maritime and aeronautical mobile services (for ships and airplanes), radio broadcasting, and amateur and experimental services.

The Modern International Telecommunication Union

On 1 January 1934, the old ITU officially became the International Telecommunication Union. This was a result of a decision made at the 1932 Madrid plenipotentiary conference to combine the two original agreements (the International Telegraph Convention of 1865 and the International Radiotelegraph Convention of 1906) into a single International Telecommunication Convention. The ITU became a specialized agency of the United Nations on 15 October 1947.

In the same year, the table of frequency allocations that had been established in 1912, which allocated to each type of radio service specific frequency bands to use so as to avoid interference with other types of uses, was made mandatory. The International Frequency Registration Board (IFRB) was set up to manage the radio frequency spectrum under the auspices of the ITU.

In 1959 the CCIR established a study group to look into space communications as a result of the launch of the Soviet *Sputnik* in 1956. In 1963 the first administrative conference on space communications was held to allocate frequencies to space services.

The ITU has divided the world into three regions for purposes of dealing with frequency allocation issues and technical standards. These regions, roughly, are Region 1, Europe and Africa; Region 2, Asia; and Region 3, the Americas. Some differences in technical standards may exist among these regions, although within each area technical standards are consistent. For instance, in Europe medium wave (or what Americans call AM) bands are 9-kilohertz wide, whereas in the Americas they are 10-kilohertz wide. This small difference allows the larger number of countries in Europe to have additional frequencies for this service that are not needed in the Americas, which are dominated by large countries (especially Brazil, Canada, Mexico, and the United States).

The ITU has a complicated organizational structure. Its permanent staff is included in the General Secretariat headed by the Secretary-General of the United Nations. Since a 1990s reorganization, there are three bureaus in this secretariat, including the Radiocommunication Bureau, the Telecommunication Standardization Bureau, and the Telecommunication Development Bureau. The members of the ITU elect a Council that oversees the secretariat between their plenipotentiary conferences, which is where major policy changes are made and regulations or resolutions are adopted concerning either radiocommunication or telecommunication. Specialized conferences are also held to deal with specific issues, such as World or Regional Administrative Radio Conferences (WARCs and RARCs), World Telecommunication Standardization Assemblies, and World or Regional Telecommunication Development Conferences. Each of these three types of conferences also has advisory and study groups that survey world practice in their subject concern, commission studies, or distill technical information that is used as the basis for determining new regulations or standards in the different areas. In addition, the WARC/RARC conferences have a Radio Regulations Board because of the more difficult matter of administrations coming to agreements and enforcing them when the communications activity in question is wireless (whether terrestrial or satellite point-to-point or terrestrial or satellite broadcasting). This is because such radiocommunication activities (including the assignment of geosynchronous satellite "parking spaces") either cross frontiers or operate in the "air space" above countries that may not derive any benefit from them.

ITU Functions

The ITU's responsibilities under this organizational structure can be grouped into three main activities. First, it coordinates radio frequency assignments, including band assignments for services and the use of the geostationary orbit for satellites, including orbital slots. Second, it recommends technical standards for international communications, including those for wired connections, bandwidths, and other technical parameters. Third, it regulates international common carrier services, including telegraphy, telephony, and data communication.

The globalization of communication that has occurred in the satellite age—including intercontinental distribution of radio and television programs, satellite-based telephony, the development of Global Positioning System (GPS) receivers and maritime communication; the explosive growth of the internet; and the need to allocate the limited number of "parking spaces" for geosynchronous satellites serving countries that are, themselves, not on the equator above which these satellites "park"—has involved the ITU in an increasing number of sovereignty-based technical issues. For instance, developed countries have argued that the allocation of orbit slots should be based on need, whereas developing countries have contended that, despite their inability to use satellites at a given time, slots should be reserved for their future use. So the question of whether technological change will enable more satellites, or more powerful satellites, to be deployed in time to meet future needs has had to be considered alongside the demands for access made by those over whose territory the satellites orbit. And because all geosynchronous satellites must orbit above the equator to maintain their positions relative to the Earth's surface (and thus appear to be stationary), the application of sovereignty arguments has had to be considered alongside the technical or economic arguments of more advanced countries. It is under the auspices of the ITU and its various technical committees, study groups, conferences, and plenipotentiaries that such issues are ultimately resolved.

One significant issue that the ITU has had to confront concerning radio has been the use and abuse of shortwave frequencies. Shortwave is the one radio service for which individual stations are not assigned particular frequencies. There are a variety of technical reasons for this, but the bottom line is that, depending on various factors, shortwave radio stations have to change their frequencies periodically. Keeping track of the use of stations' use of these frequencies is thus a major activity of the IFRB. Also, shortwave radio has historically been used for international broadcasting, and stations have purposely broadcast in multiple languages to the people of other countries. Two contentious issues developed. The first was the question of whether stations were engaged in propaganda. The second, related issue was whether a country was deliberately interfering (or jamming) the signals of a station originating outside its borders to stop the station's propaganda. Purposely interfering with another station's signal is against ITU radio regulations, so even countries that are widely known to jam others' signals have steadfastly denied that they engage in such practices. The IFRB could only register use, however, since it had no enforcement power to use in such cases.

ROBERT S. FORTNER

See also Digital Audio Broadcasting; Frequency Allocation; International Radio Broadcasting; North American Regional Broadcasting Agreement; Shortwave Radio

Further Reading

Codding, George A., Jr., *The International Telecommunication Union: An Experiment in International Cooperation,* Leiden, The Netherlands: Brill, 1952

Codding, George A., Jr., and Anthony M. Rutkowski, *The International Telecommunication Union in a Changing World,* Dedham, Massachusetts: Artech House, 1982

Fortner, Robert S., *International Communication: History, Conflict, and Control of the Global Metropolis,* Belmont, California: Wadsworth, 1993

Michaelis, Anthony R., *From Semaphore to Satellite,* Geneva: International Telecommunication Union, 1965

Tomlinson, John D., *The International Control of Radiocommunications,* Ann Arbor, Michigan: Edwards, 1945; reprint, New York: Arno Press, 1979

Internet Radio

Delivering Radio Programs Online

Internet radio involves the delivery of audio programming via digital means from one computer to other computers over the internet. It involves both simulcasts of existing over-the-air radio stations and content from internet-only stations. Internet radio was made possible by the 1995 arrival of streaming. Previously, users had to download an entire audio file before being able to listen to it. Even audio clips of short duration could take hours to download. Streaming allows the user to listen to the audio programming as it arrives in real time. This means users do not have to wait for a complete audio file to download before listening to it. Internet radio streaming can involve both live material and archived clips of audio content recorded earlier. In either case, the user must have special software that matches the software used by the station to encode and transmit the data.

Internet radio was a booming enterprise into the late 1990s, but legal decisions and a downturn in internet advertising have effectively shut down many stations today. In 2002, a dispute between internet broadcasters and the music industry came to a head when a copyright appeals board required internet radio stations to pay a per-song, per-listener fee that was prohibitively expensive for many stations. The fee was an especially great hardship for small operations, such as religious broadcasters and college radio stations, and amounted to thousands of dollars more than they made. This led to hundreds of internet-based radio stations shutting down.

By late 2002, a compromise was worked out whereby internet broadcasters could pay royalty fees on a percentage of their revenue instead of on a per-song, per-listener basis. The *Small Webcaster Settlement Act of 2002* was seen as a big victory by small webcasters and by early 2003 many small internet broadcasters were beginning to reappear.

The three leading technologies for delivering internet radio are the RealOne Player, Microsoft Windows Media, and MP3 streaming. Internet radio stations will often select one of the technologies for the delivery of their content. In some cases, stations choose to make their audio stream available in more than one of these formats, allowing listeners to choose the way they want to listen.

Streaming generally sacrifices audio quality because of the need to compress the data for delivery via narrowband (56k) telephone line modems still used by most households. Early internet radio quality was very poor, and many listeners became discouraged by the poor audio quality and problems maintaining a continuous stream. The stream would often stop and buffer (download data before it was used), inhibiting continuous delivery of the program.

Today, technological improvements and new broadband connections allow better streams and near-CD-quality sound. Listeners with cable modems or telephone DSL (digital subscriber line) services are the biggest beneficiaries. The adoption of these faster broadband connections is rising sharply. In an 18-month period between January 2001 and July 2002, the percentage of Americans with residential broadband internet access surged from 13 percent to 28 percent. As that number continues to grow, the audience for internet radio services will build as well.

Radio Stations on the Internet

Studies show that internet usage is cutting into time people would otherwise spend listening to broadcast radio. For traditional broadcast stations, delivery of programming on the internet may help recapture some of these listeners and may even generate new listeners in distant locations. Even small-market radio stations can reach the same international audiences as stations in larger markets. The concept of signal strength does not apply in the on-line world, and all stations start out on equal footing. Location is no barrier, either. It costs no more to send an internet radio program 1,000 miles than it does to send it 10 miles. On-line radio listeners say they listen more to radio stations outside their local market than they do to stations in their own locale.

One Arbitron and Edison Media Research study showed that by the middle of 2002, 35 percent of Americans had listened to internet radio, compared with 19 percent in 1998. This growth has given traditional radio broadcasters cause for concern because radio listeners now have a much greater number of listening choices on-line than they do on the radio dial.

Internet radio listeners are sometimes referred to as "streamies." As a group, streamies represent a very desirable demographic for advertisers. Streamies are among the most active group of internet users, spending more time on-line than the average internet user. Streamies are twice as likely to click on web ads and to make on-line purchases and are very interested in new devices to enable even more convenient listening. Internet radio listeners tend to be better educated and come from homes with higher incomes than regular internet users.

Capturing the internet audience and persuading listeners to revisit, however, is made more difficult with such a range of choices. Developing content worthy of repeat visits is one of the biggest challenges for internet broadcasters. Merely having a web presence to promote a station's broadcast operation is not enough. Internet broadcasters are using interactive features such as contests and live chat rooms to gain and hold on to the elusive internet audience. Concert information, celebrity interviews, and fashion information are also important content categories for the young internet radio audience.

Broadcast radio stations have traditionally had strong local identities. On the internet, some stations may decide to adopt a more national identity. Far-flung listeners with ties to a community can stay up to date with "local" news, sports, and community events from anywhere in the world.

Stations looking for a national audience may develop niche programming such as specific music genres or sports. Certain music formats may be more popular than others in the on-line world. A 1999 Arbitron and Edison Media Research study showed that 91 percent of radio listeners who have internet access prefer alternative rock. The next highest categories were Top 40 (68 percent), classical (68 percent), religious (54 percent), adult contemporary (52 percent), and news/talk (50 percent). The top-rated internet radio stations tend to be eclectic and unique-sounding outlets not commonly found on the air.

Some broadcasters remain unconvinced of the value of internet radio. For one thing, there are far fewer internet-connected computers than there are available radio sets. Internet radio also lacks the portability of broadcast radio and is not generally available in cars, at the beach, a picnic, or other gathering places outside the home or workplace. Sound quality of internet radio varies greatly, and listeners with low-speed modem connections or slow computers are often disappointed with the overall quality. Furthermore, studies have shown that many people sample on-line stations but don't return regularly. Many broadcasters are still waiting to see a return on their investment in internet radio.

Making Money on Internet Radio

Although broadcast radio is an audio-only medium, internet radio stations are free to offer interactive programming and can include images, animation, and even video. Whereas broadcast radio relies on estimating the size of audiences via ratings, internet radio can measure each time a user accesses a particular page or program and in many cases can provide detailed demographic data about the people visiting their sites.

There are three ways for internet radio stations to make money on-line; advertising, transactions, and subscriptions. Advertising is the model broadcast stations have adopted and used for decades. The ability of internet radio to reach a global audience means not only the potential for a greater number of listeners, but also that stations may be able to attract national, as well as local, advertisers. Internet radio listeners represent a desirable demographic of technology-savvy young people to advertisers as well.

Besides the standard audio-only commercials so familiar on the radio, internet radio allows stations to generate revenue through graphic advertising banners and pop-up ads as well. The "banner ad" is an easy and effective way to display advertising on a station's website. Stations charge different amounts, depending on banner size, placement, and duration on a page. The banner ad may be placed on the same page listeners go to when they want to listen to the station on-line. A greater amount can be charged for what are called "click-throughs" (money earned when users click the banner ad and go to the advertiser's site).

Other sources of revenue can be generated through classified ads and direct sales or transactions. For example, many internet radio listeners say that they would like to be able to buy music on a station's site. Advertising on internet radio stations can be tied directly to transactions conducted on-line. Whereas broadcast radio commercials depend on delayed gratification (listeners hear a commercial and will ideally buy something later), internet radio is more interactive and allows the user to go immediately from the desire to buy directly to a page where a purchase can take place. This immediacy in the on-line world changes the very nature and approach of advertising for the new medium.

Many internet radio stations have moved away from being a free service and are now charging a subscription fee. KPIG in California was the first commercial broadcaster to use the internet back in 1995. Because of copyright fees and dwindling advertising revenue, KPIG is now charging a per month fee for listeners to access their content.

Internet-Only Radio Stations

In many cases, internet radio stations exist only on the internet. Often referred to as music "channels," these ventures often play lesser-known groups and alternative music formats. In many cases, internet-only sites are providing original content, multiple channels, and fewer commercials. For artists and labels finding it difficult to get playtime on traditional stations, internet radio provides a viable option for exposure of new music.

The cost for starting up an internet radio station is far less than the cost of building or buying a broadcast station. An internet radio station can be established for less than $10,000 and does not require a license from the Federal Communications Commission.

One of the major targets of internet-only stations is the workplace. There tend to be more computers in use around offices than radio and television receivers. Internet-only stations hope to attract workers disenfranchised by traditional radio by offering more finely niched music choices, fewer commercials, and the opportunity to buy on-line. However, since internet radio uses a tremendous amount of bandwidth, workers listening to it could put a strain on a company's network and the ability to handle e-mail and other work-related applications.

Personal Internet Radio Stations

New technology and software allow anyone to become an internet radio broadcaster. Individuals can start their own stations and operate them from their homes. All it takes is a computer, an internet connection, and some free software. Users can create and customize their own radio stations on-line without the trouble of acquiring and setting up a server. Some on-line sites allow users to create their own playlists of genres and artists and to actually specify how often each is heard. The sites have large archives of music available. Once users have built their stations based on their music preferences, they can go to a webpage and listen to their own personalized station. The web address can be given to others so they can listen to the station as well.

One of the most successful audio technologies on the internet is the MP3 format. MP3 (MPEG Audio Layer 3) is a highly compressed audio format that delivers near-CD-quality sound with very small file sizes. Users can download high-quality MP3 music files even on low-bandwidth connections. On-line MP3 sites offer a great deal of free downloadable music, and one is often able to listen to new artists who promote and distribute their music on these sites. MP3 player software is available free on-line.

STEVEN D. ANDERSON

See also Audio Streaming; Digital Audio Broadcasting

Web Sites

SaveInternetRadio, <www.saveinternetradio.org>
SoundExchange, <www.soundexchange.com>
BRS Web-Radio, <www.web-radio.fm>

Further Reading

Bachman, Katy, "Not Streaming Profits," *Mediaweek* (6 September 1999)
Cherry, Steven M., "Web Radio: Time to Sign Off?" *IEEE Spectrum* (August 2002)
Graven, Matthew, "Web Radio Days," *PC Magazine* (14 December 1999)
"Internet 9: The Media and Entertainment World of Online Consumers," *Arbitron/Edison Media Research* (5 September 2002)
Kerschbaumer, Ken, "Radio Next for RealNetworks," *Broadcasting & Cable* (26 August 2002)
Kuchinskas, Susan, "Tune It to Internet Radio," *Mediaweek* (19 April 1999)
"New Media Eyes Radio's Audience," *Billboard* (23 October 1999)
Rathbun, Elizabeth A., "Clash of the 'Casters," *Broadcasting and Cable* (6 September 1999)
Shiver, Jube, Jr., "Internet Adds Dimension to Radio," *Providence Business News* (30 August 1999)
"Small Webcasters Get Break on License Fees for Transmissions of Music Recordings, as Authorized by Small Webcaster Settlement Act," *Entertainment Law Reporter* (December 2002)
"Small Webcasters, RIAA Formally Unveil Royalty Accord," *Communications Daily* (17 December 2002)
"Streaming Media Sites Expand: Downloadable Tunes Help Build Audience," *Computerworld* (19 July 1999)
Taylor, Chuck, "Webcasters Reshape Radio Landscape," *Billboard* (5 June 1999)

Ireland

Radio enjoys a special place in Irish public discourse. Ireland is a society of strong literary and verbal traditions, out of which have emerged writers such as William Butler Yeats, James Joyce, Kate O'Brien, and Samuel Beckett. The words spoken on Ireland's airwaves likewise have had considerable significance and power. From 1926, following the formation of an Irish state independent of the United Kingdom, radio came to play an important role in creating a new sense of nationalist identity. Extensive radio coverage of the Eucharistic Congress of 1932 greatly helped to consolidate close connections between the Roman Catholic Church and Irish politicians, connections that remained until recently a characteristic feature of the state.

Before 1988, the single state-owned broadcaster (Radio Eireann, later Radio Telefis Eireann [RTE]) enjoyed a monopoly of all broadcasting in the Republic of Ireland. Eamon de Valera, the predominant Irish prime minister (*Taoiseach*) of the mid-20th century, used radio effectively to disseminate his protectionist vision of an Ireland that was not only politically neutral but that he thought might also stand apart culturally from the modern world.

Listening to radio in the Republic of Ireland has often been a community experience, and people have consistently shared their relationship with the medium. In the 1950s and 1960s, men regularly gathered around a radio set on Sunday afternoons to listen to the distinctive commentaries of Michael O'Hehir on "Gaelic" football, a sport with its own special rules designed to distinguish it from "foreign" games. Today the contents of particular radio programs frequently provide a principal topic of conversation at home and in the pubs or restaurants.

Notwithstanding the fact that most householders in the Republic of Ireland now receive directly from the United Kingdom many English language television services of high quality, in addition to the four national Irish television channels, radio still has a wide listenership during the daytime and significantly influences political and media agendas. The Irish have for many years spent more time tuned to the radio than have their British neighbors.

In particular, Radio 1, the main radio service of the state-owned RTE, has provided coverage that emulates the best of the tradition of public broadcasting in Europe. During the daytime, Radio 1 creates a public space within which current and sometimes sensitive or controversial issues are discussed in a participatory fashion by well-known presenters; their guests; and members of the public, who are encouraged to phone in. During the 1990s, a wide variety of privately owned county and local radio services developed rapidly and have attracted many listeners away from RTE. Yet the main characteristic of most of these services is their local speech content. The most critically acclaimed programs on the privately owned national radio service, Today FM, are also speech centered, as opposed to music centered.

The perceived power of the spoken word in Ireland is reflected in special legislative provisions, known as Section 31, that have allowed the government to prohibit from time to time the broadcasting of interviews or reports of interviews with spokespeople for organizations that are deemed by the government to be involved in violent and undemocratic political activity, especially that relating to the conflict in Northern Ireland (the broadcasting of visual images of such organizations has never been banned).

The Republic of Ireland's transition from being a postcolonial and economically underdeveloped country to its status as the thriving "Celtic Tiger" of Europe has been reflected in part by the emergence of populist music-driven radio. Perhaps the most anomalous example of this phenomenon was Atlantic 252 (also known as Radio Tara), a long-wave station owned ultimately by RTE but managed in practice by CLT (Luxembourg). This station's diet of pop music and Americanized disc jockey patter was targeted principally at audiences across the Irish Sea and was created in 1989 to derive revenue from British advertisers, who were eager to sell their products to young English audiences. As the U.K. gradually licensed new national stations aimed at the same audience, Atlantic 252 ceased to be viable and finally went off the air on 31 December 2001.

Development

The history of Irish radio enjoys associations with several "firsts" in broadcasting. Most Irish histories of wireless point to the early radio broadcast from leaders of the Irish uprising that occurred during Easter week, 1916, when leaders of the uprising transmitted Morse code messages from their headquarters in the General Post Office on what is now Dublin's O'Connell Street. Many of the early experiments in wireless conducted by Marconi emanated from Ireland (RTE's headquarters in Dublin's Donnybrook area, the home of Marconi's Irish mother, Annie Jameson). The development of Irish radio can be seen as emerging in five phases.

Beginnings of Irish Broadcasting 1926–1945

Irish broadcasting had a modest beginning. Broadcasting began with a 1-kilowatt transmitter and an aerial system mounted on wooden sailing masts in the center of Dublin. The first broadcast was a speech by the soon to be first President of the Republic, Douglas Hyde, in January 1926. The radio service reported to the minister for post and telegraphs, and by the early 1930s the service had transmitters in Dublin, Cork, and Athlone, allowing for national coverage. Sponsored programs were part of Irish radio from the beginnings, and news coverage was provided by Irish radio reporters, rather than relying on news services. In 1937, the year that the Irish Constitution was adopted, the service became known as Radio Eireann; it held that name into the 1960s.

Quest for Independence 1945–1953

This period saw Irish radio attempt to forge a model of broadcasting that was different from that of the British Broadcasting Corporation (BBC) and one that was not mired in the bureaucracy of government. With the government of Eamon DeValera firmly in place, the radio service continued to develop in order to serve the political and economic goals of the new Free State.

Ireland did not participate in World War II, officially referring to this time period as "The Emergency." Radio Eireann featured a professional repertory company, symphony orchestra, light orchestra, news service, staff scriptwriters, outside broadcast officers, and engineers. Broadcasting during The Emergency and maintaining neutrality put a strain on what could be reported. All programming was required to be cleared by the director, and often this spilled into the Dail (parliament) debates and became a matter of public debate. Mentions of the weather, for example, were prohibited from all broadcasts, as was the coverage of sporting events. As Gorham noted, "the Government was not going to let Radio Eireann be used as an advanced weather station in the Atlantic for the belligerents of either side" (Gorham, 1967).

It was from this "emergency" situation that Radio Eireann sought independence in the years immediately following the Allied victory in Europe. The postwar move toward freedom of expression for broadcasting included the establishment of an advisory panel in 1952 that laid the groundwork for the establishment of an independent authority to oversee broadcasting that was not directly under the control of the government.

Rise and Fall of Comhairle Radio Eireann 1953–1960

The deficiencies of Irish radio were the subject of much public and government debate during this period. Some favored the continuation of direct government control, whereas others advocated following the European ethos of public service broadcasting with oversight by an independent authority. A compromise established an advisory committee, Comhairle Radio Eireann.

Comhairle Eireann's five-member council provided oversight for the radio service, subject to the approval of the minister for post and telegraphs, until the 1960 establishment of Radio Eireann. The director of broadcasting was Maurice Gorham. There was continual disagreement over the scope and function of Irish broadcasting because of lack of public funds, the need to make a decision about television, and the British electronic "invasion." There were nearly 500,000 licensed listeners for Irish radio during this period, though not all areas of the country could receive a signal. The government subsidy was augmented with the sale of program sponsorships. There were 380 employees of the radio operation, including actors, writers, and engineers, as well as the Irish Symphony Orchestra.

Radio Telefis Eireann, the "Pirates," and the Coming of Television 1960–1984

Irish radio in this period found a new regulatory environment, new sources of revenue from advertising, expanded coverage throughout the day, a Gaelic language station, authoritative news coverage, FM stations, and radio pirates. It was during this era that building a television service appeared foremost in the minds of government officials. Radio Eireann was established as an autonomous entity to be the authority over both radio and television. The first director general of the authority was an American, Edward Roth, who had been a consultant to the National Broadcasting Company (NBC) and had experience in starting up stations in Mexico and Peru. His attention, and that of others at Radio Eireann, was on television. There was a grand exodus from the radio service to the television side of "The House," which moved to new quarters in Donnybrook. Radio Telefis Eireann (RTE) was adopted by the authority as the official name of the public service broadcaster.

The rise of rock and roll and pirate radio stations had a tremendous impact on Irish broadcasting. In 1978 alone, more than two dozen pirate stations went on the air. There were numerous raids to shut them down in the early years, but these proved to be ineffective. There was little direction from the government regarding these popular broadcasters, and this, coupled with the lure of advertising dollars, allowed illegal radio operators to play popular music to the delight of their audiences. RTE's response was to launch its own music station (RTE 2) directed toward the under-25-year-olds, who represented more than half of the population. By 1984 there were more than 70 illegal radio stations operating and gaining increasing numbers of listeners and advertising dollars.

Pirate radio became a part of Irish life, it was commercially successful, the announcers had "star" status, and the government appeared helpless to do anything about it. Raids to shut down Radio Nova and Sunshine Radio in Dublin in 1983 resulted in large public demonstrations in support of these highly popular stations.

Rise of Independent Broadcasting since 1984

Ireland entered the local and private broadcasting arena at a different time and under different circumstances than most countries. The procedures and means for silencing the pirates and bringing on board replacements suitable for the audience was to be no small feat. The Irish audience had grown to depend on the illegal broadcasters during their raid on the island. The Radio and Television Act of 1988 created Ireland's independent broadcast operators and silenced the illegal operators. The Independent Radio and Television Commission (IRTC) was formed, and by the summer of 1990, new radio licenses had been awarded to 24 different ownership groups serving all counties. By 2000, 12 community radio stations, 21 independents, and a new national radio service (Today FM) were serving Irish listeners—with several new licensees set to go on the air as well. RTE introduced Lyric FM as a new service specializing in classical music and extended coverage for its Gaelic language stations.

Organizations and Stations

There are three major types of radio in Ireland, all of them claiming some allegiance to the public service broadcasting ideals that are associated with European radio generally: the state-owned RTE radio network stations, independent local commercial radio, and community radio. In addition, there are five hospital/institutional radio stations.

RTE radio channels that provide national coverage on a variety of frequencies throughout the country include RTE Radio 1; 2FM; Lyric FM; and the Irish language station, Raidio Na Gaeltachta. One local RTE radio service, Radio Cork, operates a limited schedule.

There is one independent national radio channel that provides national coverage on a variety of frequencies throughout the country, called Today FM. By 2000, there were 21 independent local commercial stations: CKR FM, East Coast Radio, Radio Kilkenny, South East Radio, WLR, Radio Kerry, Galway Bay FM, Clare FM, FM 104, LMFM, Tipp FM, Shannonside 104 FM, Highland Radio, Cork 96 FM/County Sound, 98 FM, North West Radio, Midlands Radio 3, Tipperary Mid West Radio, Northern Sound Radio, Limerick 95FM, and Mid West Radio. Ten of these stations serve the Dublin metro area, and the others serve listening areas across the countryside.

The newest radio stations on the air are the community and community of interest radio stations. Community radio stations are owned and controlled by not-for-profit organizations whose structure provides for membership, management, operation, and programming primarily by members of the community at large. These include Community Radio Castlebar, Wired FM, Dublin South Community Radio, West Dublin Community Radio, FLIRT FM, Phoenix FM, South West Clare Community Radio, Cork Campus Radio, Connemara Community Radio, NEAR FM, Community Radio Youghal, Tallaght Community Radio, Radio na Life, and the special interest station Anna Livia FM.

There are three hospital/institutional stations in Dublin, one in Waterford, and one in Cork city.

Programming

Radio programming in Ireland was long associated with extensive sports coverage of the "national" games of Gaelic football and hurling and in particular with the voice of Radio Eireann's best-known sports commentator, Michael O'Hehir. The broadcasting of popular music from outside Ireland was for decades restrained, although an antijazz campaign led by a Catholic priest in the 1930s was not entirely successful. In the 1970s young people in large numbers began to tune into foreign and domestic pirate (unlicensed) radio stations. Since 1988, local independent broadcasting has steadily grown in size and stature.

During the last quarter of the 20th century, the most influential programs on Irish radio were broadcast by RTE Radio 1 and included *Women Today,* which was feminist in concept and execution, and *Morning Ireland,* which still enjoys a very large breakfast and drive-time listenership. The main presenters associated with the great success of daytime speech programming on RTE Radio 1 during this period included David Hanly; Marian Finucane; Pat Kenny; John Bowman; Colm Keane; Myles Dungan; and Ireland's best-known broadcaster, Gay Byrne. Other well-known RTE radio personalities who present lighter programs on 2FM include Gerry Ryan, Dave Fanning, and Larry Gogan. The abiding influence and popularity of speech programming in Ireland is underlined by the fact that the most critically acclaimed program on Ireland's only privately owned national radio service, Today FM, has been *The Last Word,* which is presented in the early evening (5:00 P.M. to 7:00 P.M.) and whose establishing presenter Eamon Dunphy developed a distinctive style of in-depth discussion of controversial issues.

The programming of the independent and community radio stations reflects local content and in some cases local music. They have captured the listening ears of the contemporary audience with personalities and recorded music. Both RTE major radio services (2FM and RTE 1) and the majority of the independent stations also have digital signals that allow Irish programming to be heard worldwide via the internet.

Financial Support

Most revenue for radio comes from advertising, although RTE does receive a subsidy from a broadcast license fee, which in 2000 was 70.00 Irish punts (approximately U.S. $100) per subscribing household. In addition to RTE stations, the license fee also subsidizes the National Orchestra and three television services, among other entities. Traditionally, two-thirds of RTE's revenue has come from advertising.

The advertising and sponsorship revenue for the independent radio sector has seen steady growth, with increases of 14 to 17 percent in the years 1997 and 1998. The total radio advertising revenue in 1998 for independent stations was £32.87 million ($45 million U.S.).

Policy and Regulation

Irish radio was first established under the aegis of the Wireless Telegraphy Act of 1926, four years after a bitter civil war and six years after Ireland won its independence from Great Britain. The act defined "wireless telegraphy," in a prophetic manner, as "any system of communicating messages, spoken words, music, images, pictures, prints or other communications, sounds, signs or signals by means of radiated electromagnetic waves." Radio in Ireland operated under the portfolio of the minister for posts and telegraphs until the Broadcasting Authority Act was passed in April 1960. Under terms of this law, the Radio Eireann Authority regulated radio broadcasting; it became the Radio Telefis Eireann Authority in 1966 and is most commonly referred to as RTE. The Radio and Television Act of 1988 created Ireland's independent broadcast operators.

The Irish Constitution and its regulations regarding broadcasting pose several restrictions on freedom of expression. Irish radio, like other broadcasting, operates with a "fairness doctrine." There is an affirmative obligation on the part of all broadcasters, RTE, private, and community stations to be fair, objective, and impartial. The legacy of public service broadcasting includes statutory restrictions on several content areas, including an absolute prohibition on editorializing, the requirement that private radio broadcasting services devote a minimum of 20 percent of their airtime to news and public-affairs programming, and the absolute prohibition on broadcast advertising of religious or political advocacy.

The amount of advertising is statutorily limited to 5 percent for RTE stations and 10 percent for independents. The Broadcasting Authority Act of 1993 amended the Broadcasting Act of 1990 (passed in order to implement the European Communities Directive on Television Broadcasting) and liberalized the amount of revenue that RTE could make from advertising. A 1995 Green Paper on broadcasting examined philosophical and strategic issues for Irish radio and other electronic media.

Two separate oversight committees govern regulation of Irish radio. The Broadcasting Commission of Ireland (BCI, known as the Independent Radio and Television Commission from its foundation in 1988 until 2001) provides oversight and licensing for the independent broadcasters, and the RTE Authority sets policy and oversight for the government-owned broadcasters. Members of both these autonomous policy-guiding bodies are government appointees.

Audience Research

The Joint National Listenership Research Committee (JNLR), comprising all broadcasting organizations, the advertising agencies, and the major advertisers, conducts audience research in the Republic of Ireland. Control of the comprehensive twice-yearly survey is handled by a JNLR Technical Committee consisting of representatives of RTE (the state-owned broadcaster), Today FM (the only privately owned national radio station), the BCI (the regulator), and the Institute of Advertising Practitioners in Ireland (the agencies). The JNLR data provide very specific information on audiences and their preferences, including a detailed analysis of the social class, age, and gender of listeners over the age of 15 for each quarter hour of the day. A JNLR software package is available to permit media buyers to plan their advertising campaigns.

JNLR findings are based on personal interviews, conducted in the home, with a sample of 6,660 people over the age of 15 in the Republic of Ireland. Special adjustments are made to ensure a minimum sample of 200 for each local radio franchise area. Results of the JNLR research are divided into "listenership" and "market share." The "listenership" figures total more than 100 percent because they give equal weighting to any stations heard by the respondents on the day before the survey date, regardless of the duration of listening in each particular case. The "market share" information does total 100 percent and requests that respondents state how long they tuned in to any particular station on the day before the survey date. The latter figure is a better indicator of overall performance than is the former.

During 1999, for the first time, the combined market share of privately owned radio stations edged ahead of the combined market share of the radio services operated by the state-owned RTE. However, Radio 1, which is RTE's flagship radio service and which broadcasts mainly speech programming, continues to surpass by far any other single service, either public or private, with a national audience share around 33 percent.

As the state-owned RTE is dually funded, both from public monies and from advertising, the JNLR results are important to both RTE and the private sector in their competition for revenue.

Radio Audience and Irish Emigration

In the United States, some 30 to 35 million Americans claim Irish ancestry, quite a feat for a country of 3.5 million people. Since the beginning of Irish radio, broadcasters have considered ways and means to reach the millions of people who are connected via ancestors or interest. In 1946 the Irish Government initiated a plan to develop a high-power shortwave station that could broadcast to the United States. The project was aborted before its first broadcast, but not before a 100-kilowatt transmitter and a directional antenna that stretched over 40 miles were constructed. Today's satellite feeds and digital radio on the internet finally allow Irish broadcasters to serve their extended world audience. Irish radio stations are electronically connected via e-mail; all of the national networks and most of the local independent stations have websites; and many are "streaming" their programming, making it available to listeners worldwide. A metamorphosis of Irish radio policy and practice from an insular, nationalistic, and pastoral focus to one that includes international, local, and national constituencies has occurred over four decades. The 21st century promises continued transformations for Irish radio. The voices of Irish radio have always had something to say. They have said it with style. Today they have an expanding world audience tuning in.

COLUM KENNY AND THOMAS A. MCCAIN

Further Reading

Boyle, Raymond, "From Our Gaelic Fields: Radio, Sport, and Nation in Post-Partition Ireland," *Media, Culture, and Society* 14 (1992)

Broadcasting Commission of Ireland, <www.bci.ie>

Broadcast Ireland: Streaming Radio Stations, <www.broadcastireland.com/radio.html>
Cathcart, Rex, *The Most Contrary Region: The BBC in Northern Ireland 1924–84,* Belfast: Blackstaff Press, 1984
Clark, Paddy, *Dublin Calling: 2RN and the Birth of Irish Radio,* Dublin: Radio Telefís Éireann, 1986
Fisher, Desmond, *Broadcasting in Ireland,* London and Boston: Routledge and Kegan Paul, 1978
Gorham, Maurice Anthony Coneys, *Forty Years of Irish Broadcasting,* Dublin: Talbot Press, 1967
Hall, Eamonn G., *The Electronic Age: Telecommunication in Ireland,* Dublin: Oak Tree Press, 1993
Ireland Department of Arts, Culture, and the Gaeltacht, *Gníomhach nó Fulangach? Fáthmheas ar an Réimse Craolacháin; Active or Passive? Broadcasting in the Future Tense* (bilingual Irish-English edition), Baile Átha Cliath, Ireland: Oifig an tSoláthair, 1995
Ireland's National Broadcasting Organisation (RTÉ), <www.rte.ie/>
Kelly, Mary, and Wolfgang Truetzschler, "Ireland," in *The Media in Western Europe: The Euromedia Handbook,* edited by Bernt Stubbe Østergaard, London and Newbury Park, California: Sage, 1992; 2nd edition, London and Thousand Oaks, California: Sage, 1997
Kenny, Colum, "Section 31 and the Censorship of Programs," *Irish Law Times and Solicitor's Journal* 12, no. 3 (March 1994)
Kenny, Colum, "The Politicians, the Promises, and the Mystery Surrounding Radio Bonanza," *Sunday Independent* (21 June 1998)
McCain, Thomas A., "Radio's Oscillating Policies Jolt New Zealand and Ireland—An Applied Research Example," in *Applied Communication in the 21st Century,* edited by Kenneth N. Cissna, Mahwah, New Jersey: Erlbaum, 1995
McCain, Thomas A., and G. Ferrell Lowe, "Localism in Western European Radio Broadcasting: Untangling the Wireless," *Journal of Communication* 40, no. 1 (1990)
Mulryan, Peter, *Radio Radio,* Dublin: Borderline, 1988 (on independent, local, community, and pirate radio in Ireland)

Isay, David 1965–

U.S. Radio Producer

Independent producer David Isay's acclaimed documentaries and features, broadcast on National Public Radio (NPR), beginning in the 1990s served as the benchmark for aurally lush, compelling, and socially responsible radio. Isay, the first radio producer awarded a Guggenheim Fellowship, received a MacArthur Fellowship in 2000.

In an era in which marginalized group members are often depicted as freakish or dangerous, Isay's work paints sensitive portrayals of people living in poverty (*Charlie's Story,* among others), people with all-consuming passions, or people caught up in social struggles (*Remembering Stonewall*). Many pieces investigate spirituality; others highlight vanishing professions or fading historical icons, such as Coney Island, roadside dinosaur museums, and Jewish synagogues in the South.

Isay attended a Jewish elementary school and then a prep school in Connecticut before moving to Manhattan, where he attended high school at Friends Seminary. He graduated from New York University in 1987 and planned to start medical school, a path chosen because many of his family members are physicians. Meanwhile, however, he received grant money to produce a small documentary film about drug addiction. He notified local media about a story he felt merited attention when he met two former addicts planning to open an addiction museum. Only Pacifica station WBAI was interested, and they asked him to report the story. WBAI provided equipment and later helped edit the piece. NPR editor Gary Covino heard the broadcast and lightly re-edited it for airing on NPR; Isay earned $250 for this five-minute debut. Covino has edited most of Isay's projects since then, and Isay set aside plans for medical school to take up a brilliant career as an independent producer. He found that working in audio neatly drew together his abilities and interests.

Isay's mother, Jane, served as an early mentor; her work as a book editor influenced his finely honed editing style. From his father, Richard, a psychoanalyst, he developed an interest in people on the margins of society.

Though based in New York City, Isay has interviewed individuals across America: he has stalked poisonous reptiles with snake handlers in West Virginia and has spent time with men serving life terms in Louisiana's Angola Prison. This latter piece, *Tossing away the Keys,* resulted in the release of an inmate who had unfairly served more than 40 years in Angola.

to the Ministry of Posts and Telecommunications, called "*Yuseishou,*" which was established by combining the Ministry of Telecommunications and the Ministry of Posts in the same year.

After World War II, NHK produced many popular radio programs influenced by similar American shows, and radio broadcasting played an important role in comforting people during the often difficult transition period after the war. Some of those NHK programs included a song program, *Nodojiman Shirouto Ongakukai* (Amateur Song Contest); radio dramas such as *Mukou Sangen Ryoudonari* (My Neighbors) and *Kaneno Naru Oka* (The Hill with a Bell); and quiz programs such as *Kiminonawa* (What Is Your Name?), *Watashiwa Daredeshou* (Who Am I?), and *Tonchi Kyoushitsu* (The Wit Class). Commercial broadcasting stations also aired many entertainment programs.

Radio and Television

Following the postwar reestablishment of radio broadcasting, the first TV service was begun by both NHK and a commercial company, Nihon Television (NTV), in 1953. Because people were quite satisfied with radio broadcasting and enjoyed many favorite programs, initial deployment of the TV service was slow. However, a 1959 live broadcast triggered an increase in the sale of receivers—the wedding ceremony of Prince Akihito and Miss Michiko Shouda, who was the first ordinary citizen to become the bride of an imperial prince. Many advertisers shifted their messages from radio to television, and as a result, the radio services began to lose their major revenue.

Commercial radio stations came up with several strategies to rebuild the popularity of radio listening. They extended broadcasting hours after midnight to target individual audiences; marketed their disc jockeys' unique and interesting personalities as the appeal of a program; and reduced the number of dramas and quiz programs, which were more expensive to produce. Audience segmentation was introduced, meaning that a certain audience group, such as family, housewives, or young people, was targeted as the major audience sought during a certain time period in a day. Another strategy most commercial broadcasters began to use in 1964 was to air night games of professional baseball, especially those played by the *Yomiuri Giants*. Because of the necessity to broadcast those games across the country, commercial radio stations formed two networks, Japan Radio Network (JRN) and National Radio Network (NRN). As of December 2001, JRN had 34 affiliated stations, and NRN had 40. The networks are still mainly used to relay professional baseball games.

These strategies proved successful, and the popularity of radio broadcasting resumed by 1965. Midnight broadcasting was geared toward teenagers. Many disc jockeys played the roles of big brothers and sisters and became national celebrities among younger listeners. In addition, as the number of automobiles in Japan increased significantly in the 1960s and 1970s, radio stations succeeded in demonstrating their importance by providing timely news and information in drive time.

Although there had been strong demand for a license for FM radio broadcasting since the mid-1950s, the ministry had been worried about the commercial feasibility of FM stations. In the meantime, the ministry encouraged NHK to try to explore FM broadcasting by granting it a license as an experimental station. By 1968, NHK's FM broadcasting network covered almost 80 percent of the country. People had begun to realize the high quality of FM reception, and FM radio receivers were becoming widely diffused. Because of that circumstance, Japan's first licensed FM radio broadcasting was started by FM NHK and FM Aichi Music in 1969, and by FM Osaka Music, FM Tokyo, and FM Fukuoka Music a year later.

NHK

A 1950 Broadcast Law reestablished NHK as an independent public corporation responsible for public broadcasting service. NHK neither receives direct funding from the government nor is responsible for any governmental work. Article 7 of the law states that NHK's goal is to broadcast for the public welfare throughout the country. In order to meet this goal, Article 9 requires that NHK be responsible for domestic and international broadcasting, plus research and development to improve broadcasting in general.

NHK has a Board of 12 governors as its supreme decision-making body. Each governor represents a geographical district, and governors are chosen from various fields such as education, culture, science, and industry. Governors are appointed with a term of three years by the prime minister with the approval of the Parliament, and they are expected to operate NHK for the benefit of the people. The board is responsible for an annual budget, operating plans, and master plans for program production. A board of directors runs NHK on a day-to-day basis and consists of a president appointed by the board and a vice president and three directors who are appointed by the president with the approval of the board. In contrast with many other public corporations in Japan, in which finance, management, and personnel matters are subject to strong government influence, NHK possesses complete autonomy.

NHK is financed by fees collected from the audience. The basis of the receiving fees is included in Article 32 of the Broadcast Law, which requires that those able to tune to NHK "have to agree on a receiving contract with NHK." What this means in practice is that listeners have to pay fees to NHK regardless of whether they listen to NHK programs. Because NHK is the sole public broadcasting station independent from any government organization, receiving fees are regarded as the price necessary for its operation. In 2000 such receiving fees accounted for almost 99 percent of NHK's total operating income. A budget that includes projections of the amount of the receiving fees for the coming year, an operational report, and a settlement of accounts of the past year is annually prepared by NHK, submitted to the Minister of Posts and Telecommunications, endorsed by the Cabinet, and approved by the Parliament. Although the minister can add some comments to the budget plan, he or she is not allowed to change it.

NHK has several missions to achieve, as delineated in Article 44 of the Broadcast Law. They are (1) to broadcast quality programs in order to meet a variety of needs of the public and to contribute to the elevation of cultural standards through the broadcasting of quality programs; (2) to support the traditional culture as well as to foster new culture; (3) to provide a nationwide broadcasting network so that programs can be received at any place in the country; (4) to produce programs not only geared toward the whole country but also tailored to individual regions; (5) to conduct research and development and to disseminate the results as widely as possible; and (6) to broadcast NHK programs in foreign countries to contribute to better international understanding. Because of the public nature of NHK (and its past involvement in the government/military propaganda), the Japanese people expect NHK to produce programs that are accurate, fair, politically impartial, and neutral, with the presentation of the widest possible range of viewpoints.

NHK operates three radio services: Radio 1, Radio 2, and FM. According to 2000 statistics, Radio 1 devoted about half its time to news, about a quarter to cultural programs, and the final quarter to entertainment (a minimal amount of time, about 2 percent, went to education). On Radio 2, 66 percent was educational programs, followed by cultural programs with 20 percent and news with 14 percent. For the FM service, 41 percent was devoted to cultural programs, 37 percent to entertainment, 18 percent to news, and 4 percent to education. These data show that the primary focuses of Radio 1, Radio 2, and FM are news, education, and culture, respectively. In international broadcasting, 66 percent of the programs were news, and 32 percent dealt with various types of information. As of July 2001, Radio 1 had 214 stations, Radio 2 had 140 stations, and FM NHK had 520 stations in the country. These stations effectively cover all of Japan.

Present Radio System

All radio stations in Japan today are categorized as either general (i.e., they must air news, and more than 30 percent of programs must be education and culture related) or specialized. The latter are further divided into education or music. For an education station, more than 50 percent of programs must be educational, and more than 30 percent must be cultural. NHK Radio 2 fits into this category.

As of December 2001, there were 47 commercial radio companies, holding among them 252 AM stations; 35 of these companies were also operating TV broadcasting stations. There were 53 companies holding altogether 242 FM stations. All FM stations (except for nine outlets) were music stations and belonged to either Japan FM Network (JFN: 37 stations) or Japan FM League (JFL: five stations). Other stations included one short-wave station, one BS analog station, 10 BS digital stations, one CS analog station, six CS digital stations, and 140 FM community stations.

Sixty-eight percent of all radio programs in both NHK stations and commercial stations focused on entertainment, 14 percent was related to culture, 13 percent was news, and 3 percent was education. Advertising expenses in radio in 2000 amounted to \$1.73 billion in U.S. dollars and accounted for 3.4 percent of the total advertising expenses in Japan.

RYOTA ONO

Further Reading

20 seiki housou shi [Broadcasting History in the 20th Century], Tokyo: NHK, 2001

2002 NHK Nenkan [NHK Radio and Television Yearbook 2002], Tokyo: NHK, 2002

Housou Handbook: Bunka o Ninau Minpou no Gyoumu Chishiki [Broadcasting Handbook], Tokyo: Toyo Keizaishinpo, 1997

Inada, Taneteru, *Housou media nyumon* [Introduction to Broadcasting Media], Tokyo: Shakai Hyouronsha, 1996

Ito, Masami, et al., *Broadcasting in Japan,* London: Routledge, 1979

Kataoka, Toshio, *Housou gairon: Seidono haikei o saguru* [An Introduction to the Broadcasting System in Japan], Tokyo: Nihon Housou Shuppan Kyokai, 1988

NHK Data Book Sekaino Housou 2002 [NHK Data Book of World Broadcasting 2002], Tokyo: NHK Housou Bunka Kenkyujo, 2002

NHK Sogo Housou Bunka Kenkyujo, *50 Years of Japanese Broadcasting,* Tokyo: NHK, 1977

NHK Sogo Hoso Bunka Kenkyujo, *The History of Broadcasting in Japan,* edited by the History Compilation Room, Radio & TV Culture Research Institute, Tokyo: NHK, 1967

Nishi, Tadashi, *Zukai housou gyokai handbook* [A Diagrammed Guide to the Broadcasting Industry], Tokyo: Toyo Keizaishinpo, 1998

Tracey, Michael, "Japan: Broadcasting in a New Democracy," in *The Politics of Broadcasting,* edited by Raymond Kuhn, New York: St. Martin's Press, 1985

Tracey, Michael, *The Decline and Fall of Public Service Broadcasting,* Oxford and New York: Oxford University Press, 1998

Jazz Format

The growth of television in the early 1950s gradually replaced commercial radio as the primary family entertainment medium in the home. Radio adapted to a new role by establishing radio formats. Individual stations targeted narrow audience segments by specializing in news, talk, or any of a variety of music genres.

Although jazz music traces its roots to the formative years of the United States, it did not evolve into a bona fide radio music format until the 1950s. During the 1920s and 1930s, mainstream society viewed jazz music in the same way that rock and roll was viewed in the late 1950s: it was considered decadent. Because jazz was closely identified with black culture and affected by the racism that prevailed nationwide during that time, many of the greatest black jazz artists fled to Europe, some permanently, where they and their music were accepted openly. Paris became a cultural center of American jazz music, and the jazz music genre remains very popular there to this day in clubs and on radio. In fact, jazz radio is probably more popular in Europe than it is in the United States.

What is jazz radio? Jazz fans sometimes refer to it as music by musicians, not electricians. It includes blues, swing, bebop, fusion, Latin, and a number of other subcategories within its overall definition. Music that includes flat line piano, boring guitar, and braying saxophone is sometimes associated with jazz, but not legitimately so. New-wave music is not jazz.

Jazz grew out of blues, ragtime, and Dixieland music during the early 20th century. Following the swing era of the 1930s and 1940s, the style evolved into bebop, modern, cool, and other straight-ahead sounds by the 1960s. By this time, jazz had also been accepted as a legitimate popular art form throughout the United States. American audiences embraced the music as much as the Europeans had a decade earlier.

Jazz radio on the East Coast most likely had its roots in "Symphony Sid" Torin's live WJZ-AM radio broadcasts from the Royal Roost in New York City. Trumpeter Rex Stewart and critic and composer Leonard Feather had their own shows on AM in the early 1950s, as did Felix Grant and Ed Beech. These jazz disc jockeys are important, because early jazz formats were very much personality driven and involved a lot of talk in addition to the music.

"Sleepy" Stein, who was doing all-nighters in Chicago, moved his show to KNOB-FM in Los Angeles in 1956, and West Coast jazz radio was established. KJAZ in San Francisco had Pat Henry. KNOB and KJAZ had similar formats: one- and two-hour programs that were oriented totally toward personalities, with disc jockeys involved in lengthy announcements preceding and following each song played, which included mentioning every player or sideman, the composer, and even the record label. There were not that many jazz recordings at the time, and talkative disc jockey personalities could play virtually every current jazz record release over a 24-hour period. Other jazz radio personalities included Dick Buckley, Howard LaCroft, Frank Evans, Bob Young, Al Fox, Al "Jazzbo" Collins, Jim Gosa, Pete Smith, Dick McGarvin, and Chuck Niles. Niles, probably the "dean" of active jocks, was still doing a regular show on KLON-FM, Long Beach, as the new millennium began.

The growth of FM radio and the intrusion of album-oriented rock and underground radio during the 1970s resulted in the erosion of commercial jazz formats. For example, KKGO in Los Angeles converted to a classical music format, and KZJZ in St. Louis simply went off the air. By the mid-1990s, only a few jazz formats remained on commercial radio stations in the United States. Noncommercial radio filled the void, however, with jazz being adopted as a format at several large public radio stations and at many college stations. In 1977 National Public Radio (NPR) initiated *Jazz Alive,* a public radio network show that almost all NPR stations carried. Today, the two stations with the largest international audiences are both noncommercial: WBGO-FM in Newark, New Jersey, and KLON-FM in Long Beach, California. Both of these large public radio stations are carried on numerous cable systems and are relayed via satellite to the far corners of the globe. KLON has a special program titled *Euro Jazz* that is relayed via satellite throughout the European continent.

The format itself has changed as well. Research into audience behavior has demonstrated that most listeners want less talk and more music. This "modal music research" calls for serving a greater number of people, replacing talk with 30-second breaks, limiting the number of announcements, and not airing anything squeaky or long. The jazz format of the future will involve less talk, sharpen the focus of the music, and include memorable moments in jazz history. In all likelihood, the music will be based in the jazz styles made popular in the 1950s and 1960s (i.e., bebop, cool, straight-ahead). The big band era is blending with early rock and roll popular music into a successful commercial format.

New developments in electronic technology, such as digital audio via satellite and multiple stream audio on demand via the internet, provide new venues for jazz radio formats. For example, DirecTV carries jazz programming in Japan, and the number of internet radio stations is growing rapidly. XM Radio and Sirius are two commercial satellite radio services that offer more than 100 channels each, several of which are devoted to various forms of jazz music. These mobile services are provided to subscribers for a monthly fee and are not likely to supplant FM radio as a mainstream jazz music format. Given the fact that mainstream jazz music will always target a narrow niche, the new multi channel environment provides an ideal conduit for delivering jazz to a growing number of fans all over the world. The music itself is established as a truly American music genre; it is taught in schools and clinics and performed by prestigious musical groups throughout the world, much as classical music has been for centuries. Accordingly, the music will continue to evolve but will also remain a viable radio format for years to come.

ROBERT G. FINNEY

See also FM Radio; National Public Radio

Further Reading

Eberly, Philip K., *Music in the Air: America's Changing Tastes in Popular Music, 1920–1980,* New York: Hastings House, 1982

Gioia, Ted, *The History of Jazz,* New York: Oxford University Press, 1997

Harris, Rex, *Jazz,* London: Penguin, 1952

Kenney, William Howland, *Recorded Music in American Life: The Phonograph and Popular Memory, 1890–1945,* New York: Oxford University Press, 1999

Jehovah's Witnesses and Radio

The religious group that took the name Jehovah's Witnesses in 1931 (prior to that date the group preferred the name Bible Students) owned and operated several radio stations in the United States and Canada beginning in 1924 and used syndicated recordings on hundreds of commercial stations to supplement its broadcast outreach between 1931 and 1937. The controversial and often confrontational views of the sect involved it in frequent conflicts with other denominations and with broadcasting regulators.

The story of the Bible Students'/Jehovah's Witnesses' involvement in radio is largely the story of their second president, Judge Joseph Franklin Rutherford—a former Missouri lawyer and substitute judge with a commanding personality and a booming orator's voice. Born in 1869, Rutherford took control of the Watch Tower Society and its associated groups in 1916, following the death of the sect's founder, Pastor Charles Taze Russell. In 1917 Rutherford and several associates were convicted of sedition for their public opposition to the World War I draft; their convictions were overturned in 1919. Upon his release from prison, Rutherford took steps to revive the struggling movement, implementing a renewed program of publishing and public speaking.

As part of this effort Judge Rutherford delivered his first radio address over station WGL in Philadelphia on 16 April 1922. Soon afterward, the Bible Students acquired a plot of land on Staten Island, New York, and began construction of their own broadcasting station.

On 24 February 1924 the Bible Students inaugurated station WBBR. The noncommercial station featured classical music and hymns performed by Bible Student musicians, talks on home economics and other practical subjects, and lectures on the group's complex interpretations of Bible prophecy and chronology by Rutherford and others. The success of WBBR led to the operation of additional Bible Student stations over the next several years, including WORD in Batavia, Illinois (later WCHI), KFWM in Oakland, California, CYFC in Vancouver, British Columbia, CHCY in Edmonton, Alberta, CHUC in Saskatoon, Saskatchewan, and CKCX in Toronto, Ontario.

During 1927 and 1928 the broadcasting activities of the Bible Students began to attract the attention of government regulators. In June 1927 Rutherford testified before the Federal Radio Commission (FRC) to protest a decision denying WBBR's application to share the frequency of station WJZ, owned by the Radio Corporation of America and the flagship of the newly formed National Broadcasting Company's (NBC) Blue network, alleging that NBC was part of a religious/commercial conspiracy seeking to deny his group fair access to the radio audience. Although the FRC dismissed Rutherford's complaint, NBC offered the Bible Students free air time for the broadcast of a talk by Rutherford. The speech, entitled "Freedom for the Peoples," was delivered on 24 July 1927 and in it Rutherford denounced all other religions, the clergy, big business, and all human governments as agents of Satan. A barrage of complaints received in the wake of this address led NBC to deny Rutherford and the Bible Students any further access to its stations.

The following year the Bible Students ran afoul of broadcasting authorities in Canada. Protests from clergymen over the broadcasts of Bible Student stations in Canada were followed by allegations that the group had on two occasions sold time over its Saskatoon outlet to the Ku Klux Klan. On the strength of these complaints the licenses for the Canadian stations were revoked in March 1928.

Unable to secure time on any established network, the Bible Students turned to a network of their own—buying time on over a hundred stations from 1928 through 1930 for "The Watchtower Hour." Anchored by WBBR, and connected by American Telephone and Telegraph (AT&T) circuits, the "Watchtower Network" functioned for an hour each week, presenting talks by Rutherford and hymns performed by Bible Student musicians.

Increased costs led the organization to discontinue the live network at the end of 1930, replacing it with transcribed syndication. The Watch Tower Society purchased time on local stations and the lectures of Judge Rutherford were distributed on 16-inch shellac transcriptions manufactured by the Columbia Phonograph Company. By 1933 over 400 stations around the world were broadcasting these 12-minute talks. This "wax chain" would be supplemented by occasional live hookups from the organization's annual conventions.

As the Depression deepened, the Witnesses (adopting that name in July 1931) became increasingly combative in their attacks on organized religion, politicians, and big business and this, in turn, brought them into further conflict with station owners and the FRC. Catholic authorities, especially, took offense at Rutherford's statements and pressured station owners to discontinue the broadcasts. In 1933 the Witnesses began a nationwide petition drive for "freedom of broadcasting" and presented more than 2 million signatures to the FRC in early 1934. Allegations were immediately made that many of the signatures were forged, but the petition led Representative Louis McFadden (R-Pennsylvania) to introduce a bill that would require broadcasters to guarantee free and equal use of air time to all nonprofit organizations. Several Watch Tower Society representatives were among those testifying for this bill in March 1934. Buried under an avalanche of opposition from the National Association of Broadcasters, the established networks, and the Federal Council of Churches of Christ, the bill died quietly in committee.

Opposition to the Witnesses' broadcasts mounted steadily during the mid-1930s, culminating in Philadelphia in 1936 when Catholic leaders urged a boycott of Gimbels Department Store, owners of station WIP, which had carried the Rutherford programs for several years. Stations became increasingly reluctant to sell time to the Watch Tower Society and finally, in October 1937, the organization announced its withdrawal from commercial broadcasting, although it used special hookups for convention broadcasts until 1941.

Rutherford died in 1942 and his successors have moved away from his aggressive stances, ignoring the broadcast media in favor of direct house-to-house canvassing. Station WBBR remained in operation until 15 April 1957, when it was quietly sold. The call letters were changed to WPOW and the once-combative voice of the Watch Tower Society became a commercial station specializing in recorded music.

ELIZABETH MCLEOD

See also Religion on Radio

Futher Reading

1975 Year Book of Jehovah's Witnesses

"Bible Students Protest," *Broadcasting* (1 February 1934)

Duffy, Dennis J., *Imagine Please: Early Radio Broadcasting in British Columbia,* Victoria, British Columbia: Sound and Moving Image Division, Province of British Columbia, Ministry of Provincial Secretary and Government Services, Provincial Archive, 1983

Jehovah's Witnesses in the Divine Purpose, New York: Watch Tower Bible and Tract Society, 1959

"McFadden's Religious Time Bill Believed Defeated at Hearings," *Broadcasting* (1 April 1934)

Penton, M. James, *Apocalypse Delayed: The Story of Jehovah's Witnesses,* Toronto and Buffalo, New York: University of Toronto Press, 1985; 2nd edition, 1997

Rutherford, Joseph Franklin, *Freedom for the Peoples,* New York: Watch Tower Bible and Tract Society, 1927

Rutherford, Joseph Franklin, *Government: Speech of World-Wide Interest, Broadcast to All Continents: Hiding the Truth, Why?* Brooklyn, New York: Watch Tower Bible and Tract Society, 1935

Rutherford, Joseph Franklin, *Enemies: The Proof That Definitely Identifies All Enemies, Exposes Their Methods of Operation, and Points Out the Way of Complete Protection for Those Who Love Righteousness,* Brooklyn, New York: Watch Tower Bible and Tract Society, 1937

"Rutherford Protests to Radio Commission," *New York Times* (15 June 1927)

"Under Constant Duress," *Broadcasting* (1 April 1934)

Jepko, Herb 1931–1995

U.S. Radio Talk Show Host

Herb Jepko began what would later become the Nitecap Radio Network on KSL, Salt Lake City, Utah, on 11 February 1964 and is acknowledged by many to be the father of network talk radio.

Origins

Born in Colorado and adopted as an infant by his stepparents in Prescott, Arizona, Jepko was the only child in his family. His adoptive mother left before he was four years old and his adoptive father, Metro, was forced into Veteran's Administration hospitals on and off for the next ten years for injuries he sustained in World War I. During those times Jepko was raised by a series of foster parents until he and his father were reunited for most of his late teenage years. At age 18 Jepko entered the U.S. Army, where he learned the crafts of film production and radio broadcasting.

In the late 1950s and early 1960s Jepko pursued a series of radio jobs along the California coast, in Idaho, and eventually in Utah. He was hired by KDYL in Salt Lake City to play late-night jazz, but he soon grew bored playing music. Working a late-night shift, he began talking to his listeners about the mountains, the weather, and non-controversial news of the day. Since there were no facilities for broadcasting calls over the air, Jepko would talk to his listeners over the phone off the air while music was playing. Mail began arriving from all over the Rocky Mountain area, and it was not long before the managers of Mormon-owned 50,000 watt KSL took note of the new, talented air personality across town. While KSL had a history of hiring only Latter Day Saints (Mormons), exceptions were sometimes made for on-air talent. Jepko's nominal Catholicism and occasional drinking would later create significant tensions.

In 1962 Jepko was hired to host KSL's midday program *Crossroads*. He quickly realized that this was a different audience than the one he had been connecting with in his late-night shift at KDYL. Perhaps in part because of his unstable childhood, it was this late-night audience of the lonely, fearful, sickly, or disaffected for whom he felt the greatest affinity. Despite being a 50,000-watt clear channel, KSL was off the air from 1 A.M. to 5 A.M., and Jepko was anxious to use that time period. KSL management felt no good could come from giving listeners a reason to be awake at that hour and no advertisers would pay to reach those listeners who were awake all night. Jepko had so much confidence in his ability to make the time period work, however, that he accepted a 50 percent cut in pay, resigned from his midday job, and agreed to a six-week deadline to prove he could make the show work, or be fired.

Nitecaps Program

Just after midnight on 11 February 1964, Jepko began what would become the most successful program in KSL's history. For the first few weeks of the program, station engineers had rigged only one phone line that could be put on the air. Even on that first night, however, Jepko was seldom at a loss for callers. One of the first advertisers was for a venison cookbook, and Jepko's wife Patsy, who supported him in this late-night experiment, found herself filling orders on the kitchen table while listening to the show at night. After a night on the air, Jepko would return at 7 A.M. to shower, change, and begin making sales calls while Patsy took care of the kids. For the first year of the show, they subsisted on two or three hours of sleep each weeknight.

The program's unprecedented success was almost immediate. Within a year, local chapters of *Nitecaps,* as they became known following an on-air contest to choose a name, were meeting in five states. As the only workable method for responding to the thousands of letters received each day, Jepko began publishing *The Wick,* a monthly magazine, in June 1965. Conventions were held that drew thousands of listeners from over 30 states. KSL managers and Latter Day Saints church officials belatedly realized they had a perfect platform from which to reach those in need with proselytizing messages, but Jepko fought to keep the program away from issues of controversy and denominational religion. Some within the station expressed concerns that such a successful program was being run on church-owned facilities by a nonmember. Eventually Arch Madsen, president of KSL, would side with Jepko, allowing him to buy time from the station and operate as an independent contractor, but tensions between other KSL personnel and Jepko continued to intensify.

In January 1968 KXIV Phoenix signed on as the first Nitecap Radio Network affiliate, and network talk radio was born. The 50,000-watt clear channel KVOO, Tulsa, Oklahoma, carried the show from October 1969 to September 1971. In January 1973 WHAS Louisville, Kentucky, another 50,000-watt clear channel station, picked up the show, giving Jepko the first-ever coast-to-coast coverage for a talk show. By mid-1974, more than 80 "Nitestands" were active, with members meeting for socializing and to organize service to the elderly in their areas. While the core of the *Nitecap* audience was retired, Arbitron estimates showed a surprising number of 18–24 year olds in the midnight-to-1 A.M. hour (the only overnight hour measured by Arbitron at the time). Outside of Salt Lake City (KSL) and Louisville, Kentucky (WHAS), the audience was largely rural, because of the difficulties of receiving, in a dense metropolitan area, a clear signal from a distant AM station.

Almost all revenues from the show were derived from direct-response advertising, and Jepko insisted that *Nitecap* staff handle all order fulfillment. Too often he had seen listeners of other programs robbed by direct-response advertisers who went bankrupt without fulfilling paid orders. So Jepko required Nitecap Radio Network advertisers to ship sufficient product to his Salt Lake City offices before allowing the commercials to air. Other revenue streams included commissions earned by a travel agency and insurance company franchise operated by Nitecaps International. At its peak, a staff of 15 full-time employees handled order fulfillment, insurance policies, Nitecap tours, membership cards, and publishing of *The Wick*.

In November 1974 WBAL Baltimore affiliated with the Nitecap Radio Network, providing a clear signal into Washington, D.C., Philadelphia, and New York. This provided the first exposure to the program for many media executives living on the East Coast. On 4 November 1975, the Mutual Broadcasting System (MBS) began carrying the *Nitecap* program nationally. Thanks to the dozens of new affiliates, membership in the Nitecaps International Association (NIA) soared to over 300,000. Within a year, Arbitron estimated 10 million listeners were tuned in to the program nightly. NIA memberships likewise soared as new "Nitestands" were established across the nation. Following the affiliation with MBS, Jepko distanced himself further from KSL by building separate studios and offices for the show.

Unfortunately, as part of the agreement for MBS to carry the show, Jepko had to relinquish control over order fulfillment and sales strategy to the network. MBS sales executives attempted to sell the program on a cost-per-thousand (CPM) basis to national advertisers, but the reputation of the show as appealing primarily to an older, less-affluent, rural audience made this a difficult sale. MBS account executives quickly moved on to more profitable dayparts, and Jepko, while still enjoying the largest audience in the history of the program, began to suffer extraordinary financial losses. The combination of the increased expenses of operating his own studios and the decrease in revenue from the shift in sales strategy away from proven direct-response advertising would ultimately prove fatal.

By the fall of 1976 Jepko was under intense pressure from MBS to change the program to attract a younger audience. MBS felt more controversy and conflict would serve to attract the younger demographics, which would be easier to sell on a CPM basis to national advertisers. Jepko in turn pressured MBS to continue to serve the existing audience, to whom he

Herb Jepko
Courtesy Radio Hall of Fame

felt a strong commitment. With MBS account executives unwilling, or unable, to work deals for sufficient direct-response advertising, the show began to fail. On 28 May 1977 MBS replaced the *Nitecap* program with Long John Neble and Candy Jones. When Long John died less than a year later, MBS gave Larry King his first shot at a national audience. (Others later mistakenly cite King's program as the first network radio talk show.)

In June 1977 the Nitecap Radio Network was reborn with 10 affiliates, including KSL, but without WHAS. By 1977 14 more affiliates had joined the network, but most used low power at night and the vast majority of the core audience was unable to receive the program. Network radio, in an era before satellite distribution, required expensive American Telephone and Telegraph (AT&T) landlines, which were feasible only with the economies of scale offered by scores of stations sharing the cost. Larry King, other all-night radio programs, 24-hour broadcast television, and cable TV had fractionalized the late-night radio audience. As renewal of *Wick* subscriptions dropped, Nitestands disbanded, and orders from direct-response appeals declined, old tensions between KSL management and Jepko resurfaced, aggravated by his increased drinking. In August 1978 KSL, without advance notice to listeners or to Jepko, dropped the *Nitecap* program.

The show continued for about a year before going off the air in August 1979. Jepko made several attempts to resurrect the program in the 1980s, with the most successful involving purchasing time on WOAI, a 50,000-watt clear channel in San Antonio, Texas, which lasted for several months. The last attempt to restart the program over was made at KTKK in Salt Lake City in 1990 with Randy Jepko, Jepko's son from Patsy's first marriage, serving as co-host, but the show lasted only a few months. When Jepko's son Herb Junior died in 1992, it proved to be the final tragedy from which Jepko could not recover. He died in Salt Lake City of complications from drinking on 31 March 1995.

JOSEPH G. BUCHMAN

See also All Night Radio; KSL; Mutual Broadcasting System

Herbert Earl Jepko. Born in Hayden, Colorado, 20 March 1931. Adopted by Metro and Nellie A. Jepko of Prescott, Arizona; completed two years of college and enlisted in U.S. Army as broadcast specialist, 1953–55; worked odd jobs in radio in California and Idaho; landed late-night shift at KDYL, Salt Lake City, 1961; joined KSL, Salt Lake City, as host of midday *Crossroads;* launched *The Other Side of the Day,* midnight to 6:00 A.M., 11 February 1964, later renamed the *Nitecaps;* first network talk radio program, 1 January 1968; operated as *Nitecap Radio Network* through 4 November 1975. Died in Salt Lake City, Utah, 31 March 1995.

Radio Series

1964–77 *The Other Side of the Day* (became *Nitecaps*)

Further Reading

Bick, Jane Horowitz, "The Development of Two-Way Talk Radio in America," Ph.D. diss., University of Massachusetts, 1987

Broder, Mitch, "Small-Talk Show Is Big Draw in Wee Hours," *The New York Times* (23 February 1975)

Denton, Craig L., "Revamping the Wick Magazine," Master's thesis, University of Utah, 1976

Rose, Hilly, "The Talkers," in *But That's Not What I Called About,* by Rose, Chicago: Contemporary Books, 1978

Jewish Radio Programs in the United States

Religious broadcasts usually connote church services and evangelists. Although it is true that most radio religious programs have been directed at a Christian audience, a number of Jewish programs have been on the air since radio's earliest years.

Origins

In March 1922 New York station WJZ's radio listing announced a "radio chapel service," featuring a talk by Rabbi Solomon Foster and music by cantor Maurice Cowan. Although it had been customary for radio stations to offer short inspirational messages (usually in the morning), these were usually provided by well-known Christian clergy. In some cities with large Jewish populations, however, a rabbi was occasionally asked to speak. To the listener in 1922–23, this was something of a social revolution. Most Christians had never met or heard a rabbi before.

Rabbi Harry Levi of Temple Israel, a Reform congregation, was invited to take a turn on a daily religious program that was broadcast over Boston's WNAC in mid-1923. He got such a positive response that he was invited back. By January 1924 WNAC made arrangements to broadcast Temple Israel's services twice a month, certainly the first time most non-Jews had encountered what a Jewish worship service was like. (In March 1926 listeners could also hear a Jewish wedding, as New York's WRNY made Winnie Gordon and Julius Goldberg, along with Rabbi Josef Hoffman, radio stars for a day.) Boston's "Radio Rabbi" Harry Levi became so popular during 16 years on the air that two books of his sermons were issued, and numerous non-Jews who heard him on WNAC wrote him fan letters or came to his temple to ask for his autograph.

Radio in the early 1920s provided its audience with the chance to hear some of Judaism's biggest names, including New York's Rabbi Stephen S. Wise, long regarded by the print media as a spokesman for liberal Jews: both the *New York Times* and *Time* magazine often quoted him. As early as March 1922, Rabbi Wise gave a radio talk to encourage donations to help European Jewish refugees. On a fairly regular basis throughout the 1920s, his speaking engagements were broadcast from a number of cities; Rabbi Wise's sermons were often about the dangers of intolerance, such as a 1924 speech to protest the growing popularity of the Ku Klux Klan. A dynamic orator, Rabbi Wise developed such a following that the radio editors at major newspapers often used a bold headline and a photo to let the audience know that he was about to give another radio sermon.

Radio produced a sort of ecumenism; one listing for New York's WEAF in December 1923 featured Christmas songs for children at 6 P.M. followed by Hanukkah songs at 7. Seeing the possibilities in radio as a vehicle to promote understanding, Dr. Cyrus Adler, a scholar and president of Dropsie College in Philadelphia, helped to create a weekly program that would not be limited to sermons. In cooperation with the United Synagogues of America, an organization of Conservative Jews, the Wednesday night program went on the air in late August 1923 on WEAF, after first offering an experimental broadcast in May to see if the response would be positive. It was. This weekly Jewish program featured Jewish folk and liturgical music (sometimes sung in Yiddish, sometimes in Hebrew), discussions of Jewish holidays, and a number of famous speakers from all over the United States.

Entertainment

Jewish programming in the 1920s was not only religious in nature. Thanks to radio, listeners were able to hear two famous Cantors—the great cantor Josef (Yossele) Rosenblatt, along with the popular Jewish comedian Eddie Cantor; both appeared at a 1924 banquet to honor the Young Men's Hebrew Association's 50th anniversary. There were also programs of popular music to benefit Jewish causes; Jewish bandleaders and performers such as Irving Berlin, Leo Reisman, and Eddie Cantor were among those who participated. And by the late 1920s, most cities from the East Coast to the West Coast had rabbis on the air, usually around the major Jewish holidays. In addition to rabbis and scholars discussing Jewish customs, there was at least one popular radio show with a Jewish immigrant family as the protagonists— Gertrude Berg had created a comedy-drama called *The Goldbergs,* which began a successful run on the National Broadcasting Company (NBC) in November 1929. It was an era when many performers with ethnic names changed them to sound more "American," yet Gertrude Berg did not hide the ethnicity of her characters, nor did she hide her own ethnicity. She wrote a syndicated advice column for the Jewish press and did speaking engagements on behalf of Jewish charities. And unlike some comedy routines such as "Cohen on the Telephone," a hit record that made fun of a Jewish immigrant who was losing his battle with the English language, the Goldbergs were portrayed sympathetically, and anyone of any religion could identify with their problems. (There were also a number of popular singers and comedians on radio who were Jewish, such as Jack Benny, Fanny Brice, and Al Jolson, but at a time when anti-Semitism still flourished, most Jewish entertainers did not make overt mention of their religion.)

By the 1930s, some network programs featured Jewish themes. One of the earliest was *Message of Israel*, first heard on NBC Blue in late 1934 and hosted by Lazar Weiner, the music director of New York's Central Synagogue. Boston's Harry Levi was invited to speak on this program, which featured some of America's best-known Reform rabbis, in 1937 and again in 1938.

There were also a number of charitable and philanthropic organizations that provided radio programs, such as Hadassah, the Federation of Jewish Charities, and the American Jewish Joint Distribution Committee. Also, when a special occasion took place, such as the 50th anniversary of the Jewish Theological Seminary in March 1937, highlights from the event were broadcast. At a time when Federal Communications Commission (FCC) guidelines called for a certain amount of religion and public service, such programs served a useful purpose for the station while providing the Jewish organizations far-reaching exposure they would not otherwise have received.

And then there was the rabbi who decided to leave the pulpit to become a radio singer. Rabbi Abraham Feinburg took the air name "Anthony Frome"; known as the "Poet Prince of the Airwaves," he could be heard singing love songs on several New York stations from 1932 to 1935, at which time he gave it all up and went back to being a rabbi again. On the other hand, there was a famous opera singer who also became a cantor; his radio concerts were critically acclaimed whether he was doing Hebrew prayers or portraying the lead tenor role in "La Traviata." Jan Peerce was discovered by the famous impresario Samuel Rothafel (better known as Roxy) while singing at a hotel banquet in 1932. Soon, he was singing on NBC Blue's *Radio City Music Hall of the Air,* and by 1941 at the Metropolitan Opera, where he performed for 27 years. But throughout his life, as he had done in his neighborhood synagogue before he became famous, Peerce would chant the Jewish liturgical prayers at the High Holy days. He also made a number of recordings of sacred Jewish music, some of which have been reissued.

During the 1930s a few radio stations were airing mainly ethnic and foreign language programs, brokering out segments of the day to particular groups. One of the best known ethnic stations was New York's WEVD, where some long-running Yiddish programs made their home; thanks to WEVD, it was possible to hear anything from folk songs to entire Yiddish plays. And because WEVD had a working agreement with a Jewish newspaper called the *Forverts* ("Forward"), there were always commentators and critics who discussed the news from a Jewish perspective. WEVD also had Moses Asch, who would go on to found Folkways Records, but who in the mid-1930s hired and recorded many of the performers whose music was played on WEVD. America in the 1930s still had many immigrants who missed the culture of the old country, and radio helped to provide it. One program, *Yiddish Melodies in Swing,* went on the air on New York's WHN in 1939, and in 1941 fans were still waiting in line to get tickets to be in the studio audience.

Speaking Out

However, the 1930s became a more serious time for Jewish broadcasters as the situation in Europe worsened. As news of Hitler and the Nazis dominated newspapers, some Jewish radio programs began providing news and information that the network newscasts were hesitant to mention. Nazi military conquests were front-page news, but it was not until the 1940s that newspapers such as the *New York Times* finally decided the extermination of Jewish people was a major story. Thus, it was up to the Jewish press, and the news commentators on Jewish programs (along with a few non-Jewish commentators who spoke out, most notably NBC's Dorothy Thompson), to make sure the story was told. Rabbi Stephen Wise took to the airwaves to condemn fascism overseas while also condemning bigotry in America, as personified by Father Charles Coughlin, the anti-Semitic radio priest. The chairmen and women of many Jewish organizations, such as the American Jewish Committee and the Jewish Labor Committee, decried the persecution of Jews in Europe and tried to raise funds to help them. But Jewish public affairs programs were usually short and were seldom on the air more than once a week. Although they did call attention to the problems Jews faced, they could not compensate for the lack of coverage the rest of the week.

Ethics and Culture

The next major network program with a Jewish theme came from the Jewish Theological Seminary. Developed by seminary president Louis Finkelstein, *The Eternal Light* first aired on NBC in October 1944, featuring radio dramas about biblical personalities and famous Jewish men and women past and present. It often presented thought-provoking stories with ethical dimensions, and it was still on the air (having moved to television) in the 1980s, celebrating its 40th anniversary in 1985. Among the famous performers who were heard on this award-winning program over the years were Ed Asner, Gene Wilder, and E.G. Marshall. At the height of its popularity, this program was heard on more than 100 stations, and it won a Peabody Award for excellence.

Some Jewish celebrities began to offer their own radio shows, making use of Jewish or Yiddish culture. From 1951 through 1955, parodist and comedian Mickey Katz starred in his own radio show on KABC in Los Angeles, and in the 1960s, actor and folksinger Theodore Bikel starred in *Thedore Bikel at Home,* which aired on FM stations in New York, Chicago, and Los Angeles. Some announcers who began their careers doing Jewish-oriented programming in the 1930s or 1940s could still be heard many years later. Zvee Scooler, who had originally been an actor in Yiddish theater, did commentary on WEVD for four decades. Max Reznick, whose show *The Jewish Hour* included everything from parodies to cantorial records to the popular style of Jewish jazz known as "klezmer," was on the air on several Washington, D.C., stations from the 1940s until he retired in 1986. Ben Gailing's Jewish program *Fraylekher Kabtsen* (The Happy Poor Man) was on the air in Boston for more than 50 years. And in Chicago, Bernie Finkel's *Jewish Community Hour* celebrated its 37th anniversary in the summer of 2000.

Jewish Programs Today

Jewish music or commentary was readily available on radio as the new century began. There were radio talk shows with Jewish themes, such as *Talkline with Zev Brenner,* heard on stations in New York and New Jersey. (Brenner had even started an all-Jewish radio station, WLIR, Long Island, NY, in May 1993, and after that venture, he resumed his job as a radio and cable television talk host.) Another popular radio program with a loyal following combined requests and dedications with a wide range of Jewish music, plus commentary from an Orthodox point of view—*JM in the AM* ("Jewish Moments in the Morning") with Nachum Segal had been on the air since 1983 on WFMU in Jersey City, New Jersey. The growth of the internet enabled Jewish programs from foreign countries (including Israel) to be heard in America. The internet was also helpful to those Jewish programs that at one time were heard on small AM stations; they could now broadcast on the world wide web and gain a much larger potential audience. Newspaper owner Phil Blazer was among those who took advantage of the new technology; the *Phil Blazer Show* has been on radio in Los Angeles since 1965, but it began webcasting in 2000. Talk show host Zev Brenner also began doing webcasting, and Boston radio host Mark David, whose show *Yiddish Voice* aired on WUNR in Brookline, MA, also made his program available over the internet. National Public Radio has aired holiday concerts of Jewish music at Hanukkah (including one concert featuring Theodore Bikel), and a number of Jewish recording artists are making their music available to be heard or downloaded.

Unlike the Jewish radio programs of the early days, there are not as many radio sermons or famous radio cantors, although there certainly are Jewish programs that have a moral or ethical dimension and shows that stress Jewish theology. There are even internet programs that teach Torah (Jewish Bible) on-line and play sound files to help with singing and pronunciation.

Although much of Jewish radio programming today is oriented toward music, there are also shows about Israel and current events. And even though tolerance is much more a part of American life than it was during radio's early years, myths and stereotypes still exist; when a celebrity utters an anti-Jewish remark, or when there are questions about Jewish beliefs (such as in 2000 when U.S. Senator Joseph Lieberman, an Orthodox Jew, was the Democratic candidate for vice president), agencies such as the Anti-Defamation League or the American Jewish Congress send spokespeople to address the issue on radio and television talk shows.

One continuing controversy is the presence on the air of so-called Messianic Jewish programs, such as Sid Roth's syndicated *Messianic Vision* or *Zola Levitt Presents,* which is now heard over the internet in addition to its long run on television. Messianic programs are broadcast by Christian stations, but their intention is to convert Jews. Hosts claim they are still Jewish even though they have accepted Christianity; they use Hebrew words and Jewish terminology to disguise evangelical Christian concepts, such as referring to Jesus as "Yeshua HaMoshiah," with the hope that Jewish listeners will be less threatened by a show that says one can convert to Christianity without leaving Judaism. These shows have evoked some vehement protest from leaders of the organized Jewish community, who object to what they feel is a distortion of Jewish teachings and accuse the hosts of using deceptive tactics.

Now that there is no longer an FCC guideline encouraging religious programming, few stations are willing to give free time to a religious show. As a result, the majority of religious programs on the air are sponsored by Christian denominations or individual preachers. With the radio networks no longer providing free time, and with production costs so expensive, there are few if any religious shows of the caliber of *The Eternal Light* on radio anymore. In fact, most of the popular music stations air no religious programs at all, and yet there are still Jewish programs on radio, just as there were in the 1920s. The shows offer a wide range of styles: some are traditional, with old-timers who play the great cantorial music of the past and reminisce about the old days, but others are quite modern and exemplify the interests of the younger audience. One such show is heard on WSIA, Staten Island, New York; it features Jewish women's music and is hosted by Michele Garner, who calls herself the "Rockin' Rebbetzin" (the Yiddish word for a woman who is married to a rabbi is *rebbetzin,* and not only is her husband Eliezer a rabbi, but he too does a show on WSIA, playing Jewish rock music). Thanks to Jewish radio shows and the performers who love European Jewish culture, the Yiddish language is being kept alive; klezmer music has enjoyed a rebirth and newfound popularity with a younger demographic. Jewish radio shows are also proving helpful to Jews-by-choice, people who have converted to Judaism on their own and now want to learn more about Judaism's various customs and musical traditions. Although it is certainly true that most religious programs on radio are done by and for Christians, there continues to be a consistent Jewish presence on the air, with shows that help to create a sense of community and a sense of identity.

Another interesting trend is the resurgence of Yiddish radio programs. Although only a few American Jews speak the language, there has been great interest in reviving it, especially given its impact on Jewish music and theater in the Golden Age of Radio. Leading this effort is the Yiddish Radio Project, which, beginning in March 2002 (some excerpts appearing in late 2001), could be heard on National Public Radio. Much of the work restoring the recordings and doing the historical research was done by a New York author and musician, Henry Sapoznik. A two-volume compact disc has been issued by the Yiddish Radio Project containing the first ten episodes of the NPR program, and the Project has a website with updated information (see below).

DONNA L. HALPER

See also Cantor, Eddie; The Goldbergs; Israel; Religion on Radio; Stereotypes on Radio; WEVD

Further Reading

Fine, Joyce, "American Radio Coverage of the Holocaust," <www.motlc.wiesenthal.com/resources.books/annual5/chap08.html>

Goldman, Ari, "For Orthodox Jews, a Gathering Place on the Air," *New York Times* (24 September 2000)

Kun, Josh, "The Yiddish Are Coming: Mickey Katz, Anti-Semitism, and the Sound of Jewish Difference," *American Jewish History* 87, no. 4 (1999)

Leff, Laurel, "A Tragic Fight in the Family: The New York Times, Reform Judaism, and the Holocaust," *American Jewish History* 88, no. 1 (2000)

Shandler, Jeffrey, and Elihu Katz, "Broadcasting American Judaism," in *Tradition Renewed: A History of the Jewish Theological Seminary,* edited by Jack Wertheimer, vol. 2, New York: Jewish Theological Seminary, 1997

Wishengrad, Morton, *The Eternal Light: Twenty-Six Radio Plays from the Eternal Light Program,* New York: Crown, 1947

Yiddish Radio Project website, <www.yiddishradioproject.org>

Jingles

Station identification jingles—catchy musical motifs often accompanied by vocals—are a basic ingredient of the sound of radio stations in most regions of the world. Although they are most associated with pop and rock music commercial radio services, jingles are also common in speech-based stations and even in publicly funded radio services.

Jingles are principally used by stations to insinuate their names and slogans into the minds of listeners. In so doing, jingles help ensure that, when questioned by audience researchers, listeners will recall particular stations over their rivals. This is especially vital for commercial stations, whose advertising rates are largely dictated by the results of such surveys.

In addition to this near-brainwashing technique of implanting the station names in the minds of listeners, radio jingles can be regarded as having the following attributes:

- They provide a positive, confident, "station sound" and a general "feel-good" factor.
- They promote the most important programming elements, for example, the style and quantity of music, contests, and local information.
- They "announce" different program elements, such as news, weather, sports, and disc jockey names.
- They serve as a way of "changing gears" between different program elements, for example, between news, commercial stop-sets, and travel information.
- They serve as a way of making musical transitions, for example, between slow- and fast-paced music.

Origins

The first known singing radio jingle was done in the mid-1920s, when Ernie Hare and Billy Jones, known as "The Happiness Boys," sang songs for a number of consumer products. By the late 1930s, advertising jingles had developed to a sophisticated level of production, often involving singing choirs and full big band orchestras. The station identification jingle developed naturally from this by the 1940s: WNEW in New York, for example, asked recording artists to sing short ditties incorporating the station's call letters.

By the mid-1950s radio station management saw the jingle as part of the battle for audience ratings success in a marketplace that was becoming ever more competitive—not just with other radio stations but with the new medium of television. In this period, radio stations were rapidly moving away from the "full-service" network model with a variety of programs and toward local operations, with most adopting one of three or four basic formats, and managements had to convince listeners that their stations were in fact better than and different from the opposition. They therefore marketed and promoted in ways similar to those of any consumer product. The fact that the station jingles sounded very similar to the advertising jingles of products and services was no coincidence: they were often written, performed, and produced by the same companies.

The "founding fathers" of station jingles as a distinct entity were Bill Meeks and Gordon McLendon. The latter bought KLIF in Dallas in 1947 and appointed Meeks as his music director with a specific brief to put together live music shows. As with so many other innovations, the modern station jingle happened by accident: the jingles were used as a way of bridging the time needed in live broadcasting to set up each new vocal group.

Meeks left KLIF to start his own company, Production Advertising Marketing Service (PAMS), in Dallas in 1951. By the mid-1950s he was compiling individual station jingles into "packages"—whole series of jingles using variations of the same musical structure and slogans—to different stations across the country. By the end of 1964 the primary business of PAMS had become station jingles, and the company became the world leader in a new stratum of services to the radio industry. Meanwhile, McLendon hired another musical director, Tom Merriman, who developed and elaborated on the station identification jingles at KLIF; other stations in different parts of the United States heard these and asked for customized sets of their own. Merriman also left KLIF and in 1955 formed the Commercial Recording Corporation; his company produced some of the first jingles specifically for Top 40 radio, which was rapidly emerging as the number-one radio format.

There was a good deal of creativity and innovation in this period. PAMS is credited with being the first company to use the Moog synthesizer and the Sonovox—a device originally developed to enable people who had lost the use of their vocal chords to make intelligible speech—to create an extraordinary electronic "singing" voice. The "variable station logo" technique meant that many disparate programming elements could be linked and blended using variations on the same musical motif, often with lyrical variations of the same slogan or "positioning statement."

Despite this, the overall style of these jingles—close vocal harmonies and lush orchestrations common in the pre–rock and roll period—was outdated even by the late 1950s, and yet, curiously, the style persisted for at least another 20 years. It appears that for many years, the radio jingle was accepted by the listening public as being a musical genre of its own, with no need to bow to changes in popular music.

Occasionally, though, attempts were made to overcome this anachronism by somewhat cheekily adopting more contemporary styles: one of the PAMS jingle sets adopted the Beatles' sound, even using the group's trademark "yeah yeah." Even here, though, the jingles outlasted the creations of the form on which they were based. Stations playing more contemporary formats belatedly moved away from the traditional close harmonies and lush orchestrations, seeing these techniques as distinctly "uncool" and embarrassingly old-fashioned (U.S. FM rock stations had always taken this view of jingles and had consequently eschewed them from the start).

In 1974 former PAMS employee Jonathan Wolfert and his wife Mary Lyn set up a new company, JAM (Jon and Mary)—also based in Dallas, the world's center for station jingles—which quickly established itself as one of the leading companies in the jingles field. In 1976 they secured the contract for the most famous and imitated Top 40 radio station in the world—WABC in New York—which used PAMS jingles from 1962 to 1974. Probably the most-played jingle in the world was recorded for this station in the spring of 1976—lasting just two and a half seconds. In 1990 Wolfert bought up the rights to the jingles from PAMS (which had suspended operations in 1978 after a series of financial crises). These jingles were still in demand in many radio markets throughout the world—especially on golden oldies stations, which had their own station names and slogans sung over the original music tracks, often using many of the same singers.

Modern Era

The length of station jingles varies greatly, although most average between five and seven seconds. JAM also claims to have recorded the longest known jingle—for WYNY in the fall of 1979—which lasted three and a half minutes. "The New York 97 Song" is a vocal and musical narrative of a day in the life of a listener in New York, with repeated use of the station location, frequency, and call letters—"New York 97, WYNY."

The international ubiquity of the station identification jingle cannot simply be explained as the result of the competitiveness of the U.S. commercial radio system. Jingles are used in radio services in very different types of economic and media systems. For example, the British public service broadcaster, the British Broadcasting Corporation (BBC), which held a monopoly on domestic radio in the United Kingdom until 1973, began using PAMS jingles in 1967 on its new rock and pop network, Radio 1. This service was set up on instructions from the government after it had produced legislation outlawing the "pop pirates" of the mid-1960s—many of which had copied U.S. Top 40 and beautiful music formats and used jingles from PAMS in Dallas. Radio 1 ordered the same series of jingles that had been used by the most commercially successful of the offshore pirates, the Texan-backed Radio London. Even though the BBC carried no commercials and had no authorized competition during the late 1960s and early 1970s, its managers thought that station jingles had become an essential part of pop music programming and youth culture. Many other state broadcasters have also felt the need to invest in the "Dallas Sound" of jingles, as have the publicly funded armed forces radio networks of both the United States and the United Kingdom. The jingles have also been sung in languages other than English: the U.S. radio jingle can be fairly regarded as a form of cultural imperialism, albeit one of a benign nature.

By the mid-1980s, a new vogue for electronically produced "Sweepers," with a spoken rather than sung vocal track, became the vogue on many contemporary hit radio and adult contemporary stations. As with all commercial operations, the jingles business—dominated in the world market by JAM and TM Century—had to adapt to these new demands or face extinction. In the late 1990s, there was some indication that the fashion had moved, if not full circle, then perhaps 180 degrees, when many stations—especially those targeted to the baby boom generation, which had grown up with the Dallas Sound—began investing in the more traditional type of jingle production, albeit with a more contemporary edge.

Although fashions in station jingles will no doubt evolve still further, the one constant need for stations faced with ever more competition for listeners' loyalty is to ensure that they promote themselves on the air in the most attractive and distinctive audio fashion possible.

RICHARD RUDIN

See also McClendon, Gordon; Promotion on Radio

Further Reading

Chapman, Robert, *Selling the Sixties: The Pirates and Pop Music Radio,* London and New York: Routledge, 1992

Douglas, Susan J., *Listening In: Radio and the American Imagination,* New York: Times Books, 1999

Fong-Torres, Ben, *The Hits Just Keep on Coming: The History of Top 40 Radio,* San Francisco: Miller Freeman Books, 1998

Garay, Ronald, *Gordon McLendon: The Maverick of Radio,* New York: Greenwood Press, 1992

Hilmes, Michele, *Radio Voices: American Broadcasting, 1922–1952,* Minneapolis: University of Minnesota Press, 1997

Keith, Michael C., *Voices in The Purple Haze: Underground Radio and the Sixties,* Westport, Connecticut: Praeger, 1997

MacFarland, David, T., *The Development of the Top 40 Radio Format,* New York: Arno Press, 1979

Joyce, William. *See* Lord Haw-Haw

Joyner, Tom 1942–

U.S. Radio Disc Jockey

Tom Joyner's daily round-trip airline commute of 1,600 miles between Chicago and Dallas earned him the moniker "The Fly Jock." Beginning in 1985, Joyner continued this daily commute for eight years. It was one of several accomplishments that transformed a once small-town disc jockey into one of America's most well known on-air personalities. Another well-deserved nickname would come later, one that was just as fitting and appropriate as the earlier one—"The Hardest-Working Man in Radio."

Early Career and Life

The Tuskegee, Alabama, native started announcing during college. He had at first sung with Lionel Ritchie and the Commodores, a rhythm and blues group, but he was discouraged by his father in this pursuit. So Joyner tried announcing instead. He first worked as a local announcer at Tuskegee Institute, where he attended college, playing records in the university cafeteria and announcing at sporting events. Upon graduation, he landed an on-air position at WMRA-AM in nearby Montgomery, Alabama.

He left that job for other on-air positions in successively larger markets. For example, after WMRA, Joyner worked at Memphis' WLOK-AM. From Memphis, he moved up the river to St. Louis, where he found an announcing job at KWK-AM. From there, Joyner was hired as an announcer at Chicago's WJPC-AM. While there, he worked as a morning air personality and program director (1978–83). Moreover, he was also producer and host of the *Ebony/Jet Celebrity Showcase*, a television show. While at WJPC-AM, the station's owner and Joyner's mentor, John H. Johnson, publisher of *Ebony* and *Jet* magazines, gave Joyner his first big break. Johnson put Joyner's picture in *Jet* magazine each week as part of the promotion for the television program, and Johnson featured him in a television commercial with boxing great Muhammad Ali. In addition to his work at WJPC-AM, Joyner also worked as a disc jockey at several other Chicago radio stations, including WVON-AM, WBMX-FM, and WGCI-FM.

The Fly Jock

Joyner left WGCI-FM for a morning air personality position at Dallas' KKDA-AM. By the time his contract at the Dallas station was nearing its end, WGCI-FM management wanted him back. Joyner listened to both offers and decided to take both because neither contract had an exclusivity clause that would prevent him from working at another radio station. Joyner said "greed" got the best of him, and he signed two concurrent $1 million-plus contracts, running six years (Dallas) and five years (Chicago). Thus began the odyssey of the Fly Jock.

From 1985 to 1993, Joyner accrued 7 million frequent-flier miles, as he made the round-trip commute three days each week between Dallas and Chicago. Joyner did a morning show in Dallas at KKDA-FM (the station had acquired the license for an FM facility) from 5:30 to 9:00 A.M. In the afternoons, from 2:00 to 6:00 P.M., he did an afternoon show at WGCI in Chicago. Joyner paid $30,000 to reserve a guaranteed round-trip seat on an airline over five years.

The Hardest-Working Man in Radio

In 1994 the American Broadcasting Companies (ABC) Radio network approached Joyner with the idea of developing his radio program for national distribution and syndication. With its debut in 1994, the *Tom Joyner Morning Show* became the first live syndicated program produced by an African-American performer on radio. It is broadcast from 5:00 to 9:00 A.M. central time to a nationwide audience, Monday through Friday. Joyner's Dallas-based show combines, in more or less equal parts, rhythm and blues music, comedy, and politics. Although based primarily in Dallas, the *Tom Joyner Morning Show* is often broadcast from remote studios in various cities around the United States. It is the number-one-rated urban morning radio show in the country. Joyner's show is heard on 95 stations, reaching 5 million listeners. The show targets the 25–54 demographic group, an attractive group for advertisers. Joyner's show, when broadcast on other outlets, has catapulted many of these radio stations to number one in their respective

Tom Joyner
Courtesy ABC Radio Networks

markets. Critics complain, however, that Joyner's success has come at the expense of local disc jockeys, who are fired because they are no longer needed. Joyner, on the other hand, sees his success as an opportunity for low-rated radio stations to increase listeners and advertising revenues and to provide national programming that comes with his syndicated show.

Joyner also produces a weekend program. The *Tom Joyner Movin' on Weekend Show* can also be heard nationally and is distributed by the ABC Radio network.

Community Service

Joyner's sustained support of community service activities has endeared him to the African-American community. Among the issues he has supported include a 1980s campaign called "Drop a Dime on the Man," which identified speed traps targeting African-Americans in Dallas. In 1999 Joyner, along with political commentator and fellow team member Tavis Smiley, started a crusade to persuade retailers such as CompUSA, for example, to expand their advertising in black-owned media.

In addition, he has started the Tom Joyner Foundation, a nonprofit organization that assists college students in completing their college education at historically black colleges and universities. The foundation accepts donations from many sources, but Joyner's "Dollars for Scholars" campaign has raised more than $40,000 for students at historically black colleges, which is made available to them through his Foundation.

GILBERT A. WILLIAMS

See also Black-Oriented Radio; Disk Jockeys

Tom Joyner. Born in Tuskegee, Alabama, 1942. Bachelor's degree in Sociology, Tuskegee Institute, 1964; after graduation worked at radio stations WRMA-AM, Montgomery, Alabama; WLOK-AM, Memphis, Tennessee; KWK-AM, St. Louis, Missouri; worked at WJPC-FM (on-air personality and program director, 1978–83), WVON-AM, WGCI-FM, WBMX-FM, Chicago, Illinois; worked simultaneously at KKDA-AM, Dallas, Texas and WGCI-FM, Chicago, Illinois, earning him nickname "Fly Jock"; worked at KKDA-FM, Dallas, Texas; television producer and host, Ebony/Jet Showcase, 1982–83; host of ABC-syndicated radio show *Tom Joyner Morning Show,* 1994–; established Tom Joyner Foundation, 1998. Received four *Billboard Magazine* "Best Urban Contemporary Air Personality" awards; *Impact Magazine's* Joe Loris award for "Excellence in Broadcasting"; Mickey Leland Humanitarian Award, Congressional Black Caucus; named "Man of the Year", 100 Black Men; President's Award, NAACP; elected to the Radio Hall of Fame, 1998.

Radio Series

1980s *On the Move*
1994– *The Tom Joyner Morning Show*

Further Reading

Barlow, William, *Voice Over: The Making of Black Radio,* Philadelphia, Pennsylvania: Temple University Press, 1999
Newman, Mark, *Entrepreneurs of Profit and Pride: From Black-Appeal to Radio Soul,* Westport, Connecticut: Praeger, 1988; London: Praeger, 1989
Sherrell, Rick, "Tom Joyner, Playa Jock and King of the Hill," *Upscale Magazine* 9 (September/October 1998)
Sturgis, Ingrid, "Tom Joyner," *BET Weekend* (July/August 1998)
Williams, Gilbert Anthony, *Legendary Pioneers of Black Radio,* Westport, Connecticut: Praeger, 1998

Judis, Bernice 1900–1983

U.S. Station Manager

In the midst of network radio's golden age and on through television's first wave of success, Bernice Judis developed an inexpensive music-oriented radio program style that survived big-budget competition. Her implementation and massaging of the format over WNEW New York was analyzed and often imitated by countless media insiders from Los Angeles to Luxembourg.

Judis enjoyed admitting that she loved radio the way most men love fine cars. Reportedly, every room in her home, as well as a compartment in her handbag, was equipped with a receiver. She was always listening, or so it seemed: none of her announcers was exempt from an instructive studio hotline call, sometimes well past midnight. They affectionately nicknamed Judis "La Mama," though they never called into question their teacher's talent to motivate. Each remembered her as an executive with an instinct for hiring good people and then challenging them to be great. Even the disc jockey she fired for violating WNEW's "fashion code" by wearing red socks considered his strict boss a valued mentor.

Her entrance into broadcasting resulted from a friendship with a woman whose husband had recently purchased part interest in a radio station. In 1934 the well-heeled Judis had just returned from a European tour when her friend casually suggested that she help out the new enterprise. Within a year of Judis' agreeing to the request, her programming acumen was so evident that she graduated to general manager of fledgling WNEW.

The independent station had inadequate funding to stage the types of celebrity-oriented comedy, music, drama, or soap opera fare that was typically the province of network outlets. That never mattered, though, because Judis considered such offerings boring. She preferred a steady diet of positive pop tunes, and she maintained that others (especially young women, who often controlled the family radio dial) would share her tastes. One of her specialties was an instinct for hiring announcers who possessed the voice and ability to create a descriptive picture in the listener's mind. Judis looked for radio people ready to bloom into true "air personalities." Paramount was their potential to sell WNEW audiences the products offered by the station's growing list of advertisers.

Martin Block had a smooth spark in his delivery that Judis believed represented star quality. Her direction of the erstwhile $20-per-week announcer rocketed him to disc jockey stardom and, for years, propelled Block's *Make Believe Ballroom* program to highly salable ratings. Judis instructed him to work closer than normal to the microphone in order to generate an air of gentle romantic authority.

When authoritative columnist Walter Winchell wished (in print) for a radio station that would keep metropolitan New York's all-night work force company during the wee hours, Judis introduced the *Milkman's Matinee.* In 1936 the decision made WNEW a pioneer in 24-hour broadcasting. Although she scheduled some shows with live musicians, most of her format (including *Milkman's Matinee* and *Make Believe Ballroom*) was built around disc jockeys skillfully ad-libbing between recorded tunes. Prior to the late 1940s, most other big-city programmers considered transcriptions (in place of live bands and singers) the stuff of minor-league presentation. WNEW's shrewd manager used her sizable record library to offer people an anticipated radio staple: consistency in repeated elements. Such is the stuff that turns an ordinary song into a familiar hit recording. Literally hundreds of daily record requests flooded WNEW's switchboard. Some listeners sought to hear again and again obviously goofy selections the airstaff had spun purely as gags.

Judis is also known for adding non-musical content to what would be subsequently dubbed the "modern radio format." In the mid-1930s, Americans were gripped by the kidnapping of the infant son of transatlantic flight hero Charles Lindbergh. During the ensuing trial, the then novice WNEW official ordered her station's remote news crew to report proceedings directly from the New Jersey courthouse. Since their microphones were barred from courtroom use, Judis had her staff headquartered in the venue's nearest lavatory. Although at risk of transmitting indelicate sounds, these bathroom broadcasts were the sole electronic media offerings from inside the judicial building. Between reports, disc jockeys in the main studio played popular music. Listeners were captivated.

When the Japanese bombed Pearl Harbor, Judis was convinced that people would want more news. She scheduled round-the-clock, hourly (initially, on the half hour) newscasts by 1942—something no one else was doing. Judis' hunch quickly made "WNEW—Your station for music and news" an even more important companion for increasingly busy audiences. She rightly speculated that regular bites of pertinent information sandwiched between friendly disc jockeys conveying uplifting pop would yield a compelling soundtrack. At World War II's end, the Judis-directed WNEW targeted a young, urbane demographic, and her approach received high marks for successfully standing up to radio's postwar enemy: television.

When many radio veterans were moving to the expanding visual medium, Judis, along with her husband Ira Herbert, Rhode Island broadcaster William Cherry, and several others paid more than $2 million for WNEW. The savvy general manager and new part owner wagered that the spread of suburbs and concomitant commute of workers into cities would strengthen radio's influence, even in the face of television's quickly growing popularity. Judis met TV's evening primetime dominance by further tweaking her various 6 A.M. to 8 P.M. slots. Fourteen hours daily of popular music favorites, news briefs, and chatty disc jockeys gave ample opportunity for reaching the increasingly mobile and sophisticated thirty-something audience that corporate America's advertising agencies desperately sought to influence around the clock. People not glued to nighttime TV caught Judis' unique early-1950s counter-programming experiments. For example, she had disc jockeys spin spoken-word records, hired Milton Berle to do a serious Shakespeare performance, and tried a game show spoof in which contestants were eligible to lose their personal effects.

An attractive offer for WNEW led its owners to sell in 1954. With their share of the proceeds, Judis and her husband bought an interest in a Southern broadcast group. She oversaw the firm's Birmingham, Alabama, AM outlet. By the late 1960s, the couple had left the radio business. Judis died in Fort Lauderdale, Florida, in 1983.

PETER E. HUNN

See also Block, Martin; WNEW; Women in Radio

Bernice Judis. Born 1900. Only daughter of New York–area real estate tycoon; socialite activities and European travel, 1920s–early 1930s; New Jersey Women's Golf Champion, early 1930s; program director, WNEW New York, 1934; general manager, WNEW New York, 1935–54; part owner, WNEW New York, 1950–54; part owner of Basic Communications Stations, 1960s. Died in Fort Lauderdale, Florida, 24 May 1983.

Further Reading

Gordon, Nightingale, *WNEW: Where the Melody Lingers On,* New York: Nightingale Gordon, 1984

Passman, Arnold, *The DeeJays,* New York: Macmillan, 1971

"Radio Splits over Strategy against TV," *Business Week* (31 May 1952)

"Station WNEW Sold by Bulova and Biow for Substantially More than $2,000,000," *New York Times* (18 November 1949)

K

Kaltenborn, H.V. (Hans von) 1878–1965

U.S. Radio Commentator

Hans von Kaltenborn was radio's first news commentator and a pioneer in radio news for three decades. Prior to his broadcasting career, Kaltenborn engaged in newspaper reporting and editing and in public lecturing on current events. He is best remembered for his live broadcasts of battles during the Spanish Civil War, his marathon broadcasts of the 1938 Munich Crisis, and for President Harry S. Truman's mimicking of his 1948 election reports.

Origins

Kaltenborn was born in Milwaukee, Wisconsin, in 1878. His father, Baron Rudolph von Kaltenborn, was a former officer in the Hessian Army who came to the United States to protest the absorption of Hesse by Prussia. His mother, an American schoolteacher, Betty Wessels, died soon after Hans was born, and he was raised by a stepmother.

The family moved to Merrill, Wisconsin, when Hans was 14 years old. After only a year in high school, he ran away to work in a lumber camp. Then, for five years he worked for his father in the building material business and did odd jobs for the local newspaper, the *Merrill (Wis.) Advocate*. Later, while serving in the army during the Spanish-American War, young Hans sent articles about army life back to the *Merrill Advocate* and to the *Milwaukee Journal*. Although he did not leave the United States during the war, young Kaltenborn, who was athletic and nearly six feet tall, developed a desire for travel and adventure. After a short time as city editor of the *Merrill Advocate,* he worked his way to Europe in 1900 and spent the next two years in Germany and France. He attended the Paris International Exposition of 1900 and continued sending freelance news material back to newspapers in the United States. When he returned to the United States in 1902, he worked as a reporter and editor for the *Brooklyn Eagle,* a newspaper with which he would be associated, on and off, for about 27 years.

Lacking a high school diploma, Kaltenborn entered a special one-year program for journalists at Harvard College in 1903. After completing high school equivalency exams—he had great difficulty with the mathematics section—Kaltenborn was admitted as a regular student, majoring in political science. In 1909 he received his B.A. degree cum laude and was elected to the prestigious Phi Beta Kappa honor society. At Harvard, he debated, studied speech, won oratorical contests, ran cross country, organized a dramatic club, and perfected his distinctive German accent.

Following college, Kaltenborn traveled around the world on shipboard as a tutor to Vincent Astor. In 1910 he married Olga von Nordenflycht, a German baroness, and returned to the *Brooklyn Eagle,* for which he served as a Washington, and later, a Paris correspondent before being named the *Eagle*'s war editor. With his growing reputation as a geopolitical expert, Kaltenborn delivered, as an *Eagle* editor, a series of weekly public talks about current events in the newspaper's auditorium.

On to Radio

In 1922 Kaltenborn delivered his first radio commentary from station WYCB on Bedloe's Island in New York Harbor. The following year, he became the first regularly scheduled radio commentator when the *Eagle* sponsored Kaltenborn's regular weekly news commentaries over WEAF in New York. He remained with the *Brooklyn Eagle* until 1930, when he joined the Columbia Broadcasting System (CBS) to devote himself full-time to radio news.

During the early 1930s, when wire service reports were seldom available to radio stations, CBS and the National Broadcasting Company (NBC) employed a number of experienced newsmen to comment on the day's events. Although Kaltenborn had the advantage of being the first radio

Paul White and H.V. Kaltenborn
Courtesy CBS Photo Archive

commentator, he also benefited from his speech training, his lecturing experience, and several other unique abilities, all of which kept him in the forefront of news commentary. His spoken delivery was distinctive—clipped, precise, and easy to understand even when he spoke at a rapid pace. Among his special attributes were a firsthand knowledge of many world leaders, whom he interviewed during frequent travels abroad; fluency in the French, German, and English languages; and training in world politics. Kaltenborn also had the advantage of being able to speak without a script. For many years he delivered his radio commentaries extemporaneously with only a few notes. He also extemporaneously described political conventions, international conferences, wars, and other crises.

One of Kaltenborn's early broadcast triumphs was his live description of a 1936 Spanish Civil War battle. Perched on a haystack on high ground in France overlooking a besieged Spanish town, the commentator, wearing a steel helmet, described the battle to CBS listeners in the United States. His report was accompanied by live sounds of machine-gun fire in the background. Radio's first on-air battle report, though physically risky for 58-year-old Kaltenborn and his engineer, added greatly to his stature as a radio newsman.

Two years later, Kaltenborn was called upon by CBS to coordinate and anchor a series of radio broadcasts during the Munich Crisis in 1938. At stake was the future of the Sudetenland, as Germany's dictator Adolph Hitler, British Prime Minister Neville Chamberlain, and other European leaders met to seek a solution to the conflict in Czechoslovakia. Kaltenborn's role was to interpret and analyze the wire dispatches, shortwave reports from CBS newsmen on the scene, and speeches by the participants. Relying on his ability to understand German and French, Kaltenborn translated and provided instantaneous commentary on the speeches of various government officials received in New York by shortwave. A *Time* reporter noted that without pause, Hans von Kaltenborn translated and distilled a 73-minute speech and then for 15 minutes proceeded extempore to explain its significance and to correctly predict its consequences. Kaltenborn remained at his post at CBS in New York for some 18 days, giving about 100 separate broadcasts. He subsisted mainly on soup brought to him by his wife and napped on a couch in the studio between the frequent broadcasts.

Because of a tendency for outspokenness on controversial issues, Kaltenborn had some difficulty keeping a sponsor during the 1930s. However, by the end of the decade the Pure Oil Company became Kaltenborn's regular sponsor on CBS. In 1940 the sponsor moved Kaltenborn's commentaries to NBC to get a better time on the evening schedule. Soon Kaltenborn's keen insight into geopolitical developments was evident in his warnings about Japanese aggression, including a warning only a few days before Pearl Harbor. During World War II, Kaltenborn was exempted from the Office of War Information's requirement that commentators not deviate from scripts submitted in advance of their broadcasts. Also during the war, Kaltenborn began using the initials *H.V.* to minimize his German ancestry.

In covering the 1948 presidential election while at NBC, Kaltenborn assured listeners that even though President Truman was running a million votes ahead in the early popular vote, when the rural votes came in, the winner would certainly be Thomas E. Dewey, the governor of New York. In probably the best-known instance of presidential mimicry of a news commentator, President Truman, at a post–election dinner, told the audience that he had heard Mr. Kaltenborn's broadcast. Then the president imitated the crisp Kaltenborn's remarks and said he just went back to sleep undisturbed. The next morning, the nation learned that, as Mr. Truman had never doubted, the president was reelected to serve a full four-year term.

Kaltenborn continued to broadcast his regular commentaries until 1953. Afterward, he did occasional broadcasts on a semi-retired basis until 1958. For 33 years, H.V. Kaltenborn had covered practically all of the world's major news events and had given the public his views on the meaning of the news. He also contributed to many magazines and was the author of several books. He spent his retirement years living in Florida and died in New York City in 1965 while visiting his son.

HERBERT H. HOWARD

Hans von Kaltenborn. Born in Milwaukee, Wisconsin, 9 July 1878. Served in U.S. Army, Spanish American War; worked as city editor, *Advocate,* Merrill, Wisconsin, 1899–1900; traveled in Europe, 1901–02; reporter, *Eagle,* Brooklyn, New York, 1902; attended Harvard University, B.A. in political science, 1909; returned to *Eagle,* 1910, correspondent in Washington and Paris, later served as *Eagle's* war editor; delivered first radio commentary from experimental station on Bedloe's Island, 1922; first regularly scheduled news commentator on WEAF, for *Eagle,* 1923; hired as full-time radio commentator by CBS, 1930; broadcast battle descriptions of Spanish Civil War live on CBS, 1936; broadcast 18-day coverage of Munich Conference from CBS in New York, 1938; moved to NBC, 1940; predicted Dewey victory in presidential election, 1948. Received Phi Beta Kappa, Harvard University, 1909; DuPont Foundation Award, 1945; honorary doctorates from University of Wisconsin and Hamilton College. Died in New York City, 14 June 1965.

Selected Publications

I Broadcast the Crisis, 1938
Fifty Fabulous Years, 1950
It Seems Like Yesterday, 1956

Further Reading

Culbert, David Holbrook, *News from Everyman: Radio and Foreign Affairs in Thirties America,* Westport, Connecticut: Greenwood Press, 1976

Downs, Robert Bingham, and Jane B. Downs, *Journalists of the United States,* Jefferson, North Carolina: McFarland, 1991

Fang, Irving E., *Those Radio Commentators!* Ames: Iowa State University Press, 1977

McKerns, Joseph P., editor, *Biographical Dictionary of American Journalism,* New York and London: Greenwood Press, 1989

Taft, William Howard, *Encyclopedia of Twentieth-Century Journalists,* New York: Garland, 1986

Karmazin, Mel 1943–

U.S. Radio Executive

There was no more powerful executive in radio in the last two decades of the 20th century than Mel Karmazin. After gaining a decade of experience in radio sales and management, Karmazin formed and ran Infinity's profitable group of radio stations during the 1980s. He then merged Infinity with the Columbia Broadcasting System (CBS) in the 1990s and moved to "Black Rock," the Eero Saarinen-designed headquarters building for CBS on 6th Ave in New York City, first to run its radio operation and then to head the entire company as it merged with media conglomerate Viacom. Karmazin brought effective consolidation and cost cutting to radio station and group operation as he headed CBS's second most lucrative division and radio's top revenue-producing chain as the 20th century ended.

Born and raised in Long Island City, New York, across the East River from Manhattan, Mel (never Melvin, his given name) worked days as an account executive at the Ziowe Advertising Agency in the city while attending Pace University at night. Upon graduation, he took a position in sales, rising quickly to manager at WCBS-AM, an all-news radio powerhouse located in the heart of Manhattan. During the 1970s, Karmazin worked for Metromedia, eventually rising to general manager of its WNEW-AM and FM combination. It was from this position that he launched Infinity Broadcasting and became his own boss at the age of 37.

As a manager, Karmazin became most famous for his relentless cost cutting and format switching. Yet he was willing to try any technique to raise profits. In 1993, for example, as others were selling radio networks, Karmazin purchased Westwood One; three years later, with the merger with CBS (for $5 billion), Karmazin lorded over all the major radio networks—save American Broadcasting Companies (ABC). These networks could and did feed programming cheaply not only to stations owned by Infinity and later CBS, but also to thousands of other stations across the United States.

Karmazin never sought to own all radio stations; instead, he aimed to group stations in nearly every major city—led by stations in the top four markets of New York City, Los Angeles, Chicago, and San Francisco. Under Karmazin, CBS owned far fewer radio stations than 1999 top dog Clear Channel Communications, yet its more than 160 radio stations earned just as much in revenues, as advertisers looked to radio in major markets to couple with television and newspapers to complete their advertising plans.

Thus, through radio station operation, Karmazin was able to quietly generate about half of CBS's profits in the late 1990s. Karmazin aimed to be either number one or two in every major radio market where CBS had stations. When Mel Karmazin succeeded Michael Jordan in 1998 in CBS's top executive position, he had already formulated this plan, and thereafter he always counted on squeezing significant profits from the CBS radio group while the world looked at the ups and downs of the far more famous television network. As the 1990s ended, Karmazin operated not only flagship WCBS-AM in New York City but dozens of other stations in the country's top-twenty media market cities, including eight stations in Los Angeles (the #2 media market) and seven in Chicago (#3).

Just after Labor Day 1999, a Hollywood company most people had never heard of—Viacom—took over CBS. Although the deal was announced by Sumner Redstone, chairman and chief executive officer of Viacom, Mel Karmazin was named president and chief executive officer of the new colossus. Redstone would step aside, and Karmazin would run the day-to-day operations. This made Karmazin one of mass media's top executives, with radio now only a part of his overall responsibilities.

Redstone and Karmazin's goal in the 21st century would be to make Viacom a media conglomerate the equal of Disney, Time Warner, or the News Corporation. Redstone retained his ownership majority and thus control of the new Viacom, and Karmazin took over day-to-day operations of a corporation able to produce, promote, distribute, and present all forms of mass entertainment, from radio and television to movies and music. With a portfolio of some of the world's most recogniz-

Mel Karmazin, 25 March 1999
Courtesy AP/Wide World Photos

able brands, Karmazin and Redstone boldly proclaimed global superiority for an operation conservatively estimated as worth $35 billion.

The union of CBS and Viacom promised to test the economic theory that owning significant stakes in many mass media can provide a synergy by which the parts together can produce more profits than if they operated individually. Could CBS's radio division cross-promote and make more profitable Viacom's cable music channels (MTV, VH1, and TNN)—and vice versa? Karmazin will test his management skills in an arena wider than radio, encompassing all of mass entertainment. Owner Sumner Redstone has passed the torch, because the fine print in the merger obligates him to stick with Karmazin until 2003.

DOUGLAS GOMERY

See also Columbia Broadcasting System; Infinity Broadcasting; Ownership, Mergers, and Acquisitions; Westinghouse

Mel Karmazin. Born in New York City, 24 August 1943. Grew up in Long Island City, Queens, New York; attended Pace University, B.A. in business administration, 1965; account executive for Ziowe Advertising Agency, 1966; sales manager, WCBS, New York, 1967–70; vice president and general manager, WNEW-AM and WNEW-FM, Metromedia, 1970–81; president of Infinity Broadcasting, 1981–96; CEO of CBS Station Group, 1996–98; president and CEO, CBS, 1999–present.

Further Reading

Compaine, Benjamin M., and Douglas Gomery, *Who Owns the Media?* 3rd edition, Mahwah, New Jersey: Erlbaum, 2000

Duncan, James H., Jr., "A Second Generation Enters the Radio Industry," *Duncan's Radio Comments* 16 (June 1996)

Fahri, Paul, "Justice Department Clears Westinghouse Purchase of Infinity," *Washington Post* (13 November 1996)

Rathbun, Elizabeth A., "CBS, ABC: Two Roads to Radio," *Broadcasting and Cable* (15 September 1997)

Siklos, Richard, "Reinventing CBS," *Business Week* (5 April 1999)

Kasem, Casey 1932–

U.S. Radio Personality

For many rock and roll fans, the name Casey Kasem is synonymous with the American Top 40 Countdown, a weekly radio program of the most popular songs in the United States. It was the first nationally syndicated countdown program, and as its host and one of its founders, Kasem's impact on the landscape of American popular music is undeniable. His voice is among the most recognizable in rock and roll, joining the likes of Wolfman Jack as both a disc jockey and a celebrity.

The son of Lebanese immigrants, Kemal Amen Kasem was born in 1932 in Detroit, Michigan. Upon graduating from Northwestern High School, Kemal became an intern at WDTR, Detroit Public School's radio station. From there, he attended Wayne State University, where he landed the lead role as "Scoop Ryan, Cub Reporter," the most popular show on the campus station. As a result of his success there, he earned his own 15-minute Saturday morning show on WJR, a 50,000-watt Detroit station. He also took a position as a full-time actor on WXYZ, an American Broadcasting Companies (ABC) affiliate.

In April 1952, Kasem was drafted for army service in Korea. Because of his broadcast experience, he was assigned to Radio Station Kilroy, an Armed Forces Radio Network Affiliate in Taegu, Korea, where he started his own production team. After the war, he returned to Detroit and took a job as a newsman at WJBK, where he adopted the moniker "Casey at the Mike." After a successful stint of substituting for a disc jockey, Kasem became the station's primary disc jockey. That same year he became Krogo the Clown, hosting a children's TV program on WJBK-TV. He then left television to assist at his family's grocery store and finally relocated to Cleveland in 1959, where he took a job as a radio host at WJW.

The most prominent Cleveland disc jockey at this time was Mad Daddy (Pete Meyers), whose fast-talking, slang-laden, rhyming radio patter influenced Kasem's on-air personality. Kasem again used the nickname *Casey at the Mike.* In Cleveland, Kasem also hosted the television show *Cleveland Bandstand*, while working at WJW. The program was canceled after

Casey Kasem
Courtesy of Casey Kasem

a few months, and WJW switched formats, so he began looking for other work.

Kasem worked at WBNY in Buffalo and KEWB in Oakland before arriving at KRLA in Los Angeles in 1963. During the process, Kasem shed his comic routines and fast talk, and became a more serious, informative disc jockey following his chance discovery of a 1962 copy of "Who's Who in Pop Music," which he used thereafter to provide his listeners with information on bands and singers.

In Los Angeles, Kasem began calling himself simply *Casey Kasem,* a simpler and more serious name than *Casey at the Mike,* and he was hired by Dick Clark to host an *American Bandstand*–type program called *Shebang,* which ran on KTLA from 1965 to 1968. Kasem made his mark on the program by improvising his introductions rather than reading the prewritten cue cards. Mike Curb, a producer for Tower Records, introduced Kasem to the world of commercial voice-overs, which eventually led to his two biggest roles off the radio dial: providing the voice of Robin on the cartoon series *Batman and Robin* and also the voice of Shaggy on *Scooby Doo, Where Are You?* In addition, he provided voices for countless other cartoon shows and promotions for the National Broadcasting Company (NBC) network.

In late 1968, Kasem reunited with his old friend Don Bustany and created Kasem-Bustany productions, which they operated from Kasem's Hollywood apartment. In the summer of 1969, Kasem and Bustany met with Tom Rounds, an executive at Charlatan Productions, and Ron Jacobs, a respected program director in Los Angeles. Rounds and Jacobs had earlier formed Watermark Productions, a Los Angeles radio production company. With the financial backing of Tom Driscoll, heir to a strawberry-growing fortune, Kasem, Bustany, Rounds and Jacobs laid the groundwork for *American Top 40* (*AT40*), which would change Casey Kasem from a local disc jockey to a household name. *Casey's Coast to Coast,* in partnership with *Billboard Magazine,* was launched on 3 July 1970, a Friday night, at 7:00 P.M., by KDEO in El Cajon, a suburb of San Diego.

In order to increase its exposure, *AT40*'s creators bartered with stations; for instance, they would provide a three-hour *AT40* program, and the station would receive airtime for two advertisements in exchange. However, in late 1971 Rounds decided to begin selling *AT40* for varying amounts depending on the size of the market. (The show lost approximately $600,000 before it began to turn a profit.) In 1973 Kasem produced a special episode called "The Top 40 Disappearing Acts," a show based on one-hit wonders. This show highlighted his talent with human-interest storytelling and helped to put *AT40* on the map.

During the 1970s, musicians began to record much longer songs, which made the three-hour format difficult to maintain. So in 1978, *AT40* expanded to four hours, and Kasem introduced two other signature features of the program. First, he would begin each program by playing the top three songs from the past week's countdown. This served as a teaser for the listening audience. Second, he would introduce the "long-distance dedication," *AT40*'s most popular feature. On 26 August 1978, Kasem played Neil Diamond's "Desiree."

Kasem's work on the program was publicly recognized on 26 April 1981, when he received a star on Hollywood's Walk of Fame. His charm, storytelling ability, and embracing voice made *AT40* the most popular radio program in history, with an audience stretching from coast to coast. In 1988, he left *AT40* to begin *Casey's Top 40* for the Westwood One Radio Network, which he broadcast until March 1998, when he made a triumphant return to *AT40*. Kasem and Bustany bought the program, and revived it 38 months after it had been taken off the air.

Kasem's return to his countdown roots and the rebirth of his signature program seem to exemplify the phrase with which he ended each program: "Keep your feet on the ground and keep reaching for the stars."

ARI KELMAN

See also American Top 40; KRLA

Casey Kasem. Born Kemal Amen Kasem in Detroit, Michigan, 27 April 1932. Son of Amin and Helen Kasem, Lebanese immigrants; attended Wayne State University, 1948–52; worked at WJR and WXYZ Detroit, 1950–52; served in Korean War as broadcaster on Radio Station Kilroy, Taegu, Korea, 1952–56; host, *Krogo the Clown,* WJBK-TV Detroit, 1956; disc jockey, WJW, Cleveland, 1959; disc jockey, KRLA, Los Angeles 1963–69; host, *Shebang,* KTLA-TV, Los Angeles, 1965–68; host, *American Top 40,* 1970–88, 1998–present.

Radio Series

1970–88, 1998–present	*American Top 40*
1988–98	*Casey's Top 40*

Selected Television Series

Krogo the Clown, 1956; *Shebang,* 1965–68

Films

The Girls from Thunder Strip, 1966; *The Glory Stompers,* 1967; *The Doomsday Machine,* 1967; *Wild Wheels,* 1969; *55,2000 Years Later,* 1969; *The Cycle Savages,* 1969; *Scream Free!* 1969; *The Transplant,* 1971; *The Incredible 2-Headed Transplant,* 1971; *The Night that Panicked America,* 1975; *The Day the Lord Got Busted,* 1976; *The Gumball Rally,* 1976; *New York, New York,* 1977; *Disco Fever,* 1978; *The Flintstones Meet Rockula and Frankenstone,* 1979; *Ghostbusters,* 1984; *The Transformers: The Movie,* 1986; *Wild Wheels,* 1992; *Mr. Wrong,* 1996; *Undercover Angel,* 1999

Further Reading

Durkee, Rob, *American Top 40: The Countdown of the Century,* New York: Schirmer Books, 1999

Fong-Torres, Ben, *The Hits Just Keep on Coming: The History of Top 40 Radio,* San Francisco: Miller Freeman Books, 1998

"Top 40 Rock Show Aims for 150 U.S. Cities," *Billboard* (11 July 1970)

KCBS/KQW

San Jose, California (Later San Francisco) Station

KQW, San Jose, was one of the pioneering radio stations in the United States, the eighth to receive a Department of Commerce license, which was granted on 9 December 1921. But the major significance of KQW was that its owner, Charles Herrold, was the first person to operate a broadcast station. And although being one of the earliest stations to go on the air is important, it is the story of how Herrold got there, beginning in 1909, that makes KQW important. Today KQW is the 50,000-watt all-news KCBS in San Francisco.

Charles Herrold, after dropping out of Stanford University in 1899 and spending a decade as a freelance inventor and college instructor, decided to go into business for himself. He borrowed money from his father and in 1909 opened the Herrold College of Wireless and Engineering in a downtown San Jose bank building. The purpose of the college was to prepare young men for what was becoming a lucrative profession: wireless operator. Herrold also had a vision of inventing a new technology for a radiotelephone. Wireless was primarily Morse code based, but several inventors were just beginning to find ways to make the wireless talk. Herrold had invented and patented a system based on an oscillating DC arc, a device with its roots in the bright arc lighting of the day. His patents were for a water-cooled carbon microphone, an array of arcs burning under liquid, and a unique antenna system.

Between 1912 and 1917 Herrold, his wife Sybil, and his assistants and students at the school began a broadcasting station, regular in schedule and announced in the newspapers, with programming consisting of phonograph music and news read from the local papers. It was new, it was popular, it attracted students to the college, and it allowed Herrold to have audio content for his radiotelephone inventing. Prior to 1912, the students on the air identified the station by saying, "This is the Herrold college station broadcasting from the Garden City bank in San Jose." Later, Herrold used the call sign FN, and in 1916 he received an experimental radiotelephone license, 6XE. The evidence indicates that Herrold's small audiences began to look forward to the broadcasts, and he would have continued them were it not for the United States' entry into World War I. In April 1917, all amateurs and experimenters were ordered to cease all radio activities.

When the ban on radio activity was lifted in 1919, the arc technology once used successfully by Herrold was obsolete. So in 1919, Herrold opened a store in San Jose and built radios as a source of income. He wanted to return to the air and broadcast as before, but he lacked money for the equipment. By December 1921, Herrold had applied for and received a license as KQW, and a new transmitter using vacuum tubes was put on the air.

Like many broadcasters in the early 1920s, Herrold did not have a way to support his station. Advertising was in its infancy, and local stations had to share dial space, making it difficult for listeners to separate stations amid the interference. Many stations were sold or just went off the air, their operators giving up. In 1925 Herrold turned over his KQW license to the First Baptist Church of San Jose. In exchange for the license, the church agreed to retain and pay Herrold as its chief engineer. After a year, his contract apparently up, the church, citing financial problems of its own, fired Herrold. A headline in the San Jose paper read, "Father of Broadcasting Fired!"

In 1926 Fred Hart approached the church and offered to run KQW and make a profit with agricultural programming. In return, Hart promised the Baptists that KQW would air their Sunday morning services. The station made money as an agricultural news outlet, and Hart soon bought the station. By the end of the 1920s, broadcasting was a fully formed business, but Fred Hart still had his eye on the historical significance of his station. Contacting Charles Herrold, then a freelance sales representative for several Bay Area stations, Hart asked Herrold to try to resurrect some of the early 1909 to 1917 history and bring in materials and photos of the early station, and a promotion was developed around this information.

In 1934 Hart sold KQW to Ralph Burton and Charles McCarthy of San Francisco, and its power was soon raised to 5,000 watts. In 1942 KQW began its affiliation with the Columbia Broadcasting System (CBS). The main studios were moved to the Palace Hotel in San Francisco, although legally KQW was still licensed to San Jose, so its transmitter had to remain there. KQW was an important station during World War II, acting as a relay for shortwave transmissions for the Pacific Coast, along with airing well-known CBS network personalities. In 1945 KQW attempted one final time to publicize its pioneer history. An engineer was sent with a disc recorder to the local rest home where the aging Charles Herrold was spending his final days. A one-hour historical documentary was written, produced, and aired, with Herrold's recorded voice used at the end of the show. The actor hired to portray Herrold was Jack Webb.

After the end of World War II, KQW and CBS fought a long Federal Communications Commission battle with another station over the rights to relocate to 740 kilohertz and to increase power to 50,000 watts. In 1949 the call letters were changed from KQW to KCBS, and the transmitter was relicensed and legally moved to San Francisco; a 50,000-watt transmitter went into operation in 1951. Throughout the 1950s KCBS operated as a "full-service" station, airing a combination of news, personality, music, and CBS network offerings.

It was not until the late 1950s that CBS rediscovered that KCBS might have some status as a pioneer station. Research leading to this celebration began in 1958, when San Jose State University professor Gordon Greb discovered the Herrold history in a private local museum. The curator had pieces of the early Herrold arc transmitter technology and strong evidence that the long-forgotten broadcasts took place. Greb located the still-living witnesses to the events, including Herrold's first wife and son and former students and teachers at the Herrold school. Greb located a collection of Herrold correspondence, patents, and photographs from the important pre-1920 period. In 1959, 50 years after Herrold's 1909 beginning, a "50th Anniversary of Broadcasting" was staged by KCBS and San Jose State University's journalism department. A Herrold arc transmitter was reassembled from parts found in a local museum, and dignitaries and personalities from CBS in New York were brought to San Jose, where dinners and a parade highlighted a week of celebration. Several historical audio documentaries were aired, and a congressional resolution proclaiming KCBS the first radio station was read into the public record in Washington, D.C.

Then the most important event in the Herrold/KQW/KCBS story took place: the publication in 1959 of Greb's article in the *Journal of Broadcasting*. This story of Charles Herrold not only became the scholarly basis for the KCBS claim but also provided the historical community with evidence of Herrold's work, which has found its way into subsequent broadcast history texts.

What about the KCBS claim of "first station?" The most significant study on first broadcaster claims was published in the *Journal of Broadcasting* in 1977. In a study of four claimants by two respected historians, it was determined that KDKA in Pittsburgh could claim the title of "oldest station," because that station began on the air in 1920 and has continued, uninterrupted, to broadcast up until the present. KCBS, because of the lapse between when Herrold left the air during World War I and his return as KQW in 1921, was deemed not to be the oldest, but it could legitimately claim to be the "first station." In an ironic twist, the two large station owners, Westinghouse with KDKA and CBS with KCBS, who battled for years in the court of public opinion for the title of first broadcaster, are today owned by the same company—Viacom.

MICHAEL H. ADAMS

See also CBS; Herrold, Charles D.; KDKA

Further Reading

Adams, Mike, producer, writer, and director, *Broadcasting's Forgotten Father: The Charles Herrold Story* (videorecording), <www.kteh.org/productions/docs/docherrold.html>

Baudino, J.E., and John M. Kittross, "Broadcasting's Oldest Stations: An Examination of Four Claimants," *Journal of Broadcasting* 21 (Winter 1977)

The Broadcast Archive: Radio History on the Web, <www.oldradio.com>

The Charles Herrold Historical Site, <www.charlesherrold.org>

Greb, Gordon, "The Golden Anniversary of Broadcasting," *Journal of Broadcasting* 3, no. 1 (Winter 1958–59)

Schneider John F., "The History of KQW/KCBS," Master's thesis, San Francisco State University, 1990

KCMO

Kansas City, Missouri Station

KCMO-AM is known both regionally and nationally for two reasons: (1) it was once a high-power, 50,000 watt station that could be received across a large portion of the Midwest; and (2) because the station once employed Walter Cronkite, who has mentioned it often in discussing his early career. Originally based in Kansas City, Missouri, its studios are now located in a nearby Kansas suburb. It now operates on 10,000 watts during the day and 5,000 watts at night.

The KCMO call was first used in 1936 for a station formerly known as KWKC (which had been on the air since the 1920s) that was taken over by investors Lester E. Cox and Thomas L. Evans. By the mid-1930s, the station suffered from low power (100 watts) and no network affiliation.

Walter Cronkite, who worked for the station at that time, recalled in his biography that KCMO did not subscribe to a news wire. Cronkite, who was assigned the air name "Walter Wilcox" while at KCMO, recalls announcing sports play by play (from Western Union telegraph wire dispatches), covering news, and eventually leaving the station (but not before meeting his wife-to-be Betsy Maxwell, another station employee) when he had a dispute with management over how to react to reports of a fire at City Hall. Cronkite's version of the story is that he wanted to verify the seriousness of the fire—which turned out to be minor—with the fire department, but the station's program manager wanted to go on the air with false reports of people jumping from the building. Cronkite says that he was fired for "daring to question management's authority" and that his KCMO experience "cooled any thought I had that radio might be an interesting medium in which to practice journalism" (*A Reporter's Life,* 1996).

KCMO made major advances in solving its coverage problems by increasing to 1,000 watts in 1939 (the same year it received National Broadcasting Company [NBC] network affiliation), to 5,000 watts in 1940, and to 50,000 watts (daytime, with reduced power at night) in 1947, at which time it was assigned the frequency of 810 AM, which it occupied until the late 1990s.

The station's most stable period (and its longest continuity of ownership) began not in the golden age of radio, but as radio was being surpassed by television in 1953, when the Meredith Corporation purchased KCMO at a cost of $2 million. Under Meredith, the station heavily promoted the range and quality of signal that its 50,000 watts gave it, allowing it to reach listeners across western Missouri, much of Kansas, and parts of Iowa and Nebraska. Despite its strong daytime signal, however, its reduced power and especially its directional pattern at night often resulted in poor reception after dark in the Kansas suburbs.

Under Meredith, the station built a strong news department, and from the 1950s through the 1970s KCMO offered popular and country music formats (at one time referring to its air sound as a combination of the two). After serious losses in ratings in the late 1970s, Meredith invested heavily in a news format in 1980, at one point employing well over a dozen people in news-related capacities alone. The hoped-for surge in audience never came, however, and the news format evolved to a less expensive talk format, which also failed to move the station to a dominant position in the market. In 1983 Meredith sold KCMO to Fairbanks Communications, beginning a string of ownership changes with subsequent sales to Summit (1983), Gannett (1986), Bonneville (1993), Entercom (1997), and Susquehanna (2000).

MARK POINDEXTER

KDKA

Pittsburgh, Pennsylvania Station

KDKA in Pittsburgh began operation in 1920 and is often called the oldest regular broadcast station in the United States. The station, still owned by Westinghouse (now merged with Columbia Broadcasting System [CBS]), pioneered in many areas during its initial years on the air. Within months of its debut, the station broadcast the first regularly scheduled church services; the first program broadcast from a theater; the first on-air appearance by a Cabinet member (the Secretary of

War); and the first sporting event, a 10-round boxing match, soon followed by regularly broadcast baseball scores. KDKA developed the first orchestra exclusively used on radio, another precedent soon adopted by many other stations. The station had by then hired the first full-time radio announcer, Harold W. Arlin, a Westinghouse engineer. And a regular farm program was begun in mid 1921.

Westinghouse experimented with different means of extending and improving KDKA's signal. A shortwave station, KFKX, was placed on the air in Hastings, Nebraska, in 1923 and another followed in Pittsburgh. These and others lasted into World War II, eventually under government operation. They were one means of providing KDKA's "Far North Service" that sent the sounds of home to explorers and pioneers in northern Canada in the 1920s and early 1930s, often their only connection with the outside world.

Origins

KDKA radio began as experimental radio station 8XK in the Wilkinsburg, Pennsylvania, garage of Frank Conrad, an electrical engineer employed by the Westinghouse Corporation's East Pittsburgh plant. Conrad's experimental station was established as a point-to-point operation to test radio equipment manufactured by Westinghouse for U.S. military use in World War I. In 1919 the U.S. government canceled Westinghouse's remaining military contracts and the corporation was facing idle factories. Conrad was among the first to put his 8XK back on the air as an amateur radio telephone station and in contact with ham (amateur) radio operators. Conrad's main concerns were with the quality of his signal and the distance it would travel. He would read from newspapers and then await reports from listening posts commenting on the quality of the reception.

The people operating the listening posts soon tired of hearing news they had already seen in the newspapers and they grew weary of hearing Conrad's voice. One of them suggested that Conrad play a phonograph record. Conrad did so and soon the Westinghouse headquarters received a flood of mail requesting newer music and specific song titles. Frank Conrad had become the world's first disc jockey.

The news of Conrad's airborne music reached a department manager at Pittsburgh's Joseph Horne Department Store, who realized that people who wanted to listen to Conrad might want to purchase assembled radios. An ad was placed in the *Pittsburgh Sun*'s 29 September 1920 issue, featuring wireless sets for $10.00. That ad was seen by Harry P. Davis, a Westinghouse vice president who realized that a vast potential market could be developed for home wireless sets and that Westinghouse already had the ideal product: the SCR-70, a radio receiver made for the U.S. military in the recently concluded world war.

Decisions were made to move Conrad's station to the roof of the East Pittsburgh plant's administration building, to install a stronger transmitter, and to redesign the station for public entertainment. All was to be ready by early November 1920, a presidential election year. On 2 November, the Harding-Cox election results were broadcast by KDKA, the newly assigned call letters on its Department of Commerce license. The success of KDKA was rapidly assured, and soon many newspapers across the country were publishing the station's program schedule (usually an hour of music and talk in the evening).

According to Baudino and Kittross (1977), KDKA is the oldest U.S. station still in operation, as reckoned by the following standards: KDKA (1) used radio waves (2) to send out noncoded signals (3) in a continuous, scheduled program service (4) intended for the general public and (5) was licensed by the government to provide such a service.

Later Developments

The station's frequency shifted several times in the 1920s, between 950 and 980 kilohertz, and one final time in March 1941 (due to the North American Regional Broadcasting Agreement) to 1020 kilohertz, which it still uses. KDKA became a 50,000-watt clear channel operation by the late 1920s. In 1933, Westinghouse turned over daily management of KDKA and its other radio stations to the National Broadcasting Company (NBC) network, an arrangement that lasted until 1940.

Beginning in 1927, engineers experimented with FM transmission, using that mode for KDKA signals several hours a day. A Westinghouse commercial FM outlet was on the air in Pittsburgh by April 1942, initially programmed separately but simulcasting the AM outlet by 1948. Likewise, KDKA personnel experimented with a crude system of television in the late 1920s. However, a regular television operation appeared only when Westinghouse purchased DuMont's Channel 2 (then the only television station in Pittsburgh), dubbing it KDKA-TV, in January 1955.

As it did to most other stations, the decline of network programming brought hard times to KDKA for several years as management attempted a host of middle-of-the-road format ideas in the struggle to maintain listener loyalty, a challenge shared by other major market stations. In 1954 disc jockey Rege Cordic was hired from competing station WWSW and his huge morning drive-time popularity (in part because of his zany characters and fake commercials) helped propel KDKA up the ratings ladder over the next decade, until he left for Los Angeles. A decision in 1955 to resume broadcasts of Pittsburgh Pirates baseball games also contributed to KDKA's renewed popularity. Station newscasts expanded from several years of rip-and-read wire service-based summaries to a full news staff with substantial local presence.

The station celebrated its half-century anniversary in 1970 with considerable promotion and again laying claim to being the oldest radio station in the country. At that time it was the ratings leader in its market, reaching 50 percent more homes than its nearest competitor. In July 1982 KDKA became one of the first AM outlets to provide stereo service. A decade later the station switched from its long-time "full service" or "middle of the road" programming to take on a news/talk format that continued into the new century.

REGIS TUCCI AND CHRISTOPHER H. STERLING

See also Conrad, Frank; Group W.; Westinghouse

Further Reading

Baudino, Joseph E., and John M. Kittross, "Broadcasting's Oldest Stations: An Examination of Four Claimants," *Journal of Broadcasting* 21 (Winter 1977)

Davis, H.P., "The Early History of Broadcasting in the United States," Chapter 7, *The Radio Industry: The Story of Its Development,* Chicago: A.W. Shaw, 1928

Douglas, George H., "KDKA," *The Early Days of Radio Broadcasting,* Jefferson, North Carolina: McFarland, 1987

It Started Hear: The History of KDKA Radio and Broadcasting, Pittsburgh: Westinghouse, 1970

KDKA website, <www.kdkaradio.com>

Kintner, S.M., "Pittsburgh's Contributions to Radio," *Proceedings of the Institute of Radio Engineers* 20: 1849–1862 (December 1932)

Myer, Dwight A., "Up from a Bread-Board—KDKA's Tale," *Broadcasting,* 24 November 1941

Saudek, Robert, "Program Coming in Fine. Please Play 'Japanese Sandman'," *American Heritage: The Twenties* (August 1965)

Keillor, Garrison 1942–

U.S. Radio Humorist

Garrison Keillor is a public radio personality and writer best known for his variety program *A Prairie Home Companion,* which is produced live on Saturday evenings by Minnesota Public Radio (MPR) and distributed by Public Radio International (PRI). The program features a weekly monologue from Lake Wobegon, Keillor's fictitious hometown, where "all the women are strong, all the men are good looking, and all the children are above average." Keillor's main contribution to modern radio was to reintroduce the variety-show format. A *Time* magazine cover story called Keillor "a radio bard" and noted that his "storytelling approaches the quality of Mark Twain's."

Keillor was raised Gary Edward Keillor in a rural area eight miles outside of his hometown of Anoka, Minnesota. His family belonged to a conservative religious sect called the Plymouth Brethren, which shunned television and motion pictures but found enjoyment in a "strictly monitored" Zenith radio set. Keillor's early radio idol was Cedric Adams, star of Minneapolis Columbia Broadcasting System (CBS) station WCCO, which also carried variety shows such as *The Red River Valley Gang* and *The Murphy Barn Dance.* As a boy, Keillor would also pull in clear channel radio station signals carrying exotic rhythm and blues from far-away cities. At age 14 he discovered *The New Yorker* and dreamed of being a literary figure. In 1966 he graduated with a BA in English from the University of Minnesota, where he performed on radio station KUOM and edited the campus literary magazine.

He went to work for Minnesota Public Radio in 1969 at KSJR in the central Minnesota town of Collegeville. There he developed a three-hour morning program that evolved from classical music into an eclectic mix of musical styles that included folk, rock, and bluegrass. It was on this show that Keillor began referring on the air to the fictitious town of "Lake Wobegon." He published his first *New Yorker* fiction piece in 1970. Entitled "Local Family Keeps Son Happy," it was a short parody of small-town journalism about parents who hire a live-in prostitute to keep their teenager off the streets at night. Keillor quit radio in 1971 to concentrate on writing, but after six months he joined the staff at KSJN, MPR's St. Paul flagship station. In 1973 he again quit in order to write full-time.

During the summer of 1973, Keillor spent time listening to tapes of his boyhood WCCO radio favorites, including *The Red River Valley Gang,* which featured a regular monologue about the banjo player's visits to an uncle in North Dakota. That summer he also listened to tapes of Gene Autry programs from the "Melody Ranch" studio and to recordings of Smilin' Ed McConnell's vintage radio program, which also featured a

Garrison Keillor
Courtesy Radio Hall of Fame

story in the middle of the broadcast. When he returned to KSJN in 1974, Keillor called his morning program *A Prairie Home Companion,* its name taken from the Prairie Home cemetery in Moorhead, Minnesota. In April of 1974 he traveled to Nashville to cover the *Grand Ole Opry's* last broadcast from the old Ryman Auditorium for *The New Yorker.* During and shortly after this trip, an idea crystallized in his mind. Keillor pitched to MPR his concept for a weekly, live, old-fashioned musical variety show revolving around a monologue about Lake Wobegon. On 6 July 1974 he hosted the first broadcast of the program, which, like his KSJN morning show, was called *A Prairie Home Companion.*

Despite the growing popularity of the program, National Public Radio (NPR) declined to distribute the show nationally because it felt the show was too regional, contributing to Minnesota Public Radio's decision to set up its own network, American Public Radio (APR; today called Public Radio International). National syndication on APR began in May 1980. The program won the George Foster Peabody Award for broadcast excellence in 1981. In 1985 Keillor married Ulla Skaerved, who had been a Danish exchange student at Anoka High School and with whom Keillor became reacquainted at their 25th class reunion. Two years later, as the Twin Cities newspapers ran front-page coverage of what Keillor considered his private affairs, he abandoned *A Prairie Home Companion* and moved to Denmark, and later New York. In 1989 he began a new radio show in New York called *The American Radio Company of the Air,* which was similar in format to his earlier program. In 1992 he moved the program's production site to St. Paul, and in 1993 he reclaimed the concept and the title of *A Prairie Home Companion,* which at the beginning of 2003 was heard by some 5 million listeners weekly on more than 450 public radio stations. Keillor continues to publish works of fiction and keeps a rural Wisconsin home outside of the Twin Cities, as well as a residence in New York City.

MARK BRAUN

See also Comedy; Minnesota Public Radio; Prairie Home Companion; Public Radio Since 1967

Garrison Keillor. Born Gary Edward Keillor in Anoka, Minnesota, 7 August 1942. Married Mary Guntzel, 1965; attended University of Minnesota, Minneapolis, B.A. in English, 1966; worked for Minnesota Public Radio, Collegeville station, KSJR, 1969–71; MPR's Saint Paul flagship KSJN, 1971–73, 1974–82; broadcast *A Prairie Home Companion* on Minnesota Public Radio, 1974–87; hosted *American Radio Company* (similar to *Prairie Home Companion*), New York, 1989–92; revived *A Prairie Home Companion,* 1993–present; hosts daily five-minute radio program, *The Writer's Almanac,* 1993–present; writes biweekly romance advice column for online magazine *Salon.* Received National Humanities Medal, National Endowment for the Humanities, 1999; inducted into Radio Hall of Fame, Museum of Broadcast Communications, Chicago, 1994; George Foster Peabody Award, 1981; Edward R. Murrow Award, 1985; two ACE Awards for work in cable television, 1988; Grammy Award for recording of *Lake Wobegon Days,* 1985.

Radio Series

1974–87, 1993–present	*A Prairie Home Companion*
1989–92	*The America Radio Company of the Air*
1993–present	*The Writer's Almanac*

Selected Publications

Happy to Be Here, 1982
Lake Wobegon Days, 1985
Leaving Home: A Collection of Lake Wobegon Stories, 1987
Don: The True Story of a Young Person, 1987
We Are Still Married, 1989
WLT: A Radio Romance, 1991
The Book of Guys, 1993
Cat, You Better Come Home, 1995
The Old Man Who Loved Cheese, 1996
The Sandy Bottom Orchestra (with Jenny Lind Nilsson), 1996
Wobegon Boy, 1997
The Best American Short Stories (coedited with Katrina Kenison), 1998
Me: By Jimmy "Big Boy" Valente as Told to Garrison Keillor, 1999
Lake Wobegon Summer 1956, 2001
Good Poems (editor), 2002

Recordings

A Prairie Home Album, 1972; *The Family Radio,* 1982; *Gospel Birds and Other Stories of Lake Wobegon,* 1985; *Lake Wobegon Days,* 1986; *A Prairie Home Companion: The Final Performance,* 1987; *A Prairie Home Companion: The 2nd Annual Farewell Performance,* 1988; *More News from Lake Wobegon,* 1989; *Garrison Keillor's American Radio Company: The First Season,* 1990; *Local Man Moves to the City,* 1991; *Songs of the Cat,* 1991; *A Visit to Mark Twain's House with Garrison Keillor,* 1992; *The Young Lutheran's Guide to the Orchestra,* 1993; *A Prairie Home Companion 20th Anniversary Collection,* 1994; *A Prairie Home Christmas,* 1995; *The Hopeful Gospel Quartet: Climbing Up on the Rough Side,* 1997; *Garrison Keillor's Comedy Theater,* 1997; *Mother, Father, Uncle, Aunt: Stories from Lake Wobegon,* 1997; *The Best American Short Stories 1988* (coedited with Katrina Kenison), 1998; *Life These Days: Stories from Lake Wobegon, Vol. 2,* 1997; *A Prairie Home Companion,* 1999; *Me: By Jimmy "Big Boy" Valente As told to Garrison Keillor,*

1999; A Prairie Home Commonplace Book: 25 Years on the Air with Garrison Keillor, 1999; *A Prairie Home Companion: Pretty Good Joke Book,* 2000; *Definitely Above Average: Stories and Comedy for You and Your Poor Old Parents,* 2000; *Lake Wobegon Summer 1956,* 2001; *Good Poems,* 2002; *A Few More Pretty Good Jokes,* 2002

Further Reading

Baenen, Jeff, "Books and Authors: Garrison Keillor Spins More Tales from Lake Wobegon," *New York Times* (16 May 1990)

Fedo, Michael, *The Man from Lake Wobegon,* New York: St. Martin's Press, 1987

Karlen, Neal, "A Prodigal Son Makes His Way Home," *New York Times* (27 March 1994)

Lee, Judith Yaross, *Garrison Keillor: A Voice of America,* Jackson: University Press of Mississippi, 1991

Minnesota Public Radio: A Prairie Home Companion: Garrison Keillor, <www.prairiehome.org/cast/garrison_keillor.shtml>

Scholl, Peter A., *Garrison Keillor,* New York: Twayne, and Toronto, Ontario: Macmillan Canada, 1993

Kent, A. Atwater 1873–1949

U.S. Radio Inventor and Manufacturer

From 1921 to 1935 A. Atwater Kent's company was one of the most important U.S. manufacturers of radio receivers.

Early Years

Born in New England of an upper-middle-class family, A. Atwater Kent (he never used his first name) attended the private Wooster Polytechnic Institute from 1895 to 1897 but left before graduating to enter business. (Three decades later, he was awarded an honorary doctorate by the institution, whose laboratory he endowed in his will.) His father had been a part-time inventor and machinist before becoming a doctor, and his handiwork may have been the son's first exposure to mechanical devices. Kent's first foray into the working world came with the Kent Electric Manufacturing Company, which he formed in about 1895, with financial support from his father, and which made small electric motors, fans, and even "Amperia," a battery-powered electrical game. Although this foretold his future, it may not have been financially successful, because in about 1900 he sold his firm to Kendrick and Davis of Lebanon, New Hampshire, and briefly worked for that company.

The central part of his career came with his formation of the Atwater Kent Manufacturing Works in downtown Philadelphia in 1902. The new company made telephones, small voltmeters, and other small electrical devices. Three years later the product line was expanded to include automobile devices, including the 1906 Kent-invented spark generator ignition system, later dubbed the "Unisparker," that remained widely used into the 1970s. By the end of his life, Kent would hold 93 patents granted from 1901 to 1943.

During World War I the Kent company, by now located on Stenton Avenue in Germantown just north of center-city Philadelphia, manufactured military equipment for the U.S. Army, including a panoramic gun sight, fuse-setting equipment, a gun-training (aiming) theodolite, and a device to precisely incline a rifle. Using his trained staff, which in 1919 numbered about 125, and the manufacturing facilities developed during the war (which after wartime contracts were completed or terminated would otherwise become largely redundant), Kent's company joined the postwar bandwagon to radio receiver manufacturing. Though other firms were also entering the radio market, Kent's company enjoyed the benefits of extensive manufacturing expertise and facilities as well as an existing chain of dealerships for his automobile and electrical devices.

Radio Years

The Atwater Kent firm began selling radio receiver components in late 1921, trading on its reputation and network of dealers. Just a year later Atwater Kent radio advertising depicted fully assembled receivers in response to the growing demand for sets. The several initial models were dubbed "breadboards" because they lacked an external case and arranged their components along a wooden base. Over several years more than 120,000 steadily improved breadboard models were sold. Though successful, the radio market was changing and demanded furniture-like devices rather than experimental-looking breadboards. The company needed more space and was unable to expand in Germantown, so a new 20-acre site on Wissahickon Avenue was purchased and a routine, news bulletins, and children's bedtime stories. Three days later the station aired Easter services. The *Times* purchased the KHJ call letters (kindness, health, joy) from Kierulff in November 1922 and increased power to 500 watts. During its earliest years, KHJ stopped broadcasting for 3 minutes out of each 15-minute period in order to clear the air for distress calls. On 31 December 1922, KHJ broadcast throughout New Year's Eve, reported to be an unprecedented event.

As radio entered its golden age, KHJ became the principal West Coast affiliate of Mutual and the flagship of the regional Don Lee network, which was named for its owner, an automobile sales tycoon. The station originated numerous network programs, including the Columbia Broadcasting System (CBS) show *Hollywood Hotel,* hosted by Louella Parsons (during the years before CBS acquired station KNX). Other programming included Raymond Paige and a 50-piece staff orchestra, *Chandu the Magician,* Eddie Cantor, Burns and Allen, *Queen for a Day,* and *Hopalong Cassidy.*

Some prominent figures in mass communication passed through KHJ during its early years. Sylvester (Pat) Weaver, later president of the National Broadcasting Company (NBC), was an announcer in 1934. Helen Gurley Brown, later responsible for revamping Hearst's *Cosmopolitan,* answered listeners' letters while she was a student.

In 1950 RKO General (Tire) purchased the Don Lee broadcast properties, which included KHJ, KHJ-FM, and KHJ-TV. As music and news began to dominate local radio programming in the 1950s, KHJ featured disc jockeys and popular records. In the early 1960s, the station featured the talk personality format that had been successful at RKO General's WOR in New York. Nevertheless, by the end of 1964 KHJ had not developed a niche in the competitive Los Angeles market. In the ratings, the regional facility—on 930 kilohertz with 5 kilowatts of power and a directional antenna at night—was a lusterless 17th from the top.

The management of RKO General's radio division announced that KHJ would undergo a complete change of programming by May 1965. Although initially opposed to rock and roll or country and western, management chose to pursue an around-the-clock contemporary music format that would draw the bulk of young listeners without offending any other segment of the potential audience. That decision set in motion a chain of events that ultimately brought KHJ from virtual obscurity to legendary status.

In early 1965 RKO General retained "two men who had previously taken 'average' stations and transformed them into number one ranking in areas similar to our own." Those specialists were Gene Chenault and Bill Drake. Chenault was licensee of KYNO in Fresno, the original Top 40 station in the central valley of California. Drake had worked for Gerald Bartell's WAKE in Atlanta and KYA in San Francisco prior to joining Chenault as program director of KYNO. Drake's programming had led KYNO to victory in a tough ratings battle with KMAK, Chenault's tough Fresno competitor.

After Drake's success with KYNO, he and Chenault formed a consulting service. Their first client was KGB in San Diego, which rose from lowest to first in ratings on the 63rd day of Drake-Chenault programming. The success of KGB brought Drake-Chenault to the attention of RKO General management.

Drake and Chenault brought in Ron Jacobs, who had been program director of their Fresno opponent, KMAK, to be the new program director at KHJ. Drake and Jacobs crafted a streamlined version of Top 40 for KHJ centered on a very limited playlist of contemporary favorites aired, when possible, in sweeps of two or three songs. Most sound effects associated with Top 40 (e.g., horns, tones, beepers) were eliminated. A cappella jingles by Johnny Mann were short. Commercial loads were cut to 12 to 13 minutes per hour and clustered in strategically scheduled stop sets. News aired at 20 minutes past the hour or 20 minutes before the hour to counterprogram competitors' newscasts on the hour or at five minutes before the hour. The mix was given an on-air slogan, "Boss Radio."

Jacobs premiered a "sneak preview" of the Boss Radio format in late April 1965. Compared with other Los Angeles Top 40 stations (KFWB, KRLA in Pasadena), KHJ was noticeably uncluttered. KHJ rose to lead Los Angeles ratings during the first six months with Drake-Chenault as consultants. At the height of its popularity in the late 1960s, KHJ attracted one out of four Los Angeles radio listeners. After KHJ's phenomenal success, RKO General signed Drake and Chenault as consultants for KFRC in San Francisco, CKLW in Windsor (Detroit), WOR-FM in New York, WRKO in Boston, and WHBQ in Memphis.

By 1968 stations paid Drake-Chenault up to $100,000 annually for Bill Drake's services. Although Drake-Chenault consulted a total of only nine stations (including KAKC in Tulsa), the influence of their 1965 win in the tough Los Angeles market diffused throughout the radio broadcasting industry as managers across the nation copied the KHJ format and conservative playlist. Drake-Chenault also attracted critics who blamed the widespread imitation of the KHJ playlist for constrained promotional efforts for innovative music during the late 1960s.

The turnaround of KHJ is a classic business success story of personalities, competition, performance, and impact. In 2000, KHJ continued to thrive with a successful Spanish language format, and the Drake-Chenault sound remained popular in Los Angeles via KRTH's (formerly KHJ-FM) oldies format, which is reminiscent of KHJ during the late 1960s.

ROBERT M. OGLES

See also Chenault, Gene; Contemporary Hit Radio Format/ Top 40; Don Lee Broadcasting System; Drake, Bill; Mutual Broadcasting System; Weaver, Sylvester

Further Reading

"Adviser Becomes Boss: Drake Signs with RKO," *Broadcasting* (16 October 1972)

Fornatale, Peter, and Joshua E. Mills, *Radio in the Television Age,* Woodstock, New York: Overlook Press, 1980

"A History of Rockin' Times: A *Radionow!* Interview with Gene Chenault," *Radionow!* 2, no. 2 (1992)

Jacobs, Ron, *KHJ: Inside Boss Radio,* Stafford, Texas: Zapoleon, 2002

MacFarland, David T., *The Development of the Top 40 Radio Format,* New York: Arno Press, 1979

Puig, C., "Back When Jocks Were Boss," *Los Angeles Times* (25 April 1993)

"Rock and Roll Muzak," *Newsweek* (9 March 1970)

King Biscuit Flower Hour

Syndicated Showcase for Rock Artists

During the 1970s and 1980s, the *King Biscuit Flower Hour* presented recorded concert performances by more than a thousand artists, including the Rolling Stones, the Who, Eric Clapton, Elton John, U2, John Lennon, Elvis Costello, Aerosmith, the Beach Boys, the Fixx, Led Zeppelin, and many current and future members of the Rock and Roll Hall of Fame. It was the first live performance radio show to offer a glimpse into the daily lives of rock bands on tour. At its peak of popularity, more than 300 U.S. radio stations carried the *King Biscuit Flower Hour,* and the syndicators estimated that the weekly audience surpassed 5 million listeners.

The program's name pays homage to *King Biscuit Time* (later called *King Biscuit Flour Hour*), a famous radio program that originated in 1941 on KFFA, Helena, Arkansas. Sponsored by the makers of King Biscuit Flour and hosted by Sonny Boy Williamson and Robert Lockwood Jr., *King Biscuit Time* showcased the country blues music of the Mississippi Delta region, one of the important roots of rock and roll. Every important performer who played the honky tonks and juke joints along the "Chittlin' Circuit" from New Orleans to St. Louis appeared on the show until it left the air in 1967. Helena is now the site of the annual King Biscuit Blues Festival, which keeps the musical tradition alive.

Bob Meyrowitz and Peter Kauff were the first producers for the *King Biscuit Flower Hour.* Their company, DIR Broadcasting, began syndicating the program in 1973. The first show featured John McLaughlin's Mahavishnu Orchestra, a popular jazz-fusion band; Blood, Sweat and Tears; and an unknown group named Bruce Springsteen and the E Street band. For the next 17 years, the format remained the same. Live performances were interspersed with backstage interviews and minimal intrusion from the hosts for continuity. Venues ranged from stadiums to large auditoriums to small clubs. Every hour featured 50 minutes of music heard just as it had been performed before the live crowd.

Later, the company began a similar weekly series featuring country music artists. Production of the original *King Biscuit Flower Hour* ceased in 1990, with a library of more than 24,000 master tapes of classic rock and roll live performances. Soon thereafter, King Biscuit Entertainment bought the series and began syndication of reruns in the United States and Great Britain.

In 1996 King Biscuit Entertainment started releasing a limited number of *King Biscuit Flower Hour* performances on tapes and compact discs, using the syndicated program as a promotional vehicle. Although the classic rock radio format suffered from declining audience shares in the mid-1990s, *King Biscuit Flower Hour* retained its syndication base because its library contained material from other genres such as new wave, modern rock, blues, and alternative.

For the 25th anniversary of *King Biscuit Flower Hour* in 1998, the syndicator produced a two-hour special retrospective program, released a commemorative double compact disc, and made an important announcement. Production had begun on a new series of live performances for future *King Biscuit Flower Hour* programs. King Biscuit Entertainment also added a streaming media website (king-biscuit.com) to promote the radio series and sale of recordings and related merchandise. At the turn of the millennium, the *King Biscuit Flower Hour* could be heard weekly on nearly 200 radio stations in the United States as well as on BBC-2 in Great Britain.

ROBERT HENRY LOCHTE

See also KFFA; Rock and Roll Format

Programming History

Nationally Syndicated 18 February 1973–present

Further Reading

Cohodas, Nadine, "The King Biscuit Blues Festival: The Sonny Boy Legacy in Helena," *Blues Access* 31 (Fall 1997) <www.bluesaccess.com/No_31/kbbf.html>

King Biscuit Radio: The Legendary King Biscuit Flower Hour Radio Show, <www.king-biscuit.com/right_about3.html#kingbiscuit>

Taylor, Chuck, "King Biscuit to Observe Its 25th," *Billboard* (11 July 1998)

King, Larry 1933–

U.S. Radio and Television Talk Show Host

Larry King, who estimates he has conducted more than 35,000 interviews during his 40-year career, revitalized the possibilities of the radio and television talk show for a national as well as worldwide audience during the cable era. Although King has crafted a laid-back persona in his signature suspenders, the *Guinness Book of World Records* recognizes the indefatigable King as having logged more hours on national radio than any other talk personality in broadcasting history. From the 1980s on, King's omnipresence on radio and television and in print made him the interviewer of choice for many celebrities and politicians.

King's agreeable style and "everyman" appeal refashioned the talk show format into an international town meeting. His first interview series, on Mutual Radio, pioneered the concept of the nationwide talk show. His television series on CNN, also simulcast on Westwood One radio stations since 1994, created the first international arena for talk, reaching more than 230 million households in well over 200 countries. King credits his innate curiosity as the main ingredient for this far-reaching popularity and commercial success.

Throughout King's career, critics have questioned his ability to ask insightful and tough questions of his guests. Howard Kurtz of the *Washington Post* labeled him "a great schmoozer who makes no pretense to being a newsman." King readily admits that he never covered a news event and that, in fact, he does little research or preparation for any interview. His technique is to ask short, conversational questions, hoping to connect with his guests on a friendly level. For King, who detests confrontation, the best guest is "anyone passionately involved in what he does."

Like many broadcasters of his generation, King grew up listening to network radio, which would disappear because of television as he came of age. Born Lawrence Zieger in Brooklyn, the young King was an indifferent student, enthralled only with everything related to radio, from the escapism of *Captain Midnight* to the quiet satire of Bob Elliott and Ray Goulding. Aspiring to be the next Red Barber, the colorful sports announcer of the Brooklyn Dodgers, or Arthur Godfrey, whose on-air folksiness King would adopt with an urban twist, the frequent studio visitor was told to seek his broadcasting fortune in a smaller market. In 1957 King journeyed to Miami and got his first job, as a handyman at WAHR, a 250-watt AM station. He took over for the late-morning disc jockey and, at the station manager's suggestion, changed his name from the ethnic Zieger to King.

King was quickly noticed by larger Miami stations, and in 1958 he was hired by WKAT to anchor the valuable early-morning broadcasts. Encouraged to stand out among other drive-time disc jockeys, King created offbeat characters in the style of his comic heroes, Bob and Ray. The popularity of one such character, Captain Wainright of the Miami State Police, a *Highway Patrol* takeoff, led to his first talk show. Pumpernik's restaurant hired the entertaining upstart to host a four-hour show, live from the eatery. King discovered his talents in the ad-lib interview, first talking with waitresses and garrulous patrons, then with anyone who came by. Celebrities began to drop by the restaurant, and King flourished in the freewheeling atmosphere of the spontaneous interview. King's ability to draw people out was on display with such young performers as Don Rickles, Lenny Bruce, and Bobby Darin.

King's horizons began to expand when he was hired by WIOD in 1962. Management recognized his potential and moved his Pumpernik's show to the houseboat that served as the luxurious setting for the American Broadcasting Companies (ABC) television series *Surfside Six* for nightly broadcasts of interviews and phone calls. He realized his dream to be a sportscaster when WIOD offered him Sunday duty as color

Larry King
Courtesy Radio Hall of Fame

commentator for the Miami Dolphins. His fame opened other media doors. King began writing columns, first for the *Miami Beach Sun-Reporter,* followed by stints at the *Miami Herald* and the *Miami News.* He hosted a local talk show on Sunday nights with no time limits on WLBW in 1963, and a year later he switched to WTVJ-TV with a weekend show.

King has said he felt as if he "owned Miami" and piled up outrageous debts. Embroiled in a larceny scandal, King's career was shattered by his high-living notoriety. By 1972 he lost every media position he had accrued. During the mid-1970s King accepted any job he was offered, working as a public relations official with a horse-racing track in Shreveport, Louisiana, and as a radio commentator for the Shreveport Steamers of the World Football League. The freelance phase of his career ended when a new general of WIOD listened to archival tapes of King's best work. He was rehired to host an evening talk show, and soon the newspaper and television assignments returned. King still could not control his spending and eventually declared bankruptcy in 1978.

While getting his finances in order, King was hired to do a late-night program on the Mutual radio network. The *Larry King Show,* running from midnight to 5:30 A.M. (EST), debuted on 30 January 1978 in 28 cities and legitimized the format of the nationwide talk radio show. There had been other national talk hosts, including Herb Jebco from Salt Lake City and Long John Nebel from New York, but King demonstrated that the talk format was not a local phenomenon. Although King reveled in an uncontrolled environment, there was a distinct tripartite structure to most broadcasts. Most programs featured an hour-long interview with a guest; two hours of call-in questions for that guest; and the final hour, "Open Phone America," in which King's assorted group of "insomniacs and graveyard-shift workers" would call and chat about anything. Unlike most hosts, King did not screen calls, and he described this formula as "talk show democracy."

For the first two months, King's radio series was broadcast from Miami, and then the show was relocated to Arlington, Virginia, so that government officials could appear as guests.

The coverage grew exponentially; by the time the broadcast switched to daytime in 1992, more than 430 stations were carrying King's brand of talk. In 1982 he received the coveted George Foster Peabody Award. As he did in Miami, King used his radio fame as a calling card for other pursuits, such as writing a weekly column for *USA Today* and working for Ted Turner's CNN.

In 1985 he adapted his mostly single-guest and call-in format for cable television. *Larry King Live* has consistently been CNN's highest-rated show, emerging as a national forum on topical issues. In 1992 the major presidential candidates courted King's viewers, including H. Ross Perot, who declared his availability if "drafted" by the people. King also made headlines by arranging a debate on the North American Free Trade Agreement (NAFTA) between Perot and vice president Al Gore. Guests have ranged from world leaders and newsmakers (Mikhail Gorbachev, Margaret Thatcher, Norman Schwarzkopf, and the Dalai Lama) to entertainment and sports luminaries (Frank Sinatra, Marlon Brando, Barbra Streisand, and Arthur Ashe).

Larry King emerged as one of the dominant figures from radio's post-network era. His agreeable interviewing style attracted a wide audience across diverse demographic lines. King demonstrated that radio still has possibilities as a national medium. When music programming held sway on the FM band in the 1970s, King was one of the pioneers to conceive of the AM band as a nationwide vehicle for talk radio. The self-described "street kid from Brooklyn" helped to revive the tradition of network radio that had inspired him as a youngster.

RON SIMON

See also Mutual Broadcasting System; Peabody Awards; Talk Radio

Larry King. Born Lawrence Harvey Zeiger in Brooklyn, New York, 19 November 1933. Educated at Lafayette High School; disc jockey and interviewer at various Miami stations, 1957–71; columnist, various Miami newspapers, 1965–71; freelance writer and broadcaster, 1972–75; radio talk-show host, WIOD, Miami, 1975–78; host, Mutual Broadcasting System's *Larry King Show*, 1978–94; host, CNN's *Larry King Live*, 1978–present; host, the Goodwill Games, 1990; columnist, *USA Today* and *The Sporting News*. Received George Foster Peabody Award, 1982; National Association of Broadcasters' Radio Award, 1985; founded Larry King Cardiac Foundation, 1988; International Radio and TV Society's Broadcaster of the Year, 1989; named to Broadcaster's Hall of Fame, 1992; Scopus Award from American Friends of Hebrew University, 1994; Emmy Award, Outstanding Interviewer, 1999.

Radio Series
1978–94 *The Larry King Show*

Films
Ghostbusters, 1984; *Lost in America*, 1985; *Contact*, 1997; *Primary Colors*, 1998

Television Series
Larry King Live, 1978–present (CNN); cameos on *Murphy Brown*, *Frasier*, and *Murder One*

Selected Publications
Larry King (with Emily Yoffe), 1982
Tell It to the King (with Peter Occhiogrosso), 1990
When You're from Brooklyn, Everywhere Else Is Tokyo (with Marty Appel), 1992
On the Line: The Road to the White House (with Mark Stencel), 1993
The Best of Larry King Live: His Greatest Interviews, 1995
Future Talk: Conversations About Tomorrow with Today's Most Provocative Personalities, 1998
Anything Goes! 2000

Further Reading
Laufer, Peter, *Inside Talk Radio: America's Voice or Just Hot Air?* Secaucus, New Jersey: Carol, 1995
Munson, Wayne, *All Talk: The Talkshow in Media Culture*, Philadelphia, Pennsylvania: Temple University Press, 1993
Timberg, Bernard, *TV Talk: The History of the TV Talk Show*, Austin: University of Texas Press, 2002

King, Nelson 1914–1974

U.S. Country Music Disc Jockey

Historians have rated Nelson King among the disc jockeys who exerted the greatest influence on the commercialization of country music in the years following World War II. For almost 15 years, King served up country music to millions of listeners on his *Hillbilly Jamboree*, broadcast from 50,000-watt radio station WCKY in Cincinnati, Ohio. King's *Jamboree* transmitted strongly into the eastern United States, proving that his show could garner large audiences for country music from that populous region. He belongs to a class of early pioneering country music disc jockeys that includes Randy Blake (WJJD, Chicago, Illinois) and Rosalie Allen (WOV, New York).

After graduation from Portsmouth (Ohio) High School in 1932, King worked briefly as a stock boy at Woolworth's, but soon joined a small band, serving as its master of ceremonies. His resonant speaking voice, exhibited on various Ohio bandstands, helped him land his first radio job at WPAY in Portsmouth, where he was a staff announcer. Two years later, he became the chief announcer and musical director at radio station WSAZ in Huntington, West Virginia. At WSAZ, King's tenure included his role as part of an announcing team that covered the devastating West Virginia floods of 1937 for 381 continuous hours.

Departing soggy Huntington for Cincinnati in 1938, King began hosting a recorded music program over radio station WCPO; his *Jam for Supper* showcased the swing music of Benny Goodman, Tommy Dorsey, Artie Shaw, and others. After subsequent brief stays at radio stations WGRC in Louisville, Kentucky, and WKRC in Cincinnati, he joined WCKY in January of 1946.

It was at WCKY that management asked the new employee (then known as Charles Schroeder) to find a radio name. He was handed a Cincinnati phone book and plucked from it the pseudonym Nelson King (a change he later legalized). Soon after his arrival, he took the helm of WCKY's *Birthday Club*, *Man in the Street*, *Keep Happy Club*, and *Hillbilly Jamboree*; however, it would be *Hillbilly Jamboree,* the latter program carrying his name to national prominence.

WCKY and King had introduced the *Jamboree* in 1946, riding the growing popularity of country music during that time. King and the *Jamboree* would fuel that popularity. On the nightly show, King featured popular country music recording artists Eddy Arnold, Ernest Tubb, Red Foley, and others who were helping to establish country music as a commercial force in the late 1940s. The music of Arnold and others enjoyed burgeoning record sales and exposure on Saturday night music programs such as WSM radio's *Grand Ole Opry* and WLS radio's *National Barn Dance*. Their following would increase even more on the airwaves of the four-and-a-half hour *Jamboree*, one of the few nightly country music disc jockey programs to reach a national audience during the mid-to-late 1940s. The show also streamed into Canada, Mexico, and parts of South America.

From early in the *Jamboree*'s run, it was evident that King was attracting a significant audience during the nighttime hours. When WCKY offered to send listeners promotional pictures of the hillbilly jockey, more than 76,000 letters deluged the station over a two-week period. Products that King hawked on the air—everything from baby chicks to "genuine imitation granite" tombstones to Last Supper tablecloths that glowed in the dark—sold briskly.

King's ability to reel in a vast audience greatly impressed record companies. In an interview with the present writer, Bob McCluskey, a promotional representative for RCA Victor Records, recalled that King's show influenced tastes: "Nelson's program really controlled [the East] late at night. The people that listened to the station were very, very record conscious at that time because what was played, they bought" (see Streissguth, 1997). In light of King's influence, record companies worked diligently to court his favor. Major stars such as Gene Autry and Eddy Arnold regularly called on the powerful disc jockey, and the record companies rained refrigerators, color television sets, and crates of liquor on him. "One of the things I did . . . was I bought time on the station," said McCluskey. "In doing so, I involved Nelson . . . with a promise to play the records that I asked [him] to play." The record company practice of buying influence with popular disc jockeys was widespread during the 1940s and 1950s. However unethical the record companies' gifts and King's acceptance of them, the exchanges served to illustrate King's immense sway with listeners as well as his country music ambassadorship.

Nelson King also proved to be a boon to bluegrass music, the acoustic sub-genre of country music popularized by Bill Monroe, Lester Flatt and Earl Scruggs, and others. King loved the dynamic bluegrass sound, sitting mesmerized for hours at bluegrass performances and featuring a liberal helping of the music on his broadcasts. A major retailer of bluegrass records, Jimmie Skinner's Music Center of Cincinnati, advertised regularly on King's show and saw remarkable sales in part because of it.

Throughout the 1950s King continued to distinguish himself as an important conduit for country music. He became active in the Country Music Disc Jockey Association (which would evolve into the Country Music Association, an important trade group), and he collected *Billboard* magazine's top disc jockey award for eight consecutive years. Not even the growth of rock and roll music could wrench him from his disc jockey slot.

In the end, it would take alcoholism to mute King. Years of alcohol abuse caught up with him in 1961, when the station fired him for going on the air inebriated. He worked outside radio until 1968 when he assumed the morning disc jockey shift at radio station WCLU in Cincinnati. But his continuing alcoholism and the onset of lung cancer ended his radio career in 1970. Although he eventually gave up alcohol, he succumbed to the cancer in 1974.

Nelson King was elected posthumously to the Country Music Disc Jockey Hall of Fame in 1975, recognized as one of country music's foremost emissaries.

MICHAEL STREISSGUTH

See also Country Music Format; Disk Jockeys; Grand Ole Opry; Perryman, Tom; WSM

Nelson Charles King. Born Charles Edward Schroeder in Portsmouth, Ohio, 1 April 1914. One of three children born to Stanley and Sue Schroeder; announcer, WPAY, Portsmouth, Ohio, 1934–36; chief announcer and musical director, WSAZ, Huntington, West Virginia, 1936–38; disc jockey, WCPO, Cincinnati, Ohio, 1938–45; announcer, WGRC, Louisville, Kentucky, 1945; announcer, WKRC, Cincinnati, Ohio, 1945–46; host of *Hillbilly Jamboree,* WCKY, Cincinnati, Ohio, 1946–61; disc jockey, WCLU, Cincinnati, Ohio, 1968–70. Inducted posthumously into Country Music Disc Jockey Hall of Fame, 1975. Died in Cincinnati, Ohio, 16 March 1974.

Radio Series

1946–61 *Hillbilly Jamboree*

Further Reading

Cherry, Hugh, "Country DJs Carry Music to the People," *Music City News* (October 1980)

Friedman, Joel, "Radio, Country Music, Like Twins, Grow Big Together," *Billboard* (22 May 1954)

Kingsbury, Paul, editor, *The Encyclopedia of Country Music: The Ultimate Guide to the Music,* New York: Oxford University Press, 1998

Malone, Bill C., *Country Music U.S.A.,* Austin: University of Texas Press, 1968; revised edition, 1985

Streissguth, Michael, *Eddy Arnold: Pioneer of the Nashville Sound,* New York: Schirmer, and London: Prentice Hall, 1997

Kirby, Edward M. 1906–1974

U.S. Broadcast and Advertising Executive

Edward Kirby had a long and successful career in broadcasting, advertising, and public relations. His most important contributions to radio, however, were made while wearing his nation's uniform, working for the biggest sponsor of them all, the U.S. government.

Origins

Born in Brooklyn, New York, on 6 June 1906, Kirby was the son of a coal merchant. Sent off to boarding school at a young age, he returned to New York for a time in his high school years before going to a private military academy in upstate New York. He was later accepted at the Virginia Military Institute (VMI), where he wrote for various humor and literary magazines. Kirby received his bachelor's degree from VMI in 1926.

After graduation, Kirby worked as a reporter for the *Baltimore Sun* and later undertook economic, statistical, and market analysis for several investment banks. He eventually worked his way into advertising and public relations with C.P. Clark, Inc., of Nashville in 1929. For that firm, he directed national advertising campaigns for several important clients, including the General Shoe Corporation. He also got his first radio experience, producing nationally distributed programs for General Shoe and others.

While with C.P. Clark, Kirby came to the attention of Edwin Craig, vice president of the National Life and Accident Insurance Company of Nashville, owner of radio station WSM. Craig hired Kirby in 1933, and soon the company had increased its insurance in force by 57 percent (to $525 million) owing largely to Kirby's use of WSM as a sales tool. In 1936 Kirby married Marjorie Arnold, daughter of the dean of the

In late 1928 KNX shifted from 890 to 1050 kilohertz and moved to the Paramount Pictures lot in Hollywood. The station increased transmitter power to 5,000 watts in 1929 and doubled that in 1932. Earl sold the Los Angeles *Evening Express* but stayed in radio, running KNX under the ownership of the Western Broadcasting Company. When KNX moved its offices and studios again in 1933, to the corner of Vine Street and Selma Avenue, station power was boosted to 25,000 watts and finally to 50,000 watts in 1934. In 1936 KNX moved to Sunset Boulevard.

CBS Ownership

KNX was sold to the Columbia Broadcasting System (CBS) for $1.25 million in 1936, then the highest price ever paid for a radio station. New KNX/CBS studios were constructed and opened on 30 April 1938 at 6121 Sunset Boulevard. The Hollywood landmark station remains there today. Known as Columbia Square, the studios were home to several top-rated radio shows through the 1940s, including *Silver Theater, Melody Ranch* with Gene Autry, *Lucky Strike Hit Parade,* Jack Benny, Burns and Allen, Edgar Bergen and Charlie McCarthy, and Red Skelton. The long-running *Lux Radio Theatre* originated from the Vine Street Playhouse nearby. During World War II, *GE Radio News* with Frazier Hunt was heard. Local shows such as *The Housewives Protective League, Hollywood Melody Shop,* and *Hollywood Barn Dance* were favorites with southern California listeners. On 29 March 1941 KNX shifted its frequency one last time, to 1070 kilohertz.

In the late 1940s comedian Steve Allen worked at KNX as a disc jockey, but he soon turned his airtime into a very popular late-night interview and comedy show. The program got Allen noticed by CBS executives and was a springboard to his highly successful subsequent TV career. Bob Crane, who gained fame on TV's *Hogan's Heroes,* was a very funny morning personality on KNX from 1956 to 1965. During this time KNX had become mostly a music station with news and sports features.

In April 1968 KNX initiated an all-news format and soon operated the largest radio news department in the western United States. KNX claims to have won more awards for broadcast journalism than any other radio station in the United States. These honors include Best Newscast Award from the Associated Press and Best Newscast Award from the Radio-TV News Association (RTNA) 27 times in the past 30 years. KNX has also won more than 150 Golden Mikes from the RTNA.

JIM HILLIKER

Further Reading

Blanton, Parke, *Crystal Set to Satellite: The Story of California Broadcasting, the First 80 Years,* Los Angeles: California Broadcasters Association, 1987

KOA

Denver, Colorado Station

As a 50,000-watt clear channel station, KOA is said to stand for "Klear Over America." One of the first radio stations in Denver, KOA later became one of the West's most popular stations. KOA's powerful signal is capable of reaching 38 states in the evening hours, and the station has been heard in Canada, Mexico, and nearly every state in the United States under the right atmospheric conditions.

Changing Hands

KOA went on the air on 15 December 1924 and was authorized for 1,000 watts at 930 kilohertz. Built and operated by the General Electric Company, the station underwent many changes in operating power and dial position before settling at 50,000 watts in 1934 and at 850 kilohertz in 1941.

KOA underwent many ownership changes over the years. The station became affiliated with the National Broadcasting Company (NBC) in 1928. In 1929 NBC took over operation of KOA from General Electric, and the license was officially assigned to NBC in 1930. However, the actual change of ownership did not occur until NBC bought the transmitter from General Electric in 1934. NBC added a sister station, KOA-FM (later KOAQ-FM, now KRFX-FM), in 1948. The stations were sold to the Metropolitan Television Company (MTC) in 1952. One of MTC's principal stockholders was legendary radio, television, and motion picture entertainer Bob Hope. This group added a television station in 1953, channel 4 KOA-TV (now KCNC-TV). Bob Hope sold his interest in 1964, and General Electric repurchased the station in 1968. In 1983 General Electric sold KOA-AM and sister station KOAQ-FM to

Belo Broadcasting. In 1987 KOA was sold to Jacor Broadcasting, which merged with Clear Channel Communications in 1999.

Programming

The opening broadcast in 1924 was launched with much fanfare. With colorful prose, the station avowed its purpose "to serve with special intimacy the states that lie in the great plain—from the Dakotas and Minnesota to Texas—to the Mississippi and beyond; to spread knowledge that will be of use to them in their vast business—to further their peoples' cultural ambitions—to give wider play to their imaginations, and make melody in their ears—to bid them lift up their eyes unto these western hills whence comes new strength" (cited in *Colorado Mac News*, 1984)

In the early days, a large number of KOA radio listeners were farmers, and the station had a heavy emphasis on farm, weather, and agriculture market–related programming. *Farm Question Box* and *Mile High Farmer* were two of the longest-running and most popular agriculture-related programs on the station.

KOA claims a number of historical "firsts" in broadcasting. On 18 February 1927 KOA did a remote broadcast of the "hole-ing through" of the Moffatt Tunnel, which was at the time the longest railroad tunnel in the world. Using the railroad's telegraph circuit, which ran to the entrance of the tunnel, KOA engineers ran lines more than three and one-half miles into the tunnel to broadcast the event. On 15 November 1928 KOA engineers lugged a transmitter to the top of Pikes Peak near Colorado Springs to become the first station to originate from atop a 14,000-foot peak in the Rockies. On 6 May 1936 KOA successfully broadcast a concert nationwide from a specially equipped Radio Corporation of America (RCA) Victor train heading into Denver. Leopold Stokowski conducted Stravinsky's *The Firebird* with a portion of the Philadelphia Symphony Orchestra on board. On 27 February 1937, a ski official involved in a ski race at Berthoud Pass in Colorado was set up with one of NBC's first "pack sets" on his back and a catcher's mask with a microphone mounted inside. The idea was for him to describe what he saw and felt as he sped down the mountain. The nation waited to hear how it felt to ski down a challenging mountain course. The technical apparatus worked perfectly, but the skier forgot to talk, and all the nation heard was several minutes of heavy breathing and the rush of air.

For a number of years, the KOA Staff Orchestra, an all-string ensemble, was featured on the NBC network as a sustaining program. Such programs as *Golden Memories, Rhapsody of the Rockies,* and *Sketches in Melody* spread the fame of the KOA Staff Orchestra across the country.

The 1960s and 1970s saw the development of KOA as a news powerhouse. The station built a reputation as a leading source of news and information throughout the Rocky Mountain West.

On 18 June 1984 outspoken KOA talk show host Alan Berg was murdered in front of his condominium in Denver. The slaying, which had political and religious overtones, generated a tremendous amount of national media coverage because of the circumstances. Berg took on many groups on his high-rated talk show, including right-wing Christians, knee-jerk liberals, and the Ku Klux Klan. There were connections between Berg's death and a group called "The Order," a white supremacist group in Colorado and the Pacific Northwest. The murder weapon was later found in the home of Gary Lee Yarbrough of Sandpoint, Idaho, a member of The Order.

Today, KOA concentrates its efforts as a news/talk/sports station. The station has the largest news-gathering staff in the market and carries play-by-play coverage of the Denver Broncos National Football League team, the Colorado Rockies baseball team, and the University of Colorado football games. The station's internet address is www.koaradio.com, which includes the KOA broadcast signal.

STEVEN D. ANDERSON

Further Reading

"A Brief History of the 50,000 Watt 'Voice of the West,'" *Colorado Mac News* (3 December 1984)
KOA website, <www.koaradio.com>
Singular, Stephen, *Talked to Death: The Life and Murder of Alan Berg,* New York: Beach Tree Books, 1987

KOB

Las Cruces/Albuquerque, New Mexico Station

KOB was a pioneering noncommercial, educational radio station established in Las Cruces, New Mexico, in the years after World War I. Affiliated with the New Mexico College of Agriculture and Mechanic Arts (NMA&MA)—now New Mexico State University—and supported by the college's engineering department, KOB began as an experimental student project using equipment salvaged from the U.S. Army. Ralph W. Goddard, a professor of engineering at NMA&MA, organized the school's Radio Club on 11 October 1919. The campus organization soon acquired not only a 500-watt Marconi standard Navy spark transmitter and 60-foot tower, but also three experimental radio licenses—5CX, 5FY, and 5FZ. License 5XD, which would become KOB, was granted on 3 June 1920.

In 1922, 500 new stations began broadcasting and KOB was one of them, going on the air on 5 April. Programming included market and stock reports, live performances, recorded music, weather reports, and news accounts provided by two El Paso newspapers, the *Times* and *Herald-Post*. The college's Agricultural Extension Service was an early ally for KOB and provided ongoing support for the radio project's contributions to rural life in southern New Mexico.

Goddard, who was eventually named Dean of Engineering at NMA&MA, was a key figure in the growth and development of KOB. He was responsible for guiding the station from its beginning as a project of the Radio Club—which was largely interested in amateur and relay work—to its status a few years later as a station broadcasting to all of New Mexico. He led efforts to expand the station's programming, facilities, and wattage, negotiating for clear channel status and a power allotment of 5,000 watts with the Federal Radio Commission (FRC); both were granted on 11 November 1928. By March of the following year, the station had been approved for an allotment of 10,000 watts, making it the most powerful college radio station in the United States and the country's 13th most powerful station of any type. In the fall of 1929, Goddard applied for FRC permission to double the station's power again, which would increase its power to 20,000 watts.

Ironically, even as KOB developed as one of the country's premier radio stations, its support from the university community decreased. In spite of continuing interest in and contributions to the station by the Agricultural Extension Service, local civic groups, and segments of NMA&MA, the college administration began suggesting in late 1928 that Goddard should start looking for a buyer for the station. This diminished administrative support for KOB would ultimately prove fatal. Goddard's dream of building an educational radio station that would broadcast throughout most of the Southwest dissolved with his tragic death on New Year's Eve, 1929. While attempting to shut the station down for the afternoon, Goddard—whose shoes were soaked from a walk in the rain—was electrocuted. Although another NMA&MA professor was named station director and the application for 20,000-watt status was approved, KOB never recovered from the loss of its founder and most ardent supporter. Lacking his hands-on leadership and personal drive, station staff could muster neither adequate management expertise nor local support to keep KOB on the air. When the FRC began sending off-frequency reports and complaints that the station was not modulating, the college decided to accept a lease-purchase offer from T.M. Pepperday, owner of the *Albuquerque Journal*.

The station's assets were subsequently leased to the *Journal* in the fall of 1931; all equipment was transferred by truck more than 200 miles to Albuquerque in September 1932; and KOB's first broadcast from Albuquerque aired on 5 October 1932. The station's disposition was completed in August 1936, when NMA&MA sold KOB to the *Journal*'s newly-formed subsidiary, the Albuquerque Broadcasting Company. KOB affiliated with the National Broadcasting Company (NBC) the following year.

KOB is one example of the way in which a number of noncommercial, educational radio stations were transferred from public to private ownership in the 1930s. Lacking financial resources, institutional support, and strong leadership, this flagship western station could not endure as a publicly owned, noncommercial entity. Like many college stations, KOB's purchase and network affiliation helped to facilitate the development of a network-dominated broadcasting system in the United States. Many local listeners in Las Cruces were grieved by the station's departure, and the New Mexico State Legislature conducted an extensive debate about the station's sale to a private company. Governor Floyd Tingley, a New Dealer with close ties to Franklin D. Roosevelt, suggested that KOB become a state owned and operated public station, an idea that would be echoed in later discussions of state public networks elsewhere.

The North American Regional Broadcasting Agreement (1941) forced KOB (and many other U.S. stations) to change frequencies. In 1941 KOB moved to 1030 kilohertz, the same channel used by a Boston clear channel outlet. Later that year, just before U.S. entry into World War II, KOB moved again, this time thanks to a Federal Communications Commission (FCC) temporary permit, to 770 kilohertz, a clear channel frequency then used by what is now WABC in New York. The potential for mutual interference between the stations led to the longest legal battle in FCC history. Attorneys for each station inundated the commission with legal attempts to force the other outlet off the shared frequency. KOB was allowed to raise its power to 50,000 watts in the daytime but was required to drop back to 25,000 watts at night in an attempt to protect the New York station. In 1956 KOB was also required to install a directional antenna to reduce interference with WABC. By 1962 it had FCC permission to raise its nighttime power to 50,000 watts. Legal filings from both outlets continued to plague the FCC into the early 1980s, when KOB's final appeal to overcome WABC's primary status on the channel was turned down. Both stations remain on 770 kilohertz today, with KOB continuing to shield the New York outlet by means of directional antenna patterns.

A KOB television station was added in 1948, and in 1967 an affiliated FM outlet was added in Albuquerque. The station changed hands several times; it was purchased by Hubbard Broadcasting in 1957, which operated KOB for three decades before selling it in 1986. It changed hands again several times in the 1990s, by which time its call letters had been changed to KKOB. On 15 October 1994 the station was purchased by Citadel Communications, and by 2000 it was programmed as news/talk.

In the meantime, the university that had first supported the radio outlet lost its right to free airtime in 1951, although for a number of years that arrangement continued informally. Ralph Goddard's important role in its early years was remembered when the Las Cruces university again became a licensee in 1964 and placed KRWG (Goddard's initials) on the air as the first college/university FM station in the New Mexico. It, too, was joined by a television outlet with the same call letters eight years later.

GLENDA R. BALAS

Further Reading

Velia, Ann M., *KOB, Goddard's Magic Mast: Fifty Years of Pioneer Broadcasting,* Las Cruces: New Mexico State University Press, 1972

KPFA

Berkeley, California Station

KPFA-FM in Berkeley, California, is the first station of the Pacifica radio network and the first listener-sponsored station in the United States. Poet, philosopher, and conscientious objector Lewis Hill created the station as a means to "help prevent warfare through the free and uncensored interchange of ideas in politics, philosophy and the arts."

KPFA first went on the air 15 April 1949, and controversy dogged the non-commercial station from the start. In a postwar America known for Cold War conformity and rampant consumerism, KPFA brought many non-mainstream voices to its microphones, including African-American actor and activist Paul Robeson, Zen philosopher Alan Watts, leftist commentator William Mandel, Beat poet Lawrence Ferlinghetti, political theorist Herbert Marcuse, film critic Pauline Kael, and voices from the Bay Area's academic, pacifist, and anarchist communities.

Hill created the idea of listener sponsorship in order to fund the station's operation without having to sell commercials. This freed the station of corporate control and gave it the chance to promote political alternatives. The operating funds, Hill theorized, would come if 2 percent of the potential audience paid $10 a year to support the station.

Listener sponsorship was not the only unusual aspect of the station's operations. In the beginning, despite a bureaucratic hierarchy, everyone on staff at KPFA was paid the same salary, and decisions were made collectively. Also, clocks were removed from on-air studios so that programs could run to their natural conclusions.

Despite such innovations, financial problems and internal tensions among the staff, the volunteers, and the station's advisory board quickly came to the fore. In 1950 the station went off the air for nine months because of lack of funds. The following year, Pacifica received a $150,000 grant from the Ford Foundation, which allowed it to resume broadcasting.

In June 1953 Lewis Hill resigned because of internal political struggles at the station. A new group felt constrained under his leadership. Despite the internecine problems, KPFA continued to make history with its innovative and intellectually challenging programming. Throughout the 1950s, the station's public-affairs programs regularly addressed such hot-button issues as racial segregation, economic disparity, and McCarthyism. In April 1954 a pre-recorded radio program that advocated the decriminalization of marijuana created an uproar and led to the tape's impoundment by California's attorney general.

In August 1954 founder Lewis Hill returned to run the station. But three years later, in late July of 1957, the 38-year-old Hill, suffering from crippling rheumatoid arthritis and depression, committed suicide. He left the following note: "Not for anger or despair/but for peace and a kind of home."

KPFA and Pacifica won numerous broadcast awards for children's programming and for special programs on the First Amendment by legal scholar Alexander Meiklejohn. KPFA's public-affairs director, Elsa Knight Thompson, continued pushing the broadcast envelope despite attempts to censor the station. The Federal Communications Commission (FCC), for example, questioned KPFA's broadcasts of poets Allen Ginsberg and Lawrence Ferlinghetti, broadcasts that the government found to be "vulgar, obscene, and in bad taste."

In 1960 KPFA broadcast a three-hour documentary on the riots following House Committee on Un-American Activities (popularly known as HUAC) hearings in San Francisco. It subsequently broadcast programs on homosexuality, the blacklist, and the FBI. HUAC reacted by investigating Pacifica for "subversion," and the FCC investigated "communist affiliations" at the station, but the Commission ultimately renewed KPFA's license after a 3-year delay. KPFA later gave extensive coverage to the Berkeley Free Speech Movement, and many University of California, Berkeley, faculty members were heard regularly.

Though the public-affairs programs often stirred controversy, the station's arts programming had a significant impact on the Bay Area's cultural scene. KPFA's first music director, Americo Chiarito, boldly mixed jazz, classical, folk, and other forms of noncommercial music throughout the broadcast day. Over a 25-year period, subsequent music director Charles Amirkanian interviewed and gave exposure to the work of nearly every living composer of importance in the West, including Terry Riley, LaMonte Young, Steve Reich, Lou Harrison, Pauline Oliveros, and John Cage. Philip Elwood, later the jazz critic for the San Francisco *Examiner,* hosted various on-air jazz programs for nearly 40 years. Sandy Miranda's *Music of the World* program featured live music, interviews, and rare recordings that drew a large and devoted following that made significant financial contributions to the station. For nearly a quarter of a century, the station devoted considerable airtime to contemporary poetry and literature under the direction of Erik Bauersfeld.

Through the 1970s and 1980s there was a gradual, if fundamental, shift in how KPFA and the Pacifica network (which by then included stations in New York, Los Angeles, Houston, and Washington, D.C.) defined themselves and their target audience. Unlike the early days in Berkeley, when the audience was presumed to be an elite, educated, intellectual minority, KPFA gradually became known as a "community" radio station. Its audience, and even many on-air programmers, became increasingly defined not by ideas but by gender, ethnicity, race, and class. Third-World and women's departments were eventually created at the station. As these and other groups in the community demanded a place on the broadcast schedule, bitter arguments ensued over ideology and over questions about who speaks for whom and which groups deserve access to the microphone.

In 1999 many of these questions came to a head when the Pacifica network attempted to make programming and staff changes at KPFA, removing the station's popular general manager. The network defended the changes as an attempt to increase audience numbers and diversity. Many of the staff and volunteers, as well as activist members of the community, charged that the network was engaging in strong-arm tactics in an attempt to consolidate power and avoid accountability. What started as a personnel matter soon mushroomed into a widely publicized struggle over the station's future, prompting walkouts, strikes, and demonstrations.

Although such controversies have occurred periodically at a number of the Pacifica stations over the years, it was especially ironic that an organization that that has done so much to protect the broadcast of free speech would resort to censorship when its newscasters attempted to cover the story. When KPFA's investigative reporter Dennis Bernstein defied a network gag order by covering the crisis at the station, he was suspended from his job, then arrested along with 51 KPFA staff and activists who refused to leave the building. A live, open microphone caught the entire drama. Pacifica officials then cancelled regular programming and boarded up the station for much of July and September.

But faced with a demonstration of 10,000 angry KPFA supporters, Pacifica relented and permitted KPFA staff to go back to work. This and subsequent controversies at all five Pacifica stations sparked a grassroots campaign to democratize the network. By December of 2001 a new Board of Directors had been assigned the task of revising the foundation's by-laws to give the network's listener-sponsors more say over governance.

LARRY APPELBAUM AND MATTHEW LASAR

See also Community Radio; Hill, Lewis; Pacifica Foundation; Public Radio Since 1967

Further Reading

Hill, Lewis, *Voluntary Listener-Sponsorship: A Report to Educational Broadcasters on the Experiment at KPFA, Berkeley, California,* Berkeley, California: Pacifica Foundation, 1958

KPFA website, <www.kpfa.org>

Land, Jeff, *Active Radio: Pacifica's Brash Experiment,* Minneapolis: University of Minnesota Press, 1999

Lasar, Matthew, *Pacifica Radio: The Rise of an Alternative Network,* Philadelphia, Pennsylvania: Temple University Press, 2000

Thompson, Chris, "War and Peace: The Battle for KPFA and the Soul of Pacifica Is Over: Now the Damage Assessment Begins" (*East Bay Express,* 9 January 2002; http://www.eastbayexpress.com/issues/2002-01-09/feature.html/1/index.html)

KQW. *See* KCBS/KQW

KRLA

Los Angeles, California Station

KRLA (1110 kilohertz) is a 50,000-watt AM station serving the Los Angeles market. It was a popular Top 40 station in the 1960s, featuring stand-out radio personalities such as Casey Kasem, Bob Eubanks, Dave Hull, and Bob Hudson. KRLA associated itself strongly with the Beatles at that time, sponsoring several Beatles concerts in Los Angeles. It became the number one station in that market in 1964 and held that distinction for several years. Ironically, it was during KRLA's ratings heyday that its long legal struggle with the Federal Communications Commission (FCC) began. In 1962 the FCC denied a license renewal to station owner Donald Cooke and his company, Eleven Ten Broadcasting, over questionable on-air practices. This set off a 17-year battle, during which KRLA was operated by a nonprofit interim company. The license renewal case for KRLA is a rare instance of FCC license denial based on content and behavior rather than technical considerations.

Origins

The station's history began in 1940 when J.R. Frank Burke and other investors formed the Pacific Coast Broadcasting Company in order to apply for a broadcasting license in Pasadena, California. Burke was a fund-raiser for the Democratic Party and publisher of the *Santa Ana Register.* The FCC issued a construction permit in September 1941, just months before the United States entered World War II and placed a moratorium on all new radio construction. The new station would have the call letters KPAS and operate at 10,000 watts on 1110 kilohertz. By February 1942 an operating license was issued and KPAS went on the air as an unaffiliated station with block programming typical of the time period. Burke's accountant, Loyal King, became the station manager. In addition to music, news, and religious programs, KPAS featured dramatic sketches performed by the Pasadena Playhouse actors. Local programming and public service were hallmarks of the station in the 1940s.

In 1945, Burke sold his interest to religious broadcaster William Dumm, who in turn sold his share in the station to Loyal King two years later. The license was retained by Pacific Coast Broadcasting. The station's call letters were changed to KXLA in 1945 to suggest a more metropolitan target audience. In 1948 King began to run a popular syndicated program called *Country Crystals* and by the early 1950s, KXLA had become a country western station. Most commercial radio stations in the United States were switching to musical formats at this time, as radio talent and network financing were migrating to television.

In 1958 a major change for the station occurred when New York businessman Donald Cooke signed an agreement to purchase KXLA from Pacific Coast Broadcasting. Cooke's brother, Jack Kent Cooke, was the true interested party, but as a Canadian citizen, he was prohibited by the Communications Act of 1934 from owning a U.S. radio station. In order to help his brother finance the purchase, Jack Kent Cooke would buy the station facilities through his newly formed company, Broadcast Equipment Corporation (BEC). Don Cooke's company, Eleven Ten Broadcasting, would file for the license transfer and lease the facilities from BEC.

In March 1959, the FCC approved the transfer with the understanding that Jack Kent Cooke would have no role in managing the station. In fact, he had already been heavily involved, and in the months to come he would play a significant role in station management by making personnel decisions and planning on-air contests and promotions. The Cookes' intention was to change the format to Top 40 rock and roll and to concentrate on the teenage demographic throughout the Los Angeles market. At midnight on 1 September 1959, the station officially switched over. Nineteen-year-old disk jockey Jimmy O'Neill announced, "You have been listening to KXLA. You are now listening to KRLA-Radio for the young at heart."

Fifteen Years of "Interim" Owner Operation

In August 1959, all southern California stations were required to submit applications for license renewal according to the regular FCC schedule of renewals. KRLA did so, but in July of 1960, the FCC notified Donald Cooke that there would be a hearing for KRLA's license renewal. The FCC was concerned about several problems, including programming promises not being met, falsified logs, fraudulent contests, and the involvement of a non-citizen in station management. Although the hearing examiner recommended a one-year probationary renewal, the FCC denied renewal in 1962. The decision was upheld on appeal and the FCC ordered KRLA off the air. Comparative hearings would be held to determine the new licensee, but the FCC was concerned about leaving the frequency vacant in the meantime, as that meant the Mexican station XERB would be allowed to increase its power under provisions of an international treaty.

Oak Knoll Broadcasting, a nonprofit company, was selected to be the interim owner and licensee of KRLA. Eighty percent of their profits were to go to KCET, an educational TV station, and the other 20 percent would be distributed to other charities. Oak Knoll took possession of the station in 1964 and continued to operate it until 1979, when a new company, KRLA Incorporated, was finally selected to take over the license, ending the interim period. Oak Knoll's long period in control stemmed from the protracted FCC proceedings required to choose between more than a dozen applicants for the station license.

The early Oak Knoll years were KRLA's strongest in terms of its position in the market. In 1964, KRLA sought to distinguish itself from KFWB, its primary competitor, by focusing on a single band, the Beatles. The station played all of the cuts from Beatles LPs rather than just the hits and provided information about the music and the band members, promoting itself as "Beatle Radio." KRLA also sponsored three Beatles concerts in Los Angeles from 1964 to 1966. This strategy enabled KRLA to become a contender in the competitive and saturated Los Angeles AM market.

During the 1970s KRLA's format changed several times. In 1971 the station switched to album-oriented progressive rock in an attempt to appeal to young people in the counterculture. This move proved unsuccessful, as FM radio was already offering that format with better sound quality. In 1973 KRLA moved to soft rock, which was more successful, but not enough so to make the station profitable. In 1976, under the direction of Art Laboe, KRLA changed to an Oldies format in a successful move to recapture a segment of the dwindling AM audience.

KRLA Incorporated was actually a merger of five of the companies that had applied for the KRLA license in 1964. It operated the station until 1985, when it was purchased by Greater Los Angeles Radio, Inc. Greater Los Angeles Radio moved the studios from the Huntington Hotel in Pasadena, where they had been since 1942, to Wilshire Boulevard in Los Angeles. The transmitter was moved to Irwindale the following year. In 1997 KRLA was acquired by Infinity Broadcasting, a subsidiary of Viacom, and its format was changed to talk radio. Infinity sold KRLA to Disney-owned ABC Radio in 2001. The format was changed to sports programming with an ESPN affiliation and the acquisition of Anaheim Angels baseball games. ABC Radio relaunched the station under the new call letters KSPN and as of January 2003 had swapped frequencies with KDIS-AM (Disney Radio) on 710 kilohertz. KDIS, now on the 1110 frequency, is formatted for children with Disney product. The KRLA call letters were acquired by Salem Communications for its talk station KIEV-AM at 870 kilohertz.

CHRISTINA S. DRALE

See also Contemporary Hit Radio Format/Top 40; Kasem, Casey; Licensing; Oldies Format

Further Reading

Beem, Donald C., "Standard Broadcast Station KRLA: A Case Study," Masters thesis, California State University at Fullerton, 1980

Earl, Bill, *Dream-House,* Valencia, California: Delta, 1989; 2nd edition, 1991

KSL

Salt Lake City, Utah Station

In the early days of radio, one of the key groups of licensees was composed of churches. As is the case with many such stations, it is difficult to discuss the vision and role of KSL without considering its church affiliation and the dream to proselytize. The Church of Jesus Christ of Latter-day Saints (LDS; Mormon) acquired a radio station to broadcast general conferences of the LDS Church to people throughout the area without their having to come to the tabernacle on the city's Temple Square. Although the functions and operations of the station's mission have changed dramatically over the years, the religious influence can still be found.

KSL's predecessor, KZN, went on the air on 6 May 1922 and was among the first in the Western United States. It broadcast on 1160 kilohertz, and was designated a class A (clear channel) station in 1925. The station is still located at 1160 AM. At 8 P.M. on the first day of its broadcast, from atop the building housing the LDS-operated *Deseret News* newspaper, the station broadcast LDS Church President Heber J. Grant, who spoke of the church's mission and doctrine, quoting from the church's scriptures. Some observers felt that President Grant's remarks were the beginning of the fulfillment of a dream voiced earlier that the president of the church could deliver his sermons "and be heard by congregations assembled in every settlement of the Church from Canada to Mexico, and from California to Colorado" (see Anderson, 1922).

During its first years of operation, the station carried the voices of several famous figures. In addition to church authorities, other noted speakers used this new, fascinating, and promising medium. William Jennings Bryan delivered a ten-minute address on 25 October 1922. In 1923 President Warren G. Harding spoke over the station in a broadcast originating from the church's tabernacle. It was the first known instance of a U.S. President speaking over radio in that area.

In June 1924 *The Deseret News* sold KZN to John and F.W. Cope, who planned to overcome some of the station's engineering problems. The call letters were changed to KFPT. Later that year, in October, the station broadcast the general conferences of the LDS Church, an event that was to occur semiannually through the rest of the century. Listeners could sit at home in their own living rooms next to their radio receivers and attend to the business and spiritual matters of that faith.

In June 1923 the Mormon Tabernacle Choir began its first formal broadcast on KZN (whose call letters were changed to KFPT in 1924, then to KSL in 1925) with the program *Music and the Spoken Word,* a program that continues in the early 21st century. A few years later, when KSL joined the National Broadcasting Company (NBC) as an affiliate, it began a regular Sunday broadcast of the choir. The program continued when, in 1932, KSL moved its affiliation to the Columbia Broadcasting System (CBS). (The radio station is still a CBS affiliate, although KSL-TV is now affiliated with NBC.) It continues as the oldest continuous sustained radio program in America. The program brought fame to the station and the church, as well as a wider audience for the choir and for the tabernacle organ, from their rich acoustical setting in the century-old tabernacle.

As radio developed into a commercial medium, troubles loomed for the Mormon-owned station. Although church leaders saw nothing wrong with the business operation in conjunction with its function as "a factor in the spread of the gospel of Jesus Christ across the world" (see Hinckley, 1947), the church's standards came into conflict with some commercial practices. For example, the church advocated against the use of alcohol and tobacco. Yet network programs carried by KSL contained commercials selling beer and cigarettes, and programs contained themes or characters using these products. Not to carry such programs with their commercials could mean severance from network feeds and a drastic reduction in profits and income. Continuing to carry the programs appeared hypocritical in light of the church's teachings of abstinence from these products. With CBS's position of hard business practices guiding the decision-making process, the church was poised to lose the network affiliation and become a secondary, perhaps insignificant influence in radio in the intermountain area. Policy was established not to accept spot advertising (contracted individually with the station) for beer, wine, or tobacco, but the national network ads would continue to be carried as a necessary evil. KSL would not try to restrict network advertising or interfere with network contracts. KSL-FM was Utah's first FM station, beginning operation on 26 December 1946. Its programming was different and separate from its AM outlet.

Becoming successful as a business, KSL became the flagship station in the church's broadcast ownership, Bonneville International, established in 1964. The group owned radio and television stations in Seattle, Washington, then acquired FM outlets in New York City, Chicago, San Francisco, Los Angeles, Dallas, and Kansas City. As of 2003, Bonneville had 15 stations. From the mid-1960s to 1975, Bonneville operated an international shortwave radio station reaching various countries.

KSL quickly realized that its clear channel signal of 50,000 watts AM served more than just Salt Lake City, extending to the entire Western region of the United States. It also tried to reflect a sense of commitment to serve this extended community, as mandated by church president David O. McKay in the mid-1960s. The station moved from broadcasting high school and church basketball tournaments in the 1950s to political broadcasts of substance. Bonneville's production arm created the "Home Front" public-service messages, which sent nondenominational messages about families and values to listeners who might not otherwise tune in to religious programming. These "Home Front" features were distributed to stations throughout the country to air in a variety of programs.

The church influence in KSL's programming has brought occasional criticism of censorship and biased influence. Yet the wide range of political viewpoints, the representations of other religious denominations, and the respect garnished from its news reporting seem to quell such criticism for many observers. One KSL news director disavows any meddling in the news agenda or its coverage of stories, including those local stories critical of church policies. Although some complaints have gone to the Federal Communications Commission, none has been taken seriously enough to limit the church's operation of its stations.

Broadcast management has indicated that today KSL is a station intent to make a profit, "not to evangelize." It is a commercial broadcast enterprise owned by a religious organization "operated strictly as a business and seeking no special treatment" (see Brady, 1994). Although the station started with a dream to evangelize, the realities of commercial broadcasting make its mission for community good more general in nature.

VAL E. LIMBURG

See also Mormon Tabernacle Choir

Further Reading

Anderson, Edward H., "The Vacuum Tube Amplifier," *The Improvement Era* 25, no.3 (March 1922)

Avant, Gerry, "Major Events, 1920–1929," *Church News* (13 March 1999)

Brady, Rodney H., *Bonneville at Thirty: A Value-Driven Company of Values-Driven People,* New York: Newcomen Society, 1994

Donigan, Robert W., "An Outline History of Broadcasting in the Church of Jesus Christ of Latter-Day Saints, 1922–1963," master's thesis, Brigham Young University, 1963

Godfrey, Donald G., Val E. Limburg, and Heber G. Wolsey, "KSL, Salt Lake City: 'At the Crossroads of the West,'" in *Television in America: Local Station History from across the Nation,* edited by Michael D. Murray and Donald G. Godfrey, Ames: Iowa State University Press, 1997

Hinckley, Gordon B., "Twenty-Five Years of Radio Ministry," *Deseret News* (26 April 1947)

Lichty, Lawrence, and Malachi Topping, *American Broadcasting: A Source Book on the History of Radio and Television,* New York: Hastings House, 1975

Limburg, Val E., "An Analysis of Relationships between Religious Programming Objectives and Methods of Presentation Used by Selected Major Religious Program Producers as Compared to the Church of Jesus Christ of Latter-Day Saints," Master's thesis, Brigham Young University, 1964

Wolsey, Heber Grant, "The History of Radio Station KSL, from 1922 to Television," Ph.D. diss., Michigan State University, 1967

Zobell, Albert L., Jr., "Twenty Magnificent Years on the Air," *Improvement Era* 52, no. 9 (September 1949)

KTRH

Houston, Texas Station

A longtime news, talk, and information station, KTRH was the starting point for the careers of several national celebrities, including the two best-known anchors on Columbia Broadcasting System (CBS) television.

The history of KTRH meshes with the history of modern Houston. KTRH was a relative latecomer when it signed on the air 5 March 1930. The city had boasted an experimental radio station as early as 1919. Several commercial stations were on the air in the early 1920s, although none lasted longer than a few years. Only rival KPRC, launched by the *Houston Post-Dispatch* during a newspaper convention in the city in 1925, hinted at the promise radio would hold in the city.

The opportunity to build KTRH came as a result of the Great Depression. The economic downturn caused the regents of the University of Texas to decide against supporting their experimental station in Austin, KUT. Houston real estate magnate Jesse H. Jones, builder of Houston's emerging skyline and owner of the *Houston Chronicle,* purchased KUT and had the station's equipment boxed and transported the 165 miles to Houston.

Jones hoped to house his new station at the *Chronicle,* but the paper's editor dismissed the idea. The manager of Jones's Rice Hotel, on the other hand, was enthusiastic about a station's broadcasting from his facility and wanted the station to be "irretrievably tied to the hotel," according to Jesse Jones's nephew, John T. Jones, who would ultimately inherit KTRH from his uncle. The *TRH* in the new station's call letters stood for "The Rice Hotel."

When Ross Sterling, owner of KPRC and the *Houston Post-Dispatch,* suffered financial reverses in 1931, Jesse Jones came to the rescue and briefly controlled the *Post-Dispatch,* the *Chronicle,* and KTRH and KPRC radio stations. Jones gained such power that when he secured the Democratic Party's national convention for Houston in 1928, he also won the hearts of Texas Democrats: all 40 of Texas's electors cast their ballots for Jones's nomination for president, even though New York Governor Al Smith would ultimately win the nomination.

In 1947 the *Chronicle* established the first FM station in Houston and called it KTRH-FM. For its first two years, it simulcast the programming of KTRH-AM, and then it launched a "fine music" program of light classics. After a few years, the station returned to simulcasting. The popularity of album rock music in the late 1960s and early 1970s prompted a change of format and call letters for KTRH-FM. The letters KLOL were chosen because they resembled the 101 dial position.

On the death of Jesse Jones in 1956, the ownership of KTRH, sister station KLOL, and the *Chronicle* passed to Jones's nephew, John T. Jones, who operated the stations under the corporate name "The Rusk Corporation" (named for the downtown street where Jones's offices were located). The *Chronicle* was operated by a private foundation, Houston Endowment, that was established by the elder Jones for charitable purposes.

A graduate of Houston's San Jacinto High School, Walter Cronkite worked part-time at KTRH on his way to a journalism degree at the University of Texas in Austin. During the early 1930s Cronkite worked for the University of Texas newspaper, *The Daily Texan;* at United Press International (UPI); and at several state capital news bureaus at the same time. Dan Rather began working at KTRH in 1950, shortly after his graduation from Sam Houston State University in nearby Huntsville, Texas, where he had been a reporter for both the Associated Press and UPI. Rather's early KTRH broadcasts originated in the newsroom of the *Houston Chronicle* with the clack of wire service teletype machines in the background. "We got (the) bright idea that it would give the news program more authenticity," said John T. Jones. The younger Jones told the story of a *Chronicle* religious editor whose desk was next to Rather's broadcast desk. At the end of each newscast, said Jones, the editor would correct Rather's grammar. In 1956 Rather became the station's news director.

CBS-TV sports anchor Jim Nantz also began his career at KTRH, in 1981. While studying on a golf scholarship at the University of Houston, Nantz was an intern at the station and later host of the *Sportsbeat* call-in program.

KTRH is credited with originating the *Dr. IQ* radio quiz show during the 1930s. Ted Nabors, then KTRH program director, performed as the Doctor, and announcer Babe Fritsch took a roving microphone into the audience to choose contestants who won silver dollars when they answered questions correctly. Fritsch was the first to say, "I have a lady in the balcony, Doctor!" *Dr. IQ* was developed for national broadcast on the National Broadcasting Company (NBC) network beginning in 1939, with Lew Valentine performing as the Doctor and Allan C. Anthony as the announcer.

In the mid-1960s, KTRH began a move to the talk format under General Manager Frank Stewart. Although KTRH was not the first station to adopt the new format, it was an early entry into the talk arena. Texas farm and ranch industries prompted KTRH to establish a strong presence in agribusiness reporting, at first with information for area ranchers and later with lawn and garden programs.

In 1981 the station broadcast two live talk shows from the People's Republic of China, a first for U.S. broadcasters. Talk host Ben Baldwin and KTRH Vice President and General Manger Hal Kemp answered listener questions and described events of their travels, including a rare firepower demonstration by the People's Army Infantry. "They literally blew up a mountain for us," Baldwin reported. The station staged subsequent live broadcasts from China during the early 1980s and aired weekend features prepared by the English language staff of Radio Beijing.

Because KTRH had affiliated with the CBS Radio network in the first year of its operation, one of the stories the Houston radio rumor mill circulated in the 1970s and 1980s was that CBS had a blank check ready if John T. Jones ever decided to sell KTRH. He didn't—and the stories were never confirmed. At Jones's retirement, his son Jesse Jones III, known as "Jay," assumed operation of the stations and acquired properties in San Antonio and Austin, expanding Rusk Corporation holdings.

In 1989 Jacor Communications made a $60 million offer for KTRH and KLOL, but the deal was never consummated. Ultimately, the two stations were sold in 1993 to Evergreen Media for $51 million. Evergreen became Chancellor and later AMFM after mergers made possible by the Telecommunications Act of 1996. KTRH moved to the Clear Channel Communications roster after that company's merger with AMFM.

ED SHANE

Further Reading

Boudreaux, Phillip H., "Houston Radio: The First Sixty Years," Senior thesis, University of St. Thomas, 1982

"KTRH Stages First Live Talk Show from China," *Radio and Records* (6 March 1981)

"KTRH's Zak to Be Honored on and off the Air," *Houston Post* (5 January 1981)

"Radio Peking: The Red Rose of Texas," *Earshot* (25 October 1982)

Writers' Program, *Houston: A History and Guide*, Houston, Texas: Anson Jones Press, 1942

Kuralt, Charles 1934–1997

U.S. Broadcast Journalist

One of the most beloved of journalists, Charles Kuralt was a traveler. Most of his life was spent "on the road," a phrase that became the title of his best-known television series. But he began his career in radio, first at WUNC, the student-run station at the University of North Carolina, Chapel Hill, and later at Columbia Broadcasting System (CBS) Radio. Accompanied by a few years' newspaper experience at *The Charlotte News*, where he won a prestigious 1956 Ernie Pyle Memorial Award for his writing skills, Kuralt showed great promise as he began work at CBS Radio.

Kuralt began as a radio writer in 1957, becoming the youngest ever network correspondent just two years later at age 24 and quickly winning a reputation as a hard-news reporter. At the time he was hired by CBS, network radio's viability was uncertain. As at the other major networks, news head Dick Salant was dealing with the sweeping loss of audiences and advertisers to the upstart television medium. After dropping all its soaps and traditional programs in favor of an all-news format, CBS Radio needed unique programming to hold on to affiliated stations, and Salant's solution was a series of very short, horizontally scheduled features—"programlets" that ran four minutes or so and were embedded in commercials. The network would repeat these features as many as eight times each day, returning with fresh stories on successive weekdays. Salant picked Kuralt to do the feature series called *To Your Health*, which quickly became a popular element of CBS's successful Dimension Radio and established Kuralt as a top-notch writer and on-air reporter. After gaining television experience as host of *Eyewitness to History*, Kuralt was expected to step rapidly into the nightly network newscast—a prestigious position on the Tiffany network with the most lauded news of the day. Salant and others had tremendous expectations, touting the young Kuralt as perhaps "the next Edward R. Murrow!" Those were very big shoes for anyone to fill, but in typical fashion, Kuralt dismissed the comparison as "ridiculous."

Because Kuralt preferred the adventure of travel to the stability of a daily news job, he moved from radio to television, covering the war in Vietnam for four tours and reporting on other world trouble spots. He became CBS's chief correspondent in Latin America and later on the West Coast, but he eventually left the hard-news side of television and radio in 1967, when he found his ideal vehicle in the famed *On the Road* series. For more than two decades, Kuralt roamed America's small towns and back roads in a CBS camper, reporting on heartwarming events in ordinary people's lives. Often called the Walt Whitman (or the Norman Rockwell) of American television, he did offbeat Americana stories, telling viewers about, for example, a school for unicyclists, a gas station/poetry factory, and a 104-year-old entertainer who performed in nursing homes. He talked to lumberjacks, cooks, poets, and farmers. He told of the Franciscan Sisters of Perpetual Adoration in Wisconsin, who have been continuously praying—in shifts—for the last 100 years. Finding everyman poetry in everyday life, his offbeat human-interest stories enchanted, delighted, and touched millions of television viewers and radio listeners. These reports became part of the *CBS Evening News*, *CBS Sunday Morning*, and the weekday *Morning* program; they also aired on the CBS Radio Network. Kuralt is said to have logged more than a million miles on the road. He is quoted as saying about CBS Radio: "Going wherever I wanted to go and doing whatever I wanted to do, CBS didn't even know where I was; didn't care much where I was. I just wandered. And that was probably the best job in journalism." After more than a decade as host of the television program *CBS Sunday Morning*, he retired, only to come back as host of *An American Moment*, a 90-second series that harked back to his radio days. He also hosted for a short time the CBS Cable show *I Remember,* a weekly reexamination of significant news stories from the historical past. After a brief illness, Charles Kuralt died in 1997 of complications arising from lupus.

During his career at CBS, Kuralt wrote several best-selling books, won 11 Emmy Awards, three Peabody Awards, and numerous other prestigious broadcasting awards. At his death, Walter Cronkite called him "one of the true, greatly talented people in television." A month later, the North Carolina legislature passed a resolution honoring him, observing that "Charles Kuralt possessed a peculiar insight that enabled him to contribute substantially and effectively to the improvement and betterment of the world around him, enriching the lives of those with whom he was associated and came in contact." He was a truly great storyteller.

SUSAN TYLER EASTMAN

Charles Kuralt. Born in Wilmington, North Carolina, 10 September 1934. Son of Wallace Hamilton Kuralt and Ina Bishop Kuralt; attended University of North Carolina at Chapel Hill, B.A. in journalism, 1955; WUNC radio dramatist, 1953; editor, *Daily Tar Heel*; reporter for *Charlotte News,* 1955–57, received Ernie Pyle Memorial Award for writing, 1956; joined CBS as writer, 1957, news correspondent, 1959–94. Received 11 Emmy Awards, three Peabody Awards, George Polk Memorial Award, 1981; inducted into North Carolina Journalism Hall of Fame, 1981; International Radio and Television Society (IRTS) Broadcaster of the Year, 1985; John Tyler Caldwell Award for the Humanities, 1997 (posthumous). Died in New York City, 4 July 1997.

Television Series (many also carried on CBS Radio)

To Your Health, 1957–59; *Eyewitness to History* (later, *Eyewitness*), 1960–61; *On the Road,* 1967–80; *Who's Who,* 1977; *CBS Sunday Morning,* 1979–94; *CBS Morning News* (later, *Morning with Charles Kuralt*), 1980–82; *I Remember,* 1997; *An American Moment,* 1997

Selected Publications

To the Top of the World, 1968
Dateline America, 1979
On the Road with Charles Kuralt, 1985
North Carolina Is My Home, 1986
Southerners, 1986
A Life on the Road, 1990
Charles Kuralt's America, 1995

Further Reading

Gates, Gary Paul, *Air Time: The Inside Story of CBS News,* New York: Harper and Row, 1978
Joyce, Ed, *Prime Times, Bad Times,* New York: Doubleday, 1988
Matusow, Barbara, *The Evening Stars: The Making of the Network News Anchor,* Boston: Houghton Mifflin, 1983
McCabe, Peter, *Bad News at Black Rock: The Sell-Out of CBS News,* New York: Arbor House, 1987

KWKH

Shreveport, Louisiana Station

KWKH, a 50,000-watt clear channel station, played an important role in the commercialization of country music and rock and roll music during the 1940s and 1950s. In addition, as one of the first radio stations in Louisiana, it helped pioneer radio broadcasting in the state.

The station that KWKH would become first crackled on the air in early 1922. Engineer William E. Antony built the physical operations under the auspices of the Elliott Electric Company; in 1923 a team of investors led by a retailer of radio sets purchased the station, dubbing it WGAQ. One of the investors bought out his partners in 1925 and rechristened the station with his initials. William Kennon Henderson, who owned and operated the Henderson Iron Works and Supply in Shreveport, promptly turned KWKH into his own soapbox. At arbitrary moments during the broadcast day, he often burst into his studios and grabbed the microphone from his announcer. "Hello world, doggone you! This is KWKH at Shreveport, Lou—ee—siana, and it's W.K. Henderson talkin' to you." He railed against the national debt and chain retail stores and ridiculed over the air anybody who dared to disagree with him. He condemned the Radio Act of 1927 and sparred with both the U.S. Department of Commerce and the Federal Radio Commission, claiming that both favored chain (network) stations over independent outlets.

In the late 1920s, when Henderson applied for a power boost to 10,000 watts, federal regulators turned him down, claiming that KWKH was nothing more than a broadcaster of phonograph records. But Henderson argued that his format—which in actuality encompassed more than just record playing—satisfied his listeners' wishes. The rejection only incited the maverick's ranting resolve, and by 1930 Washington conceded to him and granted the increase in power.

As eccentric and egotistical as Henderson was, his desire to see KWKH prosper and expand its signal range helped consolidate radio's presence in Louisiana. His on-air tirades forced people to note the presence of radio, and entrepreneurs looked to the growth of KWKH as an example when they invested their own dollars in radio stations. Furthermore, it was probably Henderson's distaste for the uniformity of chain stations that led KWKH to recruit local talent to perform on its airwaves. The use of local talent, most of whom performed hillbilly music (as country music was known in the 1920s and 1930s), planted the seeds that would grow into KWKH's *Louisiana Hayride.*

In 1932 Henderson sold KWKH to the International Broadcasting Corporation, and the station changed hands again in 1935 when the *Shreveport Times,* owned by oilman John D. Ewing, took control. Under Ewing, KWKH continued the growth that Henderson had initiated, moving to modern facilities in downtown Shreveport's Commercial Building in 1936 and receiving permission to operate at 50,000 watts in 1939. (In 1934, probably much to the former owner's ire, KWKH had established a network affiliation with the Columbia Broadcasting System [CBS].) Carrying on Henderson's tradition of hiring local talent, station manager Henry B. Clay, who was Ewing's son-law; program director Horace Logan; and commercial manager Dean Upson established the *Louisiana Hayride,* which would become KWKH's most lasting mark on country music history and, indeed, on radio history.

KWKH's *Louisiana Hayride* was a country and western stage show that played weekly on the station from 1948 to 1960. Dubbed "the Cradle of the Stars," the program aired on Saturday nights from Shreveport's Municipal Auditorium and boasted among its cast members musical performers who would be the primary shapers of post–World War II country music. Important country music figures such as Hank Williams, Jim Reeves, and Johnny Cash appeared as regulars on the *Hayride* early in their careers; each used the program as a springboard to broader acceptance. In addition to providing a stage for important country music performers, KWKH and the *Hayride* would leave a lasting mark on the history of rock and roll in the mid-1950s, when singer Elvis Presley became a regular cast member on the show; the exposure he received as a cast member from 1954 to 1956 helped fuel his rise to national prominence.

KWKH's 50,000 watts of power gave the *Louisiana Hayride* its muscle. The station's signal stretched like a fan across the southwestern and northwestern regions of the United States and clipped across national borders to reach countries as near as Mexico and as far as Australia. A regular spot on KWKH's Saturday night hoedown, any aspiring country act knew, could attract recording contracts and generate bookings. The *Hayride*'s influence grew mightier in the early 1950s with its insertion into the schedule of a CBS regional network.

However, the *Hayride* would never be as mighty as the *Grand Ole Opry* on WSM in Nashville, Tennessee, and because of that, KWKH failed to hold on to its rising stars. As *Hayride* personalities gained momentum, they inevitably shifted their eyes toward the *Opry* and Nashville, where a colony that included booking agents and music publishers awaited to capitalize on the artists' successes. KWKH and Shreveport lacked such ancillary components of the music industry and therefore could not keep name artists on the show very long. The *Louisiana Hayride* became known as an "*Opry* farm club," and, largely because of the constant talent drain, it ceased regular broadcasting in 1960. The program has been reincarnated in various forms over the years, but it has never achieved the influence it enjoyed in the late 1940s and 1950s.

Today, KWKH is owned by Clear Channel Communications, which acquired the station in 1999. Although the *Louisiana Hayride* disappeared long ago from its airwaves, the station still recalls former glories with its "country gold" format, which features vintage country music from as early as the 1940s.

MICHAEL STREISSGUTH

See also Clear Channel Stations; Country Music Format; Grand Ole Oprey; WWL

Further Reading

Escott, Colin, George Merritt, and William MacEwen, *Hank Williams: The Biography,* Boston: Little Brown, 1994

Guralnick, Peter, *Last Train to Memphis: The Rise of Elvis Presley,* Boston: Little Brown, 1994

Hall, Lillian Jones, "A Historical Study of Programming Techniques and Practices of Radio Station KWKH, Shreveport, Louisiana, 1922–1950," Ph.D. diss., Louisiana State University, 1959

Logan, Horace, and Bill Sloan, *Elvis, Hank, and Me: Making Musical History on the Louisiana Hayride,* New York: St. Martin's Press, 1998

Streissguth, Michael, *Like a Moth to a Flame: The Jim Reeves Story,* Nashville, Tennessee: Rutledge Hill Press, 1998

Kyser, Kay 1905–1985

U.S. Big Band Leader and Radio Host

Largely forgotten today, Kay Kyser's band scored 11 No. 1 records and 35 top 10 hits. Kyser hosted a top-rated radio show for 11 years on the National Broadcasting Company (NBC), and he and his band starred in seven motion pictures. He has two stars set into the sidewalk on Hollywood Boulevard in Los Angeles, one for radio and another for film work. Although best known as a musical figure, in fact Kyser could not sing, read music, or play any instrument.

Origins

Kay Kyser's roots and much of his life are based in North Carolina. Born there, he originally intended to study law while attending the state university. Instead, at the suggestion of bandleader Hal Kemp, Kyser (then a university cheerleader) formed his first small band in 1926, the group consisting of six fellow students. With this initial group, Kyser soon established his role as an energetic MC, rather than as a musician. Over the next several years, the band added some professional musicians and slowly improved, but life on the road was a series of relatively short performance gigs in Depression America.

In the mid-1930s, appearances at venues in Santa Monica and then in Chicago (at the Blackhawk restaurant and club) propelled the band to national recognition and appeal. In part this was due to the music played, but much of the band's particular appeal was built around the light-hearted humor and antics of Kyser as its energetic leader. The band was something of an extension of Kyser's outgoing personality.

Radio Years

Sometime in 1937, the "college" concept was first developed to attract more customers on otherwise slow Monday evenings at the Blackhawk. Originally called "Kay's Klass," it was a kind of amateur performers night, supplemented with questions from Kyser to the musical contestants to both relax them and amuse the audience. The classroom notion may have been the idea of Kyser's young agent, Lew Wasserman. Chicago station WGN began to broadcast the band's performances as *Kay Kyser's Kampus Class*. By this point the Kyser band had developed another unique feature that would remain for many years: the "singing song title," in which a band vocalist would sing the title's words at the beginning before a full performance of the song.

In February 1938, the Mutual Broadcasting System began an eight-week series of programs carrying the Blackhawk-based band program regionally. The American Tobacco Company took up sponsorship of the show for its Lucky Strike cigarette brand and brought it to New York City and the NBC network, with the first program airing on 30 March 1938. Now known as *Kay Kyser's College of Musical Knowledge* (sometimes *College* is rendered *Kollege*) with Kyser cast as "The Old Professor," the program was an almost instant hit.

The *College* program's format was classic radio variety, built around the band's music but adding the related quiz feature along with comedy routines. Each broadcast began with Kyser's warm North Carolina–accented "Evenin' folks, how y'all?" "Diplomas" were mailed to listeners sending in music quiz questions used on the air.

Among the many instrumentalists and singers in Kyser's band, several stood out over the years. Perhaps best known was "Ish Kabibble" (Merwyn Bogue), who got his odd show business name from his comedy version of an old Yiddish song, "Isch Ga Bibble" (loosely translated, "I should worry?"), which he first performed after joining the band in 1931. He played the cornet but thrived on developing the rural "Ish" character with a "pudding bowl" hair cut, who constantly interrupted the show to recite nonsensical poems to a seemingly frustrated Kyser, becoming his onstage comedy foil. As is often the case with such performers, he was no dummy offstage, and he handled the program's payroll. He stayed with the band until the program wrapped up its television series in 1951. Among the many singers heard over the years were tenor Harry Babbitt, San Antonio native Ginny Simms, who was the band's first permanent female voice and often sang duets with Babbitt, and Mike Douglas, who would go on to become a famous variety and talk show host for television.

Kyser recorded no fewer than 11 number one songs in the space of seven years. These were "The Umbrella Man" and "Three Little Fishies" (1939); "(Lights Out) Til Reveille" and "(There'll Be Bluebirds) Over the White Cliffs of Dover" (1941); "Who Wouldn't Love You," "Jingle, Jangle, Jingle," "He Wears a Pair of Silver Wings," "Strip Polka," and "Praise the Lord and Pass the Ammunition" (1942); and "Old Buttermilk Sky" and the "Woody Woodpecker" song (1946). Overall, the band recorded some 400 songs beginning in 1935, mostly on the Brunswick, Columbia (parent of Brunswick), and Victor labels.

Kyser's band became one of the first to perform at military bases even before World War II. The first such performance took place at the San Diego marine base on 26 February 1941. Kyser became a highly successful wartime bond salesman with these and other appearances—a total of more than 500 shows at military camps and bases by the time the war ended. These

Kay Kyser
Courtesy Radio Hall of Fame

"road shows" had a somewhat different sound from the earlier broadcasts. A band bus fire in April 1942 destroyed 15 years' worth of musical arrangements, prompting a hiatus for the act. When the band reappeared, the singing song titles were gone and a newer, updated musical style became evident. Because of the pressure of the wartime appearances, the band undertook no commercial gigs (except those booked previously) for the duration, concentrating on the continuing radio show, military service shows, and the seven movies that featured Kyser and the band, all of them made within a five-year period.

Later Years

Kyser and the band moved to television with their own series, debuting in 1949. With the end of the TV series in 1951, Kyser retired at the top of his form and with nary an announcement that he was going. He was tired after years of relentless performance, had earned more than enough money, and wanted to return to North Carolina to do other things. There was also a medical reason: Kyser had begun to have serious trouble with his feet due to an arthritic-type condition in his toes, perhaps exacerbated by his years of energetic performance. Not finding help in traditional medicine, he turned to Christian Science to find relief. He became very active in the church, managing its film and broadcast department in Boston for five years in the 1970s.

He also undertook many altruistic projects, including highway safety campaigns, a health-related campaign in North Carolina, funding scholarships and providing other support for his university alma mater, and helping to establish an educational TV station for Chapel Hill, North Carolina.

CHRISTOPHER H. STERLING

James King Kern ("Kay") Kyser. Born in Rocky Mount, North Carolina, 18 June 1905. Graduated from the University of North Carolina at Chapel Hill in Commerce, 1928. Formed first orchestra, 1926; toured with orchestra, 1926–34; played Blackhawk Club, Chicago, Illinois, 1934; made radio debut in a band remote, 1937; known as host of *Kay Kyser's College [or Kollege] of Musical Knowledge,* 1938–49; retired from broadcasting, 1951; managed radio and television broadcast department, Christian Science Church, Boston, 1974–79. Died in Chapel Hill, North Carolina, 23 July 1985.

Radio Series

1937 *Kay Kyser Band Remote*
1938–49 *Kay Kyser's College of Musical Knowledge*

Television

Kay Kyser's Kollege of Musical Knowledge, 1949–50, 1954

Film

That's Right, You're Wrong, 1939; *You'll Find Out,* 1940; *Playmates,* 1941; *My Favorite Spy,* 1942; *Around the World,* 1943; *Swing Fever,* 1943; *Carolina Blues,* 1944

Further Reading

A Tribute to Kay Kyser, <http://www.ibiblio.org/kaykyser/>
Kay Kyser: A Documentary in Progress, <http://www.kaykyser.net/>
Simon, George, *The Big Bands,* New York: Macmillan, 1967; 4th edition, New York: Schirmer, and London: Collier-Macmillan, 1981

KYW

Chicago, Cleveland, and Philadelphia Station

One of Westinghouse's original outlets, KYW has been described as "the wandering radio station." Though it has served the Philadelphia area for most of its life, the call letters were also found on stations in Chicago and Cleveland. KYW's wanderlust is a result of the federal government's intermittent efforts to manage how much control large broadcasters had over America's most important cities. In the end, these labors produced voluminous litigation but little in the way of permanent results.

KYW had a Chicago address for only a dozen years, but its early start earned it a place in radio history as a pioneer radio station. In fact, it was Chicago's first radio station. Legend has it that the call letters stood for "Young Warriors."

The station first broadcast from the Commonwealth Edison office in Chicago on 11 November 1921. KYW was originally started as a partnership. Westinghouse provided the transmitter and Commonwealth Edison the broadcast location. For its part, Westinghouse rushed KYW and sister stations WBZ in Massachusetts and WJZ in Newark, New Jersey, to the air as a direct result of the success of KDKA in Pittsburgh. Westinghouse's motivation was to stimulate sales of the crystal radio sets the company manufactured. For five years, KYW was operated as a joint venture, although Westinghouse dominated the partnership. In 1926 the working relationship ended, and Commonwealth Edison eventually became associated with crosstown radio station WENR.

Three notable early programming experiments punctuate KYW's Chicago history. The first broadcast by KYW featured opera, and regular weekly opera broadcasts on the station proved an immediate success. At one point, the opera broadcasts were credited with selling nearly 2,000 radio receivers a week in the Chicago area. Second, KYW featured an early version of children's programming. Early radio personality Uncle Bob (Walter Wilson) broadcast children's stories each night, being sure to finish by 7 P.M. so as not to disturb his listeners' bedtimes. KYW also featured breaking news supplied by the *Chicago Tribune*. Seeing potential in the new medium, the Tribune Company decided to get in on the business itself and launched WGN in 1924.

The Federal Radio Commission soon grew concerned about the large number of stations in Chicago. Westinghouse offered to shift KYW out of Chicago and moved the station on 3 December 1934 to Philadelphia. Then the nation's third-largest media market, Philadelphia would prove less lucrative to Westinghouse than broadcasting from Chicago, then the nation's second most important city.

In its new hometown, KYW continued its tradition of reporting breaking news. For example, KYW covered the June 1937 *Hindenburg* disaster in nearby Lakehurst, New Jersey. From a telephone booth, a KYW reporter described to the radio audience the horrific fire, and the on-the-spot report was broadcast over both National Broadcasting Company (NBC) networks.

The association with NBC would eventually set the stage for KYW's third move—to Cleveland, Ohio. On 22 January 1956, Westinghouse, under great pressure from NBC, which wanted to upgrade its own facilities to a larger market, exchanged its Philadelphia broadcast operations for NBC's Cleveland stations. In consideration for Westinghouse receiving the smaller and less profitable Cleveland outlets, NBC also paid Westinghouse \$3 million.

Later, as a result of federal investigation into NBC's actions, it was demonstrated that NBC had forced Westinghouse into the exchange. Had Westinghouse not complied, it would have lost its valuable NBC network affiliation for its budding television operations. The Federal Communications Commission (FCC) ordered the swap undone in 1964 and found NBC culpable of abusing its network power. KYW's call letters were shifted from Cleveland, and the AM, FM, and television stations to which they were attached returned again to Philadelphia in June 1965 (the Cleveland AM station is now known as WTAM, formerly WWWE).

In October 1965, KYW became one of the first radio stations to adopt an all-news format. The format change did not bring immediate ratings success, and, reportedly, the station lost money for several years. The first decade was particularly difficult, given the popularity of crosstown AM contemporary music outlets such as WFIL and WIBG.

Today, KYW's competition comes mostly from FM stations, because KYW has long been the market's leading AM station. Like other historic AM broadcasters, KYW found that information-based spoken-word programming can be effective against music-based FM competitors. In programming its all-news format, the station uses a 30-minute news wheel and features traffic reports every ten minutes. The station does particularly well in the winter season, when the station issues snow-related closing notices for schools in eastern Pennsylvania, southern New Jersey, and northern Delaware. Arbitron reports that KYW, in addition to regularly being one of the top-three radio stations in the market, has a weekly cumulative audience of well over 1 million listeners.

Though technically the radio station is not on a full clear channel frequency, the 50,000-watt signal produced from the station's directional antenna can regularly be heard across the northeastern United States, far outside the station's primary Delaware Valley coverage area. The station has been located at 1060 kilohertz since 1941. The studios of KYW radio are housed on Independence Mall within sight of the Liberty Bell. It shares the same building as its sister television station, KYW-TV 3, now a Columbia Broadcasting System (CBS) television affiliate. The television station was Philadelphia's first when it began experimental operations just as W3XE in 1932

Despite Westinghouse's historic attempts to keep its network-owning rivals at arm's length, the company, disappointed by its prospects in manufacturing, eventually decided to concentrate on broadcast programming and merged with the CBS network in 1995. Subsequently, the Westinghouse identity disappeared. Today, the license to KYW radio is owned by Infinity Broadcasting, the radio company closely associated with CBS.

A. JOSEPH BORRELL

See also Hindenburg Disaster

Further Reading

Samuels, Rich, "It All Began with an Oath and an Opera: Behind the Scenes at Chicago's First Broadcast," *Chicago Tribune* (8 November 1993)

Shanahan, Eileen, "FCC Orders NBC to Return Station," *New York Times* (30 July 1964)

L

Landell de Moura, Father Roberto 1861–1928

Brazilian Wireless Pioneer

In the 1890s and early 1900s, Father Roberto Landell de Moura produced a series of wireless communication devices that were as original in their day as they were unrecognized.

Origins

Roberto Landell de Moura was born in Porto Alegre, the capital of Rio Grande do Sul, on 21 January 1861. Graduating as a distinguished student from a local Jesuit high school, he moved to Rio de Janeiro to study at the Polytechnic Institute. Unable to pay tuition, however, he took a job as a store clerk.

In 1881 his brother, on his way to Rome to study for the priesthood, visited him; Roberto decided to accompany him and also become a priest. He studied theology at the seminary in Rome for students from the Americas and physics and chemistry at the Gregorian University. He also became aware of scientific developments in Italy and Europe. Ordained in 1885, he returned to Rio de Janeiro and for a brief period was a temporary chaplain in the Brazilian imperial court, occasionally conversing with Emperor Dom Pedro II, who had met Alexander Graham Bell and later introduced the telephone to Brazil.

From 1887 through the 1890s Landell de Moura had a series of parish assignments in Rio Grande do Sul and then in São Paulo. He had a difficult temperament and had to be transferred several times to different parishes.

Wireless Inventions

It was during the 1890s, while posted in Campinas, São Paulo, that he formulated theories about controlled, wireless conduction of vibratory movements and light beams, believing that any sound, including the human voice, could be transmitted over land, through the air, and under the water.

During 1893 and 1894 he demonstrated these ideas in the center of the city of São Paulo. He transmitted sound without wires between two of the highest points in the city, over five miles apart, using a type of three-electrode conductor lamp. These demonstrations occurred in the presence of the British Consul and years before similar demonstrations were made by Marconi and de Forest.

Despite presenting his inventions before a representative of one of the most inventive and commercial countries in the world, however, he failed to attract interest or investment. Worse, word about his strange, "diabolical" inventions aroused the suspicions of his parishioners. They invaded his rectory and destroyed his machines.

Undismayed, he rebuilt and refined them, by 1900 obtaining Brazilian patent 3,279 for a machine transmitting sound with or without wires via space, land, or water. Continuing to find no local interest, however, he moved to the United States. Taking up residence in New York City in 1901 and surviving on a subsistence income, he remade his inventions and obtained U.S. patents 771,917 (11 October 1904), 775,337 (22 November 1904), and 775,846 (22 November 1904) for a sound wave transmitter, a wireless phone, and a wireless telegraph, respectively. The wave transmitter produced electrical oscillations of light from sound vibrations generated by the human voice or other source. The sound waves passed through a receptor with induction coils and condensers that changed them into electric or light waves, allowing their wireless transmission to a receptor that could convert them to voice or sound or light signals. His wireless phone transmitted and received voice via light waves. The wireless telegraph transmitted and received signals via various types of sound waves.

EDWARD A. RIEDINGER

See also Early Wireless

Father Landell de Moura. Born in Porto Alegre, Rio Grande do Sul, Brazil, 21 January 1861. Fourth of twelve children of Inácio José Ferreira de Moura and Sara Mariana Landell de Moura; honors graduate of Jesuit high school in São Leopoldo, Rio Grande do Sul; moved to Rio de Janeiro to study, ca. 1879; accompanied brother to study for priesthood in Rome, 1881, specialized in theology, physics, and chemistry; ordained priest, 1885; returned to Brazil, became seminary instructor in Porto Alegre, 1887–91; parish priest, Uruguaiana, Rio Grande do Sul, 1891; parish priest, Santos, Campinas, and Sant'Ana, São Paulo, 1892–99; formulated theories on and built devices for wireless sound communication; public demonstrations of inventions in capital of São Paulo, 1893–94, equipment destroyed by parishioners; received Brazilian patent for machine transmitting sound with or without wires via space, land, or water, 1900; moved to New York City, 1901–04; rebuilt his inventions and received U.S. patents, 1904; returned to Brazil, 1905; desisted from inventing; returned to parish work in São Paulo and then Rio Grande do Sul, ca. 1906. Died in Rio Grande do Sul, Brazil, 30 June 1928.

Further Reading

Almeida, B. Hamilton, *O outro lado das comunicações: A saga do Padre Landell de Moura,* Porto Alegre, Brazil: Sulina, 1983

Cauduro, Fernando, *O homem que apertou o botão da comunicação,* Porto Alegre, Brazil: FEPLAM, 1977

Fornari, Ernani, *O "incrivel" Padre Landell de Moura: História triste de um inventor brasileiro,* Rio de Janeiro: Globo, 1960; 2nd edition, Rio de Janeiro: Biblioteca do Exército, 1984

"Talking Over a Gap of Miles along a Ray of Light," *New York Herald* (12 October 1902)

Lazarsfeld, Paul F. 1901–1976

U.S. (Austrian-Born) Sociologist and Radio Research Innovator

"Paul Lazarsfeld," writes one media historian, "was undoubtedly the most important intellectual influence in shaping modern communication research" (Rogers, 1994). His pioneering radio research in the late 1930s and early 1940s helped to shape our understanding of the medium's effects and made applied radio industry research more respectable in the eyes of many academics.

Origins

Born at the beginning of the 20th century in Vienna, Austria, Lazarsfeld came of age in the lively intellectual climate of that city in the years between the world wars. Brilliant in mathematics, he earned a Ph.D. at the age of 24. But with academic jobs scarce in the mid-1920s, he taught at a Vienna high school and served as a part-time instructor at the University of Vienna while seeking a more permanent university post. He established a Research Center for Economic Psychology at the University of Vienna (where he served as an adjunct faculty member) in 1925. It became a model that Lazarsfeld would follow with later research groups in the United States. With the help of students, he directed a landmark study of unemployment in the small town of Marienthal, Austria, in 1931–32 that brought him wide attention. But because of the difficult economic times, his socialist background, and his Jewish heritage, Lazarsfeld was facing limited opportunities in Austria.

Thanks to the Marienthal study, Lazarsfeld's work reached the notice of the Rockefeller Foundation, which sponsored him for two years of research-related traveling in the United States in 1933–35. He visited many universities and research academics. When a political coup in Austria made his return to academic life there even more unlikely, he accepted a post as acting director at the University of Newark's research center in 1936. It was a tiny operation—so much so that some of his published papers appeared under a pseudonym ("Elias Smith") to disguise the fact that he was the only full-time research person working there.

Radio Research

Lazarsfeld's active importance in radio research spanned nearly two decades. This research began without him—with a proposal to the Rockefeller Foundation of a study of "The Essential Value of Radio to All Type of Listeners," proposed by Hadley Cantril of Princeton University and Frank Stanton of the Columbia Broadcasting System (CBS), to undertake experimental work about radio's audiences. Cantril, busy with his own work (as was Stanton, who by then had more than

ment; the horror multiplies when he finds out the man was his own son. "The Word" tells of a couple who descend from the Empire State Building to discover that everyone else in the world has disappeared; they conclude that God "got tired of the way [people] were doing things and destroyed them." This couple survived, to make Oboler's point that "plain ordinary people" could make the world a new and better place. Although occasionally heavy-handed, Oboler never hesitated to tackle weighty subjects within the framework of a mass horror genre, winning a new respect for the form.

As the 1930s progressed and the threat of war became clearer, Oboler left *Lights Out* and turned his writing skills to patriotic material. He revived the program for the 1942–43 season, broadcasting from New York over NBC. Oboler hosted the program himself, and its famous beginning became the one that has been most associated with the show since. To the chimes of a gong, Oboler spoke the words "It . . . is . . . later . . . than . . . you . . . think." As John Dunning (1998) has noted, the earlier hour of this series (all the broadcasts were at 8:00 P.M.), made it more accessible to people, and this became the best-remembered year of the series. Scripts were largely recycled from Oboler's previous shows, however, because he was busy doing war work. Its successful 1940s run also owes much to the sudden popularity and growth of the thriller/horror/suspense genre during the war period; *Lights Out* was the granddaddy—and model—of many of the more than 40 such programs that took to the airwaves during that time, most famously *Inner Sanctum* and *Suspense*.

The reputation and influence of the program remained strong long after television supplanted radio. This is due in part to Oboler's status as a celebrated auteur (most of Oboler's shows, unlike Cooper's, are available on tape). But *Lights Out* also stands as pivotal to radio history because it demonstrated the way in which radio programs could push listeners' imaginations to horrifying limits beyond those of reasoned vision. Radio, as horror author (and Oboler admirer) Stephen King has noted, has the ability to "unlock the door of evil without letting the monster out" (quoted in Nachman, 1998). *Lights Out* was the first program to demonstrate and exploit this aspect of radio, to the delight of its terror-stricken fans.

ALLISON MCCRACKEN

See also Horror Programs; Oboler, Arch; Sound Effects

Host
Arch Oboler

Narrator
Boris Aplon (1946)

Actors
Boris Karloff, Harold Peary, Betty Winkler, Mercedes McCambridge, Willard Waterman, Arthur Peterson, Betty Caine, Ed Carey, Sidney Ellstrom, Murray Forbes, Robert Griffin, Robert Guilbert, Rupert LaBelle, Philip Lord, Raymond Edward Johnson, and others

Writers/Producers/Directors
Wyllis Cooper, Arch Oboler, Albert Crews, and Bill Lawrence

Programming History

WENR, Chicago	January 1934–April 1935
NBC	April 1935–August 1946
CBS	October 1942–September 1943
ABC	July 1947–August 1947

Further Reading

DeLong, Thomas A., *Radio Stars: An Illustrated Biographical Dictionary of 953 Performers, 1920 through 1960*, Jefferson, North Carolina: McFarland, 1996

Dunning, John, *Tune in Yesterday: The Ultimate Encyclopedia of Old-Time Radio, 1925–1976*, Englewood Cliffs, New Jersey: Prentice-Hall, 1976; revised edition, as *On the Air: The Encyclopedia of Old-Time Radio*, New York: Oxford University Press, 1998

Harmon, Jim, *The Great Radio Heroes*, New York: Doubleday, 1967; 2nd edition, Jefferson, North Carolina: McFarland, 2001

Maltin, Leonard, *The Great American Broadcast: A Celebration of Radio's Golden Age*, New York: Dutton, 1997

Nachman, Gerald, *Raised on Radio: In Quest of the Lone Ranger, Jack Benny . . .*, New York: Pantheon, 1998

Stedman, Raymond William, *The Serials: Suspense and Drama by Installment*, Norman: University of Oklahoma Press, 1971; 2nd edition, 1977

Limbaugh, Rush 1951–

U.S. Talk Show Host

Media critic Howard Kurtz says that it is not much of an exaggeration to say that Rush Limbaugh single-handedly revived AM radio beginning in the mid-1980s and led the stampede of conservative talk show hosts that followed. By the beginning of the 1990s, Limbaugh was heard on more than 650 stations, reaching an estimated 20 million people each week. He continued to achieve similar numbers in the early 21st century. Kurtz claims that Limbaugh is more influential than other talk show hosts and commentators, even TV network giants Dan Rather, Tom Brokaw, and Peter Jennings, because he defined talk radio in a way that no one ever had before.

In addition to developing a new public forum, Limbaugh's influence has been felt in a number of other areas. His radio programs are the foundation of a multimillion-dollar empire that includes everything from conservative books to T-shirts to a monthly newsletter. In a tradition similar to that in the era of *Amos 'n' Andy,* dozens of restaurants around the country reserve "Rush Rooms" for his followers (who call themselves "dittoheads") to eat and listen to his three-hour weekday broadcasts.

In 1994 Limbaugh was credited with being the most important grassroots voice for conservative Republicans, who took control of the House of Representatives after decades of Democratic control. A *National Review* magazine cover story named Limbaugh "the leader of the opposition." His drumbeat criticism of the Clinton White House and of Hillary Rodham Clinton finally prompted the president to attack talk radio in general and Limbaugh in particular (to Limbaugh's delight). The increase in one-sided attacks over the airwaves also triggered a movement to restore the Fairness Doctrine killed by the FCC in 1987. *The Wall Street Journal* described the movement as the "Hush Rush Bill."

Tom Lewis, author of *Empire of the Air: The Men Who Made Radio* (1991), argues that Limbaugh's success speaks more to the power of radio and Limbaugh's ability to use the medium in which he has worked since he was a teen. "Better than most liberals or conservatives," says Lewis, Limbaugh is "a consummate showman who understands radio and sound, especially their ability to create a picture in the minds of listeners and their potential to capture imaginations."

To bring this about, Limbaugh borrows from the word-making skills of gossip columnist and 1930s broadcaster Walter Winchell, who coined words that not only distorted reality but evoked a wild form of humor. Limbaugh's terms, such as "feminazis" (to describe those in the women's movement) and "environmental wackos," are typical of his confrontational but engaging manner, which attracts primarily older white males. But Limbaugh also uses satire, thumping, popular recordings, and a variety of sound effects. When he talks about abortion, for example, the sound of a vacuum cleaner is heard in the background. He offers trading cards and rock songs mocking political figures.

Lewis says that Limbaugh draws from the tradition of radio humorists such as Jack Benny, Fred Allen, and even the Great Gildersleeve. "A man, a legend, a way of life," Limbaugh repeats over his Excellence in Broadcasting (EIB) radio network as well as in books, pamphlets, and newsletters. "I am Rush Limbaugh from the Limbaugh Institute for Advanced Conservative Studies. Yes, my friends, the Doctor of Democracy is on the air." According to Limbaugh, the reason for his popularity is that he's "the epitome of morality and virtue . . . with talent on loan from God."

Limbaugh's format includes lengthy monologues that sometimes meander and consist of incomplete sentences and muddled paragraphs. Along with other conservative talk show hosts, he helps to create an almost conspiratorial approach against big government and liberals in politics and the media; at the same time he celebrates the common sense of other Americans—particularly himself. Although he does not usually have guests on his program, Limbaugh is generally polite to his callers. Larry King, a talk show host who depends on guests, says Limbaugh comes to his broadcast with an agenda. "Agenda broadcasters will do anything. They'll lie. They'll fabricate, to keep the agenda going," King says.

A self-described nerd, Limbaugh got his start in broadcasting in 1967 as a disc jockey for the radio station partially owned by his father in Cape Girardeau, Missouri, where he was born and raised. After dropping out of Southeastern Missouri State University, he worked for and was fired from radio stations in McKeesport (Pennsylvania), Pittsburgh, and Kansas City, where he was also a publicist for the Kansas City Royals and announcer for its station. Because of his controversial nature, Limbaugh was also dismissed from the Kansas City station. He went to Sacramento, California, in the fall of 1984, where he replaced Morton Downey, who had been fired for telling an ethnic joke.

In California, Limbaugh found his radio audience by attacking communism, feminism, and environmentalists, developing a style that he admits is more entertainment than news or even debate. He has repeatedly said that his principal job is not so much to champion causes and proselytize but to hold an audience. After nearly tripling his California audience in four years, in July 1988 he moved to New York City's

Rush Limbaugh
Copyright 1998 Brad Trent/Premiere Radio Networks

WABC, where his program was syndicated and his loyal following multiplied and mobilized. A nationally syndicated one-half hour television program begun in 1992 was not as successful as his radio program and was dropped.

In his programs and books, Limbaugh describes and mocks various liberal movements and the effects of their agendas on society. "I believe that in order to combat the misinformation that is so prevalent on our political landscape today we have to remain informed and alert to the things that are occurring," he writes in *The Way Things Ought To Be* (1992). "The best way to do that is to read, listen to, or watch me."

An ardent admirer of Ronald Reagan and supporter of George Bush and Robert Dole, Limbaugh provided a great service for the Clinton administration when he endorsed the North American Free Trade Agreement (NAFTA) and attacked Ross Perot's anti-NAFTA crusade. However, the *New York Times* credits Limbaugh not only with the NAFTA victory but with helping sweep Republicans to Congressional victory in 1996 and with destroying the crime and lobbying bills of President Clinton. In his own words, he continues to serve as "a man, a legend, a way of life" and the "Doctor of Democracy."

ALF PRATTE

See also Commentators; Controversial Issues; Fairness Doctrine; Talk Radio

Rush Limbaugh. Born Rush Hudson Limbaugh III in Cape Giradeau, Missouri, 12 January 1951. Attended Southeast Missouri University but did not graduate; worked for radio station WIXZ, McKeesport, Pennsylvania, KQV in Pittsburgh, and KUDL, KFIX, and KMBZ, all in Kansas City, where he was also a publicist and announcer for the Kansas City Royals; hired by KFBK, Sacramento, California, 1984; signed a two-year contract with EMF Media Management, New York City, 1988; nationally-syndicated talk radio show broadcast to more than 600 stations, 2003.

Radio Series

1988–present *The Rush Limbaugh Show*

Television Series

The Rush Limbaugh Show, 1992–1996

Selected Publications

The Way Things Ought to Be, 1992
See, I Told You So, 1993

Further Reading

Franken, Al, *Rush Limbaugh Is a Big Fat Idiot, and Other Observations,* New York: Delacorte Press, 1996
Kurtz, Howard, *Hot Air: All Talk All the Time,* New York: Times Books, 1996
Laufer, Peter, *Inside Talk Radio: America's Voice or Just Hot Air?* Secaucus, New Jersey: Carol, 1995
Rendall, Steve, Jim Naureckas, and Jeff Cohen, *The Way Things Aren't: Rush Limbaugh's Reign of Error: Over 100 Outrageously False and Foolish Statements from America's Most Powerful Radio and TV Commentator,* New York: New Press, 1995
Rutenberg, Jim, "Despite Other Voices, Limbaugh's Is Still Strong, *New York Times* (24 April 2000)
Seib, Philip M., *Rush Hour: Talk Radio, Politics, and the Rise of Rush Limbaugh,* Fort Worth, Texas: Summit Group, 1993

Little Orphan Annie

Children's Serial Drama

A pioneer of the children's serial genre, *Little Orphan Annie* first bowed—or curtsied—in 1930 on Chicago station WGN. On 6 April 1931 it premiered nationally on the NBC Blue Network, later moving to Mutual Broadcasting System. Shirley Bell and Janice Gilbert portrayed Annie during the series' 11-year run; Bell from the beginning until 1940, Gilbert from 1940 to 1942.

Based on Harold Gray's popular comic strip, *Annie* featured 15 minutes of action and high adventure every weekday afternoon or early evening, initially based primarily in her adopted hometown of Tompkins Corners, and later in more exotic, faraway places. She fought all forces of evil, including gangsters and criminals, reminding her faithful listeners at the end of the show to "be sure to drink your Ovaltine." Indeed, the premium toys offered by *Little Orphan Annie* and its long-time sponsor seemed at times to compete with the stories themselves, taking up four to six minutes of the 15-minute broadcast. Children who tuned in were urged to get their own "swell Ovaltine shake-up drinking mug" by sending ten cents and the proof of purchase from an Ovaltine can.

In addition to Bell and Gilbert, other cast members included Allan Baruck and Mel Torme (Joe Corntassel, Annie's best friend), Henry Saxe, Boris Aplon, and Stanley Andrews (Oliver "Daddy" Warbucks), Henrietta Tedro and Jerry O'Mera (Ma and Pa Silo, the farm couple who cared for Annie when Daddy Warbucks was away on business), and Pierre Andre (Uncle Andy, the announcer). The voice of Sandy, Annie's dog, was provided by Brad Barker. Among Annie's favorite expressions were "Leapin' lizards" and "Jumpin' grasshoppers."

The program's writers employed a simple but very effective technique to keep listeners, especially young children, returning to hear the next installment of Annie's adventures: the cliff-hanger. Episodes seldom ended with finality or resolution. Instead, story lines "flowed" from one episode to the next, occasionally reaching a conclusion but never without the development of a new story line to take its place. Beginning with *Annie,* this open-ended approach—leaving listeners in suspense at the end of each daily broadcast—was particularly evident for decades to come in children's radio and television programming and motion picture serials. In addition, Annie's radio adventures appealed to youngsters because the episodes often articulated childhood dreams of experiencing the glamour of the adult world.

NBC's radio network connections were not completed for regular U.S. coast-to-coast broadcasting until 1933, two years after *Little Orphan Annie*'s network premiere. As a consequence, *Annie* in its infancy was actually two different programs—one originating in Chicago, the other in San Francisco. Listeners in the eastern and central areas of the U.S. heard Shirley Bell in the lead role, while listeners in the far west heard Floy Hughes. Identical scripts ensured some consistency between the two productions, but west coast listeners were no doubt startled to suddenly hear different actors after the program's operations were consolidated in Chicago in 1933.

For over five years *Little Orphan Annie* aired six times a week, going to five times weekly beginning in 1936. The series moved from NBC to Mutual in 1940, at which time Ovaltine, the show's original sponsor, decided instead to put its advertising dollars in *Captain Midnight,* a new children's suspense show. Taking the chocolate drink mix's place as *Annie*'s sponsor was the breakfast cereal Puffed Wheat Sparkies, but by this time other adventure shows were outgunning Annie at her own game. Faced with declining ratings, *Little Orphan Annie*'s last broadcast was on 26 January 1942.

DAVID MCCARTNEY

Cast

Little Orphan Annie	Shirley Bell, Floy Hughes, Bobbe Deane, Janice Gilbert
Joe Corntassel	Allan Baruck, Mel Torme
Oliver "Daddy" Warbucks	Henry Saxe, Stanley Andrews, Boris Aplon
Mrs. Mary Silo	Henrietta Tedro
Mr. Byron Silo	Jerry O'Mera
Uncle Andy (announcer)	Pierre Andre
Sandy (Annie's dog)	Brad Barker
Aha	Olan Soule
Clay	Hoyt Allen

Producer/Creator

Based on the comic strip by Harold Gray

Programming History

WGN Chicago	1930
NBC Blue	1931–October 1936
NBC	November 1936–January 1940
Mutual	1940–42

Further Reading

Boemer, Marilyn Lawrence, *The Children's Hour: Radio Programs for Children, 1929–1956,* Metuchen, New Jersey: Scarecrow Press, 1989

Dunning, John, *Tune in Yesterday: The Ultimate Encyclopedia of Old-Time Radio, 1925–1976*, Englewood Cliffs, New Jersey: Prentice-Hall, 1976; revised edition, as *On the Air: The Encyclopedia of Old-Time Radio*, New York: Oxford University Press, 1998

Swartz, Jon D., and Robert C. Reinehr, *Handbook of Old-Time Radio*, Metuchen, New Jersey: Scarecrow Press, 1993

Terrace, Vincent, *Radio Programs, 1924–1984: A Catalogue of over 1800 Shows*, Jefferson, North Carolina: McFarland, 1999

Localism in Radio

U.S. Regulatory Approach

The concept of "localism," or serving a specific community, has always been central to the practice of radio programming and to government policies concerning broadcasting in the United States. In contrast to most of the rest of the world, American radio stations were allocated to local communities and licensed to serve audiences defined by the boundaries of those communities. The Federal Communications Commission (FCC) has described its radio allocation priorities as (1) providing a usable signal from at least one station to everyone and diversified service to as many persons as possible, and (2) creating sufficient outlets for local expression addressing each community's needs and interests. That system of license allocation remains the foundation of American broadcasting. As for programming, local service has frequently been a key element in a station's ability to survive and prosper. Research and experience have consistently demonstrated that local content is one of the things listeners value most highly.

However, economic realities have usually impelled broadcasters toward centralized program distribution. Networks began developing in broadcasting's earliest days, and although the traditional radio networks have long been reduced to providing news and sports for radio, a new generation of networks offering full-time formats appeared as the increased availability of satellite service in the 1980s made such a service viable. By 2000, the emergence of distribution technologies that no longer rely exclusively on nearby transmitters to reach individual audience members—such as direct-to-home satellites and the internet—and the ability of large radio groups to program clusters of stations from a central location led some to suggest that localism is an idea destined to be little more than a quaint relic of a bygone age. Larry Irving (then head of the National Telecommunications and Information Administration) told the 1999 National Association of Broadcasters convention in Las Vegas that localism has "gone the way of the buffalo." Yet other industry observers continue to argue forcefully that the most successful radio stations are those that do the best job of connecting with the localized needs and interests of their audiences. This view holds that localism will be even more important in a future of ever-greater competition from sources such as the satellite-based Digital Audio Radio Services (DARS), offered nationwide since November 2001 by XM Satellite Radio and since 2002 by Sirius.

These contradictory assertions may all, in fact, be accurate, depending on one's vantage point. Much like the phrase "The public interest," the meaning of localism has always been in the eye of the beholder—typically either the FCC or a station licensee. The very vagueness of the term has enabled a variety of regulators, industry spokespersons, and public service advocates to laud the importance of localism in different situations.

Localism in U.S. Broadcast Regulation

As a matter of policy, localism is closely tied to a number of regulatory goals. These are generally expressed as the need to limit centralized (program) power or authority in order to create more diverse content—the robust and varied "marketplace of ideas" central to the American understanding of free speech. For several decades, the FCC has pointed to the importance of localism as a means of providing diverse program content for the furtherance of the public interest. Some policy makers also argue for the need to protect local communities and smaller interests from being overwhelmed by programs developed by (and for) larger national interests. The desire to diffuse political power has been a running theme throughout American history (the federal system of government is perhaps the most obvious result). Added to the widely accepted notion that the media are capable of exerting great influence on society, the decision to dilute the power of a single broadcast entity—station or especially network—seems an obvious choice.

The commission has a specific charge in the Communications Act to "encourage the larger and more effective use of radio in the public interest." The FCC has generally interpreted this to mean that it should try to allocate the maximum technically feasible number of stations around the country. Thus, structural definitions of localism (in a geographic or spatial sense) have most often guided policy makers. This understanding of localism assumes that stations licensed to transmit to a geographically restricted area will focus their programming on the specific needs and interests of the citizens residing in that area. In this context, localism as policy has been put into regulations affecting the distribution of licenses to various communities. Localism is also seen in the bedrock obligation of all broadcast licensees to serve the needs and interests of their community of license (which at one time involved an elaborate process to ascertain the needs of that community) and in the preference that was granted to active local ownership when, prior to 1996, the comparative hearing process was used to choose a licensee from among mutually exclusive competing applicants.

In one of the earliest examples of localism, the Radio Act of 1927 divided the country into five "zones" and, in the case of competing license applications, directed the Federal Radio Commission (FRC) to distribute stations among the zones according to frequency, power, and time of operation, with concern for fairness, equity, and efficiency. The Davis Amendment, added one year later, required the FRC to provide *equality* of service, in terms of both transmission and reception, in each of the five zones. These sections, with slight modification, passed into the Communications Act of 1934. Although the zone system was repealed in 1936 and the law was modified to require once again that the FCC simply provide a fair, efficient, and equitable distribution of radio service to each state and community, localism was (and is) undeniably a powerful concern in Congress.

Another significant, and more recent, example of the FCC's structural concern with localism was the decision to drastically restructure the system of FM station allocation in order to increase the number of available stations in the early 1980s. Generally referred to by its FCC docket number (80-90), this order authorized three new classes of stations and modified the interference and operational rules with the goal of allowing first (and sometimes second) FM stations in communities where none had been possible before under the original 1962 Table of Assignments. As a result, the number of FM stations in the United States grew from around 3,000 in 1980 to slightly under 6,000 by the end of 1990 and continued to climb to more than 8,500 by 2003.

Localism and the Business of Radio

However, the growth was not solely good news for the industry or the audience. Many observers lamented the increase in interference in the FM band. The rise in the number of stations combined with the simultaneous deregulation of radio to rapidly escalate station values. Many owners found themselves too far in debt, particularly during the economic downturn in the late 1980s and early 1990s. Sometimes as a result of the significant economic hardships resulting from the combination of increased competition and debt load, sometimes in response to a perceived change in the desires of the audience, stations cut back or eliminated local air staffs and news operations. If the goal of the FCC's restructuring was to create more local content for more communities, the result could best be described as mixed. Although many stations continue to thrive by providing their audience a programming diet heavy with local content and involvement, satellite-provided formats, other syndicated product, and the ever-increasing ease of automating a station combine to create a significant economic incentive for many licensees to reduce localism to commercials and weather forecasts.

This illustrates the inherent conflict between policy rhetoric and the changing economic realities facing licensees. Though frequently lauding localism in policy pronouncements, the FCC has seldom promulgated, and even more infrequently enforced, local program guidelines or requirements. Nearly all of those that ever existed, such as the fairness doctrine, the ascertainment primer, commercial guidelines, and news and public-affairs guidelines, disappeared by the mid-1980s. The reasons behind this deregulatory trend (critics would term it failure) have been hotly debated because they are so complex. The regulatory problem is one of accommodating the various interests—licensees, program producers, audiences, networks, regulators, advertisers—in a rule that comports with common understanding of the First Amendment. Although this task was difficult under the public trustee model of regulation that guided broadcasting's first 50 years, it is practically impossible to set firm local content guidelines under the current regulatory philosophy, which moves much of the control from government policy to marketplace competition.

A further complicating factor is the constant evolution of the media environment. From the system envisioned at the time of the Communications Act in 1934, rigidly structured along relatively narrow geographic lines, radio in the United States moved almost immediately to a distribution system with a wider geographic frame (regional and national sources feeding the majority of programming on "local" stations). From a few hundred local stations linked with relatively new networks at the time the Communications Act of 1934 was written, radio grew decades later to a business of thousands of "local" outlets providing a relatively few national music or talk program formats. Development of internet and satellite distribution has merely enhanced the trend to national program types provided through local outlets. The degree to which any one of those outlets wishes to be truly local (reflecting and projecting its own community) is left to the discretion of the licensee.

The vague nature of localism itself is a final complication. One's understanding of concepts such as "local" or "community" colors any practical application of localism. The term *can* mean full-scale involvement of a station with its community, or (as is more usually the case) it can mean mere mention of local weather (and local commercials) within a syndicated music format heard on hundreds of stations across the country. As traditionally viewed by the FCC, real localism is probably somewhere in the middle, but much closer to the former—and many argue that it's also good business. In his book *Radio Programming: Tactics and Strategy,* programmer and consultant Eric Norberg asserts that localism and human contact are the elements that listeners value most in a station and that therefore the core of what makes a station successful is the relatable local person on the air. This viewpoint takes on the air of common wisdom in the industry trade press, particularly in advice given to programmers and air talent. Researchers repeatedly find that local information (weather forecasts, traffic information, event news) is one of the top reasons people tune in to radio.

Localism as a Social Construct

Critics, and occasionally even the FCC, have suggested that the goals of localism can be addressed in a different fashion, recognizing that communities frequently form around shared tastes, interests, and ideals without specific reference to a geographic boundary. For example, in the rulemaking that eliminated many of the radio programming guidelines, the commission noted that

> communities of common interests need not have geographic bounds. . . . The economics of radio . . . allowed that medium to be far more sensitive to the diversity within a community and the attendant specialized community needs. Increased competition in large urban markets has forced stations to choose programming strategies very carefully. (FCC, "Deregulation of Radio," *Notice of Inquiry and Proposed Rulemaking,* 73 F.C.C.2d 457 [1979], at 489)

This alternative view of community can also be seen in the FCC's decision to approve satellite-delivered DARS. Despite the diversity of programming alluded to above, many program interests go unfulfilled by traditional terrestrial radio because the audience for a particular type of music or information is simply too small within the service area of a single station or is otherwise unattractive to advertisers. Beginning in late 2001, however, satellite radio services included program channels that would not be economically viable on a single station in a given market (e.g., five separate jazz channels). The technology can aggregate widely separated audiences in a fashion that does not serve traditional localism but surely adds to content diversity.

Trials of Modern Localism

Sometimes, competing concerns such as spectrum efficiency have prevailed over localism. Prior to 1978, the FCC issued Class D FM licenses to college and community stations, permitting low-power operation (a maximum of 10 watts, with a tower height less than 100 feet) in the noncommercial part of the FM band. These stations represented a variety of operational styles, from student-run stations at colleges or high schools to stations licensed to civic groups and generally run by a largely volunteer or all-volunteer staff. By their very nature, these operations were strongly committed to their community and would appear to personify the localism ideal, often featuring material not available through full-power stations.

In 1972 the Corporation for Public Broadcasting (CPB) petitioned the FCC to explore several issues related to more efficient use of the FM channels set aside for noncommercial educational stations. In comments to the commission, they argued (with the support of others, including the National Federation of Community Broadcasters) that the 10-watt broadcasters were effectively blocking more efficient use of the spectrum. Essentially, CPB and its supporters were arguing for more stations that met their qualifying guidelines for size and professionalism, at the expense of smaller operations. Supporters of Class D stations, primarily the licensees themselves, countered with various arguments for retaining the service as it was, including the point most relevant here—the truly local nature of the service.

The FCC recognized that the Class D stations were indeed meeting discrete local needs. But in this case, the commission put the emphasis on the efficiency argument put forth by CPB, announcing new rules that effectively forced existing Class D stations to upgrade their facilities to Class A minimums or else become a secondary service, facing interference or being bumped from their assignments. In the FCC's view at that time, there was not sufficient spectrum available for both full-power, larger coverage area stations and low-power operations (although the commission has long accepted the need for low-powered translator and booster stations that extend the coverage of existing FM and TV stations but are prohibited from originating any programming themselves).

It is somewhat ironic that the FCC issued rules in 2000 that will create a new class of low-powered FM or microradio stations. The rulemaking comes in response to petitions arguing that, in the wake of industry consolidation following the 1996

Telecommunications Act, radio ownership and content are insufficiently diverse, and that current stations often fail to address local needs. Although congressional intervention curtailed the number of LPFM stations that could be licensed, the FCC had issued more than 400 construction permits by the end of 2002, and 73 LPFM stations were on the air in January 2003

GREGORY D. NEWTON

See also Australian Aboriginal Radio; Canadian Radio and Multiculturalism; College Radio; Community Radio; Deregulation; Licensing; Low-Power Radio/Microradio; Pacifica Foundation; Ten-Watt Stations

Further Reading

Barrett, Andrew C., "Public Policy and Radio—A Regulator's View," *Media Studies Journal* 7 (1993)

Collins, Tom A., "The Local Service Concept in Broadcasting: An Evaluation and Recommendation for Change," *Iowa Law Review* 65 (1980)

Federal Communications Commission, *Commission en banc Programming Inquiry, Report and Statement of Policy,* 44 FCC 2303 (1960)

Federal Communications Commission, *Deregulation of Radio, Notice of Inquiry and Proposed Rulemaking,* 73 FCC 2d 457 (1979)

Federal Communications Commission, *The Suburban Community Policy, the Berwick Doctrine, and the De Facto Reallocation Policy, Report and Order,* 93 FCC 2d 436 (1983)

Jones, Steven G., "Understanding Community in the Information Age," in *CyberSociety: Computer-Mediated Communication and Community,* edited by Steven G. Jones, Thousand Oaks, California: Sage, 1995

Kelley, E.W., *Policy and Politics in the United States: The Limits of Localism,* Philadelphia: Temple University Press, 1987

Kemmis, Daniel, *Community and the Politics of Place,* Norman: University of Oklahoma Press, 1990

McCain, Thomas A., and G. Ferrell Lowe, "Localism in Western European Radio Broadcasting: Untangling the Wireless," *Journal of Communication* (Winter 1990)

Milam, Lorenzo W., *Sex and Broadcasting: A Handbook on Starting a Radio Station for the Community,* 2nd edition, Saratoga, California: Dildo Press, 1972; 4th edition, as *The Original Sex and Broadcasting,* San Diego, California: MHO and MHO, 1988

Norberg, Eric G., *Radio Programming: Tactics and Strategy,* Boston: Focal Press, 1996

Stavitsky, Alan G., "The Changing Conception of Localism in U.S. Public Radio," *Journal of Broadcasting and Electronic Media* 38 (1994)

Local Marketing Agreements

Brokered Agreements among Stations

As part of the deregulation of radio ownership initiated by the Federal Communications Commission (FCC) during the late 1970s and early 1980s, radio station owners and operators began to sign program service and/or marketing agreements, known in the industry as local marketing agreements. In 1992 the FCC formally approved this form of ownership and operations agreements, and they became commonplace, particularly in the four years leading to the easing of ownership rules in the 1996 Telecommunications Act.

Alliances for local marketing agreements may be located in the same market, in the same region, or in the same service (AM or FM). The allied owners and operators draw up and sign legal agreements defining financial control over their allied properties, but the owners still maintain their separate licenses and studios. After 1996, it often became simpler to simply purchase a station, but local marketing agreements were still used to make transitions to new owners cheaper, easier, and more cost-effective.

A local marketing agreement thus has become a time-brokering agreement between stations that can address either programming or advertising time. Basically, the originating or principal station in the local marketing agreement pays the "affiliate" a monthly fee either to partially simulcast programming or to air original satellite-delivered programming. This type of agreement differs from a satellite format network affiliation arrangement, wherein the affiliate pays for programming. The originating station in a local marketing agreement can strike an arrangement with the leased station for either handling or sharing advertising sales.

The benefits of local marketing agreements to the originating station include expanded coverage area and thus the potential for increased sales of advertising. For example, during the early 1990s, owners of many struggling AM stations signed local marketing agreements to stabilize their flow of profits rather than take on the risks and costs involved in trying to establish a new format.

By 2000, although the number of stations and formats seems endless in major markets, in fact, local marketing agreements allowed two separate radio stations to operate jointly, and so the number of operators (or voices) was actually far fewer than the number of stations (or outlets). Usually the financially strong station reaches a combined operation and sales agreement with a financially troubled station in the same community to oversee programming and advertising time sales for a percentage of the advertising sales. Although the parties exercising a local marketing agreement are not required to file the agreement with the FCC, the licensee of the weaker partner station is still required to meet the station's maintenance and community standards (although in practice these requirements have become so minimal during the 1990s that this threat of losing a licensee over such a deal offers no risk).

The local marketing agreement policy helped redefine the institutional relationship that had formerly been restricted to affiliation. Indeed, once the FCC in 1992 formally relaxed regulations to allow local radio owners to own and control more than one station in the same service market, a wave of deals took place. One scholar calculated that over 50 percent of the commercial radio stations became involved in some aspect of consolidation between 1992 and 1996, including local marketing agreements. The pace of consolidation has increased tremendously since the Telecommunications Act of 1996 became law, with local marketing agreements used to make the transition to combinations of radio stations.

Local marketing agreements permit the parties to take advantage of cost and organizational structural efficiencies; to dominate a market with variations of one format; to eliminate redundant jobs; to develop broader marketing plans and solutions for advertisers; and, in the end, to increase profits for stockholders and investors. One can simply buy a station, or, to be more flexible in the short run, one can set up a local marketing agreement to test whether a formal alliance might work better. Some owners delay formal merger decisions until they figure out how to consolidate personnel and facilities.

There have been, therefore, numerous examples of different uses of local marketing agreements, as owners have utilized combinations of acquiring stations to form new and, they hope, more profitable alliances. Consider a top 50 market, Charlotte, North Carolina, where in spring 1992 the market was being served by ten radio owners and operators, who owned 3 AM and 12 FM stations. Two years later, those ten had dwindled to six owners and operators. By fall 1996, Charlotte had consolidated to the point of having only four viable radio owners and operators controlling the same 15 stations. Local marketing agreements created much, though not all, of this consolidation.

DOUGLAS GOMERY

See also Licensing; Ownership, Mergers, and Acquisitions

Further Reading

Albarran, Alan B., *Management of Electronic Media,* Belmont, California: Wadsworth, 1997

Chan-Olmsted, Sylvia M., "A Chance for Survival or Status Quo? The Economic Implications of the Radio Duopoly Ownership Rules," *The Journal of Radio Studies* 3 (1994)

Compaine, Benjamin M., and Douglas Gomery, *Who Owns the Media?* White Plains, New York: Knowledge Industry, 1979; 3rd edition, Mahwah, New Jersey: Erlbaum, 2000

Ditingo, Vincent M., *The Remaking of Radio,* Boston: Focal Press, 1995

Federal Communications Commission, Mass Media Bureau, Policy and Rules Division, *Review of the Radio Industry, 1997,* MM Docket No. 98-35 (13 March 1998, as part of 1998 Biennial Regulatory Review)

opponents in the state legislature impeached but failed to convict Governor Long in 1929.

Huey Long launched his second gubernatorial campaign in a 3 August 1927 speech carried over radio station KWKH in Shreveport. The station, owned by Long's friend William K. Henderson, played a major role in many of Huey Long's future radio addresses, all of which—like his 3 August speech—were carried free of charge and for unlimited time.

Huey Long's radio skills were honed to perfection during his years as governor. His unique oratorical style fit perfectly with the expectations of his listeners. Long often spoke effortlessly and without notes for two or three hours at a time. His rapid-fire monologues freely moved from quoting Bible verses, to joking, to harsh name calling of political foes. He even resorted to deliberately mispronouncing words and ignoring accepted rules of grammar in order to identify himself more closely with his audience.

Many of Huey Long's radio addresses were broadcast over KWKH by remote control from the Governor's Mansion in Baton Rouge or from a hotel room. He often began his remarks by telling listeners that Huey Long (or the "Kingfish"—a sobriquet that he began using in the late 1920s that came from a character's name on the popular *Amos 'n' Andy* radio show) was going to be talking and that they should phone a friend who also might tune in. He then would spend several minutes in small talk while calls were made and listeners gathered.

Huey Long was elected to the U.S. Senate in 1930 while serving as Louisiana governor. He retained the governorship for another year and was finally sworn in to the U.S. Senate in January 1932. Soon after his arrival in Washington, Long made waves by proposing what he called his "Share-Our-Wealth" plan as a means of redistributing wealth in America. He told a Depression-weary America of his plan during a 30-minute February 1934 nationwide radio broadcast on the National Broadcasting Company (NBC) network. Time for the broadcast was donated, following the practice of both NBC and the Columbia Broadcasting System (CBS) to supply such time to members of the U.S. Congress. The enormous number of letters that flowed into Senator Long's office following his radio address required some 48 secretaries and typists to answer, and by 1935 Long had a bigger office and more employees than any other U.S. Senator. Although economists berated his "Share-Our-Wealth" plan as unworkable, the notoriety that Long was gaining nevertheless persuaded the radio networks to allow the senator more airtime.

Network radio proved to be more than just an excellent forum to push his economic plan, though, when in August 1935 Senator Long announced his break with President Roosevelt over policy issues and his plans to run for the presidency himself. His chances of defeating Roosevelt seemed good, according to knowledgeable observers who placed Huey Long's political strength as second only to the president's. Long departed from his short-lived campaign in September 1935 to be present during a special session of the Louisiana legislature. While hurrying through a hallway of the state capitol on 8 September, he was gunned down by an assassin, a Baton Rouge physician named Carl Weiss. Other than a presumed dislike for Long's dictatorial politics, no definite reason for the assassination has ever been determined. Huey Long remained conscious for two days after the shooting but died on 10 September at age 42.

RONALD GARAY

See also Controversial Issues; KWKH; U.S. Congress and Radio

Huey Long. Born in Winnfield, Louisiana, 30 August 1893. One of ten children of Huey Pierce and Caledonia (Tison) Long. Studied law at Oklahoma School of Law (1912–13) and Tulane University (1914–15); admitted to bar and began practicing law, Winnfield and Shreveport, Louisiana, 1915; member of Louisiana State Railroad Commission, 1918–28; governor of Louisiana, 1928–32; U.S. Senator, 1932–35; announced candidacy for U.S. President, 1935. Died (assassination) in Baton Rouge, Louisiana, 10 September 1935

Further Reading

Bormann, Ernest, "A Rhetorical Analysis of the National Radio Broadcasts of Senator Huey Pierce Long," *Speech Monographs* 24 (November 1957)

Bormann, Ernest, "This Is Huey P. Long Talking," *Journal of Broadcasting* 2 (Spring 1958)

Brinkley, Alan, *Voices of Protest: Huey Long, Father Coughlin, and the Great Depression,* New York: Knopf, 1982

Hair, William Ivy, *The Kingfish and His Realm: The Life and Times of Huey P. Long,* Baton Rouge: Louisiana State University Press, 1991

Williams, T. Harry, *Huey Long,* New York: Knopf, 1969

Lord Haw-Haw (William Joyce) 1906–1946

World War II Propaganda Broadcaster

During World War II, William Joyce, known to his British listeners as "Lord Haw-Haw," was one of the most famous expatriate radio propagandists for the Axis powers (others included "Axis Sally," Ezra Pound, and "Tokyo Rose"). After the war, Joyce was found guilty of treason and executed by the British.

Origins

William Joyce was born in 1906 in New York City to Irish parents who were naturalized American citizens; he was thus a citizen himself, an issue of import at the end of his life. When he was three, his family returned to Ireland, eventually settling in Salthill, Galway.

Joyce was educated at the College of St. Ignatius Loyola in Galway, but his education was interrupted by the bitter struggle for Irish independence. By his own later admission, Joyce worked as a spy for the British "Black and Tans" in 1920–22. This was a despised military force sent to Ireland in the summer of 1920 to reinforce the Royal Irish Constabulary. Joyce's support of and cooperation with such an intensely despised organization was enough to force the Anglophile family to move to England after the Irish achieved independence in 1922. There Joyce first attended Battersea Polytechnic and later Birkbeck College of University of London, from which he received a first class honors degree in English in 1927.

While in college, Joyce became politically active in conservative and then far-right circles. He eventually dropped out of graduate study and in 1933 joined Sir Oswald Mosley's British Union of Fascists, rising to be director of propaganda and later deputy leader. He later fell out with Mosley and was dropped from the party. In 1933 he made what became a fatal mistake: declaring he was a British citizen (and one born in Galway when Ireland was part of Britain) to obtain a British passport to ease his ability to travel as political tensions rose. He renewed the passport (and in essence, the citizenship vow) in August 1939. The fraudulently acquired passport was later a central element in the post-war charge of treason.

Broadcasting for Germany

In August 1939, Joyce and his second wife, fearing arrest as avowed fascists while war loomed, fled England and went to Germany. He soon found a position with German radio, broadcasting anti-British propaganda beginning on 18 September 1939. In 1940 he became a naturalized German citizen. Both he and his wife worked for the Nazi regime for the rest of the war.

His voice was beamed to Britain on a program called *Views on the News*. Broadcasts featured war-related news and ad hominem attacks on Allied leaders, Jews, and communists. His signature opening, "Germany calling, Germany calling," became well known to the British public. The "Lord Haw-Haw" appellation came first from a *Daily Express* correspondent who referred to his "speaking English of the haw-haw, dammit-get-out-of-my-way variety." British listeners were at first amused at his somewhat affected high-class and sneering tone as he attacked Jews and other enemies of Germany.

Through 1942, as the Third Reich enjoyed a number of military successes, Joyce related events to his overseas listeners with considerable candor. As war-related news was then being censored by the British government and the BBC, many Britons turned to Joyce to ascertain the facts. As the war turned against Germany, however, Joyce's broadcasts—nearly always made from studios in Berlin—grew more pointed and shrill, and the declining number of British listeners no longer trusted what they heard.

Joyce's final and arguably most dramatic broadcast was recorded at a transmitter near the Danish border in April 1945 after Hitler's suicide. Though never broadcast, in it Joyce railed yet again against the dangers of Bolshevism and reflected upon Germany's role in the war. He was quite clearly agitated and may have been intoxicated. A few weeks after the war ended in May, he was captured by British troops as he tried to flee Germany.

Trial and Execution

Joyce was returned to England and tried for treason in a three-day London jury trial in September 1945. He was quickly found guilty and sentenced to death. The trial and appeals of the judgment were widely followed, and transcripts were published in two contemporaneous books. The tricky citizenship issue was over-ridden by arguments that in fraudulently seeking a British passport before the war, he had placed himself under the protection of—and thus in a position of loyalty to—the British crown. After all appeals were exhausted, in January 1946 Joyce was hanged at London's Wandsworth prison.

Joyce's propaganda role is significant. For one thing, his broadcasts served as something of a radio role model for Ezra Pound, who later performed analogous services for Italy. Second, Joyce's broadcasts (and those of both "Axis Sally" and "Tokyo Rose," who were Americans and were also tried for treason) raise the question of whether broadcast polemics and editorializing necessarily constitute treason. Joyce's execution poses a third question: is it possible for the victors in any given struggle to try the vanquished in such a way as to ensure fair treatment for the defeated?

WILLIAM F. O'CONNOR AND CHRISTOPHER H. STERLING

See also Axis Sally; Fritzsche, Hans; Propaganda; Tokyo Rose; World War II and U.S. Radio

William Brooke Joyce. Born in Brooklyn, New York, 24 April 1906, son of Michael Joyce, an Irish-born naturalized American citizen, and Gertrude Emily Brooke. Moved to Galway, Ireland, 1909. Attended College of St. Ignatius Loyola, Galway, Ireland; family moved to England, 1922; attended Battersea Polytechnic, London, England, 1922–23; Birkbeck College of University of London, B.A. 1927; Deputy Leader, British Union of Fascists, 1933–37; broadcast Nazi propaganda, Berlin, 1939–45. Executed at Wandsworth prison, London, England, 3 January 1946.

Selected Publications

Twilight over England, 1940

Further Reading

Cole, John Alfred, *Lord Haw-Haw and William Joyce: The Full Story,* London: Faber, 1964; New York: Farrarr Straus and Giroux, 1965

Doherty, M.A., *Nazi Wireless Propaganda: Lord Law-Haw and British Public Opinion in the Second World War,* Edinburgh: Edinburgh University Press, 2000

Hall, J.W., editor, *Trial of William Joyce,* London: William Hodge, 1946

Roberts, Carl E. Bechhofer, editor, *The Trial of William Joyce,* London: Jarrolds, 1946

Selwyn, Francis, *Hitler's Englishman: The Crime of Lord Haw-Haw,* London: Routledge and Kegan Paul, 1987

West, Rebecca, "The Revolutionary," in *The New Meaning of Treason,* by West, New York: Viking Press, 1964

Lord, Phillips H. 1902–1975

U.S. Radio Actor, Writer, Producer

SOUND EFFECTS: [Marching feet, machine-gun fire, siren wail.]
VOICE: Calling the police! Calling the G-Men! Calling all Americans to war on the underworld!
ANNOUNCER: *Gangbusters!* With the cooperation of leading law-enforcement officials of the United States, *Gangbusters* presents facts in the relentless war of the police on the underworld.

Loud and urgent openings such as this signaled another episode of the radio program *Gangbusters,* which aired for more than 20 years. This fast-paced drama was renowned for snappy dialogue and realistic sound effects, particularly at the show's beginning. (These features gave rise to the expression "coming on like gangbusters" to describe anything with a rapid start.) The creator of *Gangbusters* was Phillips H. Lord, a radio actor, writer, producer, and developer who was one of the industry's most successful artists during the "Golden Age" of radio in the 1930s and 1940s.

As Lord explained when the show (called *G-Men* for its first year) debuted on 16 September 1935, the series was based on actual FBI cases provided by Director J. Edgar Hoover. Lord and his assistant Helen Sioussat had an office in the Department of Justice next to Hoover, who allowed them unprecedented access to FBI case files. "I went to Washington, was graciously received by Mr. Hoover, and all of these scripts are written in the department building," Lord told listeners during the inaugural program, which was about the killing of John Dillinger. "Tonight's program was submitted to Mr. Hoover, who checked every statement, and made some very valuable suggestions."

However, Hoover almost backed out of the project after the first broadcast, in the wake of scattered criticism about this new style of radio entertainment. Lord hastily wrote Hoover to mollify him, vowing "There is NOTHING that I won't do to win your respect and confidence." What Hoover required Lord to do was to use only the material provided by the FBI and nothing more. Moreover, all scripts had to be approved by Hoover or one of his aides, and all had to show that every criminal was punished, either with death or a long jail sentence.

Lord endured these stringent guidelines because he recognized the importance of the new show—as a new genre of radio entertainment and a legacy for his career. "This series, Mr. Hoover, means more to me than anything else in the world," Lord explained in private correspondence. "My whole future will be based on the success of this program, and there isn't a stone I'll leave unturned toward making it the finest thing in radio."

Before he created this crime drama, Lord's future had seemed secure in a different broadcast format. He was one of the brightest stars on radio with a loyal nationwide audience for *Sunday Evening at Seth Parker's*. Lord played the elderly preacher from Jonesport, Maine, who discussed local happenings with his family and neighbors in the parlor. The *Seth Parker* broadcasts conjured up images of white, steepled churches, modest cape-style homes, and vessels plying the harbor, as the friendly folks prayed, sang, and discussed morality. The dialogue and plots hearkened to a simple, old-fashioned way of life during the uncertainty of the Depression. Lord created and wrote the series, and even wrote some of the hymns. It spawned a 1931 movie, *Way Back Home*, which starred a 29-year-old Lord in heavy makeup.

Lord based *Sunday Evening at Seth Parker's* on his own experiences. The son of a Connecticut Congregational minister, Lord developed his narrative skills from his grandfather, Hosea Phillips, a traveling salesman "whom none could beat in story telling, witticisms, and good common sense," Lord later wrote. After education at Bowdoin College and stints as a high school principal and candy maker, Lord got his break when the National Broadcasting Company (NBC) approved his proposal for a radio show about small-town life. "An old-fashioned cottage gathering of friends and neighbors coming together to sing hymns and discuss news of the town," Lord described it. The show became a huge success, as ten million listeners tuned in weekly to the cottage by the sea.

After five years Lord decided to have the Parkers and their Jonesport neighbors take an around-the-world cruise while Lord and his cast did so in real life. They departed on 8 December 1934, in the four-masted schooner *Seth Parker*, with sophisticated shortwave broadcasting equipment aboard that enabled them to continue their weekly broadcasts. This remarkable and highly publicized voyage was cut short two months later when a hurricane disabled the boat in the South Seas, 300 miles from Tahiti. The schooner's passengers were rescued by the Australian heavy cruiser *HMS Australia*, which carried one of King George's sons. Convinced that the ship's SOS calls were a publicity stunt, Australian officials demanded that Lord be brought up on charges. He was exonerated after the cruiser's captain asserted that Lord's maritime distress was real.

Embarrassed by this international incident, Lord abandoned the *Seth Parker* show and changed career paths. He developed the fast and violent action show *G-Men* in 1935 as a sharp contrast to the leisurely and gentle *Seth Parker*. After renaming the show *Gangbusters*, Lord and his production company, Phillips H. Lord Inc., developed other programs. The factual human-interest show *We the People* debuted in 1936 and aired for three seasons. It was soon followed by the serial drama *By Kathleen Norris*, the aviation adventure *Sky Blazers*, and the crime dramas *Treasury Agent* and *Counterspy*. Lord even revived *Seth Parker*. Before he turned 40, Lord had become one of the most successful writers, actors, and producers in radio.

Amid the flurry of development, Lord produced a new courtroom drama set in New York, entitled *Mr. District Attorney*. Horror-fiction writer Alonzo Deen Cole sued Lord for appropriation of literary property, claiming that he had submitted the idea to Lord but was never compensated or acknowledged. Lord won in trial court, but the verdict was overturned on appeal. The New York Court of Appeals ruled that Lord had indeed stolen the idea from Cole.

Lord tried to make the transition to television as a producer, but after two failed series he retired to coastal Maine, where he found contentment promoting community dances and sing-alongs. ("I'm exceedingly ambitious," he once told a reporter, but then added, "I'd swap all I have for contentment.") After suffering from myasthenia gravis, he died on 19 October 1975 in Ellsworth, Maine, just a short drive from Jonesport.

RALPH FRASCA

See also Gangbusters

Phillips H. Lord. Born in Hartford, Connecticut, 13 July 1902. Played title role on *Sunday Evening at Seth Parker's*, 1929–36, 1938–39; embarked on around-the-world cruise, broadcasting from the ship *Seth Parker*, 1934; shipwrecked near Tahiti, 1935; creator, producer, narrator for *Gangbusters*, 1935–57. Died in Ellsworth, Maine, 19 October 1975.

Radio Series

1929–36, 1938–39	*Sunday Evening at Seth Parker's*
1932–33	*The Country Doctor*
1934–45, 1951–52	*The Story of Mary Marlin*
1935–57	*Gangbusters* (*G-Men* during first year)
1936–39	*We, the People*
1941	*Great Gunns*

Films

Way Back Home, 1931

Selected Publications

Seth Parker and His Jonesport Folks: Way Back Home, 1932

Further Reading

Cole v Phillips H. Lord Inc., 28 New York State Supp. 2nd series 404 (1st Dept.) (1941)

DeLong, Thomas A., *Radio Stars: An Illustrated Biographical Dictionary of 953 Performers, 1920 through 1960*, Jefferson, North Carolina: McFarland, 1996

Helen J. Sioussat Papers, Library of American Broadcasting, University of Maryland, <www.lib.umd.edu/LAB/COLLECTIONS/sioussat.html>

Menken, Harriet, "Radio Personalities," *New York American* (9 February 1936)

"Radio Writer and Producer Phillips H. Lord Dies," *Ellsworth American* (23 October 1975)

Low-Power Radio/Microradio

Small Community Radio Stations

Microradio is a political movement with the goal of putting low-power FM transmitters into the hands of community activists, minority groups, and those with no hope of getting a traditional Federal Communications Commission (FCC) license to broadcast. Under the leadership of Free Radio Berkeley (FRB) founder Stephen Dunifer, instructions are readily available to anyone who wants a low-cost transmitter kit, programming help, and legal representation. Microradio broadcasters are often students and street people without property, and when ordered by the FCC to cease operations, they simply move to another location. FCC enforcement is uneven and has been complicated by recent court rulings. In *United States v Dunifer* (July 2000), the court ruled against Dunifer's right to broadcast without a license. That decision led activists to pressure the FCC more strongly for a licensed low-power FM service.

Origins

The history of unlicensed radio goes back to the early radiotelephone experimenters who simply went on the air without asking anybody's permission. By the end of the 1920s, the radio spectrum had been divided between commercial, amateur, and experimental users. All were required to have licenses, first from the Federal Radio Commission and after 1934 from the FCC. But there have always been scofflaws, mostly referred to as "pirate broadcasters." From ships anchored offshore with powerful transmitters to radio hobbyists broadcasting entertainment on amateur radio frequencies, there is a long history of unlicensed broadcasting. The most blatant of those illegal operators were usually caught and fined, and their equipment was confiscated. Unlike the modern microradio movement, most such "pirates" were not political activists.

For a time, however, small radio stations were not only allowed, they were actively encouraged. With the inception of FM radio, the FCC encouraged noncommercial operations. When frequencies went begging, the FCC in 1948 initiated a low-power category of stations, the so-called Class D outlets, that might use as little as ten watts of power and cover a very small area with a usable signal. But they were broadcast outlets, often held by nonprofit groups unable to afford anything larger. By the 1970s there were several hundred such stations on the air.

Low-power stations were increasingly resented by the FM radio business, which was rapidly expanding in the 1970s and 1980s. Tiny stations took up valuable frequencies that full-power outlets coveted. In 1978 the FCC began to reverse course, requiring these stations to use at least 100 (and more likely 1,000) watts or give way to full-power stations that would provide more services to more listeners. By the mid-1990s, only a handful of the old Class D stations remained, and most of those were preparing to use the required higher amounts of power.

A model for what could develop with radio was borrowed from television. As a result of experiments originating in Canada in the 1970s, the FCC became interested in and eventually approved a class of low-power television stations in the early 1980s. These were to use very low power on VHF channels or up to about 1,000 watts on UHF channels that would be "dropped in" among already allotted full-power channels in such a way as to keep interference at a minimum. After years of legal wrangling over how best to handle the mountain of conflicting applications, the FCC was granted the right to hold lotteries among mutually exclusive applications. By 2000 there were hundreds of low-power television outlets on the air, some providing original programs but many connected by satellite and offering typical entertainment programs otherwise not receivable in isolated rural areas.

Ted Coopman (1999) tells of the beginning of the modern microradio movement:

> The modern micro broadcasting movement began on November 25, 1986 in a public housing development in

Springfield, IL. Put on the air for about $600, the one-watt station broadcast openly on 107.1 FM as Black Liberation Radio (now Human Rights Radio). The operator, Mbanna Kantako, a legally blind African-American in his mid-thirties, started the station because he felt that the African-American community in Springfield was not being served by the local media. Kantako felt that because the African-American community had a high illiteracy rate, radio would be the best way to reach this community.

According to FRB founder Dunifer, the goal of the microradio movement was to have so many transmitters in use across the country that the FCC would be overwhelmed, finding itself in a situation similar to one it faced in the early 1970s, when the FCC was unable to control the widely popular citizens band radio service. Licensing for that service was eventually dropped. Unlike the "pirates" of the past, who hid from the FCC, microradio proponents act in open defiance of the law, challenging the government to arrest them, to shut them down. They believe that they are entitled to the airwaves; that they are shut out of the current allocation of FM licenses; and that, because of scarcity and resulting high cost, licenses are available only to the wealthy.

Dunifer was successful in convincing the National Lawyers Guild that the right to broadcast was a civil rights issue, one of giving access to all people, especially those disenfranchised by licensed media. In 1993 Dunifer's ten-watt radio station started operation at 104.1 FM, offering music and political commentary while actively challenging the FCC. In June 1993, the FCC issued a *Notice of Apparent Liability* to Dunifer for unlicensed broadcasting and fined him $20,000. Dunifer was represented by the Lawyers Guild, and in 1995 a U.S. district court in Oakland, California, heard arguments on constitutional issues in Dunifer's case, arguments stating that the FCC had not proved that his broadcasts caused harm to licensed broadcasters. The FCC responded, saying that siding with Dunifer would cause thousands more to go on the air, and the resulting interference would be chaotic.

The attorney arguing the case for Dunifer raised other issues in the hearing that would eventually cause the FCC to study the possibility of a licensed low-power FM service. One such issue was that the 1978 elimination of the Class D ten-watt educational license in favor of licenses for stations over 100 watts was overly restrictive and violated Dunifer's First Amendment right to free speech. It was argued that the commission's failure to provide a low-power service did not provide for the public interest, convenience, and necessity required under the Communications Act of 1934. It was further argued that the financial qualifications required for an FM station license violated the equal protection clause of the Constitution. The FCC disagreed on all counts.

In November 1997 Federal District Judge Claudia Wilken ruled in favor of Dunifer, saying the FCC had failed to prove Dunifer had done harm to existing broadcasters. Buoyed by this victory, Dunifer continued to help others get on the air and to promote the creation of a low-power service. But in June 1998 the FCC prevailed in court, and Dunifer was taken off the air, based on the fact that he had applied for neither a license nor a waiver to broadcast and therefore lacked standing to challenge the FCC. In July 1999 Dunifer's attorneys appealed, arguing that the FCC had not acted in good faith when dealing with microradio, citing as evidence a San Francisco applicant's request for such a license, which had been ignored by the FCC.

Development of the microradio movement has been based on a common belief that stations very low in power escape the jurisdiction of the FCC. This misconception is based on Part 15, Subpart D of the FCC rules, which permits unlicensed operation of very low power transmitters. This rule only allows, however, for an effective service range of 35 to 100 feet in the FM band. On AM frequencies, unlicensed transmitters cannot cover a radius larger than about 200 to 250 feet. The FCC argues that if microradio stations are able to reach listeners, they are almost certainly operating illegally.

Low-Power FM

As the number of microradio stations grew and complaints from the National Association of Broadcasters (NAB) increased, the FCC closed a number of unlicensed microradio operations. Both the protests and the microbroadcasting movement continued to grow, however. Media reports suggested that between 500 and 1,000 microradio stations were on the air in 1998, although five years earlier only a handful existed.

In spite of NAB protests, FCC Chair William E. Kennard suggested that proposals to establish a legal microbroadcasting service were worthy of consideration. The FCC issued a *Notice of Proposed Rulemaking* on 28 January 1999 to authorize an LPFM broadcast service. The FCC's *Notice* cited concern about the increasing concentration of ownership of media properties, with concomitant loss of diversity, in addition to suggesting that smaller communities outside metropolitan areas were often deprived of local focus in programming.

In response to its *Notice,* the FCC received thousands of comments representing the views of community groups, labor unions, religious organizations, state and local government, and others. The commission announced that response to the petition had indicated a broad interest throughout the country in the LPFM proposal. Further, the commission's webpage on "Low Power Broadcast Radio Stations" was accessed more than 15,000 times in 1998 alone. The FCC received thousands of additional phone and mail inquiries each year regarding the legality of low-power broadcasting.

The commission's decision was released 20 January 2000 with the announcement that such a service would, in fact, be established. In spite of aggressive protest and threatened legal action by the NAB, the FCC voted to create an entirely new type of radio station, with the intention of enhancing service to underrepresented groups and local communities. The FCC's restrictions were as follows:

> These (LPFM) stations are authorized for noncommercial educational broadcasting only (no commercial operation) and operate with an effective radiated power (ERP) of 100 watts (0.1 kilowatts) or less, with maximum facilities of 100 watts ERP at 30 meters (100 feet) antenna height above average terrain (HAAT). The approximate service range of a 100 watt LPFM station is 5.6 kilometers (3.5 miles radius). LPFM stations are not protected from interference that may be received from other classes of FM stations (www.fcc.org).

Several years after the initial announcement of the Low Power (LPFM) service, the FCC had received several thousand applications. Most of the applicants have been from religious organizations, the rest from community foundations and educational entities. A current list is on the FCC database (<www.fcc.gov/fcc-bin/fmq?state=&serv=FL&vac=&list=2>). Approximately 200 Construction Permits (CP) had been issued and 40 applicants had received actual licenses. Most of these were in the L1 category, 100 watts or less. Only a few have been under the L2 designation, 10 watts or less. As of the beginning of 2003, the FCC was not taking any additional applications.

While at least some of the original impetus for the LPFM service was to either legitimize or remove from the air the so-called pirates or micro broadcasters, based on the list of applicants, CPs and licenses issued, many observers suggest that the following will likely happen: former big city pirate/micro broadcasters will not receive licenses due to interference concerns in crowded FM markets; most licensees will be religious organizations, schools, and community foundations. And if the experience from the educational FM "boom" of the 1970s is repeated, most of these LPFM broadcasters will eventually lose interest and funding and abandon their licenses, many of which will be taken over by larger NPR broadcasters and well-funded networks and religious organizations.

MICHAEL H. ADAMS AND STEVEN PHIPPS

See also British Pirate Radio; College Radio; Community Radio; FM Radio; Licensing; Localism in Radio; Ten-Watt Stations

Further Reading

Coopman, Ted M., "FCC Enforcement Difficulties with Unlicensed Micro Radio," *Journal of Broadcasting and Electronic Media* 43, no. 4 (Fall 1999)

The Free Radio Network: Pirate Radio Top 20 Links of All Time <www.frn.net/links>

New York Free Media Alliance Microradio Documents <artcon.rutgers.edu/papertiger/nyfma/str/links.html>

Rogue Radio Research: Research and Resources on Micro Radio/Low Power FM <www.rougecom.com/rougeradio/>

Sakolsky, Ron, "Anarchy on the Airwaves: A Brief History of the Micro Radio Movement in the USA," *Social Anarchism* 17 (July 1992)

Sakolsky, Ron, and Stephen Dunifer, *Seizing the Airwaves: A Free Radio Handbook,* Edinburgh and San Francisco: AK Press, 1998

Soley, Lawrence, *Free Radio,* Denver, Colorado: Westview, 1999

Yoder, Andrew R., *Pirate Radio Stations: Tuning In to Underground Broadcasts,* New York: McGraw Hill Professional, 1990

Lum 'n' Abner

Comedy Show

A party line rings, two "backwoods" voices respond ("I-grannies, Abner, I believe that's our ring." "I-doggies, Lum, I believe you're right."), and the announcer gently brings us up to date with the latest events here in Pine Ridge, Arkansas, home of the "Jot 'Em Down Store" run by Columbus "Lum" Edwards and Abner Peabody. On *Lum 'n' Abner,* Chester Lauck (Lum) and Norris "Tuffy" Goff (Abner) performed their homespun country characters for more than 20 years, starting on a local Arkansas station and eventually airing for a time on every major radio network.

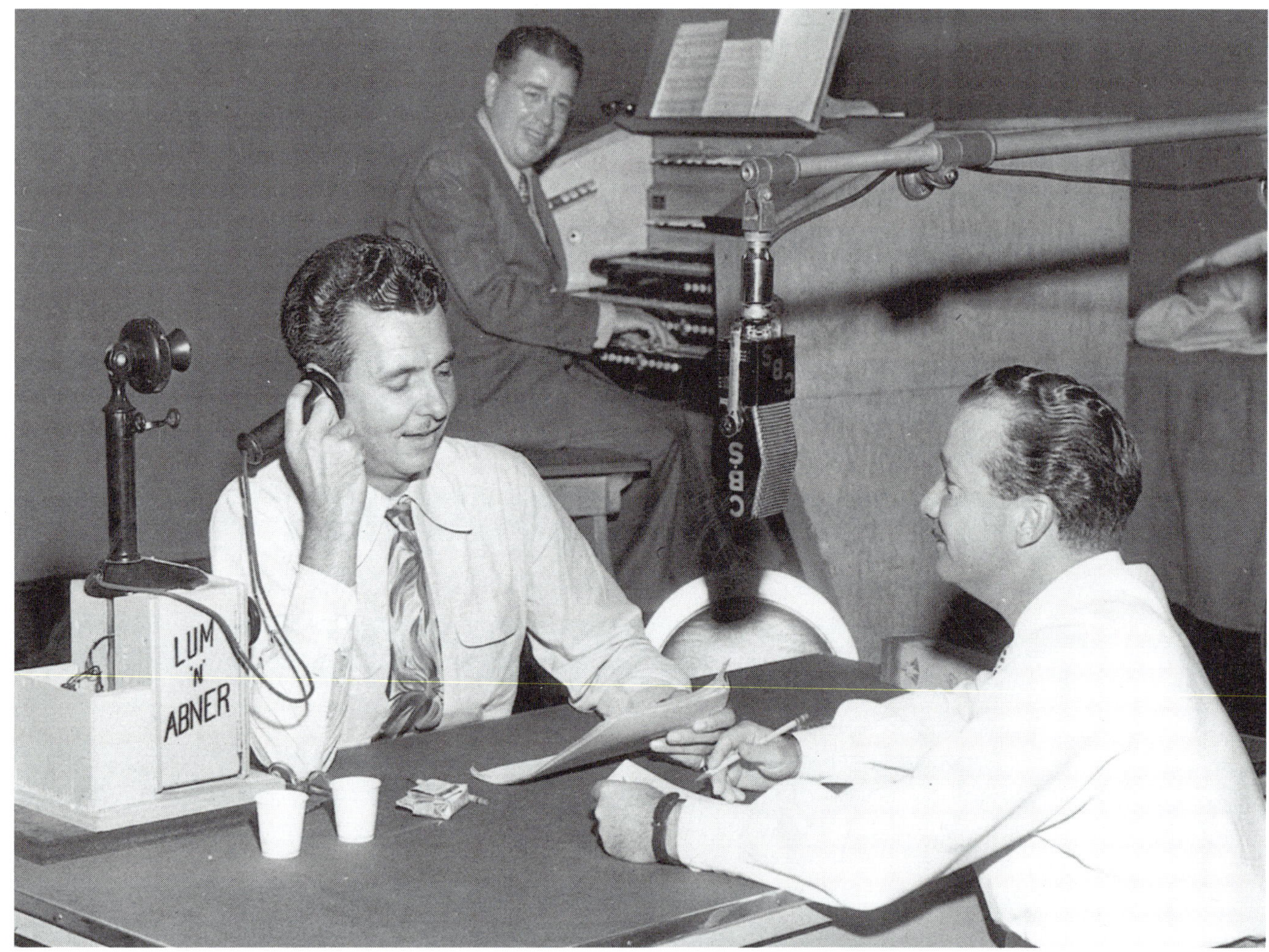

Chester Lauck and Norris Goff broadcast another episode of *Lum 'n' Abner* in 1947–48 as organist Ralph Emerson awaits his cue
Courtesy National Lum & Abner Society

Lauck and Goff met as boys in a Mena, Arkansas, grade school, and they were performing impressions and blackface comedy together by the early 1920s. In 1931 an Arkansas radio station invited locals to perform for a flood-relief benefit, and Lauck and Goff planned to do their blackface act. Seeing the station overrun with other blackface teams, they instead performed an "Ozarkian humor" routine to great listener response. They were signed to a weekly 15-minute program on KTHS in Hot Springs, Arkansas, and by late summer of that year they had both a sponsor (Quaker Oats) and a foothold on a regional National Broadcasting Company (NBC) station, Chicago's WMAQ. From then on, *Lum 'n' Abner* became an increasingly hot property; whenever Lauck and Goff were not pleased with their time slot, pay, or sponsor, another network gladly wooed them away.

The show's style was wry and folksy, deriving much of its humor from a combination of misfiring schemes and "countrified" misunderstandings of aspects of the everyday lives of their listeners, as when Lum and Abner try to do their taxes or when the town blacksmith, Cedric Weehunt (Lauck), tries to become a ventriloquist without realizing that ventriloquists are supposed to change the dummy's voice and try not to move their lips. The characters were not stooges, though (despite Abner's regular catchphrase—"Huh?"—often uttered after Lum's long and involved explanation of his latest plan): the folks of Pine Ridge always ended up on top. There was a gentle tone to the humor and a relaxed pace to the steadily rolling plot arcs that made the program a long-lived and reliable performer in the ratings.

Some plotlines were self-contained in a single episode, as when Lum and Abner struggle to do their 1942 taxes, first concluding that they owe "the givverment" 8,912 sacks of sugar, then realizing that the government actually owes them (though they decide instead to send along the extra cash in the till to help the war effort). However, more often the plots wove their way onward for weeks, as when the town's richest man, Diogenes Smith, holds a campaign to discover and reward the kindest person in town, which begins a complex train of events in which Lum eventually becomes an unwitting courier of counterfeit money.

Lauck and Goff performed the voices of most of the program's major characters. In fact, for the first six years they performed all of the characters, and those they did not perform (the female characters in particular) were only talked about in the store and never actually appeared on the show. NBC executives were continually pushing to broaden the program's scope; one 1933 memo read, "I have impressed upon the boys the necessity of more action and other characters. Their scripts from now on will have both. They are afraid of women characters, for they feel they can't write the dialogue. However, we will work hard to accomplish this." In the late 1930s female characters did begin to appear occasionally, most often voiced by Lurene Tuttle.

The program was particularly popular in its home setting, rural Arkansas. In 1936, to celebrate *Lum 'n' Abner*'s fifth anniversary on the air, the unincorporated Arkansas town of Waters officially changed its name to Pine Ridge at a ceremony attended most notably by the governor of Arkansas, Lauck, Goff, and town resident Dick Huddleston, a fictionalized version of whom was voiced by Goff on the program. To this day one can visit the "Jot 'Em Down Store" in Pine Ridge, which serves as a *Lum 'n' Abner* museum and which is where the National Lum and Abner Society holds conventions for aficionados. There is also a country store in Kentucky's Fayette County that styles itself the "Jot 'Em Down Store"; it dates back to a 1937 visit by Lauck and Goff on their way through the area to buy some horses.

The show's popularity was parlayed over the years into a series of seven feature films starring Lauck and Goff (with appearances by performers such as Zasu Pitts, Grady Sutton, Franklin Pangborn, and Barbara Hale) and produced by RKO, beginning with *Dreaming Out Loud* in 1940 and ending with *Partners in Time* in 1946. One additional film was released by "Howco Productions" much later, in 1956, an odd installment that found Lum and Abner out of the familiar territory of Pine Ridge, traveling to Paris and Monte Carlo. It was actually an edited-together version of three *Lum 'n' Abner* television pilots that the Columbia Broadcasting System (CBS) had originally produced in 1949 to no great acclaim.

In late 1948, during its second stint on CBS, *Lum 'n' Abner* changed from its 15-minute serial-comedy format to a half-hour comedy variety program, including for the first time such trappings as an orchestra and a studio audience. This was how the program seemed to end its days, going off the air in the spring of 1950. But three years later it reappeared once more in its traditional 15-minute format, first for a limited 13-week series and then for a final six-month run.

Lum 'n' Abner is one of the radio programs for which many recordings and scripts still survive. Some tapes are available from commercial sources, but there are also thriving collections circulating in the hands of ordinary fans. *Lum 'n' Abner* may not have the instant modern name recognition of programs such as *Amos 'n' Andy*, but it has made its mark—not only among radio fans, but also on the map of the United States.

DORINDA HARTMANN

Cast

Columbus "Lum" Edwards, Cedric Weehunt, Grandpappy Spears, Snake Hogan	Chester Lauck
Abner Peabody, Squire Skimp, Dick Huddleston, Mousey Gray, Doc Miller	Norris Goff
Ellie Conners, Sgt. V.W. Hartford, Nurse Lunsford	Lurene Tuttle
Diogenes Smith, B.J. Webster, Mr. Sutton	Frank Graham
Detective Wilson, Dr. Roller, Pest Controller, Mr. Talbert, FCC Man	Howard McNear
Duncan Hines, W.J. Chancellor	Francis X. Bushman
Ira Hodgekins, Caleb Weehunt	Horace Murphy
The Baby, J.W. Tiffin	Jerry Hausner
Mr. Talbert's Father	Ken Christy
Dr. Samuel Snide (dentist)	Eddie Holden
Doc Ben Withers (veterinarian)	Clarence Hartzell
Lady Brilton	Edna Best
Rowena	Isabel Randolph
Otis Bagley	Dink Trout

Announcers

Tom Nobles (1931), Charles Lyon (1931), Del Sharbutt (1931–33), Gene Hamilton (1933–34), Carlton Brickert (1934–38), Lou Crosby (1938–44), Gene Baker (1944–45), Forrest Owen (1945–48), Wynden Niles (1948–50), Bill Ewing (1953–54)

Producer

Larry Berns

Directors

Bill Gay, Robert McInnes, Forrest Owen

Writers

Betty Boyle, Norris Goff, Chester Lauck, Roz Rogers, Jay Sommers, Howard Snyder, Hugh Wedlock, Jr.

Programming History

26 April 1931–7 May 1954 (many changes in network and time slots)

Further Reading

"Amid the Native Corn," *Newsweek* (6 October 1947)
Crawford, Byron, "Setting Store by a Name,"*Courier-Journal* (Louisville, Kentucky, 24 May 1991)
Lum 'n' Abner: Frigidaire Announces Sponsorship," *New York Times* (18 August 1948)
"*Lum 'n' Abner* to End," *New York Times* (30 June 1948)
Salomonson, Terry G.G., *The Lum 'n' Abner Program: A Radio Broadcast Log of the Comedy Drama Program,* Howell, Michigan: Salomonson, 1997

Lux Radio Theater

Anthology Drama

One of the most popular and prestigious radio programs for two decades, *Lux Radio Theater* was a dramatic anthology that mainly presented movie adaptations with big-name Hollywood stars.

Lever Brothers had been using celebrities to endorse its Lux Toilet Soap in magazine ads throughout the 1920s, and in 1934 the J. Walter Thompson Advertising Agency proposed to Lever Brothers an extension of this promotional tactic, the sponsorship of a radio drama presenting stars of the stage and screen. The resulting program, *Lux Radio Theater,* aired on the National Broadcasting Company (NBC) and originated from downtown Manhattan, premiering on 14 October 1934. Given its New York locale, the program presented mostly Broadway talent and properties in this early period. During the first season, the show's host was a fictitious character named Douglass Garrick, played by John Anthony and billed as the show's producer. Peggy Winthrop, another fictional character played by Doris Dagmar, supplied commercials. In addition to commercials, the show's framework included a scripted chat session between Garrick and each particular episode's stars.

By the end of its first season, *Lux Radio Theater* was a critical success, but given their sizable investment in the show, Lever Brothers had hoped for higher listener ratings. When the Columbia Broadcasting System (CBS) offered them the advantageous 9 P.M. Monday night slot, *Lux* made the switch to that network and time. The second season brought a few other changes: a new actor, Albert Hayes, played Douglass Garrick; Art Millett was added as announcer; and the Peggy Winthrop character was eliminated.

By the end of the 1935–36 season, ratings still were not where Lever Brothers wanted them. Under the assumption that film stars would draw larger national audiences than Broadway stars, in 1936 *Lux* moved to Hollywood, specifically to the Music Box Theater on Hollywood Boulevard. Producers also hoped to bring in a famous Hollywood name to replace the Douglass Garrick figure as host. They settled on famed movie director Cecil B. DeMille, who, although he was assigned the title of producer, was simply a host figure and often appeared only for dress rehearsals and the actual recordings. It was hoped that DeMille's famed persona as creative tour de force would help lend a prestigious, glamorous image to the show. Additionally, by specifically presenting him as a producer, essentially an authorial voice, *Lux Radio Theater* posed the show as a first-rate cultural experience on par with DeMille's epic film productions. Equally important, framing DeMille as a creative force helped to elide the fact that the show was really created by an ad agency concerned mainly with advertising revenue, rather than solely a culturally beneficial endeavor.

As a result of these changes, the 1936–37 season brought a plethora of stars in movie adaptations, such as Errol Flynn and Olivia de Havilland in "Captain Blood," Irene Dunne and Robert Taylor in "Magnificent Obsession," and Fredric March and Jean Arthur in "The Plainsman." Cooperation between J. Walter Thompson and the Hollywood studios made this possible. The studios would offer the broadcast rights to film properties for a fee (usually no more than $1,000), though sometimes the rights would be offered for free in exchange for publicity on the air. The show's most substantial expense came from talent costs. DeMille made $2,000 per show in this period, and each headlining star received $5,000. (The average yearly salary in the U.S. in 1937 was $1,327.00) Given that a minimum of two headliners per episode was the general rule, talent costs could reach as high as $20,000 per episode.

In 1940 *Lux Radio Theater* changed recording facilities to the Vine Street Playhouse. This venue offered a more spacious

Lux Radio Theater, Spencer Tracy, Fay Wray, and Cecil B. DeMille
Courtesy CBS Photo Archive

stage, and the actors would thus all stand on the stage throughout the whole program, even if their parts were completed in the first act. DeMille sat off to the side at a card table, and an offstage mike provided sound effects and some commercials. The New York version of the show had not been presented in front of an audience, and a change in this policy upon the move to Hollywood resulted in a furious weekly demand for tickets.

The next period of upheaval for *Lux Radio Theater* came in 1944, when DeMille left the show. His departure stemmed from a dispute with the American Federation of Radio Artists (AFRA). A proposition on that year's state election ballot would have allowed a Californian the right to obtain employment without first gaining union affiliation. AFRA strongly opposed this measure and charged all members one dollar to fund a battle against its passage. The right-wing DeMille resolutely objected both to AFRA's stance and to their demand that members fund it, and he refused to pay his dollar. This resulted in a suspension of his AFRA membership, meaning he was also barred from any radio work. *Lux* first turned to guest hosts in the interim, including Brian Aherne and Lionel Barrymore, but when it became clear that neither AFRA nor DeMille would relent, the producers settled on a permanent host in William Keighley, a lesser-known Hollywood director. Keighley took on DeMille's role as host and faux-producer; however, he could never match the famed director's prestigious presence and vaunted image.

Keighley retired in 1952 and was replaced by Irving Cummings, a fellow Hollywood director. However, *Lux Radio Theater's* ratings began a precipitous slide, particularly into 1954, as Lever Brothers and J. Walter Thompson were giving more attention and money to the television version of the program, *Lux Video Theater* (1950–57). In 1954, NBC reclaimed *Lux Radio Theater* and tried to resurrect the show's stately image with a marketing campaign and presentation of "twenty of the greatest Hollywood pictures" during the 1954–55 season. However, this did little to stem the show's decline, and the show aired its final broadcast on 7 June 1955. The television version carried on until 1957, but it never reached the popularity of the radio version, particularly because it could not offer the caliber of stars the radio show had. Only the radio version could tout a history of 926 episodes starring the most famous talent of the era.

CHRISTINE BECKER

Cast

Host "Douglass Garrick"	John Anthony (1934–35)
Other Hosts	Cecil B. DeMille (1936–45), William Keighley (1945–52), Irving Cummings (1952–55), Lionel Barrymore (1945), Walter Huston (1945), Mark Hellinger (1945), Brian Aherne (1945), Irving Pichel (1945)
Announcers	Melville Ruick (1936–40), John Milton Kennedy (1940s), Ken Carpenter (later years)

Directors

Antony Stanford (1934–36), Frank Woodruff (1943), Fred MacKaye (1944–51), Earl Ebi (1951–55)

Programming History

NBC Blue	October 1934–June 1935
CBS	July 1935–June 1954
NBC	September 1954–June 1955

Further Reading

Billips, Connie J., and Arthur Pierce, *Lux Presents Hollywood: A Show-by-Show History of the Lux Radio Theater and the Lux Video Theater, 1934–1957,* Jefferson, North Carolina: McFarland, 1995

DeMille, Cecil B., *The Autobiography of Cecil B. DeMille,* edited by Donald Hayne, Englewood Cliffs, New Jersey: Prentice-Hall, 1959; London: Allen, 1960

Hilmes, Michele, *Radio Voices: American Broadcasting, 1922–1952,* Minneapolis: University of Minnesota Press, 1997

Jewell, Richard B., "Hollywood and Radio: Competition and Partnership in the 1930s," *Historical Journal of Film, Radio, and Television* 4, no. 2 (1984)

M

Magazines. *See* Fan Magazines; Trade Press

Make Believe Ballroom. *See* Block, Martin

Mankiewicz, Frank 1924–

U.S. Broadcast Executive, President of National Public Radio

Frank Mankiewicz, the third president of National Public Radio (NPR), led the organization from obscurity to national significance. Mankiewicz took over leadership of the seven-year-old NPR in August 1977.

Public radio, as represented by NPR and the Association of Public Radio Stations, had just emerged from two years of internal strife by merging to create a "new," more powerful National Public Radio to provide political, promotional, technical, and programming leadership. The fresh start required a new president, someone who was "somebody," preferably a politician, a journalist, a showman, a promoter, and a celebrity whose phone calls would be returned. The new NPR board found that individual in Frank Mankiewicz, scion of the Hollywood family, press secretary to the late Senator Robert Kennedy, director of George McGovern's 1972 campaign for president, and a syndicated columnist and author. Along the way, Mankiewicz had spent six years as a Hollywood lawyer, had directed Peace Corps operations in Latin America, and had run unsuccessfully for the California state legislature and the U.S. Congress from Maryland. His name was also to be found on the enemies list of President Richard Nixon, about whom Mankiewicz had written two books.

Mankiewicz did not disappoint. He transformed public radio in four critical areas: publicity, programming, politics, and satellite delivery.

Publicity

The news media finally discovered NPR after having virtually ignored it for its first seven years. Much of the publicity centered on Mankiewicz as a personality. He was good copy, always ready with the quick quip and the memorable sound bite. More fundamentally, his programming and political exploits gave the media something interesting to write about.

Programming

Mankiewicz persuaded President Jimmy Carter to participate in a two-hour national call-in program from the White House on a Saturday afternoon, 13 October 1979. For those two

Frank Mankiewicz
Courtesy AP/Wide World Photos

hours, all media focused on the previously unknown NPR. In addition to publicity, the president's exclusive appearance gave the obscure network much-needed credibility. Similarly, Mankiewicz persuaded the Senate leadership to allow NPR to broadcast live its debate on the Panama Canal Treaty. Although the Senate had always excluded cameras and microphones from its chambers, Mankiewicz broke that barrier, creating what he described to the press as a historic broadcast of a historic event.

In entertainment, Mankiewicz reached back to his boyhood, when his father, Herman Mankiewicz, teamed with Orson Welles on the *Mercury Radio Theater* and its most famous production, "War of the Worlds." NPR would collaborate with George Lucas on a radio adaptation of the *Star Wars* film trilogy. The series brought new attention and new listeners to NPR.

Of most lasting consequence, however, was Mankiewicz's decision to move NPR news into radio's prime time, morning drive. *Morning Edition*'s debut in November 1979 more than doubled NPR's news output, its budget, and its staff, and it focused those resources where they could reach the most potential listeners. *Morning Edition* indirectly transformed both NPR and the programming on its member stations. NPR became a competitive 24-hour news organization for the first time, changing its public image and its internal psychology from that of an ancillary alternative service to that of a primary source of breaking news. To make room for *Morning Edition* in prime morning hours, stations were forced to rethink and refocus their entire approach to programming. For the most part, they eliminated their weakest program elements and concentrated on their strongest. A combination of these changes doubled public radio's cumulative audience from 4 million to 8 million a week between 1979 and 1982.

Politics

After almost a decade of struggle between public radio and public television over the division of federal funds between the two media, Mankiewicz boldly persuaded Congress to earmark 25 percent of its funding for public radio. Public television had argued that radio should receive only 10 percent; prior to Mankiewicz, public radio had felt victorious when it won 17 percent. The extra money made possible *Morning Edition* and other dramatic improvements in the public radio service.

Satellite Delivery

Congress also agreed in 1979 to fund a satellite delivery system for public radio, the first radio network to use that technology for network distribution of programs. The satellite not only improved the technical quality of NPR programming but also provided the flexibility to produce programming from a variety of locations around the country, making possible such important innovations as the live broadcasts of Garrison Keillor's *A Prairie Home Companion* from Minnesota each Saturday night.

Despite all his success, the Mankiewicz presidency did not end happily. The election of Ronald Reagan in 1980 resulted in a 25 percent reduction in federal funding for public broadcasting in 1983. With typical brashness, Mankiewicz proposed to meet that challenge not by reducing service but by expanding it. He planned to raise the necessary money from the private sector, through increased "underwriting" of programming and through business ventures that would utilize NPR's satellite capacity to provide a variety of commercial services to the public. The concepts had merit, but Mankiewicz did not have the time or the money to realize them. None would produce significant revenue for several years, and all required significant investments right away.

By early 1983 NPR found itself running a $3 million deficit, which increased to $6 million and threatened to reach $9 million if drastic action were not taken. NPR was insolvent. Ultimately, member stations and the Corporation for Public Broadcasting bailed NPR out of its difficulties, but a condition of that bailout was the firing of Frank Mankiewicz in April 1983. Although he had hoped to leave NPR to become commissioner of baseball, Frank Mankiewicz instead formed an alliance with Republican political operative Robert Gray to spend the rest of his career in one of Washington's most influential lobbying organizations.

JACK MITCHELL

See also National Public Radio

Frank Mankiewicz. Born in New York City, 16 May 1924. Son of Herman J. and Sara (Aaronson) Mankiewicz; attended University of California, Los Angeles, A.B., 1947, Columbia University, M.S., 1948, and University of California, Berkeley, LL.B, 1955; served in U.S. Army Infantry, 1943–46; journalist, Washington, D.C. and Los Angeles, 1948–1952; admitted to Bar of California, 1955; practiced law in Beverly Hills, 1955–61; director, U.S. Peace Corps, Lima, Peru, 1962–64, and Latin America regional director, 1964–66; press secretary for Senator Robert F. Kennedy, 1966–68; syndicated columnist with Tom Braden, 1968–71; national presidential campaign director for Senator George Mc Govern, 1971–72; author and journalist, 1973–77; president, National Public Radio, 1977–83; public relations executive and consultant, Hill and Knowlton, 1983–present.

Selected Publications

Perfectly Clear, Nixon from Whittier to Watergate, 1973
U.S. v Richard M. Nixon: The Final Crisis, 1975
With Fidel: A Portrait of Castro and Cuba (editor, with Kirby Jones), 1975
Remote Control: Television and the Manipulation of American Life (with Joel Swerdlow), 1978

Further Reading

Collins, Mary, *National Public Radio: The Cast of Characters*, Washington, D.C.: Seven Locks Press, 1993
Engelman, Ralph, *Public Radio and Television in America: A Political History*, Thousand Oaks, California: Sage, 1996
Looker, Thomas, *The Sound and the Story: NPR and the Art of Radio*, Boston: Houghton Mifflin, 1995
McDougal, Dennis, "The Public Radio Wars," *Los Angeles Times* (8–11 October 1985) (four-part series)
Zuckerman, Lawrence, "Has Success Spoiled NPR?" *Mother Jones* 12, no. 5 (June–July 1987)

Ma Perkins

Serial Drama/Soap Opera

A widow whose homespun wisdom guided her family, friends, and neighbors in the fictitious rural community of Rushville Center, *Ma Perkins* was also one of radio's most enduring soap operas, captivating American and, at times, overseas audiences for 27 years, from 1933 to 1960. Throughout its entire run of 7,065 episodes, Virginia Payne portrayed the leading character.

A 15-minute daytime serial drama, *Ma Perkins* premiered on NBC radio on 4 December 1933, three months after its local debut on WLW in Cincinnati, Ohio. Cincinnati-based Procter and Gamble Company, makers of Oxydol soap flakes, initially sponsored the broadcast. The drama was so popular it was picked up by competing networks, CBS and Mutual, at various times, while continuing its broadcasts on NBC as well. CBS acquired exclusive rights to the show in 1949, where it remained until its final broadcast in 1960.

Ma had three children, John, Fay and Evey, whose husband Willie Fitz (portrayed by Murray Forbes, also for all 27 years of the broadcast) managed Ma's lumber yard business. Shuffle Shober, Ma's business partner, was portrayed by Charles Egleston for 25 years, until his death in 1958. Edwin Wolfe succeeded him for the remaining run of the show. Writer Orin Tovrov wrote the scripts for more than 20 years.

Ma Perkins' roots can be traced not only to Cincinnati, but also to Chicago, where advertising executives Frank and Anne Hummert created the program. It was in Chicago that the Hummerts also originated other popular daytime drama serials in the early 1930s, including *The Romance of Helen Trent* and *Just Plain Bill.* Though each of the serials offered different characters with different story lines, they were all products of the Hummerts' desire to present familiar themes that had proven popular with Depression-era listeners seeking some measure of assurance and security in an unstable world. According to Marilyn Matelski, the Hummert approach was a simple formula. "[It] combined fantasies of exotic romance, pathos and suspense with a familiar environment of everyday life in a small-town or rural setting. Combined with an identifiable hero or heroine, this formula produced an overwhelming audience response" (see Matelski, 1988).

The story lines for *Ma Perkins* often reflected the turbulence of the times. In one broadcast during World War II, for example, Ma learns that her son John was killed in combat in Europe. The "news" prompted a flood of letters from devoted listeners expressing their sympathy. Throughout its run, *Ma Perkins* conveyed the vulnerability of life in ways not depicted before on radio. It did so while reaffirming the belief that people could solve their problems so long as they believed in one another. In 1938, Ma offered this observation: "Anyone of this earth who's done wrong, and then goes so far as to right that wrong, I can tell you that they're well on their way to erasing the harm they did in the eyes of anyone decent."

Virginia Payne was only 23 years old when the serial drama premiered, and, remarkably, she never missed a performance in its 27-year run. A Cincinnati native and graduate of the University of Cincinnati, she faithfully dressed the part at countless public appearances and even personally answered many of her listeners' letters. Her down-home language included such expressions as "I ain't sure I understand it" and "Land sakes!"

U.S. audience interest in radio soap operas declined during the 1950s because of the growth of television, and *Ma Perkins* was no exception. By 1960 CBS radio cut three of its ten serials, and NBC dropped its last surviving soap opera; ABC had ended all of its daytime serials the year before. The sponsors that owned the programs abandoned radio in favor of television because of its increasing audience appeal. The transition to television created a vacuum in radio advertising sales, and by 1960 only one quarter of network radio advertising time was being sold. Local radio, meanwhile, was blossoming: disc jockeys replaced radio network programs in large numbers by the end of the 1950s, and radio station managers discovered that the locally hosted music program proved cheaper for advertisers and more popular to listeners as well. By the 1955–56 season, *Ma Perkins'* radio audience share had fallen to one quarter of its all-time high in 1944–45. On 25 November 1960, the show's final broadcast featured the family at its traditional Thanksgiving meal:

"I look around the table at my loved ones and to me the table stretches on and on. Over beyond the other end past Shuffle I see faces somehow familiar and yet unborn, except in the mind of God... Someday, Fay will be sitting here where I'm sitting, or Evey, or Paulette... They'll move up into my place and I'll be gone, but I find right and peace in that knowledge... I give thanks that I've been given this gift of life, this gift of time to play my little part in it," Ma said at the table.

Virginia Payne was only 50 years old when *Ma Perkins* ended. During the 1960s and 1970s she remained active in show business, appearing in radio commercials and touring in such productions as *Life With Father, Becket,* and *Oklahoma!* Shortly before her death in Cincinnati on 10 February 1977, she appeared on radio one last time on *The CBS Radio Mystery Theater.*

DAVID MCCARTNEY

See also Hummert, Frank and Anne; Soap Opera

Cast

Ma Perkins	Virginia Payne
Fay Perkins Henderson	Rita Ascot; Marjorie Hannan; Cheer Brentson; Laurette Fillbrandt; Margaret Draper
Evey Perkins Fitz	Dora Johnson; Laurette Fillbrandt; Kay Campbell
John Perkins	Gilbert Faust
Shuffle Shober	Charles Egleston (1933–58); Edwin Wolfe (1958–60)
Willie Fitz	Murray Forbes
Junior Fitz	Cecil Roy; Arthur Young; Bobby Ellis
Paulette Henderson	Nannette Sargent; Judith Lockser
Augustus Pendleton	Maurice Copeland
Mathilda Pendleton	Beverly Younger
Gladys Pendleton	Patricia Dunlap; Helen Lewis
Paul Henderson	Jonathan Holoe
Gregory Ivanoff	McKay Morris
Gary Curtis	Rye Billsbury
Charley Brown	Ray Suber
Tom Wells	John Larkin; Casey Allen

Announcers
Bob Brown, Jack Brinkley, Dick Wells, Marvin Miller, Dan Donaldson

Producers/Creators
Frank and Anne Hummert; Robert Hardy Andrews

Writers
Robert Hardy Andrews, Lee Gebhart, Lester Huntley, Natalie Johnson, and Orin Tovrov

Programming History

WLW, Cincinnati, Ohio	August 1933–December 1933
NBC	1933–49
NBC Blue	February 1937–December 1937; June 1938–November 1938
CBS	January 1938–May 1938; September 1942–November 1960
Mutual	1935–36

Further Reading

Allen, Robert Clyde, *Speaking of Soap Operas,* Chapel Hill: University of North Carolina Press, 1985

Cox, Jim, *The Great Radio Soap Operas,* Jefferson, North Carolina: McFarland, 1999

Edmondson, Madeleine, and David Rounds, *The Soaps: Daytime Serials of Radio and TV,* New York: Stein and Day, 1973

LaGuardia, Robert, *From Ma Perkins to Mary Hartman: The Illustrated History of Soap Operas,* New York: Ballantine Books, 1977

Matelski, Marilyn J., *The Soap Opera Evolution: America's Enduring Romance with Daytime Drama,* Jefferson, North Carolina: McFarland, 1988

Stedman, Raymond William, *The Serials: Suspense and Drama by Installment,* Norman: University of Oklahoma Press, 1971; 2nd edition, 1977

Stumpf, Charles K., *Ma Perkins, Little Orphan Annie, and Heigh-Ho Silver,* New York: Carlton Press, 1971

The March of Time

Network Docudrama Series

The March of Time, a radio forerunner of today's television "docudramas," was a widely heard and imitated news dramatization program—a "radio newsreel"—that lasted 14 seasons on network radio (1931–45) and led to a famous motion picture documentary series of the same name. *The March of Time* is best remembered by the words of the title spoken by the mellifluous narrator Westbrook Van Voorhis, who also spoke on the newsreel version and ended both by saying "Time marches on."

Origins

The idea for a digest of the news for radio broadcast originated with Fred Smith, the first station director at WLW in Cincinnati. Smith wrote and directed dramas for WLW and introduced many program ideas at the station. In 1925 he hit upon the novel idea of reading various items taken (without permission) from newspapers and magazines. Smith called the program *Musical News;* after each story, a brief musical piece was played by the staff organist.

About the same time, a similar idea was put into print in a new kind of weekly magazine. Henry R. Luce and Briton Hadden, friends since prep school and Yale, quit their jobs at the *Baltimore News* in 1922 to found a magazine that summarized the week's news—an idea they had discussed since their days in boot camp during World War I. The magazine was *Time,* and its first issue appeared 3 March 1923. (Luce ran *Time* alone after the death of Briton Hadden in 1929.)

In 1928, Fred Smith at WLW got permission from *Time* to use an advance, or "makeready," copy of the magazine sent to him by airmail (then just begun) so that he could rewrite items for a weekly news summary on WLW. Soon Smith was hired by Time Incorporated and traveling the Midwest, signing up stations for a daily news summary the magazine would syndicate to radio stations. Beginning on 3 September 1928, 10-minute scripts were delivered by airmail to be read by local announcers on more than 60 stations. In New York, WOR called the program *NewsCasting* and broadcast it from 5:50 P.M. to 6:00 P.M., Monday through Friday. Smith himself was WOR's reader for the first year.

The radio program's title was apparently the first use of the word *newscast,* and it was listed in the *New York Times* radio logs by that title. While at WLW, Smith had coined *radarios,* after *radio* and *scenario,* for original radio plays he wrote and produced. (*Time* was even better known for its neologisms, coining *cinemaddict, newsmagazine,* and *socialite,* among others.) By the spring of 1929, the ten-minute summaries were being carried on as many as 90 stations (up from 60 stations just a half-year earlier)—the first large-scale regular daily news broadcast carried in the United States—but it was never a true network program because each station developed its own script.

Fred Smith next conceived of dramatizing the news. In September 1929 he produced a five-minute "news drama" in cooperation with *Time* and submitted his audition program to a number of stations with the title *NewsActing.* Although *newsacting* did not become a household word, the program idea caught on. By December 1929 Smith and a crew of six to eight actors were producing a weekly five-minute drama for distribution by electrical transcription. These programs were not full-scale dramatic productions, but they included sound effects, occasional music, and the portrayal of the voices of actual people involved in the news stories. Within a few months the *NewsActing* records were being broadcast over more than 100 stations nationwide. At the time, other network programs were dramatizing history, but none had tried a weekly presentation of current news.

Henry Luce wanted to advertise *Time* on the radio networks, and on 6 February 1931, an experimental program was sent via telephone wires to the home of a Time Inc. executive where a small group was gathered that included CBS president William S. Paley. Exactly a month later, on Friday, 6 March 1931, *The March of Time* was fed from CBS's New York studios and carried on 20 of the network's affiliates (there were then about 80) at 10:30 P.M. EST. (The program's title was taken from a Broadway show tune of the same name.) After a five-second fanfare, the announcer said:

> The March of Time. On a thousand fronts the events of the world move swiftly forward. Tonight the editors of *Time,* the weekly newsmagazine, attempt a new kind of reporting of the news, the re-enacting as clearly and dramatically as the medium of radio will permit some themes from the news of the week.

Network Years

This first program dramatized the reelection of William "Big Bill" Thompson as mayor of Chicago, the sudden death of the *New York World* by merging with the *New York Telegram,* and shorter segments on French prisoners sent to Devil's Island, revolution in Spain, prison reform in Romania, a roundup of news of royalty, an auction of Czarist possessions in New York, and the closing of the 71st United States Congress.

During the first season the program ran 13 weeks. It returned on 8 September 1932, but as a sustaining feature because *Time* executives had decided that they could not afford advertising, which they said the magazine no longer needed. In the magazine the editors justified the radio show's cancellation by asking, "should a few (400,000 *Time* subscribers) pay for the entertainment of many (9,000,000 radiowners)?" In the 29 February 1932 issue, *Time* also argued, "For all its blatant claims to being a medium of education, radio contributes little of its own beyond the considerable service of bringing good music to millions." In November 1932 the magazine resumed its sponsorship. *The March of Time* did its part to promote magazine sales: at the end of each program, listeners were reminded that they could find more details in the issue of *Time* magazine soon to be on newsstands.

On 1 February 1935, the *March of Time* newsreel began as a monthly film series in movie theaters. It began as a typical newsreel with several items in each issue, but the January 1938 issue focused on a single topic, "Inside Nazi Germany." After October 1938, single subjects were being treated exclusively as the series became a documentary rather than strictly a news series. The documentary series ran until 1951. Time Inc. also produced historical television documentary series, such as *Crusade in Europe,* based on the book by General Dwight D. Eisenhower (1948). The title *March of Time* was also used for a syndicated series of television documentaries produced by David Wolper.

Broadcast of *The March of Time*
Courtesy CBS Photo Archive

During the 1933–34 season, *March of Time* was sponsored by Remington-Rand, and Westbrook Van Voorhis became the voice of the program. In 1935 there was a variety of sponsors when a daily 15-minute version was tried for one season. In 1938 the program was sponsored by Time Inc.'s *Life* magazine (purchased by Luce in 1936), and the announcer worked that title in to the opening of the program: "Life! The life of the world, its conflicts and achievements, its news and fun, its leaders and its common people." The program was not aired during the 1939–40 and 1940–41 seasons. After seven years on CBS, it moved to the NBC Blue network. In July 1942 the format was changed to only one or two dramatized segments and many more live, on-the-spot news reports. By the 1944–45 season, *March of Time*'s last, listeners were hearing the actual voices of newsmakers on many network programs.

At the century's end much was being made of the "synergy" of cross-media ties, the idea being that the interaction of a media corporation's units (say, a magazine division feeding story ideas to a film company) would encourage the making of products and profits greater than the sum of the corporation's parts could make by acting alone. *The March of Time* was one of the first examples of synergy. Fred Smith's idea of reading a few news items accompanied by musical selections that became *Musical News, NewsCasting, NewsActing,* and *March of Time* led to the many "dramatized news" programs—documentaries and docudramas—on television today.

LAWRENCE W. LICHTY

See also Documentary Programs; News

Narrators
Ted Husing, Westbrook Van Voorhis, Harry Von Zell

Programming History

CBS 1931–39
NBC Blue 1941–45

Further Reading

Bohn, Thomas W., and Lawrence W. Lichty, "The March of Time: News As Drama," *Journal of Popular Film* 2, no. 4 (Fall 1973)

Fielding, Raymond, *The American Newsreel, 1911–1967*, Norman: University of Oklahoma Press, 1972

Fielding, Raymond, *The March of Time, 1935–1951*, New York: Oxford University Press, 1978

Lichty, Lawrence W., and Thomas W. Bohn, "Radio's March of Time: Dramatized News," *Journalism Quarterly* 51, no. 3 (Autumn 1974)

Marconi, Guglielmo 1874–1937

Italian Wireless Inventor

Marconi's daughter Degna Marconi published an account of the prophetic revelation given at the future inventor's birth: as the household servants joined family members in crowding into the birth room, a servant tactlessly commented, "What big ears he has!" Guglielmo's mother, Annie, although spent from a difficult labor, is said to have defended her newborn by saying, "He will be able to hear the still, small voice of the air." Little did she know that 21 years later, he would be sending messages across distances of more than a mile with no wires, but only air between the transmitter and receiver.

Youth

Marconi's early years afforded little formal education. He was born in the Italian countryside to an Italian father and an Irish mother. Although he attended an occasional school, he rarely got along well in such settings. Most of his education was gained either under his mother's tutelage, lessons that included readings from the Bible and English language instruction, or from books in his father's library. In the library, he was introduced not only to Greek mythology and history, but also to heroes of electricity such as Benjamin Franklin and Michael Faraday.

Marconi's father was impatient with his son's excessive interest in electricity and his constant tinkering with mechanical apparatuses. While constructing experiments based on Franklin's writings about electricity, the young Marconi succeeded in getting high-voltage electricity to vibrate a series of dinner plates to the point of destruction. Thereafter, his father began to systematically ruin Guglielmo's experimental equipment whenever he found it. It was his mother who worked to nurture her son's interest in exploring science. She helped him protect and hide the scientific devices he built.

The First Wireless Transmissions

At about the age of 20 Marconi was given the attic in the family's house as a private workspace where he could dabble in his hobby. He spent many secluded hours there working on experiments inspired by an article about the electromagnetic wave experiments of Heinrich Hertz. Marconi realized the possibilities for wireless communication if he could improve on Hertz's work. Although Hertz had showed that an alternating electric current in a closed electrical system could cause sparks to result in a totally separate loop of wire a few feet away, he had not conceived of the communication possibilities. However, others were making a communication connection. Even before Marconi had read about Hertz's experiment, others were building more powerful electromagnetic demonstrations and showing that the waves traveled for many feet without wires. None of the other experimenters, however, actually created anything practical in a communication sense.

Marconi began by duplicating Hertz's experiments in his attic. One night in 1894 he woke his mother and led her to the attic. There he tapped a telegraph key on one side of the attic. From the far end of the attic came the sound of a bell ringing every time the telegraph key was struck. While Marconi was said to lack the theoretical capacity of others experimenting with electromagnetic waves, his persistence and his ability to make his mechanical apparatuses do what he wanted them to do made him successful where others were not.

A neighbor, Professor Augusto Righi of the University of Bologna, provided informal tutelage on electromagnetic waves and gave suggestions for improving Marconi's early experiments. But Righi was also one of those who discouraged Marconi, suggesting that he would not succeed in making wireless communication practical. Such discouragement served to

Guglielmo Marconi
Courtesy Marconi International Fellowship Foundation

solidify Marconi's resolve to find a solution. Marconi's signals soon began to reach from the attic to the lowest corner of the home's interior. By the spring of 1895, Marconi began to experiment with communicating over longer and longer distances outdoors. By summer, he could send signals several hundred yards.

Much of the ability to advance the distance at which signals could be received was based simply on increasing the electrical power used by the transmitter, a very inefficient mode of sending signals over distances. It was at this point that Marconi made one of his most important early breakthroughs. In an attempt to transmit radio waves over greater distances, he used two large sheets of iron, placing one on the ground and holding the other in the air. Suddenly the distances the waves traveled were vastly multiplied. Marconi realized that placing part of the transmitter in the ground reduced the electrical resistance, allowing the signal to travel up to a mile and a half. He had invented the grounded antenna. Soon he was sending signals more than two miles, even over and around hills and other obstacles.

Selling the Wireless Idea

Because Marconi did not have the high-level connections that a formally educated scientist would have developed, he and his mother relied on the advice of the parish priest and the family physician to decide how to pursue promoting his new invention. They chose to send a detailed document describing the invention to the Italian Minister of Post and Telegraph in early 1896. The invention, however, was rejected, and Marconi took this as a rejection by his country. (He may have fared better had he submitted the invention to the Italian Navy, which may well have recognized the invention's tremendous potential for ship-to-shore communication.) Based on that initial rejection, and probably also at the urging of his Irish mother, they decided that England, the greatest maritime power of that time, was most likely to be interested in the invention.

After applying for a patent in London, Marconi presented his invention to William Preece, an official of the British postal system. Preece arranged for Marconi to present his invention to many prominent scientists, who were impressed with his achievements. As the commercial possibilities began to be realized, Marconi was asked to sell his invention, but instead he formed his own company, the Marconi Wireless Company, and sold only a minority interest to investors. He achieved a wireless link across the 8.5 miles of the Bristol Channel in 1897 and began to install radio towers on the coast to warn ships in dangerous waters.

Marconi was also invited to demonstrate his invention to the Italian Navy, a right that he had reserved for his invention when his patent was filed. In his demonstrations and tests in Italy, he first proved that his signals could pass beyond the horizon through, it would seem, a wall of ocean. This finding would prove crucial in forecasting his eventual transatlantic communication.

Marconi was not only an inventor, but also a shrewd businessman. When asked to set up a transmitting station on a ship off the coast of England to transmit reports on a yacht race, he was quick to take up the challenge. Although this move did little to advance his scientific knowledge, he knew it would provide immense publicity for his invention. Queen Victoria picked up on the idea of ship-to-shore communication and adopted its use for the royal family between Osborne House on the Isle of Wight and the royal yacht.

In 1899 Marconi established a wireless link across the 32 miles of the English Channel. But simply conquering greater and greater distances became somewhat less important for Marconi, who had realized that the nature of wireless communication allowed only one transmitter at a time to operate in an area without interference. If two stations operated simultaneously, they would cancel each other out. Therefore, Marconi began to fix his attention on both focusing the direction of the signal and tuning the signal. Shortly after his Channel link was established, he wrote to Preece that he had made some advances in tuning the radio signal. He received a patent for this tuning invention a few years later.

While Marconi was in the process of experimenting with tuning in 1899, he took his first trip to the United States at the invitation of American businessmen who encouraged him to set up wireless systems for the America's Cup yachting races. In addition to these publicity-oriented transmissions, Marconi also met with the U.S. Navy and Army for demonstrations.

During these tests, Marconi experimented with the tuning components of his system but refused to let navy personnel in on the full details, because he had not yet applied for a patent. The navy, although only mildly enthusiastic, ordered several ships to be outfitted with wireless devices, but a deal with the Marconi company later fell through because of the high cost—ten cents per word—for using the devices. Shortly after his departure after his two months in America, an associate filed papers to charter the American Marconi Company.

On the return voyage from the United States, Marconi established a couple of other firsts in wireless transmission. The *St. Paul,* on which he sailed, was outfitted with a wireless to contact the English coast upon their approach. From 66 nautical miles away, the wireless connection was made. This allowed enough time for news of the South African War to be telegraphed to the ship, copies of the *Transatlantic Times* printed with the news, and the copies sold to passengers so that they could catch up on the news hours before they reached port.

Crossing the Atlantic

Marconi continued to focus for the next couple of years on developing his tuning device, but he was also aiming toward his next major target, transatlantic communication. Many great minds in the scientific community had determined that transatlantic wireless would be impossible because of the inability of radio signals to follow the curvature of the Earth. Marconi had of course already demonstrated to himself that he was sending signals that would bend around the Earth over bodies of water. To make this transatlantic leap, he first built a transmitting station at Poldhu, on the coast of southwestern England. He then tried to build a receiver on Cape Cod, Massachusetts, in early 1901, but he was beset by storms that destroyed his stations.

He then took a shortcut to prove his theory correct. He sailed to Newfoundland and, using a kite, hoisted a wire 400 feet in the air. At 12:30 P.M. on 12 December 1901, Marconi and one assistant heard the three distinct clicks of the letter *S* telegraphed according to a preset schedule from Poldhu to St. Johns, Newfoundland, a distance of more than 2,100 miles. The transmission was received repeatedly that afternoon. Two days later, Marconi released his results to the press, results that were heralded by some and belittled by others (critics credited static electricity as the cause of the signals).

Though Marconi initially planned to build a permanent receiving station in St. Johns, threats from transatlantic cable companies to institute legal action persuaded him to move south to Nova Scotia. There, the Canadian government worked to help him build a permanent receiving station. The funding they provided also helped finish the facility being completed on Cape Cod. Still, there were many unknowns in the technical aspects of transatlantic signals. In 1902, on a trip from England back to North America, Marconi discovered that signals traveled better at night than during a clear day. This was an important, if misunderstood, discovery. Try as he might, Marconi could not overcome the daytime impediment to his long-distance transmissions for several years, because he did not understand the impact of various frequencies on the propagation of electromagnetic waves.

Later Life

The Marconi companies faced considerable financial pressures through these years because the promise of transatlantic wireless seemed uncertain owing to problems with dependability in the system. To make these transmissions commercially viable, they had to be available on demand at any time. It was 1907 before commercial transmissions began in earnest between Glace Bay in Nova Scotia and a new station in Clifden on the west coast of Ireland. The Marconi company continued to concentrate on wireless telegraph service, with all ships required to have telegraph service by 1911.

All his work on creating a practical wireless method of international communication earned Marconi the Nobel Prize for physics in 1909. When the liner *Republic* went down after a midocean collision that same year, all passengers were saved thanks to wireless distress calls. The *Titanic* similarly used its wireless in 1912 to summon help that saved one-third of the passengers. Marconi sought further developments, including a portable wireless that could be used from lifeboats at sea, automatic warning alarms for untended wireless equipment, and triangulation techniques that would pinpoint a ship's position more accurately.

The early 1910s were filled with patent fights and business battles, but they were also largely profitable ones for the Marconi companies. Incorporating the inventions of others, Marconi's companies even began sending some wireless telephony messages of voice and music, a practice becoming known as "radio." After Italy entered World War I in 1915, Marconi was made an officer on the engineering staff responsible for wireless. Because of the need for secrecy in communication, he worked on developing shortwave communication and direction finders for the Italians and secured funding to update and expand the insufficient quantity of wireless equipment.

After the war, in 1920, the British rescinded a Marconi experimental license for broadcasting an opera performance, saying that it interfered with legitimate services. Thus, the U.S. got a two-year head start on Britain in regular radio broadcasting and the mass production of receivers. In 1922 Marconi was again allowed to begin limited broadcast service. By this time he was performing many of his radio experiments on his yacht, *Elettra,* which he had purchased in 1919.

Marconi's personal experiments continued to improve on his wireless inventions. He was eventually able to send directional signals and could make them circle the globe using shortwave technology. He capitalized on this achievement by establishing permanent communication links between territories of the British Empire. He also helped to establish Vatican Radio on shortwave so that the pope could speak to the whole world at once.

By 1927 Marconi's health began to fail. He continued to work on experiments in radar and with microwaves. He again overcame the doubts of theoreticians about the ability of microwaves to travel long distances, and he set up the first microwave communications link at the Vatican. Marconi died in 1937.

STEPHEN D. PERRY

See also Early Wireless; Fleming, John Ambrose; Hertz, Heinrich; Vatican Radio

Guglielmo Marconi. Born in Bologna, Italy, 25 April 1874. Self-educated and tutored by scientist Augusto Righi; invented methods of propagating electromagnetic waves over short distances, 1895, eventually reaching across Atlantic ocean, 1901, and around the world, 1922; invented method of tuning signals to avoid interference, 1904; worked on developing radar and microwave systems, advancing shortwave systems, and implementing permanent wireless communication between British colonies; set up Vatican radio service enabling Pope to speak to world, 1931. Received Nobel Prize in physics, 1909. Died in Rome, Italy, 20 July 1937.

Selected Publications

Wireless Telegraphy, 1899
Signals Across the Atlantic, 1901
The Progress of Electric Space Telegraphy, 1902
Recent Advances in Wireless Telegraphy, 1906
Wireless Telegraphic Communication, 1909
Radiotelegraphy, 1911

Further Reading

Baker, W.J., *A History of the Marconi Company,* New York: St. Martin's Press, 1971
Carter, Alden R., *Radio: From Marconi to the Space Age,* New York: Watts, 1987
Dunlap, Orrin Elmer, Jr., *Marconi: The Man and His Wireless,* New York: Macmillan, 1937; reprint, New York: Arno Press, 1971
Gunston, David, *Marconi: Father of Radio,* New York: Crowell-Collier Press, 1965; as *Guglielmo Marconi: Father of Radio,* London: Weidenfeld and Nicolson, 1965
Ivall, Tom, and Peter Willis, "Making Continuous Waves," *Electronics World* 103, no. 2 (February 1997)
Jolly, W.P., *Marconi,* New York: Stein and Day, and London: Constable, 1972

Kreuzer, James H., and Felicia A. Kreuzer, "Marconi—The Man and His Apparatus," *The A.W.A. Review* 9 (1995)
Marconi, Degna, *My Father, Marconi,* New York: McGraw Hill, and London: Muller, 1962
Masini, Giancarlo, *Marconi,* New York: Marsilio Publishers, 1995
Simons, R.W., "Guglielmo Marconi and Early Systems of Wireless Communication," *GEC Review* 11, no. 1 (1996)

Market

Radio markets are defined in geographic, demographic, and psychographic terms. Often, the definition of a radio market involves all three factors.

The primary definition of a market for radio is geographic—the area served by the transmitter's coverage, whether one city, a group of counties, or an entire region. Within that "market," the station establishes its listener base and sells advertising to attract those listeners to area retailers.

The second definition for radio derives from the medium's ability to target individual audience segments. Demographically drawn markets allow stations to focus programming on specific age groups: young adult women, for example, or teenagers. National radio programs define their markets using demographics.

Some markets defy geography and demography and consist of people of similar interests and tastes—psychographics. These similarities are often referred to as "lifestyle characteristics." However, consultant George A. Burns cautions radio marketers to understand the differences: "A group of nude skydivers may have only that in common. Their radio tastes can vary widely. While there may be broad commonalities among listeners to an individual radio station, it seems almost impossible to define them." Burns (1980) suggested that lifestyle characteristics of a particular group "*converge* at an individual radio station, *as far as radio listening is concerned.*"

As radio emerged in the 1930s as a viable—and valuable—medium for advertisers, geographic divisions were the most effective and most often used because of network radio's national reach. A 1939 promotional flyer from the National Broadcasting Company (NBC) showed the percentage of radio ownership in each of nine regions of the United States. The same flyer divided radio families by each of the four standard time zones, by size of city, and by whether listeners were in rural or urban locations.

Geography

The specific geographical definition of a market begins with guidelines set by the U.S. government's Office of Management and Budget (OMB). Radio's ratings services base their market areas on OMB's "Metropolitan Statistical Area" (MSA), "Primary Metropolitan Statistical Area" (PMSA), and "Consolidated Metropolitan Statistical Area" (CMSA). The government assigns each county surrounding a major population area to a specific MSA, PMSA, or CMSA.

For purposes of radio and television ratings, the definitions are modified by the Arbitron Company for radio and Nielsen Media Research for television, based on the needs and desires of their subscribers. Stations subscribing to the ratings service vote on which counties are included in a ratings report and which are excluded or assigned to another market area. Modern media markets are typically defined in three ways: for radio, the Metro Survey Area ("Metro") and the Total Survey Area (TSA); for television, the Designated Market Area (DMA).

As an example, the San Francisco, California, Arbitron ratings report contains data from Sonoma County listeners, and Sonoma County is considered part of the San Francisco Metro in Arbitron ratings reports. Sonoma County radio station operators, however, elected to define their county as a radio market, too. The result was a ratings report for San Francisco and an additional report for Santa Rosa (the largest city in Sonoma County). The ratings for Santa Rosa are duplicated in the San Francisco report, creating what is called an "imbedded market."

In contrast, Philadelphia and Wilmington, Delaware, are similar in that they are geographically side by side. Just as Sonoma County listeners hear San Francisco stations as easily as local outlets, Wilmington listeners can hear Philadelphia stations. However, the two are separate and distinct radio markets, as elected by Arbitron's subscribers in each area. The Philadelphia report and the Wilmington report have no duplication.

The Federal Communications Commission (FCC) generally defines a market area based on the signal contours of individual stations and overlapping signals. The FCC definition thus often differs from the Arbitron definition.

Targeting

Targeting was first found in advertising texts of the early 20th century. There was clear awareness of the use of different periodicals to target various populations: children, farmers, college students, and religious people, for instance. There were "trade" or "class" magazines such as those aimed at plumbers or Masons. Small-town newspapers targeted their audiences specifically, giving local influence to the advertisements. In 1915 Ernest Elmo Calkins suggested "canvassing consumers" in different cities around the country in order to gather information for an ad campaign. Calkins was a pioneer of targeting specific types of people as audiences and creating non-geographic markets.

General Motors advertising was an early example of targeting and segmentation. Trying to work out of a sales slump in the 1920s, the company reorganized its strategy based on price segments. Chevrolet, Pontiac, Buick, and Cadillac automobiles were priced differently and advertised to different markets based on socio-economic criteria.

This led manufacturers to support magazines and radio stations that reached the consumer segments they wanted for their products. Radio at the time was more mass than segmented; however, it became an ideal demographic and segmentation medium, creating communities of like-minded listeners. Writing in *American Demographics* magazine, Joseph Turow recognized the benefit of targeting to specific communities: "Target-minded media help advertisers [target a specific audience] by building primary media communities formed when viewers or readers feel that a magazine, radio station, or other medium resonates with their personal beliefs, and helps them chart their position in the larger world."

Just as media has changed since targeting and segmentation began, so has research. The statistical tools available to the researcher have grown tremendously over the years. The plunging cost of computation makes it both economically and logistically feasible to merge large databases and create new analyses. Researchers are now able to uncover relationships among demographic, attitudinal, behavioral, and geographic elements of the population. Those relationships are called "clusters."

Clusters

The saying "birds of a feather flock together" represents the idea of clustering. By combining demographic data, a market can be grouped into clusters, also known as "geodemographic segmentation systems."

Cluster systems take many demographic variables and create profiles of different individual or household characteristics, purchase behaviors, and media preferences. Most clusters used in media sales and analysis have catchy, descriptive names in an attempt to make them easier to remember. Examples are "Elite Suburbs," "2nd City Society," "Heart Landers," and "Rustic Living"—all from Claritas' PRIZM cluster system.

Marketing Tools magazine claims that cluster systems

> are especially powerful when used in conjunction with business mapping. Sophisticated mapping software programs easily link demographics to any level of geography (a process called "geocoding"). Some software can pinpoint specific households with neighborhoods from . . . customer data and then create schematic maps of neighborhoods by cluster concentrations.

The geographic element distinguishes clusters from psychographic segments. Another difference is that cluster categories are based on socio-economic and consumer data, not attitudinal data.

When mapping and media mix, Zip codes are often used as a targeting tool. Radio stations tend to use Zip codes to target potential Arbitron diary keepers, but this is not an exact science. In *The Clustering of America* (1988), Michael Weiss introduced the use of Zip codes as a clustering device. His work introduced marketers to age, education, and buying segments originally developed by Claritas Corporation for their PRIZM database. As effective as Weiss was in describing clustering, the net result among media sellers was his Zip code analyses.

With more than 36,000 Zip codes in the United States, precise segmentation is difficult. A Zip code does not constitute a segmented market, even though a single Zip code can contain 35,000 addresses or more. In New York City, because of the density of the population, Zip codes are somewhat cohesive in terms of ethnic and socio-economic mix. In the smallest towns, there may be only one Zip code, thus defying segmentation. That is why the cluster systems were developed.

Lifestyles

Cluster analysis is often confused with psychographics, and the words *psychographics* and *lifestyles* tend to be used interchangeably. There is a difference, though: *psychographics* usually refers to a formal classification system that categorizes people into specific types based largely on psychological characteristics; *lifestyle* is more vague and generally refers to organizing people by attitudes or consumer behavior—"politically

Eugene F. McDonald, Jr.
Courtesy of Zenith

many of his competitors, McDonald also established a design and planning unit early in the war to design the initial postwar commercial lines. This planning unit also allowed Zenith to maintain employment for a number of engineers who had families in occupied Europe and thus could not qualify for war production work. This kind of generosity coupled with enlightened self-interest was typical of McDonald's approach to life.

Commander McDonald's lifelong passion for television began in 1933. Zenith went on the air with W9XZV in 1939, and in 1941, the first color broadcast was transmitted. Because of McDonald's interest in FM radio and his friendship with Edwin Howard Armstrong, Zenith's FM station went on the air on 2 February 1940, operating under the call letters W9XEN (later W51C and WEFM). McDonald is probably best remembered today as the first and most vocal advocate of what was to become cable and pay-per-view television. Thanks to his personal persistence and political acumen, Zenith received permission to conduct a limited commercial test of "Phonevision" in 1951. Also under his direction and guidance, Zenith invented the first wireless television remote control in 1956, named Space Commander in McDonald's honor (later Space Command).

From his earliest days in radio, McDonald's close associates remarked on his uncanny ability to predict the future of this very innovative and volatile industry. McDonald was inducted posthumously into the Broadcast Pioneers Hall of Fame on 4 April 1967, nine years after his death. Among the accomplishments listed in the citation were his roles as founder, president, and first chairman of the board of Zenith Radio Corporation; his dynamic merchandising strategies; his inventions (29 patents) and innovations; his role as explorer; and his role as the first president of NAB. He was also cited for having established one of the nation's pioneer radio stations (WJAZ) and for fostering the development of shortwave radio, international communication, ship-to-shore, FM, VHF and UHF television, radar, and subscription television. In 2000 the Commander was an inaugural inductee into the Consumer Electronics Hall of Fame by the Consumer Electronics Association.

HAROLD N. CONES AND JOHN H. BRYANT

See also National Association of Broadcasters; Zenith Radio

Eugene F. McDonald. Born in Syracuse, New York, 11 March 1886. Served as Lieutenant in World War I; formed and became president of Zenith Radio Corporation with Karl Hassel and R.H.G. Mathews, 1923; founded WJAZ radio station, Chicago, 1923; formed and became president of National Association of Broadcasters (NAB), 1923; developed shortwave radio receivers, 1925; pioneered early television with Zenith's W9XZV, 1939; helped create Zenith's FM radio station, WEFM, 1940; posthumously inducted into Broadcast Pioneers Hall of Fame, 1967; posthumously inducted into Consumer Electronics Hall of Fame, 2000. Died in Chicago, Illinois, 15 May 1958.

Selected Publications

Youth Must Fly: Gliding and Soaring For America, 1942

Further Reading

Bensman, Marvin, "The Zenith-WJAZ Case and the Chaos of 1926–27," *Journal of Broadcasting* 4 (Fall 1970)

Bryant, John H., and Harold N. Cones, *The Zenith Trans-Oceanic: The Royalty of Radios,* Atglen, Pennsylvania: Schiffer, 1995

Bryant, John H., and Harold N. Cones, *Dangerous Crossings: The First Modern Polar Expedition, 1925,* Annapolis, Maryland: Naval Institute Press, 2000

"Commander McDonald of Zenith," *Fortune* (June 1945)

Cones, Harold N., and John H. Bryant, *Eugene F. McDonald, Jr.: Communications Pioneer Lost to History,* Record of Proceedings, 3rd International Symposium on Telecommunications History, Washington, D.C.: Independent Telephone Historical Foundation, 1995
Cones, Harold N., and John H. Bryant, *Zenith Radio: The Early Years: 1919–1935,* Atglen, Pensylvania: Schiffer, 1997
Cones, Harold N., and John H. Bryant, "The Car Salesman and the Accordion Designer: Contributions of Eugene F. McDonald, Jr., and Robert Davol Budlong to Radio," *Journal of Radio Studies* 8, no. 1 (Spring 2001)
Zenith Radio Corporation, *The Zenith Story: A History from 1919,* Chicago: Zenith Radio Corporation, 1955

McGannon, Don 1921–1984

U.S. Broadcasting Executive

Don McGannon was the long-time head of Westinghouse Broadcasting (Group W) and a respected member of the industry's leadership in the 1960s and 1970s. He sought quality programs and public service as well as profits. He is best known for Group W's successful introduction of the all-news format to radio and for developing a stronger commitment to public affairs by the Group W stations. He was a strong believer that broadcasters should serve in the public interest.

Origins

Donald McGannon was born in New York City in 1921. One of four children, he attended Fordham University (which would decades later name a communications research center after him), earning a B.A. in 1940. As with many others of his generation, he served in the U.S. Army during the war (becoming a major), returning to Fordham to earn a law degree in 1947. Admitted to the Bar in both New York and Connecticut, McGannon became a practicing attorney for several years. He worked briefly for a Connecticut democratic congressman, but when the politician suffered a fatal heart attack, McGannon was out of work. Here family ties helped out. McGannon's dentist brother had a patient who was a DuMont television executive, seeking some help. On learning of the out-of-work attorney, he agreed to interview McGannon and hired him.

Thus McGannon's broadcasting management career began in 1951 when he became the assistant director and then general manager of the several owned-and-operated television stations for the struggling DuMont Television Network. Just months before that network closed operations in 1955, he moved to Westinghouse Broadcasting Company as vice president and general executive, and he was elected president later that year.

Developing Public Service

One of McGannon's first major actions at Westinghouse—masterminding the retrieval of former Westinghouse property KYW-AM-TV (then in Cleveland, Ohio) from the National Broadcasting Company (NBC)—was indicative of the confrontational stance he would take with the titans of the industry. The Federal Communications Commission (FCC) found that Radio Corporation of America (RCA) Chairman David Sarnoff, who had been quoted in print threatening to withdraw his network's affiliation with Westinghouse, had coerced the company into swapping for NBC facilities in the smaller market of Cleveland; the FCC ordered the trade undone.

In 1956 McGannon took the bold step of severing the network affiliations of all Westinghouse radio stations and beginning to build an independent news-gathering operation. A well-staffed Washington bureau was soon established. Facilities for foreign operations began building in 1961.

Under McGannon, Westinghouse Broadcasting Company (by 1963 known as "Group W") was both admired and disliked in the industry. It was admired because the company was a leader in public service. To many it was the model of what a broadcasting group should be. The company was disliked because this willingness to do more than the law required put the company at odds with most of the industry and with the National Association of Broadcasters (NAB), which fought the FCC on every rule it imposed. McGannon had a keen awareness of Westinghouse's historic role in the birth of commercial radio (KDKA), and he imbued his employees with a sense of mission and social responsibility. As a corporation, Westinghouse believed that the right of the federal government to regulate the airwaves was legitimate. The broadcast spectrum was indeed a public resource to be managed for the good of all, and McGannon cast himself as the "good steward" who obeyed all

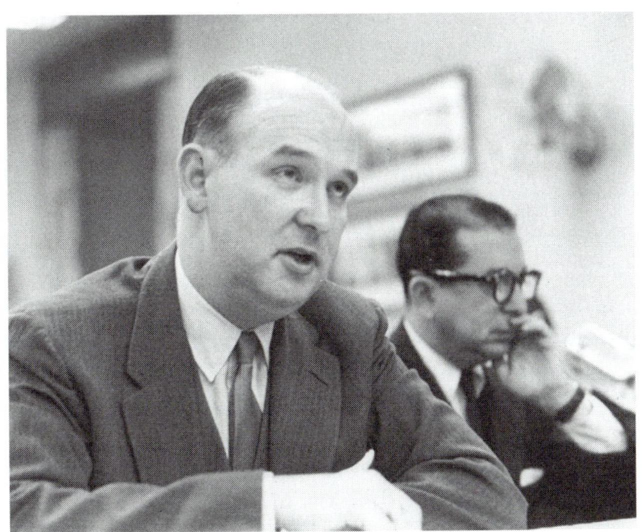

Donald H. McGannon
Courtesy AP/Wide World Photos

the rules—who in fact did more than the rules required—and yet managed to give an enviable return to his stockholders. This undercut the argument propagated by the NAB, *Broadcasting* magazine, and others that the fairness doctrine, license renewal requirements, "equal time" regulations, and other government stipulations were onerous and financially burdensome.

McGannon's position was simple. The station licenses were what made Westinghouse broadcasting possible, so they must be protected. If the FCC wanted half an hour of community-oriented programming per week, then McGannon would give them two as insurance. McGannon often stated that the more public service his company did, the more money it made. It was almost embarrassing, he joked.

In keeping with this approach, McGannon also instituted something called the Public Service Conferences. These were held regularly beginning in 1956 in cities where Westinghouse owned stations. Multi-day events, the conferences brought together radio and television personnel, scholars, and politicians from around the country to discuss the role broadcasting should play in dealing with issues of concern to the nation. The first conference was held in Boston and featured addresses by Senator John F. Kennedy and Vice President Richard Nixon. These conferences were important in their own right and enhanced the reputation of Westinghouse immeasurably.

Instituting All-News Radio

In July 1962 Westinghouse Broadcasting Company purchased WINS-AM for a reported $10 million. Westinghouse at first experimented with various music formats in an unsuccessful effort to top market leader WABC-AM. Then, on Monday, 19 April 1965, WINS-AM went to an all-news format, becoming one of the first stations to make what was then considered a radical format transition. Importantly, WINS-AM pioneered the all-news format in the largest media market in the United States.

WINS-AM became the place on the dial where New Yorkers tuned to learn about breaking news. In November 1965, when a major blackout darkened northeastern cities, WINS-AM kept millions informed during the crisis as they listened on battery-powered portable radios. Eventually, three of Group W's radio stations adopted the format. WINS-AM continued to broadcast "all news, all the time" in 2000.

McClendon's all-news radio stations broadcast mostly hard news in 15-minute blocks. Each block featured its own reporter, who compiled and edited wire-service copy and read it on the air. McGannon's version of the all-news format differed in that it made extensive use of Westinghouse's growing news-gathering capabilities, which included Washington and foreign bureaus as well as news staff in the large cities where Westinghouse owned outlets. An established network of stringers was also available when needed. Because of this, Westinghouse presented a much more varied slate of material, including local, national, and international features; sports; education; religion; finance; and "actualities." The Westinghouse/Group W approach to the all-news format was an immediate success.

Impact

During his long tenure heading Westinghouse, McGannon developed and refined the broadcasting group as a recognized entity within the industry. Previously, the two main organizing principles of radio and especially of television were the network and the individual station, which was almost always a network affiliate. Westinghouse Broadcasting Company was neither fish nor fowl. Although its television stations were network affiliates, they were first and foremost Westinghouse stations. This group identity was even more pronounced in Westinghouse's radio properties, which dropped all network ties in 1956. The previously mentioned Public Service Conferences were actually seen as a way of establishing the "Group" concept in the mind of the industry and the public.

McGannon's Westinghouse was ahead of its time on several issues. In a mostly white male world of button-down shirts and blue suits, McGannon sought diversity. To this end, he founded a job-training program to help minority workers get a start in the industry. Called the Broadcast Skills Bank, its name was later changed to the Employment Clearinghouse. McGannon also ordered all Westinghouse properties to drop cigarette advertising several years before legally required to do so.

Graham McNamee (right) seated with Phillips Carlin
Courtesy Library of American Broadcasting

Justice William Howard Taft's oath of office and inaugural address has McNamee's voice in the background, saying "We are ready." Another classic recording includes McNamee's breathless description of Charles Lindbergh's return to U.S. soil after his historic 1927 solo flight across the Atlantic.

It was as a sportscaster, though, that McNamee was most widely known. He covered World Series games from 1923 to 1935, major college football games (including the first Rose Bowl broadcast in 1927, which was also NBC's first coast-to-coast broadcast), significant tennis matches, and momentous prize fights, including the second "fight of the century" between Gene Tunney and Jack Dempsey in 1927. His colorful broadcast of that match for the 73-station NBC Red network reached an estimated 50 million people, a staggering number given the relatively few radio sets then in use.

As was common for early journalists, McNamee tended to embellish the moment to create emotional impact, but his style fit with the times and made the announcer as much a star as the athletes he covered. His mere presence conveyed to listeners that an event had importance. In describing the players, he cultivated heroic images that outshone many athletes' actual achievements. The mythic aura around such legendary giants as Babe Ruth and Jack Dempsey was largely the creation of a generation of sportscasters known for their hyperbole. McNamee and his contemporaries spun a web of adulation for sports heroes that endures among fans today.

After covering a dozen World Series between 1923 and 1935, as well as hundreds of boxing and tennis matches and football games, McNamee's sports announcing ended with the rise of a new group of radio announcers. The times demanded terse and factual reporting of sports in the style of such greats as Red Barber and Lindsey Nelson, and NBC pulled him from covering the 1935 World Series and moved him to other tasks. McNamee had become dated and controversial because he talked so much and because many sportswriters and fans found his vivid game dramatizations unwarranted. At the same time, subsequent sportscasters idolized him for his great voice, his historic sportscasts, and his ongoing achievements in developing play-by-play and color sports announcing. McNamee's role in creating the celebrity sportscaster and building the national fan base for major sports has not been forgotten.

His final years at NBC were largely given over to announcing for Rudy Vallee's variety program and appearing as a straight man for Ed Wynn. He occasionally did spot news reporting, as with the maiden voyage arrival in New York of the *Queen Mary* in 1936 and a pier-side description of the burning of the French liner *Normandie* in February 1942. On 24 April 1942 he announced *Elsa Maxwell's Party Line*—his final broadcast. He died shortly thereafter of an embolism of the brain, just two months before his 53rd birthday. Obituaries and editorials on his death appeared across the country; "Mr. Radio is Dead" is how the *Cleveland Plain Dealer* headlined its story.

McNamee was one of the first celebrities of the new medium, touted as "The World's Most Popular Announcer" on a cup from *Radio Digest* in 1925 after winning a national competition over 132 other announcers (listeners were fickle: two years later he was ranked ninth in another poll; in 1931 he was ranked fourth, but second among sports announcers, after Ted Husing). Long after his death, McNamee was recognized with the first group of notables inducted (in 1977) into the National Association of Broadcasters Hall of Fame. A national award in his honor was established for a sportscaster who achieved success in a second field of endeavor, just as he had gone from singing to announcing. The Graham McNamee award has gone to such noted individuals as President Ronald Reagan, Walter Cronkite, Bryant Gumbel, and Larry King, all of whom were sportscasters at some point in their careers. Heywood Hale Broun wrote that "McNamee justified the whole activity of broadcasting," adding that he "has been able to take a new medium of expression and through it transmit himself—to give it vividly a sense of movement and of feeling. Of such is the kingdom of art" (cited in Smith, 1987, 1992).

SUSAN TYLER EASTMAN

See also National Broadcasting Company; Sportscasters; WEAF

Graham McNamee. Born in Washington, D.C., 10 July 1888. Trained as baritone, New York City, 1907–21; made professional debut, 1921; radio announcer, WEAF, 1923–26; color announcer for first broadcast of World Series, 1923; chief announcer for live news and sports, NBC Radio Red Network, 1926–34. Inducted posthumously into National Association of Broadcasters Hall of Fame, 1977. Died in New York City, 9 May 1942.

Selected Publications

You're on the Air, 1926

Further Reading

Gorman, Jerry, Kirk Calhoun, and Skip Rozen, *The Name of the Game: The Business of Sports,* New York: Wiley, 1994

Harwell, Ernie, *Tuned to Baseball,* South Bend, Indiana: Diamond Communications, 1985

Rader, Benjamin G., *American Sports: From the Age of Folk Games to the Age of Televised Sports,* Englewood Cliffs, New Jersey: Prentice Hall, 1983; 2nd edition, 1990

Smith, Curt, *Voices of the Game: The First Full-Scale Overview of Baseball Broadcasting, 1921 to the Present,* South Bend, Indiana: Diamond, 1987; revised edition, as *Voices of the Game: The Acclaimed Chronicle of Baseball Radio and Television Broadcasting—From 1921 to the Present,* New York: Simon and Schuster, 1992

McNeill, Don 1907–1996

U.S. Radio Morning Show Host

For 35 years, many Americans woke up to the easygoing banter of Don McNeill, the genial host of the Chicago-based *Breakfast Club* program. To a considerable degree, he created the morning radio format that became widely popular throughout the industry.

Origins

Born in Illinois in 1907 and raised in Wisconsin, McNeill moved around to a number of radio stations in the Midwest, San Francisco, and New York for several years. He struck it

Don McNeill
Courtesy of family of Don McNeill

big (though that was anything but clear at the time) when NBC-Blue asked him to take over *The Pepper Pot,* a struggling morning show in Chicago that had no advertisers and precious few listeners. Starting at $50 a week, McNeill began hosting the program on 23 June 1933 and renamed it *The Breakfast Club* a few months later.

Breakfast Club Years

Right from the start, McNeill made changes that greatly helped turn the program into a huge success. For its hour-long slot, he fairly quickly developed what he called the "four calls to breakfast," which divided the program into quarter-hour segments, each of which emphasized something different. It started with an opening song that usually began:

> *Good morning, Breakfast Clubbers, we're glad to see ya!*
> *We wake up bright and early just to howdy-do ya*

The first segment offered interviews with the studio audience (added in 1938); the second featured "memory time" with sentimental poetry and prayer time (added in 1944 and retained after the war, along with hymns provided in a nonsectarian fashion); the third was "march time"; and the fourth and last was dubbed "inspiration time" and also often included more sentimental poetry. Although the audience never knew it, no breakfast was ever consumed on the program.

After the first few years, McNeill sought and was granted permission to move from a scripted program to a variety program that was both unrehearsed and spontaneous. *Breakfast Club* took on greater charm at that point, as McNeill's ability to ad-lib in almost any situation came through. So did his genial nature and easygoing tone, along with that of the rest of the regular cast, which sometimes included his wife and young sons. One singer on the program, Charlotte Reid (known on the air as Annette King), later became a member of the Federal Communications Commission and of the U.S. Congress. Other regulars at one time or another included Fran Allison and singers Patti Page, Johnny Desmond, and Anita Bryant. Some cast or orchestra members stayed around for decades. While the jokes were often bad and the program corny, it was appealing for its ability to seemingly speak to each listener individually. Many of the segments were enlightened by listener mail that amounted to thousands of weekly letters by the 1940s. And on occasion, the program left Chicago to go on tour, broadcasting from other cities.

The show's huge popularity became very evident with a 1944 promotional offer. As John Dunning writes: "When the sponsor offered *Breakfast Club* membership cards, 850,000 people wrote in. No more than 15,000 were expected, and McNeill had to go on the air and beg out of the promotion. It still cost the sponsor $50,000 to retreat" (see Dunning, 1998). Luckily by then there *were* sponsors. Through the 1930s the program lacked advertising support. During most of the war years, only about a quarter of the program was sponsored. Indeed, it was not fully supported by advertising until 1946—13 years after McNeill had taken it over and demonstrated he could build big and loyal audiences. Saturday broadcasts were dropped in 1945, by which time McNeill was making $1,000 a week.

The show was clearly aimed at middle America, just as it was broadcast *from* middle America. The program was first broadcast (until 1948) from the 19th-floor NBC studios in Chicago's huge Merchandise Mart building, then for five years from the Little Theater in the Civic Opera Building. In 1953 it moved to the Terrace Casino of the Morrison Hotel and two years later to the College Inn at the Sherman House, and finally in 1963 to specially rebuilt Clouds Room on the 23rd floor of the Allerton Hotel, all in downtown Chicago. But its heart and content were clearly not in the city, but rather in rural America. And the do-good nature of the program came through in constant appeals to visit those in retirement homes or hospitals, to collect food for the starving refugees in postwar Europe, and to help the poor at holiday time at home. Listeners and often studio audiences came through every time.

Final Years

In 1950, by then making $100,000 a year, McNeill signed a 20-year contract with the network to continue the program, one that gave him two months of vacation per year. *Don McNeill's TV Club* aired on ABC-TV in 1950–51 and a brief attempt to simulcast the successful radio program on television as *The Breakfast Club* failed after just one season on ABC in 1954–55. In 1957, the program began to be tape recorded a day before broadcast. Toward the end of the program's run, McNeill was receiving upwards of 10,000 letters each month. The program was still being carried on more than 220 stations when it ended its 35 year run on 27 December 1968, the last regularly scheduled network radio broadcast originating in Chicago. The old-fashioned sound of *The Breakfast Club* no longer fit a radio industry largely devoted to popular music formats, or a country facing the height of the Vietnam War and political assassinations.

CHRISTOPHER H. STERLING

See also American Broadcasting Corporation; Godfrey, Arthur; Morning Shows

Donald Thomas McNeill. Born in Galena, Illinois, 23 December 1907. Graduated from Marquette University, 1930: B.Phil. Entered radio with work at WISN, Milwaukee, 1928; WTMJ, Milwaukee, 1929; WHAS, Louisville, 1930. NBC *Breakfast Club* host, 1933–68. Taught communications at

Marquette and Notre Dame Universities, 1970–72. Spokesperson for Deltona Corporation, 1970–80. Inducted into Broadcaster's Hall of Fame, 1979; Radio Hall of Fame, 1989. Died in Evanston, Illinois, 7 May 1996.

Radio Series

1933–68 *The Breakfast Club*

Television

Don McNeill's TV Club, 1950–51; *The Breakfast Club,* 1954–55 (simulcast)

Selected Publications

20 Years of Memory Time; The Most Popular Selections since 1933, 1953

Mr. Don McNeill, esq. Presents his Breakfast Club Family Album, 1942

Further Reading

The Don McNeill Collection is housed at Marquette University's Memorial Library.

"'Breakfast Club' Ending its 35-Year Stay on Radio," *New York Times* (27 December 1968)

Doolittle, John, *Don McNeill and His Breakfast Club,* Notre Dame, Indiana: University of Notre Dame Press, 2001

Dunning, John, "The Breakfast Club," in *On the Air: The Encyclopedia of Old-Time Radio,* New York: Oxford University Press, 1998

Froman, B., "The Man Who Came to Breakfast," *Colliers* (13 May 1950)

McPherson, Aimee Semple 1890–1944

U.S. (Canadian Born) Radio Evangelist

The Pentecostal evangelist Aimee Semple McPherson, known as "Sister Aimee" to her followers, was an early celebrity of mass media religion. She was the first woman both to preach by radio (1922) and to hold a station license (1924). She founded the third radio station in Los Angeles, the still-extant KFSG, as the mouthpiece for her new denomination, the International Church of the Foursquare Gospel. At the time of her death in 1944 she had made plans for the United States' first televangelist network.

For this extraordinary career as one of the first modern media-savvy evangelists, her early preparation seems rather unremarkable. She was a dutiful missionary wife and later was a little-known itinerant preacher. Born Aimee Elizabeth Kennedy in 1890 in rural Ingersoll, Ontario, she was converted to Pentecostalism by her future husband, Robert Semple, during a 1907 revival. The Semples set off in 1910 to evangelize China, but Robert died shortly after their arrival and just weeks before the birth of their first child, Roberta Star Semple. Aimee Semple then left the ministry for about five years, during which time she went to the United States, married again, and had a second child, Rolf. By 1915, however, she was back on the road working full-time as an itinerant preacher on the Pentecostal tent circuit. Her travels and spreading fame eventually took her to Los Angeles in 1918, which remained her home for the rest of her life.

By 1921, when McPherson broke ground for Angelus Temple on West Sunset Boulevard, she was already famous for her faith-healing meetings ("stretcher days," as she called them), which packed stadium-sized crowds in San Diego, San Jose, and Denver. Her fast-growing organization, now called the International Church of the Foursquare Gospel, was the largest private relief charity in Los Angeles during the Depression years (and has since become a thriving worldwide church). At Angelus, to the accompaniment of a huge organ, a 14-piece orchestra, a brass band, and a 100-voice choir, she performed "illustrated sermons" every Sunday night for 20 years. These illustrated sermons were stage spectacles and Biblical dramas with elaborate costumes rented from Hollywood studios and with huge sets and special effects.

McPherson had first preached on Oakland's Rockridge radio station (now KNEW) in April 1922. She preached occasional services over *Los Angeles Times* station KHJ while raising $75,000 to construct a 500-watt broadcasting facility inside Angelus Temple. Her own radio station, KFSG, began broadcasting on 6 February 1924. A Class A 500-watt transmitting station, with unlimited broadcast time, the station could be heard over much of the western United States and the Pacific, even as far away as Hawaii. Like other early radio pioneers in the 1920s, McPherson occasionally "wandered" the airwaves in search of a clearer broadcast frequency. In reply to Secretary

Aimee Semple McPherson, 21 August 1930
Courtesy AP/Wide World Photos

of Commerce Herbert Hoover's 1927 directive to keep to her assigned frequency, she fired off an infamous telegram that read, "PLEASE ORDER YOUR MINIONS OF SATAN TO LEAVE MY STATION ALONE. YOU CANNOT EXPECT THE ALMIGHTY TO ABIDE BY YOUR WAVELENGTH NONSENSE. WHEN I OFFER MY PRAYERS TO HIM I MUST FIT INTO HIS WAVE RECEPTION. OPEN THIS STATION AT ONCE." Eventually Sister Aimee learned to be more diplomatic in her dealings with the federal regulators; renewal of Sister Aimee's license to operate KFSG was never denied.

KFSG featured nearly round-the-clock programming: there were Sunday devotional and revival services simulcast in their entirety from the main auditorium, programs of sacred music, programs designed for children, a "family altar call," civic talks, and weekly divine healing and baptismal services. Later, McPherson added travel shows, a religious news program, and even live studio serials patterned after popular radio dramas of the time, with titles such as *Jim Trask: Lone Evangelist.* Perhaps more so than any other fundamentalist media preacher, she duplicated the genres of commercial radio. But faith healing, in particular, set McPherson apart as a radio phenomenon. Listeners were invited to kneel by their radios, touching the speakers to receive cures for their ailments. Prayers for healing were a regular part of the programming of KFSG and brought heavy mail response claiming cures for every known disease. Speaking in tongues was also a regular feature of KFSG programming, except during World War II, when McPherson stopped airing speaking in tongues because of wartime prohibitions against coded messages.

Sister Aimee's regional following was numbered at 30,000 by 1930, and her radio audience was certainly much larger. In 1937, on her license renewal application, McPherson claimed that KFSG constituted a "church of the air" with a registered membership exceeding 50,000. Most of her adherents and regular listeners came from the ranks of the lower middle class, including many recent migrants to the West Coast. As Sister Aimee herself put it, perhaps slightly tongue in cheek, "I bring spiritual consolation to the middle class, leaving those above to themselves and those below to the Salvation Army." Fundamentalist churches in the area drew close to 80 percent of their membership from people who had lived in Los Angeles for less than ten years. Eager for the familiar and positive message Sister Aimee radiated, these recent arrivals were nonetheless embedded in the emerging consumer and entertainment culture of the time, and radio was a natural way for her growing denomination to tap a huge potential audience.

The latter half of her ministry was dogged by controversy, although McPherson's media outlets and her rigorous schedule of services and appearances did not falter. During the lean Depression years, when many smaller religious radio stations and programs went off the air for lack of funds, KFSG continued to broadcast (although with a shared-time arrangement with KRKD Los Angeles), and McPherson opened additional branches of her church in dozens of cities. She died in Oakland, California, just before her 54th birthday. Her son Rolf continued the ministry until 1988. After her death, the Church operated one of Los Angeles' early FM stations, KKLA FM (97.1) until the early 1950s and later assumed control of KRKD's AM and FM outlets.

McPherson's radio career permanently linked religious broadcasting with show business and successfully bridged old-time revivalism and modern mass communication. Her imprint remains not just on Pentecostalism itself, but on the exuberant, colorful, and sometimes all-too-human face of media evangelism throughout the century.

TONA J. HANGEN

See also Evangelists/Evangelical Radio; Religion on Radio

Aimee Semple McPherson. Born in Ingersoll, Ontario, Canada, 9 October 1890. Only daughter of James Kennedy and Mildred Ona Pearce; married Irish Pentecostal itinerant preacher Robert Semple, 1908; widowed in Hong Kong, 1910, their daughter born later that year; married Harold McPherson, 1912; became Pentecostal itinerant, 1915, often traveling with her mother and two children; settled in Los Angeles, 1918; established Angelus Temple in Hollywood, 1923; founded radio station KFSG, 1924; established the International Church of the Foursquare Gospel, early 1920s (name incorporated 1927); married David Hutton, 1931; broadcast at least weekly over KFSG for 20 years. Died in Oakland, California, 27 September 1944.

Selected Publications

Bridal Call, 1917–1934
This Is That: Personal Experiences, Sermons and Writings of Aimee Semple McPherson, Evangelist, 1923
Aimee Semple McPherson: The Story of My Life, 1973

Further Reading

Blumhofer, Edith Waldvogel, *Aimee Semple McPherson: Everybody's Sister,* Grand Rapids, Michigan: Eerdmans, 1993

Epstein, Daniel Mark, *Sister Aimee: The Life of Aimee Semple McPherson,* New York: Harcourt Brace Jovanovich, 1993

Mavity, Nancy Barr, *Sister Aimee,* Garden City, New York: Doubleday Doran, 1931

McLoughlin, William, "Aimee Semple McPherson: Your Sister in the King's Glad Service," *Journal of Popular Culture* 1 (Winter 1967)

McWilliams, Carey, "Aimee Semple McPherson: 'Sunlight in My Soul,'" in *The Aspirin Age: 1919–1941,* edited by Isabel Leighton, New York: Simon and Schuster, 1949; London: Bodley Head, 1950

Media Rating Council

Industry Self-Regulatory Group

Originally known as the Broadcast Rating Council (BRC) and then as the Electronic Media Rating Council (EMRC), the official sanctioning body for ratings services now goes by the name Media Rating Council (MRC). The MRC works to maintain confidence and credibility in ratings through its self-stated goal of setting standards that ensure that surveys of media audiences are conducted in a manner that encourages quality, integrity, and accurate disclosure of the research process. A nonprofit agency sanctioned by the U.S. Justice Department, the MRC consists of 70 members from broadcast and cable trade associations, media owners, advertising agencies, cable networks, print and internet companies, and national networks.

The MRC is a nonprofit industry organization that is run on its membership fees. Each member organization provides one person to serve on the MRC board of directors, which makes the final decision as to whether reports that have been audited will receive accreditation from the council. Among the members of the New York–based MRC are the National Association of Broadcasters, the Television Bureau of Advertising, the Radio Advertising Bureau, and the Cable Advertising Bureau.

The MRC was established in 1964, when it was known as the BRC. A self-regulatory agency, the BRC was formed in response to an investigation by the House Subcommittee on Communications of the Interstate and Foreign Commerce Committee, under the chairmanship of Oren Harris. The Harris Committee, as it was known, held hearings in 1963 to investigate ratings and audience research. The hearings arose both from an increased focus on television ratings after the quiz show scandals of 1961 and from complaints by advertisers that they couldn't obtain upfront information about research methodology from the Nielsen Company. Essentially, the Harris Committee was concerned that if ratings were defective or deceptive, they would affect programming selections by stations and work in a manner that was not in the public interest.

The credibility and validity of ratings became a growing industry concern as U.S. representatives questioned executives from broadcast-measurement companies about the quality of their research. Faced with the possibility of government interference in the ratings business, broadcast industry leaders obtained permission from the U.S. Justice Department (to avoid any perception of antitrust violations) to set up the BRC, thereby ensuring a means of self-regulation.

The BRC changed its name to the Electronic Media Rating Council in 1982 to include all electronic media, such as radio, television, and cable. The "Electronic" in the council's name was dropped in 1997 when the council started performing audits on print services as well as broadcast services. On 4 September 1996 the *Study of Media and Markets* (a national survey of over 20,000 adults performed by Simmons Market Research Bureau) became the first multimedia research study with a primary focus on print media to receive accreditation from the council. Today the MRC audits organizations such as Mediamark Research (which provides research to all forms of advertising media collected from a single sample) and J.D. Power and Associates (which publishes the annual *Car and Truck Media Studies* to assist with marketing and media strategies).

Audience measurement services voluntarily submit their studies to the MRC for review and possible accreditation. The MRC then commissions audits by an independent accounting firm (currently the Ernst and Young Corporation) to review the data. The ratings services pay the cost of the audit to the MRC, which in turn pays the auditors. This system allows for some separation between the parties and establishes that the MRC, not the ratings service, is supervising the audit.

Even though the auditing process is voluntary, many organizations still seek accreditation so that they will be considered legitimate in the industry and will therefore be better able to sell their ratings. Organizations seeking accreditation must agree to conduct their service as represented to users and subscribers, undergo MRC audits, and pay for the costs of the audits. The results of the audits are reviewed by the council's board of directors to determine if the ratings service will receive accreditation. Should a report receive accreditation, the organization submitting the report still must re-apply the following year and have the report reviewed on an annual basis. The MRC accredits syndicated services and individual reports, not entire companies. Accredited services and reports carry the MRC double-check logo.

In 1993 the council made an unprecedented move when it voted to suspend its accreditation of the spring Atlanta Arbitron survey, citing an on-air promotional campaign by Atlanta broadcasters that "hyped" (aimed to increase response rates) the survey by urging listeners to cooperate with Arbitron's diary-based system. The council decided that the effort could have an adverse effect on methodology and thus distort the survey results. The move to suspend accreditation met with sharp criticism from the Radio Advertising Bureau (RAB), which was upset with what it believed was an unfair bias toward television. Despite the controversy, the RAB currently holds membership in the MRC.

The MRC has expanded its role over the years but still functions primarily in an effort to maintain rating confidence and credibility. Melvin A. Goldberg, then EMRC executive director, explained in 1989 that obtaining accreditation required adherence to specific minimum standards that outlined basic objectives of reliable and useful electronic media audience measurement research. Acceptance of those standards was voluntary and was one of the conditions of EMRC membership.

According to Goldberg, the minimum standards fell into two groupings: (1) "Ethical and Operations Standards," and (2) "Disclosure Standards." The Ethical and Operations Standards governed the quality and integrity of the overall process of producing ratings. Meanwhile, the Disclosure Standards specified which information a ratings service had to make available to users, to the EMRC, and to its auditing agent. The overall effect of the standards was to assure anyone using EMRC-accredited ratings that the ratings actually measured what they said they did.

Thus, the minimum standards established professional codes of conduct that ratings services had to agree to in order to gain accreditation. For example, a ratings service was required to submit complete information on its survey methodology, including sampling techniques, recruiting procedures, weighing, tabulations, coding and computer software, and the eventual ratings. The standards that Goldberg referred to are still in effect today. However, the MRC has also added electronic delivery requirements that govern the proper way for ratings services to deliver data to a third party electronically. The MRC has also incorporated internet ratings reports into its auditing processes.

MATT TAYLOR

See also A.C. Nielsen Company; Arbitron

Further Reading

Beville, Hugh Malcolm, *Audience Ratings: Radio, Television, and Cable,* Hillsdale, New Jersey: Erlbaum, 1985; 2nd edition, 1988

Goldberg, Melvin A., "Broadcast Ratings and Ethics," *Review of Business* (Summer 1989)

Mergers. *See* Ownership, Mergers, and Acquisitions

Mercury Theater of the Air

Anthology Radio Drama Series

Mercury Theater of the Air was an offshoot of Orson Welles' successful theater company that had catapulted him to Broadway fame. Using many of the same actors, he put on a series of radio plays in 1938 under the title *First Person Singular,* although the *Mercury Theater* name was better known. At first the show had no sponsor and few listeners, but the success of the legendary "War of the Worlds" episode in 1938 persuaded Campbell's Soup to back it. The radio plays gave Welles national fame and allowed him to branch into films with his seminal *Citizen Kane* in 1941. The originality and technical flair that marked that film applied equally to Welles' radio productions.

By 1938 Welles, an accomplished Shakespearean actor, was becoming a noted radio performer, having worked on *March of Time;* adapted, directed, and starred in the seven-part *Les misérables;* and, most notably, by becoming the voice of *The Shadow.* He was approached by William Lewis, head of programming at Columbia Broadcasting System (CBS), to make nine one-hour adaptations of famous books. Welles' budget of $50,000 for the nine shows was not much, given that he had been earning $1,000 per week, and the shows were slated for an unpromising timeslot against the popular *Edgar Bergen and Charlie McCarthy.* However, the deal did offer Welles creative carte blanche; he did not need to worry about pleasing a sponsor or appealing to any target audience. (CBS made the offer at a time when Congress was threatening legislation aimed at raising radio standards, so owner William S. Paley wanted programs that could turn CBS into a veritable patron of the arts.)

Broadcasts began on 11 July 1938 on the WABC network with "Dracula." Like subsequent performances, it followed Welles' view that "the less a radio drama resembles a play the better it is likely to be." This represented a major departure from many previous programs that had tried to recreate live theater down to the last detail, even including intermissions and chatting patrons. Instead, Welles introduced an omnipresent narrator, himself, who played several roles. This not only allowed Welles to take center stage but also changed the narrator into a storyteller, all the other characters effectively becoming projections of himself. Welles was equally innovative in his use of music. He asked Bernard Herrmann (head of music for CBS) for an unprecedented amount of musical scoring for each drama: up to 40 minutes in 57-minute-long performances. Similar demands were made with regard to sound effects, which overlapped the dialogue instead of occurring at the end of a speech as had been the practice. Welles demanded that even the faintest rustle of leaves be reproduced, despite the fact that very few listeners would have been able to hear these effects on their crackling AM radios. (They can now be appreciated on compact disc recordings, however.)

Nearly all of the dramas performed by the Mercury Players, which included "Treasure Island," "The Thirty-Nine Steps," and "The Count of Monte Cristo," were classics with family appeal. Their tone was not patronizing, but neither were they too advanced for children to appreciate. Welles decided which story to perform each week, but despite opening credits saying that each play was "produced, directed, and performed by Orson Welles," his participation was strictly limited to reviewing the script, making last minute changes, and performing. His long-time collaborator John Houseman and experienced radio man Paul Stewart oversaw the script writing and rehearsals. For the cast, Welles was able to call upon members of the Mercury stage theater such as George Coulouris, as well as other experienced radio actors including Ray Collins, Agnes Moorehead, and Martin Gabel.

Despite this range of talents, the programs rarely attracted more than four percent of the national radio audience until "War of the Worlds" greatly increased Welles' fame. The resulting sponsorship by Campbell's Soup caused the title of the series to be changed to *The Campbell Playhouse Series* in midseason 1939. This new name also reflected the demise of the Mercury Theater, which had fallen apart following a number of unsuccessful theater productions.

Ostensibly, *The Campbell Playhouse Series* was the same program as the *Mercury Theater of the Air*, and the dramas continued to be hour-long adaptations. However, the new and bigger budget allowed Welles to cast star names, including Katharine Hepburn ("A Farewell to Arms"), Laurence Olivier ("Beau Geste"), Gertrude Lawrence ("Private Lives"), and Walter Huston ("Les misérables"), as his co-stars. Former members of the Mercury stage theater continued to work on the show, but they were now reduced to supporting roles. The presence of a star also affected the show's format, as the play would now be sandwiched between segments of talk show style patter as Welles chatted with his guest star. This would invariably include some banal reference to the joys of Campbell's soup. Campbell's also inserted commercial breaks for soup ads into the plays themselves. The *Mercury* broadcasts had been uninterrupted, but now cliff-hangers had to be created to insure that listeners did not tune in a different program during the commercials.

The *Campbell*'s plays were also based upon noticeably different books. The *Mercury*'s eclectic mix of classics was forsaken in favor of more populist and more modern works, primarily best-sellers from the previous decade. There were also reworkings of previous Welles productions, which were an indicator of the extent to which the program lost much of the *Mercury*'s originality and innovation. Welles' contribution also dropped off considerably as he began to concentrate more on theater before relocating to Hollywood. During this period, he would fly to New York on the day of the performance, make the broadcast, and then fly back to Hollywood. The production of the plays was thus left to Houseman and Stewart, with almost no input from Welles at all. However, *Campbell Playhouse* was one of the most popular shows on radio until Welles finally pulled the plug in March 1940 to fully concentrate on cinema. He had considered moving the show to Los Angeles so that his actors could be employed while he worked in films, but Campbell's refused to give up the Broadway panache that the show's New York connection provided. The *Mercury* name did make some sporadic returns to radio whenever Welles needed to raise some quick cash, but these later programs were mainly rehashes of previous performances that added little to the originals.

NEIL DENSLOW

See also Drama; Playwrights on Radio; War of the Worlds.

Cast

The "Mercury Players"	Orson Welles, Ray Collins, Agnes Moorehead, George Coulouris, Frank Readick, Georgia Backus, Bea Benaderet, Everett Sloane, Edgar Barrier

Producers/Creators

Orson Welles and John Houseman

Programming History

WABC (CBS)

First Person Singular/Mercury Theater of the Air	July 1938–December 1938
Campbell's Playhouse Series	December 1938–March 1940; 1946

Further Reading

Brady, Frank, *Citizen Welles: A Biography of Orson Welles,* New York: Scribner, 1989; London: Hodder and Stoughton, 1990

Callow, Simon, *Orson Welles: The Road to Xanadu,* London: Jonathan Cape, 1995; New York: Viking, 1996

Houseman, John, *Run-through: A Memoir,* New York: Simon and Schuster, 1972; London: Allen Lane, 1973

Leaming, Barbara, *Orson Welles,* New York: Viking, and London: Weidenfeld and Nicholson, 1985

Naremore, James, *The Magic World of Orson Welles,* New York: Oxford University Press, 1978

Thomas, François, "Dossier: La radio d'Orson Welles," *Positif* (October 1988)

Metromedia

Group Owner of Radio Stations

Corporate executive John Kluge made his mark as the founder of Metromedia, a media conglomerate that operated through the 1960s and 1970s both with independent television stations in major U.S. cities and with owned and operated major-market radio stations as well. Although not well known to the general public, Kluge emerged in this period as one of the most powerful media moguls. Kluge proved that independent television stations and big-city radio stations could make millions of dollars in profits by counterprogramming. In 1985, when Australian Rupert Murdoch offered Kluge nearly $2 billion for Metromedia's seven television stations, he sold out and began to reinvent Metromedia. He was out of big-city radio during the late 1980s.

John Werner Kluge surely represents the American success story. Kluge grew up poor in Detroit, but in 1933 he won a scholarship to Columbia University, where he earned a degree in economics. Serving U.S. Army Intelligence during World War II, he returned with little taste for resuming a career in the employ of others. He looked for ways to make money, including buying and selling radio stations.

After World War II, Kluge came to radio (and television) for its advertising power in the growing market of Washington, D.C., where he had served in World War II. Kluge bought and sold radio stations; his first was WGAY-FM in Silver Spring, Maryland, a Washington, D.C., suburb. As radio reinvented itself as a format medium, Kluge bought and sold stations across the United States, with early investments in radio groups including the St. Louis Broadcasting Corporation, Pittsburgh Broadcasting Company, Capitol Broadcasting Company (Nashville), Associated Broadcasters (Fort Worth–Dallas), Western New York Broadcasting Company (Buffalo), and the Mid-Florida Radio Corporation (Orlando).

Kluge became aware of television as an investment possibility when he ran into an acquaintance on a street in Washington, D.C.; the acquaintance casually mentioned that the failed Dumont television network was going up for sale. In January 1959 Kluge acquired Paramount Pictures' share of what remained of DuMont, the television stations of Metropolitan Broadcasting, for $4 million. He then consolidated his radio and television holdings and later bought and sold interests in restaurants, outdoor and direct-mail advertising, and magazines.

Indeed, Kluge never stopped trading radio—if he figured he could make a profit. So in 1982, for example, he sold WMET-FM (Chicago) and KSAN-FM (San Francisco) and acquired KHOW-AM and WWBA-FM in Tampa, Florida. Yet with Federal Communications Commission rules permitting Metromedia to own only seven AM and seven FM radio stations, Kluge held on to stations in top markets because they made the most money. Metromedia held WNEW-AM and -FM in New York City, KLAC-AM in Los Angeles (acquired in 1963), and WIP-AM in Philadelphia (acquired in 1960) for the longest amount of time. Once a station was acquired, Kluge assigned managers to squeeze maximum profits, not caring what format was used. His stations employed all formats: adult contemporary, beautiful music, all-news—any format that worked in that particular major market.

In his heyday, Kluge grew famous for cutting costs and maximizing revenues; indeed, once he had assembled Metromedia, he moved the operation's headquarters out of expensive Manhattan across the Hudson River to Secaucus, New Jersey, where rents were lower. He secured the cheapest possible programming, and then, even with small audience shares, Metromedia could, with bare-bones costs, make a profit. But not every well struck oil. One disastrous misstep was Kluge's purchase of the niche magazine *Diplomat.* Another was his vision of forming a fourth television network, a venture in which Kluge only lost millions of dollars.

In April 1984 Kluge took Metromedia private, and so he possessed three-quarters of Metromedia stock when he sold the seven television stations to Rupert Murdoch a year later. The eventual sale of the radio stations in the late 1980s would make Kluge more than $100 million, a great deal of money, but little compared to the billions made from the sale of the television stations.

But by the 1990s Kluge was again at work building a new corporate empire. He again took a qualified risk by sinking his fortune into paging devices such as beepers and mobile telephones. He bought licenses for operation in major markets, waited as the market evolved, and then sold out at a profit. He also went global, forming Metromedia International Telecommunications to bring wireless cable and communications businesses to the emerging markets of Eastern Europe and the former Soviet republics. In 1994 Kluge began to break into the radio broadcasting business once again by acquiring Radio Juventus in Hungary. By 1999 Metromedia International was a leading operator of radio stations in Eastern Europe and the former Soviet Union, with 15 stations in ten markets.

DOUGLAS GOMERY

Further Reading

Compaine, Benjamin M., and Douglas Gomery, *Who Owns the Media?* New York: Harmony Books, 1979; 3rd edition, Mahwah, New Jersey: Erlbaum, 2000

Gelman, Morris J., "John Kluge: The Man with the Midas Touch," *Television Magazine* (July 1964)

Gomery, Douglas, "Vertical Integration, Horizontal Regulation—The Growth of Rupert Murdoch's Media Empire," *Screen* 27, nos. 3–4 (May–August 1986)

Kluge, John W., *The Metromedia Story,* New York: Newcomen Society in North America, 1974

Metropolitan Opera Broadcasts

Bringing Opera into the Home

Performances of New York's Metropolitan Opera have been broadcast regularly on radio since 1931 and occasionally on television since 1977. Milton Cross was the voice of these broadcasts for more than four decades. The continuous support of Texaco (ChevronTexaco after 2000) from 1940 to 2004 formed what was probably the world's longest-running commercial broadcast sponsorship.

Origins

New York's Metropolitan Opera Association was formed in October 1883, and soon the city was presented with an annual season of fine opera performances by top-drawer orchestras and singers. But for decades the only way to hear and see a "Met" performance was to purchase an expensive ticket and attend a program in New York.

The first hint of an alternative means of delivering opera came on 13 January 1910. With the permission of the Opera's assistant director, wireless inventor (and opera lover) Lee de Forest set up one of his transmitters high in the attic above the stage with a temporary bamboo antenna on the roof. Several microphones were placed on the stage. That first transmission included scenes from *Cavalleria Rusticana* and Enrico Caruso singing in *Pagliacci* and was heard primarily by other radio operators and some reporters. And what they heard was anything but a clear signal, given the crude equipment of the time. The poor results did not endear the company's management to the rising medium of radio.

Despite radio's later development, opera director Giulio Gatti-Casazza resisted further experimentation with radio microphones for two decades out of a fear of lost ticket sales. He also felt mere *listeners* would lose the visual aspect of opera. Public reasons given for the lack of radio coverage included technical problems with placement of microphones and contracts with lead singers that forbade such transmission. Yet the Met was being bypassed by others.

The Chicago Civic Opera went on the air in 1922 when station KYW debuted with a focus on opera broadcasts. And other performing companies were heard in other cities. So were some performances by the Manhattan Opera Company that then competed with the Metropolitan Opera. The fan magazine *Radio Digest* began an editorial campaign to get the Met to change its mind that same year. Station WEAF wasn't waiting—they formed their own in-studio opera companies, one for grand and one for light opera—and broadcast their performances for several years. The original language of the composer was used, but performances were cut to fit one-hour time slots. And they took place in a studio, not on an opera stage, limiting what could be accomplished. Broadcasts of opera from Europe could occasionally be picked up by U.S. listeners tuning shortwave. The broadcast sound quality left much to be desired, but at least the operas were being heard.

What finally turned the tide was the Met's need for new sources of operating funds during the depression. NBC secured the broadcast rights for $120,000 for the first season, outbidding rival CBS. Broadcasts began with *Hansel and Gretel* on Christmas Day of 1931 with Deems Taylor providing the initial commentary. He received howls of protest when he timed his comments to appear over the music and soon changed his approach. The first broadcasts also featured announcer Milton Cross, who would remain as host until 1975, doing more than 800 broadcasts and missing only two in all those years.

Metropolitan Opera Auditions of the Air was developed as a separate program (1935–58; on NBC until 1937, then Blue/ABC) and also featured commentary by Milton Cross with Edward Johnson, managing director of the Met, as host. Each week aspiring operatic performers would do their best to earn audience support and a contract from the Met.

Despite the interest of a small but vocal audience, sponsorship for the broadcasts was difficult to arrange and harder to perpetuate. Several backers (American Tobacco, RCA, and Lambert drugs) came and went, and by the late 1930s, opera broadcasts from New York were threatened by a lack of continuing advertiser support.

Texaco Sponsorship

Beginning on 7 December 1940, the Texaco oil company took up sponsorship of the weekly broadcasts. Though at first this seemed merely the latest in a changing parade of financial backers, Texaco stuck with the program, pleased with the highbrow audience it attracted. More than six decades later, ChevronTexaco continued to support the Metropolitan Opera broadcasts, forming what was probably the longest continuing relationship of an advertiser and a program in radio history. The broadcasts under Texaco at first continued on NBC, then moved to Blue (which became ABC) until 1958, at which point CBS carried on the series for two years. But Texaco became unhappy with declining network interest and decided to create its own specialized network of stations to carry the Saturday matinees. So in 1960 the Texaco-Metropolitan Opera Radio Network was created, with Texaco arranging for the AT&T connecting lines to link the slowly growing number of stations carrying the broadcasts.

As microphone and other radio technologies improved, so did the sound of the opera on the air. By the late 1940s a more complex multiple microphone technique was being used. The opera company understandably insisted that no microphone be placed where it could be seen, so broadcasts utilized four microphones placed near the stage footlights. These were aimed at the floor to receive a more equal (reflected) sound of the varied singing voices and spoken lines and were supplemented by two more microphones suspended above the orchestra. Stereo transmissions were added in the 1960s. The entire radio operation was upgraded with the move of the Met into its new opera house in New York's Lincoln Center in 1966. The opera network was connected by satellite in the 1980s.

Over the years the opera broadcasts were supplemented with the "Texaco Opera Quiz" and "Opera News on the Air," features that became very popular. On Milton Cross' death in 1975, Peter Allen took over the host role and continued it into the 21st century. In 1977 several operas were televised on the Public Broadcasting Service, which still does three or four such performances each season (with English subtitles). Texaco also supported formation of a Media Center to archive past broadcasts.

Over the years the audience for the programs grew, first in the United States and then beyond. The 20-week (November–April) season of Saturday matinees has long been transmitted throughout the United States and Canada (more than 325 stations across North America), and beginning in 1990, to 27 European countries. Australia and New Zealand joined the network in 1997, and in 2000 the Texaco network welcomed Brazil and Mexico. By the end of the 2000–2001 season, Texaco had sponsored 1,212 Metropolitan Opera broadcasts of 144 different operas.

In May 2003 ChevronTexaco (as the sponsor had become after a 2000 merger) announced that the 2003–04 season of matinee broadcasts would be the last they would sponsor, ending a nearly 65-year run, the longest continuous sponsorship in radio history. The Metropolitan Opera said it would continue the broadcasts and seek new sponsors. ChevronTexaco said the series had been costing about $7 million a year, reaching some 10 million listeners in 42 countries. Their announcement to terminate sponsorship came after the oil company saw a drop in both profits and stock price.

CHRISTOPHER H. STERLING

See also Classical Music Format; Cross, Milton; De Forest, Lee; KYW; Music; Taylor, Deems

Programming History

NBC Red or Blue (sometimes both)	1931–40
NBC Red	1940–43
Blue Network/ABC	1943–58
CBS	1958–60

Texaco-Metropolitan Opera Radio Network	1960–2000
ChevronTexaco-Metropolitan Opera Radio Network	2000–2004

Further Reading

De Forest, Lee, *Father of Radio: The Autobiography of Lee De Forest,* Chicago: Wilcox and Follett, 1950

DeLong, Thomas A., *The Mighty Music Box: The Golden Age of Musical Radio,* Los Angeles: Amber Crest Books, 1980

Jackson, Paul J., *Saturday Afternoons at the Old Met: The Metropolitan Opera Broadcasts, 1931–1950,* Portland, Oregon: Amadeus Press, and London: Duckworth, 1992

Jackson, Paul J., *Sign-Off for the Old Met: The Metropolitan Opera Broadcasts, 1950–1966,* Portland, Oregon: Amadeus Press, and London: Duckworth, 1997

LaPrade, Ernest, *Broadcasting Music,* New York: Rinehart, 1956

Metmaniac Opera Archive Online <www.metmaniac.com>

Pogrebin, Robin, "Chevron Texaco to Stop Sponsoring Met's Broadcasts," *New York Times* (21 May 2003)

Mexico

Radio reaches more people in Mexico than any other electronic medium of communication. Despite the broad reach of television and the growing role of cable and the internet, radio broadcasting continues to be a central arena of commerce, culture, education, and politics. Beginning in the 1920s and 1930s, radio's dynamic sound format made it an ideal medium for Mexico's orally based popular cultures, including music, verbal humor, and melodrama. Despite the strong influence of the U.S. commercial broadcasting model, radio practices in Mexico reflect the interaction of a variety of social forces: local popular cultures, the national state, transnational commercial interests, and civic groups.

Broadcasting Begins: The 1920s and 1930s

Beginning with Mexico's first radio transmission in 1908, radio was the domain of amateur operators who used the medium primarily for point-to-point communication. With the authorization of the first non-experimental broadcasting stations in 1923, however, state and commercial interests began to develop radio as a broadcast medium. State-sponsored stations were operated by specific branches of government (e.g., the Navy, the Ministry of Education) with limited public funding. Commercial stations were financed by entrepreneurs who drew on both transnational capital and domestic capital accumulated in ancillary fields (e.g., newspaper publishing, electronics, retail sales). The first commercial station, CYL, was launched by Luis and Raul Azcárraga, distributors of U.S. radio parts and receivers, in partnership with the Mexico City newspaper *El Universal.* Another early station, CYB (later XEB), was started by a cigarette company, El Buen Tono, with French financial backing. Sixteen stations were broadcasting by 1926, and nineteen by 1929, the year Mexican stations received the "XE" and "XH" call letter designations.

Beginning in the early 1920s, the U.S. government and U.S. media interests made a concerted effort to open broadcasting markets in Mexico and Latin America to U.S. commercial interests. At the 1924 Inter-American Conference on Electrical Communications held in Mexico City, however, Latin American delegates resisted this offensive and argued that a strong governmental role in radio was essential to economic self-determination. Although they conceded to the principle of commercial competition where feasible, Latin American delegates resolved that electronic communication media were public services over which national governments held direct control. These resolutions became the basis for Mexico's first radio law of 1926, the Law of Electrical Communications (LCE) and the subsequent Law of General Means of Communication (LVGC).

Despite considerable U.S. pressure, Mexico's radio regulations were strongly influenced by state activism and economic nationalism. The 1926 LCE declared the airwaves to be a national resource, allowed only Mexican citizens to own or operate radio stations, and prohibited any transmissions that attacked state security, public order, or the established government. The LVGC established a system of 50-year concessions for commercial broadcasters. Concessionaires were prohibited from making political or religious broadcasts and were required to carry government transmissions free of charge. Further regulations prohibited radio studios from being located on foreign soil and required broadcasting in Spanish. In practice, this regulatory framework ensured the rapid development of commercial broadcasting while giving the state a privileged position of access and control within a highly nationalistic broadcasting system.

Commercial broadcasting grew enormously during the 1930s with an infusion of advertising revenue from transnational companies like Colgate-Palmolive and Proctor and Gamble. Although many prominent regional stations started in the early 1930s, Mexico City became the undisputed power center of commercial broadcasting. The centerpiece of Mexico City broadcasting was XEW, "La Voz de América Latina desde México" ("The Voice of Latin America from Mexico"), founded by Emilio Azcárraga in 1930. XEW soon became the most powerful and most popular radio station in the country. By the late 1930s Azcárraga had organized two national radio networks anchored at stations XEW and XEQ in Mexico City. Along with the rise of the Azcárraga group, the late 1930s witnessed the consolidation of the commercial broadcasting industry and the formation of a powerful industry group that became the Radio Industry Chamber (the CIR, later CIRT) in 1942.

As commercial broadcasting advanced, the central state also expanded its efforts to harness the new medium. The number of government-operated radio stations reached a peak of 14 during these years. Important stations were operated by the National Revolutionary Party (PNR, later the PRI), the Ministry of Public Education, and the Autonomous Department of Press and Publicity. Beginning in 1937 the government produced a "National Hour" program of official information and national culture that all stations were required to broadcast weekly. At the same time, radio regulations promoted national culture by requiring all broadcasters to include 25 percent "typical" Mexican music in their broadcasts. The 1930s also saw the beginning of state-sponsored radio education projects, which included the distribution of radio sets to rural schools and working class neighborhoods.

Whereas comedies and drama serials dominated U.S. broadcasting during the 1930s, music was the mainstay of Mexican radio. This was due, in part, to an explosion in the production and circulation of popular music in Mexico and Latin America beginning in the 1920s. Musical programs included boleros, mambos, and rancheras performed by orchestras, bands, and soloists, often interspersed with comedy interludes in a "musical variety" format. In addition, radio regulations created a climate that promoted national musical forms over the radio medium. Commercial broadcasters found that Mexican singers and musicians performing popular Mexican tunes provided the ideal radio content to satisfy state nationalism and capture the national broadcasting market. Mexican orchestras featuring performers like Agustín Lara and Toña la Negra dominated the XEW network schedule.

Although dramatic series were less prominent than music during the 1930s, broadcasters developed a number of dramatic and nonfiction program formats by modifying and adapting U.S. soap operas, game shows, and news and sports programs. Radionovelas, for example, combined the techniques and formats of U.S. soap opera production with the melodrama of Latin American music, film, and theater.

Radio listening grew considerably over this period as the number of stations grew, the quality and consistency of programming improved, and the price of radio sets dropped. Listeners had been scarce in the 1920s, but between 1930 and 1935 the number of radio receivers grew to an estimated 200,000 sets and reached an estimated 600,000 sets by the end of the decade. Surveys from Mexico City indicate that well over half of the urban population had regular access to the medium by the late 1930s, although radios were still rare in rural areas and smaller regional cities.

A Golden Age: The 1940s and 1950s

The 1940s and 1950s were characterized by steady growth and consolidation in the radio industry and the increasing popularity of the radio medium. Between 1941 and 1950 the number of radio sets grew from more than 600,000 to almost 2 million. Assuming approximately six listeners for every radio receiver, a 1943 survey indicates that over 90 percent of the populations of Monterrey and Torreón had regular access to radios, compared to 79 percent in Guadalajara, 68 percent in Mexico City, and only 33 percent in Puebla and Morelia. In 1941 Azcárraga and his partner Clemente Serna Martínez consolidated their radio holdings into a single company, Radio Programs of Mexico (RPM). By the mid-1940s RPM networks encompassed nearly half the stations in Mexico and included more than 30 affiliates in Central America, the Caribbean, and the northern rim of South America. Several other national networks emerged in the 1940s, including networks anchored by stations XEB and XEOY-Radio Mil.

Mexican broadcasters' regional power became evident in 1946 when the chief organ of the radio industry, the CIR, organized the Inter-American Association of Radiobroadcasters (AIR) to promote commercial broadcasting throughout Latin America. In 1948 AIR delegates drew up the *Bases of the AIR*—a set of principles to guide broadcasting legislation in the region. These principles declared broadcasting to be a fundamentally private activity undertaken in the public interest. The document argued that government regulations should be limited to the technical aspects of broadcasting and that the state should be prohibited from competing with private broadcasters for advertising dollars. The unstated goal of the AIR was to ensure that the commercial model of broadcasting would guide television development in the region.

World War II had a significant impact on the consolidation and expansion of the Mexican radio industry. During the war the U.S. government sponsored a massive propaganda recognition and spread to stations throughout the United States and Canada.

Over the next five years on KHJ, Morgan earned unparalleled shares in the Arbitron radio ratings, with over 20 percent of Los Angeles listeners tuned to him each morning. He became the voice of Boss Radio and introduced a cast of characters, slogans, and routines that made him even more popular. His "Getting Morganized," a machine he created to give his listeners a zap of extra luck or energy, became so popular that he took it with him when he left KHJ.

During these years, Morgan did all of KHJ's promotions and images, and he began to take on work outside of Los Angeles. He became known as the quintessential boss jock, and his style was frequently copied by morning disc jockeys throughout the country. He was one of the best-known broadcasters in the country, and in 1967 he received *Billboard* magazine's "Air Personality of the Year Award."

In 1969 Morgan narrated and coproduced (with Ron Jacobs) the first ever "rockumentary," the 48-hour *KHJ History of Rock and Roll.* It aired worldwide, and when it was broadcast in the Los Angeles market, it received a 60 share in the ratings.

During this time, Morgan also began appearing on television, hosting and narrating for several shows, including *Boss City, Morgan's Alley,* the American Broadcasting Companies' (ABC) *In Concert,* the National Broadcasting Company's (NBC) *The Helen Reddy Show,* and the local KHJ-TV channel 9's *Groovy Show.*

Morgan left KHJ in 1970 to move to Chicago to work at WIND, but, not liking the cold climate, he returned to Los Angeles in 1972 to his morning drive slot at KHJ. He remained there until 1974, when he took a job at FM station K-100 (KIQQ) another Drake-Chenault programmed station. The team re-created their AM radio success on the FM dial, and soon it became the number-one morning program in Los Angeles.

In 1975, Morgan took a job at KMPC-AM as a staff announcer, working a split shift on the weekends and filling in for the weekday staff, including legendary Dick Whittinghill. In 1979 Morgan took over for Whittinghill when he retired. At KMPC Morgan formed his "Good Morgan Team," a group of radio professionals that specialized not only in music, but also in news, weather, sports, and traffic. He stayed at KMPC through the early 1980s despite several format changes and finally a transition to talk radio. Even this did not hurt his popularity: he had the best ratings on the station and became highly successful at the talk format as well.

During this time, Morgan also cohosted a local television show on KNBC-TV called *The Everywhere Show* and became the announcer on the nationally syndicated musical series *Solid Gold.* He also hosted several nationally syndicated radio shows, including *Record Report* and the ABC/Watermarks *The Robert W. Morgan Special of the Week,* on which Morgan interviewed the most popular music personalities in the world. He even recorded a program for TWA's in-flight radio called *Morgan's Manor.* During this period, he also recorded thousands of radio and TV commercials, movie trailers, and documentaries.

In 1982 Morgan left KMPC to return to a music show and took a job at Emmis Broadcasting's KMGG Magic 106 FM. In 1985 he returned to KMPC when they returned to a music format, and in 1986 he gave the last broadcast of 93 KHJ when it signed off the air (the station was sold and subsequently became a Spanish format).

In August 1992 Morgan took over as the morning disc jockey at KRTH (K-Earth) 101-FM, oldies radio, and was reunited with The Real Don Steele. He remained there until 1998, when lung cancer forced him to retire. When K-Earth 101 held a retirement tribute for him at the Museum of Television and Radio in Beverly Hills, it was hosted by Dick Clark, who spoke at great length of Morgan's influence on Top 40 radio. In 1993 Morgan received a star on the Hollywood Walk of Fame, and he was inducted into the National Broadcasters Hall of Fame in 1994. He was also inducted into the Ohio Broadcaster's Hall of Fame. He died of lung cancer in 1998 and was posthumously inducted into the Radio Hall of Fame at Chicago's Museum of Broadcast Communications. Samples of his work are on permanent display at the Museums of Television and Radio in Beverly Hills and New York and at the International Broadcasting Congress Archives in Brussels.

JUDITH GERBER

See also Disk Jockeys; Drake, Bill; KHJ; Radio Hall of Fame

Robert W. Morgan. Born in Mansfield, Ohio, 23 July 1937. Oldest child of Arthur and Florence Morgan; attended Wooster College and worked at WWST, 1955; worked at several California radio stations, such as KMAK, Fresno, and KEWB, San Francisco; morning disc jokey at KHJ, Los Angeles, 1965–71 and 1972–73; disc jockey, various Los Angeles stations 1973–98; retired, 1998. Inducted into Broadcasters Hall of Fame, 1994, and Radio Hall of Fame, 1999. Died in Los Angeles, California, 22 May 1998.

Radio Series

1969	*KHJ'S History of Rock and Roll*
1976–81	*Robert W. Morgan Special of the Week*
1977	*Record Report*

Television Series

Morgan's Alley, 1968; *Boss City,* 1968; *Groovy Show,* late 1960s; *In Concert,* 1973–75; *Everywhere Show,* 1979

Further Reading

Barrett, Don, *Los Angeles Radio People: Volume 2, 1957–1997*, Valencia, California: db Marketing, 1997

Fong-Torres, Ben, *The Hits Just Keep on Coming: The History of Top 40 Radio*, San Francisco: Miller Freeman Books, 1998

MacFarland, David, *Development of the Top Forty Radio Format*, New York: Arno Press, 1979

Mormon Tabernacle Choir

U.S. Choral Group Featured in Radio Broadcasts

The Mormon Tabernacle Choir's broadcast program, *Music and the Spoken Word*, began soon after The Church of Jesus Christ of Latter-day Saints began broadcasting in 1922 on KZN (today's KSL) radio in Salt Lake City. Only rehearsals were broadcast occasionally at first, in part because the musicians were skeptical of the fidelity of radio. The first formal broadcast came on 26 June 1923, when President Warren G. Harding also spoke. The choir began its regular live performance schedule when KSL became an NBC affiliate, and on 15 July 1929 the first regular network Tabernacle Choir program was aired nationally. Today, it claims the title, "longest running continuous network radio program in America" with more than 3,600 broadcasts to its credit. The performance is delivered live when every Sunday morning more than 2,000 radio, television, and cable operations broadcast the program worldwide.

Choir History

The Mormon Choir first sang in Utah's Salt Lake Valley more than 150 years ago, coming into existence in 1847. There was no "Tabernacle" or organ at that time so the group sang in an improvised bower of trees where adobe blocks and poles supported the roof of leaves and branches. From this beginning it played a central part in the early church's commitment to celebrate culture—both sacred and secular events. The historic auditorium, the Tabernacle, is a dome-shaped building in Salt Lake's Temple Square, first used in 1867. The Tabernacle organ was installed that same year and it has become the most recognized symbol of the choir.

In 1849 John Perry, a Welshman, became the choir's first regular director. Perry was followed by others who brought formality and discipline to the choir organization. Evan Stephens, also from Wales, conducted the choir for 27 years. Under his leadership, the choir grew from 125 singers to over 300. Stephens directed the choir on its first major concert tour in 1893 and is credited with laying the foundation for the choir's growing international acclaim.

The choir program tradition consists of song, organ recitals, and a short non-denominational sermon dubbed the "Spoken Word." For more than 40 years, Richard L. Evans provided the latter—thought-provoking, inspirational messages usually two to three minutes in length, all eventually published in a series of books. Upon his death, Evans was replaced by J. Spencer Kinard, who worked for 19 years. He was replaced by the current voice, Lloyd Newell.

The Choir on the Air

The choir began its national broadcast history with the NBC network in 1929, when KSL became an affiliate. When KSL switched to the CBS network in 1933, the choir followed. The "Spoken Word" unit of the program was added in 1936.

It is no wonder that the fidelity of these first programs was questioned by the musicians. As the story goes, KZN/KSL apparently owned only one microphone in those early days. Thus, according to a prearranged plan, the station briefly went off the air as a courier dashed across the street from the station to the Tabernacle carrying that one microphone. There, a tall stepladder was installed near the organ console, and the announcer climbed to his precarious perch atop that ladder holding the microphone that was to pick up both the music and the announcer's words.

Coverage of the choir has grown with technology. In 1948, television broadcasts began in Salt Lake and the choir was among the station program lineup. In 1961, it became a part of the church's launch into international shortwave radio. In the 1970s, the choir took part in the first satellite broadcasts. Today, choir presentations use radio, television, cable, satellite, motion picture recording technology, and the internet in its world-wide distribution. The choir has made more than 150 recordings (some of them CDs), five of which have attained gold status with sales of over 500,000, and two have received platinum awards for sales of more than 1 million. Perhaps best known is their classic recording of "Battle Hymn of the Republic," recorded with the Philadelphia Orchestra, for which the choir was awarded a Grammy in 1959.

ELDEAN BENNETT

See also KSL

Further Reading

Evans, Richard L., *Unto the Hills,* New York: Harper, 1940
Evans, Richard L., *This Day and Always,* New York: Harper, 1942
Godfrey, Donald, Val E. Limburg, and Heber Wolsey, "KSL, Salt Lake City: At the Crossroads of the West," in *Television in America: Local Station History from Across the Nation,* edited by Michael D. Murray and Donald G. Godfrey, Ames: Iowa State University Press, 1997
Hicks, M., *Mormonism and Music: A History,* Urbana: University of Illinois Press, 1989
Jeffery, Charles, *The Mormon Tabernacle Choir,* New York: Harper and Row, 1979
Limburg, Val E., "Mormon Tabernacle Choir," in *Historical Dictionary of American Radio,* edited by Donald G. Godfrey and Frederic A. Leigh, Westport, Connecticut: Greenwood Press, 1998
Mormon Tabernacle Choir website, <www.musicandthespokenword.com>

Morning Edition

U.S. Newsmagazine Program

For millions of Americans, weekday consciousness begins with *Morning Edition:* the clock radio turns itself on to the razzle of B.J. Liederman's theme music or the resonant calm of host Bob Edwards.

The program began on 5 November 1979. After National Public Radio (NPR) scored a success with its afternoon newsmagazine, *All Things Considered,* the network's member stations wanted a morning service. They were not necessarily looking for a program. Most member stations at the time ran classical music in the morning, but they wanted news and feature elements they could drop into those programs. Unlike *All Things Considered,* which had evolved on the air, *Morning Edition* was to be a planned creation, conceived by a committee with input from stations and guidance from audience researchers. The format was to be a two-hour series of segments, with a mix of news, sports, arts, and features. None of the segments was to be longer than nine minutes, with fixed times so that stations could take what they wished and cover the rest with local news, weather, and traffic. But production of the pilot was handed over to two morning newsmen from the immensely popular Washington commercial station WMAL. As former producer Jay Kernis recalls, "they knew how to do *AM* drive-time radio, but they broke the promise of public radio. They ignored the audience we'd been building." The pilot failed. Host Bob Edwards remembers them as sounding "like a bad talk show in a small market."

NPR's news director, Barbara Cohen, fired the producers and the first two hosts. Kernis, the arts producer, was promoted to senior producer and "given ten days to re-invent the show and teach the staff how to produce it." Frank Fitzmaurice came in as executive producer to oversee the program's news content. Bob Edwards, who had spent five years as cohost of *All Things Considered* with Susan Stamberg, was recruited to fill in as host for 30 days, until the network could find someone to take over.

Jay Kernis recalls how he and Edwards thought through the role that began Edwards' more than two decades as the solo host of the program. "A host is not an announcer," Kernis says. "A host is the glue that holds the show together. There should be this vortex of information, and in the calm center, there should be the host, Bob, steady Bob, carrying it all back to you."

Kernis also preached to his staff that the show's strict format should be liberating rather than restricting, a concept he says he got from the late arts producer Fred Calland, who pointed out that *All Things Considered* had to create a new architecture in its relatively free-form 90 minutes every day. The program could succeed or fail, depending on whether that structure was successful. With *Morning Edition,* the architecture was a given. As Calland put it, "you know the perimeters of the canvas, now you can paint."

One example of working within the structure came when Kernis had to write a short piece of advance copy for Bob's

U.S. Congressmen who criticized rock and roll as being repulsive African jungle music. Later in the year, WINS-AM, perhaps unable to compete with Murray the K's institutional imitators WABC and WMCA, was sold to Westinghouse, Inc., and its format was switched to all news. After eight years, Murray the K left WINS radio.

By early 1966 Kaufman had established "Murray the K's World" in Garden City on Long Island, New York—a multilevel and multimedia entertainment complex converted from an abandoned airplane hangar. Although short-lived, the nightspot is said to have influenced music venues such as the Plastic Inevitable, Manhattan's Cheetah, and the Fillmore Auditorium in San Francisco. Unfortunately, Kaufman suffered huge financial losses from his intended month-long run of rock shows, due to a subway workers strike. According to his lawyer, the strike prevented performers and audiences from coming to the theater after the first few performances. He sought $25 million in damages.

The determined Murray Kaufman then turned his efforts to another television concept for the Office of Economic Opportunity, this time at the local level, producing at least two highly rated television programs for New York's Channel 5. His first was a 90-minute combination of in-concert and on-location clips of popular recording artists, with drop-in talk segments by other noted celebrities. The first show drew an estimated 3 million viewers to two broadcasts. The second show was entitled *Murray the K's Special for the Year 2000*. It aired in May 1966 and sought to explain the vast differences in language, music tastes, and dress that would exist between youth and adults at the turn of the century.

Early in 1966 WOR Radio hired Kaufman as its program director. However, in July 1966 management shifted Kaufman and three other disc jockeys from WOR-AM to WOR-FM and cut their wages by 50 percent. FM radio was still developing its economic base, but the American Federation of Television and Radio Artists represented the four disc jockeys in their efforts to retain their AM salaries. Kaufman and the others went on strike until October 1966 when the matter was resolved, although they remained at lower salaries.

The low frequency of commercials on FM radio allowed Kaufman to initiate an indelible radio programming innovation by pioneering progressive radio. He emphasized the perspective that albums contained many good songs that were appropriate for airplay. He also encouraged exposing the full artistic expressions of recording artists by playing their extended-time productions. Nonetheless, a very proud Murray the K cut short his tenure at WOR-FM in 1967 when program consultant Bill Drake was brought in by management to institute programming guidelines and policies. Before departing, Kaufman went on the air and blasted WOR's decision, just as he had done previously at WINS.

From 1967 to 1976 Kaufman worked short stints at various radio stations but never achieved the high points of his earlier career. He began with CHUM (Canada) from 1967 to 1968. Later in 1968 he took a weekend slot on his old WMCA station from 2:00 to 7:00 P.M. on Sundays. After several other brief stints, in the late 1970s he moved to California with his sixth wife, soap opera actress Jacklyn Zeeman. He had three sons from previous marriages. His last job in radio was hosting a syndicated radio program for Watermark.

Kaufman, a smoker, had battled cancer since 1973 and was finally forced to relinquish the job with Watermark in 1981 due to his failing health. After retirement he was financially strained but lived in comfortable surroundings. (Tony Orlando provided him with a 24-hour nurse and housekeeper.) Murray Kaufman died of cancer in Los Angeles on 21 February 1982.

LAWRENCE N. REDD

Murray the K. Born Murray Kaufman in New York City, 14 February 1922. Song plugger for Bob Merrill, 1950s; popularized Merrill's "How Much Is That Doggie in the Window," 1950s; president, National Council of Disc Jockeys, 1950s; spearheaded council's relief effort during Hungarian Revolution, 1956; late-night radio disc jockey, WMGM, 1957; hosted *Swingin' Soiree* for WINS, 1958; spokesperson for government youth program "New Chance," 1965; hosted "Murray the K's Special for the Year 2000" for television, 1966; worked short stints at various radio stations, 1967–76. Died in Los Angeles, California, 21 February 1982.

Publication

Murray the K Tells It Like It Is, Baby, 1966

Further Reading

"Dropout TV Show Irks Republicans," *New York Times* (30 June 1965)

"Jockeys Finally Catch up with WOR-FM," *New York Times* (7 October 1966)

Price, Richard, "Going Down with Murray the K," *Rolling Stone* (15 April 1982)

"Youth Wants to Uh Uh Uh," *New York Times* (15 May 1966)

Murrow, Edward R. 1908–1965

U.S. Radio Journalist

Journalists from broadcast and print media alike consider Edward R. Murrow one of the greats of his time. His legacy was to be among the first to put radio news into a league of respectable journalism. He later helped to launch serious television journalism and started the first TV newsmagazine program. Most important, he established traditions of courage and integrity in the profession of broadcast journalism at a time when radio was still developing the tone of its news function.

Murrow's work spanned from the onset of World War II until the Kennedy administration, from his radio reporter years to president of Columbia Broadcasting System (CBS) News, to director of the U.S. Information Agency. His on-air radio reports of events helped shaped the nation's ideas of what was occurring in the world. His dramatic language in radio formed aural word pictures that were rare and striking to his listeners, and he carried his style into the new medium of television. Murrow became a mentor to many in broadcast news, who carry his ethics and style into today's news efforts. His legacy can be found in the most prestigious journalism awards, in schools named after him, and in historical accounts of corporate and government integrity.

Origins

Murrow was born on 25 April 1908 in Polecat Creek, North Carolina, as Egbert Roscoe Murrow, the son of a farmer. He had two older brothers, Lacey and Dewey. When Egbert was six years old, the family migrated by train to the Northwest. They ended up near Blanchard, in the northwest corner of Washington State, where the family struggled to make ends meet.

As a teen, Egbert changed his name to Edward. He became active in sports, debate, and drama, and he even served as a bus driver for the scattered students of his small school. After his graduation from high school, he worked as a logger in the virgin woods of the Olympic Peninsula. It was in this setting that he learned both virtues and vices, appreciation for the rugged beauty of the Olympic Mountains, and the language and smoking habits of the rough brotherhood of lumberjacks, which tempered the Quaker traditions of his family.

After a year and two summers of working in the woods, Murrow enrolled in 1926 at Washington State College (WSC; now Washington State University) in Pullman, on the east side of the state. His intentions were to pursue a curriculum in prelaw and then go to law school. But circumstances led him to a public-speaking course taught by a dedicated and inspiring teacher, Ida Lou Anderson. She was demanding, but she recognized a talent in Murrow and helped him to develop his potential. Eventually she became his mentor and instilled in him a love of the use of language, a somber introspection regarding his own values, a love of philosophy, and a flare for the dramatic. He went on to enroll in the first broadcasting course, "Radio Speaking."

After his first year of working for his room and board, he became involved in the Greek fraternity system at WSC. Through the political influence of his peers, Murrow was elected student body president for his senior year. He graduated in 1930. He then became an officer for the National Student Federation of America in 1932, traveling to New York City with a meager living expense to run the national office. Soon, Morrow became active in encouraging student exchange among various countries, working for the International Institute of Education.

Honing Radio Journalism

In 1935, Murrow was hired as director of talks at CBS. In this position, he traveled to Europe to line up speakers, and he was there in 1938 at the time of Hitler's invasion and annexation of Austria. The event was described by Murrow in a radio report from Vienna (the beginning of his journalistic career), in which he described the mood of the people, the political setting, and the street scenes in vivid details that made his verbal descriptions come to life. This was to be the first of many descriptions he would later give of the effects of World War II.

Murrow's observations from the European front often came from mere notes used when he dictated his descriptions from his mind's eye. His descriptions and dramatic use of language captured the ear of American listeners, who were compelled not only by the events but by Murrow's style as well.

As the war intensified, Murrow began to broadcast about war events from the basement of the British Broadcasting Corporation (BBC)'s Broadcast House. Murrow turned to shortwave radio for his 1940–41 London "Blitz" broadcasts. The broadcast quality was questionable, with fading and fidelity dependent on the earth's atmosphere. Yet it was those broadcasts, often sent late at night London time for the convenience of listeners on the U.S. East Coast, that led to his first real fame. They were timely, compelling, and immediate for a world breathlessly awaiting word about the quickly growing war. And each one began with his standard and quite dramatic signature opening: "This . . . is London." (In later years he said that Ida Anderson had suggested to him that he insert the pause after the first word.)

Edward R. Murrow
Courtesy CBS Photo Archive

His descriptions of the English people under siege carried the insights of a sociologist:

> This is a class conscious country. People live in the same small street or apartment building for years and never talk to each other. The man with a fine car, good clothes and perhaps an unearned income doesn't generally fraternize with the tradesmen, day laborers and truck drivers. His fences are always up. He doesn't meet them as equals. He's surrounded with certain evidences of worldly wealth calculated to keep others at a distance. But if he's caught in Piccadilly Circus when the sirens sound, he may have a waitress stepping on his heels and see before him the broad back of a day laborer as he goes underground. If the alarm sounds about four in the morning, as it did this morning, his dignity, reserve and authority may suffer when he arrives half-dressed and sleepy, minus his usual defenses and possessed of no more courage than those other who have arrived in similar state. . . . Maybe I'm wrong . . . but I can tell you this from personal experience, that sirens would improve your knowledge of even your most intimate friend (Broadcast, 4 September 1939).

Murrow reported not only on the war situation, but on the people. He spoke of girls in light dresses, boys sobbing, old toothless men, and women clutching their belongings as they left their bombed-out homes. His characterizations were vivid and insightful:

> I'm standing tonight on a rooftop looking out over London. . . . Out of one window there waves something that looks like a white bed sheet, a window curtain swinging free in this night breeze. It looks as though it were being shaken by a ghost. There are a great many ghosts around these buildings in London. . . . Down below in the streets I can see just that red and green wink of the traffic lights; one lone taxicab moving slowly down the street. Not a sound to be heard. As I look out across the miles and miles of rooftops and chimney pots, some of those dirty-gray fronts of the buildings look almost snow-white in this moonlight here tonight (Broadcast, 22 September 1940).

Sometimes Murrow's courage bordered on foolhardiness. On one notable occasion, and ignoring direct network orders that he not place himself in harms way, he accompanied a B-17 bomber crew on a bombing mission over Berlin. His descriptions became some of his most memorable writing:

> The clouds below us were white, and we were black. D-Dog [the plane] seemed like a black bug on a white sheet. The flack began coming up. . . . The small incendiaries [we dropped were] going down like a fistful of white rice thrown on a piece of black velvet. As Jock hauled the Dog up again, I was thrown to the other side of the cockpit, and there below were more incendiaries, glowing white and then turning red. The cookies—the four-thousand-pound high explosives—were bursting below like great sunflowers gone mad. And then, as we started down again, still held in the lights, I remembered that the Dog still had one of those cookies and a whole basket of incendiaries in his belly, and the lights still held us. And I was very frightened (Broadcast, 3 December 1943).

As the war ended, Murrow accompanied the troops into the concentration camps of Germany. He told his listeners:

> If you are at lunch, or if you have no appetite to hear what Germans have done, now is a good time to switch off the radio, for I propose to tell you of Buchenwald. . . . The prisoners crowded up behind the wire. We entered. . . . There surged around me an evil-smelling horde. Men and boys reached out to touch me; they were in rags and the remnants of uniform. Death had already marked many of them, but they were smiling with their eyes. . . . When I entered [one of the barracks] men crowded around, tried to lift me to their shoulders. They were too weak. Many of them could not get out of bed. I was told that this building had once stabled eighty horses. There were twelve hundred men in it, five to a bunk. The stink was beyond all description (Broadcast, 15 April 1945).

Postwar Broadcast Journalism

After the war, Murrow returned to the states and to CBS, which was now a much different organization than the one he had left in 1937. William Paley, head of CBS, persuaded Murrow to take over the network's news organization. His radio war broadcasts had become famous, and his fame would certainly help enhance the network's programs. In his new position, Murrow tried to reflect American life, with both its shortcomings and strengths. He created the radio programs *As Others See Us,* a report on how the foreign press viewed the United States, and *You Are There,* a recreation of historical events. And, since radio had been looked down upon by the print media, Murrow produced the program *CBS Reviews the Press,* in which he ensured that the criticism of radio would become a two-way street, and that "mutual criticism will benefit both."

Eventually Murrow was joined by a young producer, Fred Friendly, and together they produced *I Can Hear It Now,* a 45-minute record that was soon followed by others. The 1950–51 network program *Hear It Now* evolved from that project. It was a sound documentary, a kind of magazine on the air, covering several subjects each week in its hourly network format.

Murrow reluctantly entered into television with Friendly. They produced a 1952 TV counterpart to their radio series entitled *See It Now*. Early programs appear rough and unpolished compared to today's network magazine shows, but the focus was not on appearance but on substance.

Not all of Murrow's work was flattering to the news efforts of his day. In 1958, in an address to his colleagues at the meeting of Radio and Television News Directors Association, he told a startled audience:

> So far as radio—that most satisfying and rewarding instrument—is concerned, the diagnosis of its difficulties is rather easy. . . . In order to progress, it need only go backward. To the time when singing commercials were not allowed on news reports, when there was no middle commercial in a 15-minute news report, when radio was rather proud, alert and fast. . . . If radio news is to be regarded as a commodity, only acceptable when saleable, then I don't care what you call it—I say it isn't news.

The same speech was highly critical of television, which helped to further sour Murrow's already poor relations with CBS.

After the election of John F. Kennedy as president in 1960, Murrow was chosen to head the U.S. Information Agency. Once critical of government positions, he now found himself spokesperson for the Kennedy administration and its relay of news throughout the world.

In the fading days of his career, Murrow's contributions were recognized with the Medal of Freedom, the country's highest civilian honor. He was also knighted by Queen Elizabeth for his vital role to England during World War II. Ill health, likely brought on by his chain smoking, forced Murrow into early retirement. He died of cancer in 1965 at the age of 57, leaving his wife, Janet, and one son, Casey.

VAL E. LIMBURG

See also Columbia Broadcasting System; Friendly, Fred; Hear It Now; News

Edward R. Murrow. Born Egbert Roscoe Murrow in Polecat Creek, North Carolina, 25 April 1908. Attended Stanford University and the University of Washington; graduated from Washington State College (now Washington State University), 1930. Served as assistant director of the Institute of International Education, 1932–35; began career with CBS as director of talks and education, 1935; became director of CBS' European Bureau in London, 1937; during World War II, hired and trained distinguished corps of war correspondents, including Eric Sevareid, Howard K. Smith, Charles Collingwood, and Richard C. Hottelet; returned to U.S. as CBS vice-president and director of public affairs, 1946; resigned to return to radio broadcasting, 1947; narrated and produced *Hear It Now* radio series, 1950–51; brought series to television as *See It Now*, 1952–58; began *Person to Person* television program in 1953; moderated and produced *Small World*, television series featuring discussions among world figures, 1958–60; appointed by President John F. Kennedy to head U.S. Information Agency (USIA) in 1961, and remained in post until 1964. Recipient: nine Emmy Awards. Died in New York, 27 April 1965.

Radio Series

1950–51 *Hear It Now*

Television Series

See It Now (1952–58); *Person to Person* (1953–59); *Small World* (1958–60)

Selected Publications

This Is London, edited by Elmer Holmes Davis, 1941

In Search of Light: The Broadcasts of Edward R. Murrow, edited by Edward Bliss, Jr., 1967

Further Reading

Bernstein, Mark, and Alex Lubertozzi, editors, *World War II on the Air: Edward R. Murrow and the Broadcasts that Riveted a Nation*, Naperville, Illinois: Sourcebooks/MediaFusion, 2003

Cloud, Stanley, and Lynne Olson, *The Murrow Boys: Pioneers on the Front Lines of Broadcast Journalism*, Boston: Houghton Mifflin, 1996

Kendrick, Alexander, *Prime Time: The Life of Edward R. Murrow*, Boston: Little Brown, 1969

Persico, Joseph E., *Edward R. Murrow: An American Original*, New York: McGraw-Hill, 1988

Smith, Robert Franklin, *Edward R. Murrow: The War Years*, Kalamazoo, Michigan: New Issues Press, 1978

Sperber, Ann M., *Murrow: His Life and Times*, New York: Freundlich Books, 1986

Museums and Archives of Radio

Repositories of Radio History

There are many public, private, and academic archives and museums whose sole purpose is to preserve radio broadcast documents and programs. Most of them have audio and visual recordings, books, periodicals, pamphlets, oral histories, interviews, and other documents that trace the history of radio programming and radio broadcasting. In addition, several archives and museums are devoted to the history of radio technology itself and the development and advancement of this technology. Still others trace the important figures and individuals in radio broadcasting, radio technology, and radio history. There are also dozens of old-time-radio collector's clubs with less extensive collections of radio broadcasts.

Origins of Museums and Archives of Radio

The idea of establishing formal radio museums and archives began in the 1940s, and such institutions were actually developed in the 1970s. The Broadcast Pioneers Library was the first organized library of radio history; it was begun in the 1960s and formally opened in 1972. However, some less formal collections began much earlier. For example, in 1949 the Library of Congress began to collect and preserve some radio programming. At the same time, the National Archives started collecting and preserving programming from governmental sources and began receiving donated new programs and material from radio stations and networks throughout the U.S.

During World War II, the Armed Forces Radio Services began to produce discs in order to bring radio programs to U.S. troops during the war. These discs would later become the basis of privately traded material. During the same period, a few network and syndicated programs were distributed on discs as well.

In the 1950s the Broadcast Pioneers organization unsuccessfully attempted to establish a museum of broadcast history. During the same decade, individuals began seriously recording and collecting radio programs with the introduction of home-recording equipment and the demise of network radio. Reliable and affordable reel-to-reel recorders were first introduced into the consumer market during this time.

However, radio program collecting did not become truly popular until the 1960s, when classic radio programming began to change dramatically and, many felt, to disappear. As a result, many individuals began to realize that preserving such programs was essential for documenting the history of radio. Individuals began to organize for the purpose of exchanging radio programs, information, and resources. As expected, as radio formats rapidly changed, radio stations began to discard their old stored material and programs. These informal groups began collecting such materials, and a collectors movement started to grow. These groups also created newsletters on radio program collecting.

One of the earliest and most influential of the collectors groups was the Radio Historical Society of America, founded by Charles Ingersoll in the 1960s. Ingersoll also started one of the first newsletters for collectors of old-time radio programs, and *Radio Dial* set the standard for those to follow. One of these was *Hello Again*, started in 1970 by Jay Hickerson. It remains the most popular of old-time-radio collector group newsletters. This newsletter was also part of the formation of the Friends of Old Time Radio. *Hello Again* was successful because it brought together more than 100 of the most active program collectors. According to Professor Marvin R. Bensman of the University of Memphis Radio Archives, "Today, approximately 160-plus active collectors comprise the mass of privately collected broadcast material available."

Another factor that helped spawn the radio collector movement was the sale of radio programs to private individuals. J. David Goldin, a former Columbia Broadcasting System (CBS), National Broadcasting Company (NBC), and Mutual engineer, first mass-marketed and sold radio programs. In the late 1960s, Goldin formed Radio Yesteryear, a company that sells audio recordings of classic radio programs. He also started an album subsidiary of the company called Radiola.

A big boost came for the establishment of the first broadcast history library when William S. Hedges, a former NBC executive, began collecting items for the Broadcast Pioneers History Project between 1964 and 1971. This collection, which consists of nearly 13,000 items, including correspondence, articles, and speeches in 540 different subject categories, formed the core collection of the Broadcast Pioneers Library, which opened in 1972 and led to the establishment of the Library of American Broadcasting.

Public and Academic Museum and Archive Collections

Library of American Broadcasting (University of Maryland)

The first formally established institutional radio archive was the Broadcast Pioneers Library, begun in the 1960s and formally opened in 1972. It was housed in the headquarters of the National Association of Broadcasters (NAB) in Washington, D.C., until 1994. It then became part of the library system at the University of Maryland, College Park, and became known as the Library of American Broadcasting, one of the most extensive collections of the history of broadcasting. The collection consists of audio and video recordings, books, periodicals, pamphlets, oral histories, photographs, personal collections, and scripts that pertain to the history of broadcasting. The library features more than 8,000 volumes ranging from engineering manuals to programming histories. It is particularly strong in its book collection from the early part of the 1920s and 1930s, tracing the evolution of broadcasting.

The library's audio holdings include 1,000 interviews, speeches, news broadcasts, special events, and oral histories (with many accompanied by transcripts) of such important radio figures as Edgar Bergen, Norman Corwin, Leonard Goldenson, Lowell Thomas, and William Paley. There are also thousands of recordings in many formats, including more than 8,300 recorded disks, 25,000 photographs, and 10,000 CDs of commercials in its Radio Advertising Bureau Collection. Also housed here are many specialized collections from radio performers, executives, broadcast engineers, writers, producers, and magazine publishers. Highlights include political speeches from Franklin D. Roosevelt, Winston Churchill, and Harry S. Truman in the Donald H. Kirkley collection; more than 160 recordings of congressional hearings, political speeches, and other media events of the 1960s and 1970s in the Daniel Brechner Collection; and The Center for Media and Public Affairs Collection, which contains talk radio programs.

Some of its other holdings include some 7,000 pamphlets, ranging from 1920s Bell Laboratories radio engineering bulletins to promotional materials from broadcast networks; the Westinghouse News Collection (1958–82), which consists mainly of raw feeds from the Washington bureau; the Associated Press Radio Competition Collection (1967–68), which contains samples of radio journalism, almost exclusively from California.

There is also a collection of government documents that includes the Navigation Bureau List of Radio Stations (1913–27), Federal Radio Commission (FRC) and Federal Communications Commission (FCC) decisions, and congressional reports and hearings.

The museum acquired the Chester Coleman Collection of the NAB Library and Historical Archive in June 1998. This collection includes more than 4,000 books and periodicals. The NAB collection also includes historical meeting and convention minutes, newsletters, promotional materials, and scrapbooks.

National Public Broadcasting Archives (University of Maryland)

An additional archive housed at the University of Maryland is the National Public Broadcasting Archives (NPBA). The archives originated as a cooperative effort between both educational institutions and broadcasting organizations, including the Corporation for Public Broadcasting (CPB), Public Broadcasting Service (PBS), National Public Radio (NPR), the Academy for Educational Development, and the University of Maryland. The idea was spearheaded by Donald R. McNeil, a former PBS board member who was concerned that the history of public broadcasting was at risk.

The archives, which opened 1 June 1990, form part of the Archives and Manuscripts Department of the University of Maryland Libraries. They consist of historical materials from the major organizations of U.S. noncommercial broadcasting. These include PBS, Children's Television Workshop, CPB, NPR, Agency for Instructional Technology, America's Public Television Stations, Association for Educational Telecommunications and Technology, Public Service Satellite Consortium, and the Joint Council for Educational Telecommunications.

The NPBA also has personal papers from many influential public broadcasting figures and a reference library containing basic studies of the broadcasting industry, rare pamphlets, and journals on relevant topics. The archives also house a collection of audio and video programs from public broadcasting's national production and support centers and from local stations. There is also a collection of oral history tapes and transcripts from the NPR Oral History Project.

Museum of Broadcast Communications (Chicago)

The Museum of Broadcast Communications is devoted solely to radio and television broadcasting and is housed on two floors of the Chicago Cultural Center. (The Museum was scheduled to relocate to new premises in Chicago's River North area as of Spring 2004.) The museum's purpose is to educate the "public, teachers, and students about the profound influence of radio, television, and advertising in our world." It does this via hands-on exhibits, broadcasting memorabilia, a public archives collection, and educational outreach programs.

The Museum was founded in 1987 by Bruce DuMont, the nephew of television pioneer Allen B. DuMont, using private contributions. It consists of changing exhibits, radio and television archives, a Radio Hall of Fame, an Advertising Hall of Fame, the Lynne Harvey Radio Center, and a gift shop. The museum's public archive, the Arthur C. Nielsen Jr. Research Center, contains over 85,000 hours of television and radio broadcasts, commercials, and newscasts, with 13,000 television programs, 4,000 radio broadcasts, and 11,000 television commercials, all of which can be screened on site in one of 26 study suites. All programs in the archive's collection are cross-referenced and cataloged in a fully computerized retrieval system. The collection focuses on Chicago television news, talk/interview programs, documentaries, political broadcasts, programs of its Radio Hall of Fame inductees, sports programming, and "Golden Era" television dramas.

Included in its archives is an extensive historic radio program collection, the Chuck Schaden Radio Collection, which contains more than 50,000 programs and is considered to be the largest of its kind in the United States. The Lynne Harvey Radio Center features a live, weekly broadcast of *Those Were the Days,* by radio historian Chuck Schaden, complete with a live studio audience, as well as other live broadcasts. The museum also hosts many special events, including an annual induction ceremony into its Radio Hall of Fame, which pays tribute to the legends of radio. The Hall of Fame was founded by the Emerson Radio Corporation in 1988 and was taken over by the Museum of Broadcast Communication in 1991. There is also a collection of vintage radio and television sets from local donors.

Museum of Television and Radio (New York and Los Angeles)

The bicoastal Museum of Television and Radio (New York and Los Angeles) is devoted to radio and television broadcast history, particularly focusing on the individuals and programs that make up that history. The museum was founded in New York in 1975 by William S. Paley, chairman of CBS, as the Museum of Broadcasting. The museum changed its name and moved to a larger headquarters in September 1991. The New York museum's holdings include some 100,000 radio and television programs, as well as 10,000 commercials. It also includes two screening rooms, two theaters, a group listening room, 96 individual booths equipped with television and radio consoles, a research library, and a gift shop. In addition, there are three public galleries that display broadcast industry artifacts.

In March 1996 the Los Angeles branch of the museum opened in Beverly Hills. It has the same features as its East Coast predecessor. Because Los Angeles is the number-one radio market in the United States, the Los Angeles museum offers more of an emphasis on radio than the New York branch (which focuses more on television). In addition, when radio was in its heyday during the 1930s and 1940s, many shows were made in Los Angeles. According to Norm Pattiz, a trustee of the museum and chairman of Westwood One, "We're now in the No. 1 and No. 2 radio markets, with exactly the same material available at both museums."

Both locations offer seminars by critics, directors, producers, performers, journalists, and writers, including University Satellite Seminars, and both offer a wide variety of programs from the collection in two screening rooms and two main theaters, as well as constantly changing special exhibits. Programming from current series and exhibitions is shown throughout the day.

American Library of Radio and Television (Thousand Oaks, California)

The American Library of Radio and Television is part of the Special Collections Department of the Thousand Oaks Library System. Its holdings focus specifically on the history of radio rather than on the individuals in the profession. The library was founded in 1984 after the Thousand Oaks Library System broke away from the Ventura County System. The newly formed Library Foundation and the Friends of the Library decided that they wanted the library to focus on larger programs and a research collection. Specifically, they were anxious to fill a niche in the Los Angeles area by focusing on a special historical collection. They chose broadcasting because several of the library organizers had extensive contacts in the radio broadcasting industry.

Along with Maryland's Library of Broadcasting, the American Library of Radio and Television offers one of the largest collections of broadcasting documents in the United States, and it has an extensive reference collection of radio materials including 23,000 radio and television scripts, 10,000 photographs, 10,000 books on the history of radio and television broadcasting, pamphlets, sound recordings, periodicals, 200 maps and charts, manuscripts and personal papers, 5,000 audio recordings, and 50 oral history tapes.

In addition, the library contains archives of such notable individuals and stations as Norman Corwin, Bob Crosby, Monty Masters, Carlton E. Morse, Rudy Vallee, and KNX AM. Their Radio Series Scripts Collection contains scripts from 1930 through 1990, and their Radio Sound Recordings Collection contains recordings from 1932 to 1994.

The George Clark Radioana Collection at the Smithsonian Institution

The George H. Clark Radioana Collection is a part of the National Museum of American History of the Smithsonian Institution. The collection was assembled by George Clark of the Radio Corporation of America (RCA) and is one of the most extensive collections of documents and publications on the history of wireless and radio in the United States. It was transferred from the Massachusetts Institute of Technology to the National Museum of American History in 1959. The collection occupies more than 276 linear feet of shelf space, but it has not been fully indexed.

The collection is particularly strong from 1900 through 1935. There is extensive biographical information on the men who developed the technical aspects of radio and the industry; information on the inception, growth, and activities of radio companies, most notably the National Electric Signaling Company and RCA; and photographs of all aspects of radio.

The United States Library of Congress and the National Archives

Both the United States Library of Congress and the National Archives in Washington, D.C., have collections of voice recordings and radio programs. The Library of Congress has received donations of transcriptions of old radio shows. There are over 500,000 programs in their collection, including a large number of British Broadcasting Corporation (BBC) and Armed Forces Radio and Television Service (AFRTS) recordings. The library also has a large collection of radio-related items, such as early folk and regional programs, as well as a large selection of NBC Radio's broadcast discs from 1935 to 1970, which cover the Depression, World War II, postwar recovery, and radio comedy and drama programs. Other radio collections include the WOR-AM collection, United Nations recordings, Library of Congress concerts and literary recordings, and the Armed Forces Radio Collection. There is also an extensive collection from the U.S. Office of War Information (OWI), which, between 1944 and 1947, transferred thousands of items used to support the war effort to the Library of Congress. These items include OWI sound recordings, photographs, and a small number of research papers. In addition, the Motion Picture, Broadcasting, and Recorded Sound Division holds nearly 50,000 acetate disc recordings of foreign and domestic radio broadcasts.

The National Archives of the United States also features a broad collection of radio-related material. Most of these are housed at the Special Media Archives Services Division's Motion Picture, Sound, and Video unit at Archives II in College Park, Maryland. The holdings include 150,000 reels of film, 160,000 sound recordings, and 20,000 videotapes. These materials were obtained from both public and private sources. The sound recordings catalog includes radio broadcasts, speeches, interviews, documentaries, oral histories, and public information programs. The library indicates that the earliest recording they have dates from 1896, with the bulk of their recordings coming from between 1935 and the present.

Some of the specialized catalogs in the holdings include the NPR catalog, which contains NPR news and public-affairs broadcasts from 1971 to 1978, and the Milo Ryan Photoarchive Collection, which includes 5,000 recordings, primarily of CBS-KIRO radio broadcasts from 1931 to 1977. These materials were originally kept at the University of Washington and contain news and public-affairs programs, speeches, interviews, wartime dramas, and daily World War II news programs. The library also features the American Broadcasting Companies (ABC) radio collection, which consists of 27,000 radio broadcasts of news and public-affairs programs from 1943 to 1971.

Duke University Library Advertising History Archive

Duke University Library has a special Advertising History Archive that is part of the John W. Hartman Center for Sales, Advertising, and Marketing History. The advertising history collection located in the Hartman Center is the J. Walter Thompson Company Archives. The J. Walter Thompson Company is one of the world's oldest, largest, and most innovative advertising firms. The collection documents the history of the company and, as part of this, its role in radio broadcasting. The J. Walter Thompson Company's Radio Department produced some of the most popular radio shows on the air during the 1930s and 1940s. These include *Kraft Music Hall, Lux Radio Theater,* and *The Chase and Sanborn Hour.* These and other Thompson programs are housed in the collection.

In 1979 the J. Walter Thompson Company Archives were formally established in the company's New York Office. In 1987 Chief Executive Officer Burt Manning authorized the gift of the entire collection to Duke University. The archives contain over 2,000 linear feet of printed and manuscript materials, nearly 2 million items in all, half of which are advertisements. The archives house the Radio-Television Department files, which include microfilm of scripts of most of the agency-produced radio and television shows from 1930 to 1960, including *Kraft Music Hall, Lux Radio Theatre,* and *Lux Video Theatre.* Most of the holdings in the archives are open to researchers except for recent and unprocessed materials.

The Pavek Museum of Broadcasting (St. Louis Park, Minnesota)

The Pavek Museum of Broadcasting, located in St. Louis Park, Minnesota, a suburb of Minneapolis, houses a large collection of antique radios, televisions, and other broadcasting memorabilia and equipment, including an actual old-time radio studio. The mission of the museum is to provide a broader knowledge of how pioneers in electronic communications affected the evolution of society, to stimulate a new recognition of the practical and real contributions that exploring science and the communication arts can bring, and to provide a permanent and living repository for the preservation of these historic items.

The museum opened in 1988, and most of its collection comes from the original Joseph R. Pavek Collection. Pavek, an electronics instructor for Dunwoody Institute, started his collection in 1946. He also had his own electronics business, and he began storing his collection at his business. By the 1970s he began to look for someone to take over the collection, house it, staff if, and make it available to the public. In 1984, unable to find such a person, he was set to sell the collection, but Earl Bakken, the inventor of the pacemaker, stepped in and, with Paul Hedberg of the Minnesota Broadcasters Association, formed the nonprofit organization that became the umbrella for the museum.

The Pavek collection consists of over 1,000 radio receivers, transmitters, and televisions from the first half of the 20th century. Highlights of the collection include a working 1912 rotary spark-gap transmitter, crystal radios of the early 1920s, a collection of vacuum tubes (including several original de Forest Audions), and a large collection of radio literature. Additional donations from radio and television stations and from other collectors have greatly increased the size of the original collection.

Included at the museum is the Charles Bradley Collection, which has examples from over 60 Minnesota radio and television manufacturers from the 1920s and 1930s. There are also many examples of historic broadcast equipment on display, such as cameras, consoles, and microphones. The museum also houses the Jack Mullin Collection, which documents the history of recorded sound, with over 125 years of audio recording technology, starting with the earliest days of the phonograph. Mullin is credited as being the person who brought back two tape recorders from a German radio station while serving in the Signal Corps at the end of World War II. At the time, tape recording was an unknown technology in the United States, and Mullin was immediately hired by Bing Crosby to tape-record his popular radio program for broadcast, the first use of tape recording in American broadcasting.

Also featured in the archives is the Pioneer Broadcaster Series, which preserves videotaped interviews with radio pioneers. The museum also has an educational program with classes, workshops, and exhibits for both children and adults. There is also a library of technical and service information on electronics and electronic communication. Besides the permanent collection, the Pavek also displays items on loan from other private collections.

University of Memphis Radio Archive

The University of Memphis Radio Archive is a collection of broadcast programs that was started over 30 years ago by Dr. Marvin R. Bensman of the Department of Communication. Bensman began his radio collection from original transcriptions, private collectors, and other institutional collections. The collection is intended to be a representative sampling of most series and shows.

The collection is housed in the Microforms Department of the McWherter Library at the University of Memphis. Individuals may request audiocassettes of these radio programs. Programs have been selected because they give a sense of the history and development of broadcasting. They feature the key events that influenced the regulation of broadcasting and of broadcasting programming. Some highlights of the archive include Westinghouse's 50th Anniversary program; the history of broadcasting from the 1920s to the 1970s; 50th anniversary shows about the development of the BBC, NBC, and CBS; early pioneer broadcasters and/or inventors; Aimee Semple McPherson's broadcasts; the American Society of Composers, Authors, and Publishers' (ASCAP) Cavalcade of Music concert in 1940 consisting of live performances by musical stars including Berlin, Handy, and others; *Year-end Reviews;* CBS Radio Workshops; World War II broadcasts; numerous movie dramatizations; and classic comedies.

Private/Personal Museum and Archive Collections

American Museum of Radio

The Bellingham Antique Radio Museum is a nonprofit museum located in Bellingham, Washington. It is a private collection gathered over the past 25 years by Jonathan Winter, who started collecting radios when he was a child. The collection spans the history of radio from the time it began through the early 1940s and features over 1,000 antique radios on display. The Bellingham Antique Radio Museum, as it was originally called, opened in 1988 in a small room and moved to larger quarters in downtown Bellingham in 1990. In 2001 the museum moved to a new, larger facility.

In addition to its collection of antique radios, there is other material on display highlighting the history of radio technology, including historical photographs; books and magazines from radio's early days; microphones, coils, tubes, speakers, and other parts; biographies of people involved in radio history; audio clips of some of the more historic broadcasts; and clips of radio entertainment shows from the early days of broadcasting.

U.S. National Marconi Museum (Bedford, New Hampshire)

The U.S. National Marconi Museum was created by the Guglielmo Marconi Foundation in 1995 to help publicize the name of Marconi, the "Father of Wireless." The museum is located in Bedford, New Hampshire, and the collection features equipment, historical literature, and audiovisual presentations on the development of radio communications. It features displays of early Marconi wireless equipment, along with the progression of radios up to a current cellular telephone exhibit. The museum also features a restoration room for repairing vintage radios, a machine shop, and a facility room for educational lectures to school groups and for meetings of electronic-oriented organizations.

The John Frey Technical Library contains thousands of radio communication periodicals, some in a series dating from 1920. All the publications are indexed and cataloged on CD-ROM and can be accessed by internet on the library computer. The library also features hundreds of engineering, text, and reference books.

Museum of Radio and Technology (Huntington, West Virginia)

The Museum of Radio and Technology is a small, private collection consisting of old radio and television sets, and it is staffed exclusively by volunteer museum members. It features several displays, including a radio shop of the 1920s and 1930s that has a variety of radios from that era, including battery radios; horn speakers; a wind-powered generator; a radio-television sale room featuring radios, television sets, and wire recorders; a Gilbert toy display; a vintage hi-fi room with tube-type audio equipment and related components such as amplifiers, tuners, tape recorders, receivers, microphones, and turntables. The highlight of the display is the Western Electric transmitter, a 1930s 5,000-watt AM transmitter complete with power supply components and studio equipment.

The Radio History Society's Radio-Television Museum (Bowie, Maryland)

The Radio History Society is a nonprofit organization dedicated to the preservation of radio and television history. In June 1999 the society opened its new Radio-Television Museum in Bowie, Maryland, housed in a fully restored turn-of-the-century building.

The Radio Historical Society owns a large collection of old literature and radio artifacts relating to the history of radio and television broadcasting. Some of their collection includes radio sets from the 1920s through the 1960s plus local broadcast memorabilia. Their permanent and changing exhibits include home receivers, novelty radios, broadcast microphones, and communication and ham radio equipment. They also maintain a display area at George Washington University's Media and Public Affairs Building in downtown Washington, D.C., with changing displays.

Society to Preserve and Encourage Radio Drama, Variety, and Comedy

The Society to Preserve and Encourage Radio Drama, Variety, and Comedy (SPERDVAC) is an organization of old-time radio enthusiasts that has assembled one of the most important and well-maintained radio program archives in the world. There are over 20,000 original transcription discs, as well as a large library of printed materials and scripts. In addition, there are over 2,000 reels of old-time radio available only to its members. SPERDVAC also produces a monthly newsletter and a catalog listing the thousands of shows in its collection, and it hosts monthly meetings and annual conventions in the Los Angeles area.

JUDITH GERBER

See also Nostalgia Radio; Peabody Awards; Radio Hall of Fame

Further Reading

Godfrey, Donald G., compiler, *Reruns on File: A Guide to Electronic Media Archives,* Hillside, New Jersey: Erlbaum, 1992

Hedges, William, and Edwin L. Dunham, compilers, *Broadcast Pioneers History Project; Third Progress Report and Historical Inventory,* New York: Broadcast Pioneers, 1967

Hickerson, Jay, *What You Always Wanted to Know about Circulating Old-Time Radio Shows (But Could Never Find Out),* N.p.: 1986

Lichty, Lawrence W., Douglas Gomery, and Shirley L. Green, *Scholars' Guide to Washington, D.C., Media Collections,* Baltimore, Maryland: Johns Hopkins University Press, and Washington, D.C.: Woodrow Wilson Center Press, 1994

Pitts, Michael R., *Radio Soundtracks: A Reference Guide,* Metuchen, New Jersey: Scarecrow Press, 1976; 2nd edition, Metuchen, New Jersey, and London: Scarecrow Press, 1986

Sherman, Barry, et al., "The Peabody Archive and Other Resources," *Journal of Radio Studies* 1 (1992)

Smart, James Robert, compiler, *Radio Broadcasts in the Library of Congress, 1924–1941: A Catalog of Recordings,* Washington, D.C.: Library of Congress, 1982

Swartz, Jon David, and Robert C. Reinehr, *Handbook of Old-Time Radio: A Comprehensive Guide to Golden Age Radio Listening and Collecting,* Metuchen, New Jersey, and London: Scarecrow Press, 1993

Music on Radio

Music has been a staple of radio programming since the medium's creation in the early 1920s. Indeed, David Sarnoff's historically fabled memo—real or not—foresaw radio's potential future as a "music box." Before radio, to be able listen to music one had to play an instrument (most often a piano), purchase a poorly recorded disc, or pay to attend a live performance. Radio broadcasting changed that by offering frequent free musical performances for the simple purchase of a radio receiver.

Radio music history can be divided into two eras, divided by a short but confusing transition period. During the first (to 1950), most music was broadcast live as a part of a variety of radio formats, both network and local. The second era (since 1955) followed a brief and difficult transition but soon saw station programmers regularly playing music using specific short lists of recordings. This focus on specific formats has defined radio music, with only the conversion from various disc formats to audiotape and then back to digital discs and tapes changing the means of recording and playback. Indeed, technical change underlies any historical analysis of music on radio. The phonograph record as a means of listening to music preceded radio, but it was radio broadcasting that vastly expanded the musical recording industry—first on 78-rpm records, then, after the war, on 33 1/3-rpm long-playing records and 45-rpm records into the 1960s. Thereafter came a decade or so of analog audiocassettes, and finally, at the end of the 20th century, compact discs and other digital formats.

Network Tin Pan Alley Era (to 1950)

Music as a popular radio program genre started when many advocated the new medium as a means to bring high-art music such as opera and orchestral recitals to the mass public. But although European classical music never disappeared from radio's schedule as radio entered the network era during the late 1920s, its presence quickly gave way to popular music and in particular to variety shows starring musical talents such as Rudy Vallee and Al Jolson. New York City's Tin Pan Alley created the music that big bands and their singers offered radio listeners.

By the early 1930s, both the National Broadcasting Company (NBC) and the Columbia Broadcasting System (CBS) had discovered genres of musical programs that the public preferred. The networks tried classical music; varieties of popular music; and what might be called light or background music, which was designed for listeners involved in activities other than dedicated listening. But although broadcasting classical concert music suggested that radio was providing a "good" to the masses, comedy and variety shows created the mass audiences advertisers sought. Broadcasts of the Chicago Civic Opera on NBC Blue (Saturday), the Cities Service Orchestra on NBC Red (Friday), the Edison Electric Orchestra on NBC Blue (Monday), and the Paramount Symphony Orchestra on CBS (Saturday) maximized prestige but drew small audiences.

Variety shows proved to be the most successful means of creating a profit with music programming. These shows varied depending on how pop music was emphasized—from a comic host with a musical guest to a musical host with a comic as guest. The latter—the musical variety program—became the most popular network radio genre during the 1930s. Top attractions centered more and more on name bands, including Guy Lombardo's Orchestra on CBS or the Paul Whiteman Orchestra on NBC, both broadcast on Monday nights.

Through the 1930s, so-called light music offered the second-largest musical category of radio shows; for example, Jesse Crawford played the pipe organ on CBS on Sunday nights, Lanny Ross (later of *Your Hit Parade* fame) soothed his audiences on NBC on Saturday nights, and—in a rare case of sponsor naming—The Wheaties Quartet performed as intended background music on CBS on Wednesday nights.

By the mid–1930s, NBC and CBS were offering some of the most popular free musical entertainment during those hard times. Indeed, sales of phonograph records plunged during the Great Depression as fans substituted listening to music on the radio for the relatively expensive purchasing of individual phonograph records. Radio became the place where new popular tunes were introduced, and their creators and players became musical stars.

Although during the day local stations still offered non–network live music from the community, prime time had become big time for radio listeners and programmers. Yet stations in large cities did maintain orchestras to play for the local programming. In reality the music that most fans sought came primarily from New York City and then in small doses from Los Angeles–based studios that used musical talent associated with movie making.

By 1940 classical concert music still offered prestige, but on fewer and fewer programs. A star system developed as NBC put together its own classical orchestra, led by Arturo Toscanini. At CBS, William S. Paley signaled that his star was Andre Kostelanetz, who by 1940 was on the air not one night, but two. NBC continued to hire a classical orchestra in order to identify itself as the higher-class network, and by the early 1940s they had scheduled the Boston Symphony, the Firestone Concert, the Minneapolis Symphony, and the Rochester Philharmonic Orchestra.

Judging by the number of shows offered in 1935, radio listeners seemed to prefer a named band with an identifiable sound to a group with the name of its sponsor—even if that was a full classical orchestra. Yet some names in the light music category could and did become pop music stars, such as Kate Smith, who had high ratings in 1935 despite being on the air for only 15 minutes on CBS on Saturday nights.

By 1935 variety shows—which had always had a popular music component—reigned as the most popular of radio's genres. No list can be complete, but the big bands of the day could be found throughout the schedules of NBC and CBS—including the Bob Crosby orchestra, Fred Waring, Horace Heidt, Paul Whiteman, and the "waltz king" Wayne King. Guy Lombardo's orchestra remained a fixture on CBS on Monday nights, symbolizing more and more that the name was in the band and its singers, not in some amalgamation fashioned directly by the sponsor.

In short, the popular mainstream music of the 1930s and 1940s was found primarily on network radio. Orchestras were hired to perform live to generate a studio-made "high-fidelity" sound before the innovation in the late 1940s of 33 1/3-rpm and 45-rpm records. In-house studio orchestras were formed to provide background music for dramatic shows as well.

Big bands played remotes for dances in such ballrooms as the Aragon in Chicago and the Pacific Square in San Diego, at beach or other waterside attractions (the Steel Pier in Atlantic City and the Glen Island Casino in New Rochelle, New York), at restaurants (the Blackhawk in Chicago, the Copacabana in New York City), and at major hotels in most big cities. Such remotes offered popular venues for radio broadcasting through the 1940s and symbolized the hot new sounds for dancers of the era.

The rise of "name" singers was another emerging trend. Through the 1930s singers, led by Bing Crosby, learned to use the microphone for effect, not simply as a means of broadcasting. Ratings spiked when Crosby and Frank Sinatra—as well as Rosemary Clooney, Ruth Etting, Helen Kane, Peggy Lee, and Doris Day—were scheduled. Soloists hardly represented the lone form of popular radio singing. There were duos, trios, and quartets—from the Ink Spots to the Mills Brothers, from the Andrews Sisters to the Boswell Sisters. Singing intimately and in a number of styles, all based on Tin Pan Alley arrangements, became a true art form through radio broadcasting.

The war years proved to be the final hurrah for the musical variety show. National defense bond rallies often functioned as all-star radio variety shows, meant to outdo all other radio extravaganzas. Programs such as *Music for Millions, Treasury Star Parade,* and *Millions for Defense* not only drew needed bond sales but also were beamed overseas or recorded for later playback for the troops fighting in Europe and the Pacific. The top stars of network radio toured for the United Service Organizations (USO) and went abroad to entertain soldiers near the fronts. Indeed, radio star and big band leader Glenn Miller was killed while traveling from one such show to another. The war years also proved the crest for big band singers on network radio. Kate Smith and Dinah Shore, for example, starred in some of the most popular shows on radio.

This system of making live music came apart, however, because of the demands of its most famous star, Bing Crosby. Crosby hated the necessities of live broadcasting, which demanded a rigid schedule that included doing shows twice (once for the Eastern and Central time zones and then a second time for Mountain and Pacific time zones). In 1946 Cosby moved his top-rated show from NBC to ABC to obtain relief. ABC, desperate for ratings, allowed Crosby to prerecord his *Philco Radio Time* using newly developed audiotape technology. He did not have to be in the studio when his show debuted (on 16 October 1946), nor weekly as it ran on ABC until June 1949. At that point William S. Paley, head of CBS, also gave into recorded music programs and as a part of his famous "talent raids," offered Crosby more money than ABC could afford.

Even though Tin Pan Alley and its allies in Hollywood largely dominated music played on the radio through the 1940s, there were alternatives. In particular, hillbilly music shows were becoming hits on the networks and on many local stations, particularly on small-town outlets in the South and West.

NBC led the way on the network level with *The National Barn Dance* and *The Grand Ole Opry,* both on Saturday nights. Numerous Southern stations offered live music, particularly during early morning hours. The demand for hillbilly music exceeded the supply, and so border stations based in Mexico blasted at 1 million watts music by hillbilly favorites such as the Carter Family, Jimmie Rodgers, Cowboy Slim Rinehart, and Patsy Montana.

The Carter Family—a trio composed of A.P. Carter, Sara Carter, and Maybelle Carter—was the first family of country music, and their famed 1928 Bristol, Tennessee, recording sessions kicked off a new genre of popular music. Jimmie Rodgers was also at those Bristol sessions and should be counted among the creators of hillbilly music. Nolan "Cowboy Slim" Rinehart, "the king of border radio," was a singing cowboy who, because of border stations' power, was heard across the nation as much as his more popular rival, singing cowboy Gene Autry. Cowgirl Patsy Montana teamed up with Rinehart for a series of transcribed duets during the 1930s and became so popular that her 1935 recording "I Wanna Be a Cowboy's Sweetheart" became the first million-selling record by a female hillbilly artist.

Ethnic artists found it more difficult to gain access to even local radio. In particular, although African-Americans were developing rhythm and blues music, the genre could rarely be heard on the radio during the 1930s and 1940s. Race records and juke joints offered the sole outlets, but the music was there and rich in form and style for the great change that was about to happen to radio music broadcasting.

Transition (1948–55)

Beginning in the late 1940s, NBC and CBS committed themselves to network television. They transferred their big bands and pop singers—plus some symphonic music—to TV and used profits from network radio to fund their new, and in the future far more profitable, medium. This worked well for the networks, and Bing Crosby, Kate Smith, Tommy Dorsey, and particularly Dinah Shore became mainstays of network television programming of the 1950s.

Their departure—and that of most other network programming—left radio stations looking for something new. Stations would find their salvation and reinvention in rock, an amalgamation of country and race forms. As rock was developing through the early and middle 1950s, Todd Stortz and Gordon McLendon pioneered Top 40 radio. They developed a short list of top tunes and played them over and over again. Teenagers of the 1950s—not interested in the big band, Tin Pan Alley music of their parents—embraced Top 40 radio. AM radio stations—looking for something to fill their time as network programs migrated to television—looked to Top 40 as their salvation.

There were sizable vested interests in keeping the live musical variety show going. These included the performing music societies, the American Society of Composers, Authors, and Publishers (ASCAP) and Broadcast Music Incorporated (BMI). Even more concerned was the American Federation of Musicians, the performer's union which tried to slow adoption of the innovative recording techniques—tape and discs—which union leaders and members feared (correctly) would lessen the demand for their live services. These parties, in addition to many parents and religious leaders, found rock music subversive and threatening. Even NBC tried to keep the variety musical show alive on radio with *The Big Show* on Sunday nights in 1951, but with no success.

Format Radio (Since 1955)

The symbol of the Top 40 revolution in radio was singer Elvis Presley. Gone were the big bands, dominated by brass and woodwinds, with dozens of players; these had been replaced by combos of a drummer and a couple of guitars. The electric guitar gave the necessary amplified sound and beat. The singer, who had been just one part of the big band, was now moved to the forefront, and with Chuck Berry, the singer sang his or her own compositions (so the songwriters of Tin Pan Alley were no longer needed). And, most important, the music of the margins—hillbilly and race—moved to the forefront as the amalgam labeled rock and roll. Elvis was the "hillbilly cat." Chuck Berry grew up in St. Louis listening to both *The National Barn Dance* and *The Grand Ole Opry*. After more of a struggle, blues music, later dubbed rhythm and blues in its urban form, came to mainstream rock in the form of Detroit's "Motown Sound."

Rock, country, and rhythm and blues formats spawned a myriad of newer sub-formats for radio stations that wanted to be more than just "the other" Top 40 station in town. Taking but a single example, the history of country symbolizes how one marginal form became mainstream in the last half of the 20th century—indeed the top format in all radio by century's turn.

As rock splintered into many subtypes, each with devoted audiences, country rose to become the music that many white Americans listened to, in part because during an era of civil rights unrest, country recognized and appreciated that its roots were not tinged by music with more direct African-American roots. Country had its origins in the folk and hillbilly music that was so marginal during the network radio era—save for the popular "barn dance" programs. For advertisers, country attracted white, middle-class, suburban America—the audience they most wanted to reach.

Entrepreneurs provided a new name, and "country and western" was used into the 1960s. With the rise of Nashville (Tennessee) as an important recording center, however, the "western" was dropped, and by the time country format radio took off, the name was simply "country." What would become known as the "Nashville sound" worked as Hank Williams made country songs popular as pop music—an approach also heralded by Jim Reeves and Patsy Cline. By the 1960s, country emerged as an alternative genre, with stars such as Johnny Cash, Jimmy Dean, Loretta Lynn, and Dolly Parton. As rock seemed to lose its roots in the 1970s, country became an even more popular radio format. By the 1980s many surveys found country to be the most popular format on radio. A once marginal music style had become a dominant form of pop music, all made from a central location in Nashville.

With the innovation of portable and automobile radios, radio listening moved out of the home and became ubiquitous, particularly with the advent of the Walkman. The average person listened more than three hours per week. Advertisers targeted ethnic groups (principally African-Americans and Latinos), different age groups, income classes, and genders with different types of music. Adult contemporary music worked best for those 25–34 years old, whereas album-oriented rock was aimed at teenagers; their college-aged cousins seemed to prefer classic rock and contemporary hits radio. Country generally appealed to an older audience.

By the middle 1990s, many argued that radio had become too formulaic. Virtually all radio sought female suburban adults 18–34 who listened to radio on their way to and from work. Artists (in any format) who did not fit that pattern of attraction were simply not played. In the 1990s, for example, Top 40 morphed into "contemporary hits radio" and largely abandoned those who had once helped to create it: teenagers.

Creating a complete listing of these format formulas is almost fruitless—the annual *Broadcasting and Cable Yearbook* by the late 1990s listed more than 75 formats—starting with adult album alternative or AAA, moving on to urban contemporary, variety (four or more formats), Vietnamese, and finally women.

In the 1990s country music was among the most popular formats on U.S. radio. In turn, country spun off the gospel music format, and later the Christian contemporary music format. Its composers and stars were influenced by rock stylings they grew up with; superstar Garth Brooks recalled the group Kiss as his key influence. Indeed, during the 1990s one could more easily find a Willie Nelson or Loretta Lynn "classic" tune covered and then played on an adult contemporary format, a beautiful music format, or an easy listening format station than on a country station.

Other formats enjoyed great popularity during this turn of the century period as well, among them what was termed Contemporary Rock, Rhythmic Oldies, Urban, and Hot AC. Perhaps the most tuned by young listeners was Hip Hop. It inspired the newest incarnation of the Top 40/CHR format because the Hip Hop sound dominated the best selling music charts nationally, if not globally.

DOUGLAS GOMERY

See also, in addition to individual formats discussed above, American Federation of Musicians; American Society of Composers, Authors and Publishers; Broadcast Music Incorporated; Canadian Radio and the Music Industry; Classical Music Format; Crosby, Bing; Formats; Grand Ole Opry; McClendon, Gordon; Metropolitan Opera; National Barn Dance; Recordings and the Radio Industry; Singers on Radio; Storz, Todd; Talent Raids; Vallee, Rudy; Walkman

Further Reading

DeLong, Thomas A., *The Mighty Music Box: The Golden Age of Musical Radio,* Los Angeles: Amber Crest Books, 1980

Eberly, Philip K., *Music in the Air: America's Changing Tastes in Popular Music, 1920–1980,* New York: Hastings House, 1982

Fornatale, Peter, and Joshua E. Mills, *Radio in the Television Age,* Woodstock, New York: Overlook Press, 1980

Fowler, Gene, and Bill Crawford, *Border Radio,* Austin: Texas Monthly Press, 1987

Hickerson, Jay, *The Ultimate History of Network Radio Programming and Guide to All Circulating Shows,* Hamden, Connecticut: Hickerson, 1992; 3rd edition, as *The New, Revised, Ultimate History of Network Radio Programming and Guide to All Circulating Shows,* 1996

Joyner, David Lee, *American Popular Music,* Madison, Wisconsin: Brown and Benchmark, 1993

Landry, Robert John, *This Fascinating Radio Business,* Indianapolis, Indiana: Bobbs-Merrill, 1946

MacDonald, J. Fred, *Don't Touch That Dial! Radio Programming in American Life, 1920–1960,* Chicago: Nelson-Hall, 1979

Malone, Bill C., *Country Music U.S.A.,* Austin: University of Texas Press, 1968; revised edition, 1985

Routt, Edd, James B. McGrath, and Frederic A. Weiss, *The Radio Format Conundrum,* New York: Hastings House, 1978

Sanjek, Russell, and David Sanjek, *American Popular Music Business in the 20th Century,* New York: Oxford University Press, 1991

Whetmore, Edward Jay, *The Magic Medium: An Introduction to Radio in America,* Belmont, California: Wadsworth, 1981

Music Testing

Determining Radio Audience Preferences

There has never been as great a need for accurate music research data in the radio industry as there is today. Not only are many more entertainment options available to potential listeners, but the expectations for ratings and profit performance continue to increase. As a result, programmers of music-oriented stations have adopted a variety of research methods to better understand their listeners' attitudes toward particular songs.

Requests

Perhaps the most easily overlooked source of music data is a station's request line. Many programmers recognize requests as an inexpensive and simple way to collect music information. Instructing disc jockeys to tally songs that people care enough about to request is an easy and cheap way to obtain a daily glimpse of titles that excite listeners. However, programmers should be careful not to place too much confidence in request data. Listeners with enough spare time to place requests may not best represent a station's audience. To better ensure that a station's entire audience range is represented, programmers rely on more scientific methods.

Callout

The primary method of music testing is callout research. Callout consists of trained interviewers telephoning randomly selected listeners of a particular station and having them use a pre-established scale to rate 15 or 20 "hooks" from songs the station plays. A hook is a brief lyrical segment, often the title or chorus, that captures the essential quality of the song.

According to Tony Novia of *Radio and Records* magazine, callout began in the 1970s when broadcasters believed they could predict which new songs would become popular by having listeners rate hooks from the very latest releases. Unfortunately, because the songs were so new and had not received any airplay, respondent unfamiliarity resulted in unreliable data. Beginning in the 1980s, programmers realized that callout was an effective tool for obtaining data about familiar music. The hook, in effect, was just long enough to "jog the memory about a song in question" (Novia, 2000). Today, most users of callout recommend that songs not be included in research until they reach a high level of familiarity through airplay. For example, radio consultant Guy Zapoleon reports that a general rule is to have a song play at least 100 times on a station before placing it into callout.

During callout, respondents provide data after each hook is heard over the phone line. First, they are asked if they recognize the song. If they do, a favorability-scale question is generally asked next. For example, listeners may rate the song on a scale of 1 to 10, where 1 means they dislike the song very much and 10 means they like the song very much. Another type of data often obtained during callout is a fatigue or burnout measurement. Listeners are asked, "Are you tired of hearing this song on the radio?" Especially in contemporary music formats, fatigue data is important for determining when to decrease airplay of a popular song title.

Perhaps the biggest benefit of callout research is the ability to gather music data quickly, easily, and inexpensively. These benefits result in the ability to generate weekly reports on current music. Drawbacks to callout include the reliance on hooks, the brevity of which sometimes fail to capture the essence of a song; the comparative low fidelity of telephone lines, which may negatively bias results; and the high refusal rates of respondents, which can be expected any time researchers make unscheduled telephone calls. Two newer and less prevalent music testing techniques, the personal music test and call-in research, have been developed to address these shortcomings.

Call-in and Personal Music Tests

As is implied by its name, call-in research consists of listeners telephoning a station's research department to complete music tests. This method allows listeners to provide information at their convenience. A similar method involves invitations to visit the station's website, where listeners can participate in a music test in which audio of hooks (or even of entire songs) is streamed. There are several drawbacks to these two methods that must be kept in mind. First, just as with those who request songs, listeners who have the time or interest to phone a station or visit its website to participate in a music survey may not be representative of listeners in general. Second, there is no way to adequately prevent one listener from providing opinions more than once, thereby biasing the results.

A personal music test attempts to combine the scheduling convenience of call-in with the representativeness and quality control of callout. Using this music testing method, telephone interviewers call a random selection of station listeners and schedule an appointment for them to visit a research facility at a convenient time. Upon their arrival, listeners are given a hook tape of current music and a personal cassette player with headphones. Listeners work through the hooks, providing ratings for each hook at their own pace. The personal music test ensures that respondents devote the undistracted time required to provide valid data. Furthermore, because telephone lines are not involved, the fidelity of the music being tested is much closer to what is actually heard over the air. A major drawback to the personal music test, however, is cost, because most stations employing the method have found that a financial incentive is necessary to increase participation.

Auditorium Music Tests

Although the methods mentioned above tend to be used to collect opinions of fewer than 30 current songs, auditorium music tests (AMTs) are generally employed to test between 350 and 700 older songs. Familiarity with the titles is assumed; the goal here is to determine the best-liked "gold" music among the station's target audience. For oldies formats, the method is often used to determine the entire playlist; therefore, oldies programmers conduct AMTs each quarter, whereas more contemporary music stations can afford to do them only once or twice a year. AMTs consist of inviting between 75 and 150 randomly selected listeners to an auditorium and playing a hook tape for them. The shared sense of purpose and controlled environment are key benefits to this method. One drawback is the cost of auditorium rental and respondent incentives. Another is the possibility of respondent fatigue, which can be lessened by scheduling breaks periodically during hook presentation.

ROBERT F. POTTER

See also Audience Research Methods; Auditorium Testing

Further Reading

Eastman, Susan T., Douglas A. Ferguson, and Timothy P. Meyer, "Program and Audience Research," in *Broadcast/Cable Programming,* 5th edition, edited by Eastman and Ferguson, Belmont, California, and London: Wadsworth, 1997

Kelly, Tom, *Music Research: The Silver Bullet to Eternal Success,* Washington, D.C.: National Association of Broadcasters, 2000

Lynch, Joanna R., and Gillispie, Greg, "Creating an Image," in *Process and Practice of Radio Programming,* by Lynch and Gillispie, Lanham, Maryland: University Press of America, 1998

MacFarland, David T., *Contemporary Radio Programming Strategies,* Mahwah, New Jersey: Erlbaum, 1990; 2nd edition, as *Future Radio Programming Strategies: Cultivating Leadership in the Digital Age,* 1997

Norberg, Eric, G., "Promoting Your Station," in *Radio Programming: Tactics and Strategy,* by Norberg, Boston and Oxford: Focal Press, 1996

Novia, Tony, "Reach Out and Touch Some Listeners," *Radio and Records* (11 February 2000)

Porter, Chris, "Music Testing: Pros and Cons," *Radio and Records* (28 August 1998)

Zapoleon, Guy, "When to Put a Song into Callout," *Radio@Large* (15 July 1999)

Mutual Broadcasting System

U.S. National Radio Network

The Mutual Broadcasting System was unique among the four national radio networks. Whereas the other networks originated most of their programming from studios in New York City and Hollywood, Mutual was a cooperative program-sharing venture whose member stations around the country provided most of the programming. As the last major network to be established, Mutual's stations tended to be the ones the other networks did not want: low-powered rural stations with limited listening areas. Thus, although Mutual was eventually to proclaim itself the nation's largest radio network based on the number of affiliates it served, it was continually mired in last place in a four-way race.

Mutual and its affiliates created many memorable programs, such as *The Adventures of Bulldog Drummond, Buck Rogers, Double or Nothing, 20 Questions, The Falcon, The Green Hornet, The Shadow, Sherlock Holmes, The Lone Ranger, Dick Tracy, Queen for a Day,* and *Captain Midnight,* and featured personalities such as the controversial Father Charles E. Coughlin, Dick Clark, Merv Griffin, Mike Wallace, and, in later days, Larry King. But the network's fourth-place status and chronically weak financial position often resulted in its best programs being lured away to the deeper-pocketed competing networks.

Creating a Fourth Network

Because local radio listening areas, or markets, varied widely in both population and number of stations locally available, the three-network system (National Broadcasting Company [NBC] Red, NBC Blue, and the Columbia Broadcasting System [CBS]) worked well in some places and not as well in others. Markets with three local stations willing to affiliate with a network (despite the advantages, not all stations desired affiliation) were ideally suited to the status quo. Markets with fewer than three stations frequently saw a station affiliated with more than one network, with one network considered the station's primary affiliation and another network constituting a "secondary" affiliation. In markets with four or more stations desiring network affiliation, somebody, obviously, was going to be disappointed. In a competitive environment with four or more stations, the affiliation contracts usually went to the more powerful stations.

The early 1930s saw several attempts to start a fourth radio network, from the Amalgamated Broadcasting System (headed by popular radio comedian Ed Wynn, often billed as "the perfect fool": the network folded in five weeks) to an American Broadcasting Company (no relation to today's ABC) that lasted a few months. Among the many reasons for the high failure rate, two deserve special consideration, because they were to resurface continually as formidable challenges to anyone trying to compete with NBC or CBS. The first was the fact that the three major networks already had solid relationships with the best advertisers, and they still had much airtime to sell. In many ways, a sustaining program represented an unsold commercial slot. Ideally (for the network), the entire schedule would be commercial. Thus, the sales representatives at NBC and CBS aggressively went after advertisers to buy more time, often offering discounted rates to large advertising accounts. Any start-up radio network was going to have a tough time convincing advertisers to stray from the majors. A second problem was the ragtag nature of most of the stations not already signed with NBC or CBS. As much as these stations wanted network affiliation, this accumulation of largely low-powered and/or rural stations would not be very attractive to national advertisers.

Ironically, the company that was finally to establish a fourth network started life with no national network intentions. In 1934 four powerful independent (non-network) stations banded together to form the Quality Group. The purpose of the group, which consisted of WOR (New York), WGN (Chicago), WLW (Cincinnati), and WXYZ (Detroit), was twofold. First, they would share their better sustaining programs among themselves. Second, they would offer an alternative to the producers of commercial programs who wanted access to four major metropolitan markets without going through one of the established networks. As a Quality Group spokesman stated, "We will endeavor to make suitable time arrangements for advertisers seeking to broadcast in important markets through the use of a few stations having high power and a vast listening audience. . . . Each station will remain independent and make its own decision in accepting programs. . . . Several programs are now broadcast over this group of stations by mutual agreement." The "mutual" nature of the cooperative venture apparently struck a chord, because the organization was almost immediately renamed the Mutual Broadcasting System (MBS). By the time it celebrated its first anniversary in 1935, Mutual carried 20 hours of commercial broadcasts and 40 hours of sustaining broadcasts per week. The anniversary was bittersweet, however. The one non-stockholding partner in the venture, WXYZ, had just jumped ship to NBC Blue. Mutual was able to replace WXYZ in the Detroit market by signing CKLW, an across-the-border Canadian station that had served as the area's CBS affiliate. CBS had dumped CKLW in favor of yet another Detroit station, WJR, when it increased its power to 50,000 watts.

More significantly, a major schism regarding the future of the company was developing among the three owner stations. Desiring to increase the operation's revenue, WGN and WOR wanted to open Mutual up into a broader network serving more stations. WLW was opposed to this plan. Whereas the metropolitan locations of WGN and WOR (Chicago and New York City) gave them local access to millions of listeners, WLW got most of its audience through the far-flung reach of its nighttime 500,000-watt signal. If Mutual began to sign affiliates in the cities reached by WLW's signal, the station reasoned, WLW would lose much of its audience to these local stations. WLW wanted MBS programming to remain exclusively available to the original four markets to preserve its own unique appeal to its geographically widespread audience.

WLW was outvoted. By early 1936 some individual Mutual programs (but not the complete network schedule) were being carried on a network of nine stations. By the fall of 1936, Mutual announced expansion to the West, signing affiliation agreements with the Don Lee regional network in California and with several Midwest stations along the American Telephone and Telegraph (AT&T) line, which was leased to carry the network's signal to the West Coast. At the same time, WLW announced that it was turning in its MBS stock. It remained an MBS affiliate for many years, even continuing to supply Mutual with some original programming. Its own schedule, however, became increasingly a mix of MBS and NBC. The Federal Communications Commission (FCC) eventually decided that the "superstation" experiment was a failure, placing WLW at too much of a competitive advantage over other stations, and WLW became a regular 50,000-watt clear channel station. The Don Lee network picked up WLW's stock, as well as one-third of the cost of the expanded network operations. Before the network hookup to California was in operation, Mutual signed a Washington, D.C., station and another regional network, Colonial, as a New England affiliate. Thus by the end of 1936, Mutual was a true transcontinental network, albeit one with huge gaps (most significantly, the southern half of the United States). Despite this expansion, MBS executives remained committed to the network's unique vision. Company President W.E. Macfarlane emphasized the independent nature of Mutual's affiliates by noting, "The Mutual Broadcasting System was organized with the purpose of presenting better programs, allowing stations to maintain their independence, and creating a network of stations which would serve the country's listening audience and still allow stations to fulfill obligations to their various local communities."

By 1937 Mutual was serving 51 affiliates. The complete network schedule consisted of 30-3/4 hours of commercial programs and 93-1/4 hours of sustaining programs per week. Within a year, the total number of affiliates was up to 51. Yet increasingly, Mutual was finding itself frustrated by the major networks. Many of its new affiliates were only secondary Mutual stations. These stations owed their primary allegiance (and best broadcasting hours) to one of the major networks. On these stations, Mutual only got the broadcast times the major networks did not want: the hours with the fewest listeners that were the most difficult to sell to national advertisers. Because of its weak position relative to the other networks, Mutual became an early practitioner of "counter-programming." If the most popular program on radio in a given time slot was a drama, Mutual would schedule a musical show opposite it.

The FCC Network Probe

Early in 1938, in part responding to growing complaints from Mutual about its difficulty in competing with the entrenched New York-based networks, the FCC initiated a probe of possible network monopolistic practices. Data soon confirmed some of Mutual's complaints—CBS and NBC, with three networks between them, dominated the strongest stations across the country. Mutual was having trouble getting a competitive foothold in the business.

Mutual's winning of the rights to provide the baseball World Series broadcasts in 1938 and 1939 brought other network practices into sharp contrast. CBS and NBC ordered their affiliates to stick with their own network programs, even when those stations wanted to carry the highly popular games (and Mutual was willing to provide them). The closed-door approach of the New York networks certainly helped to underline Mutual's anti-competitive arguments.

Based in part on information provided by Mutual, the FCC issued its final report on chain broadcasting in May 1941, calling for a host of changes in the relationship between networks and their affiliates. After a fierce legal battle, and several long congressional hearings, the rules were upheld in a landmark Supreme Court decision in 1943. NBC and CBS were forced to modify many of their affiliation contracts, somewhat evening the playing field for Mutual.

The Decline of Mutual and Network Radio

The MBS continued to expand through the 1940s, reaching 400 affiliates in 1947. It became the first network to include FM stations in its lineup, although these affiliations were plagued in the beginning by a dispute with the American Federation of Musicians that prohibited any musical programs from being carried over FM stations without an additional fee. The expanded Mutual network now reached 84 percent of the nation's radio homes, although only 60 percent of the network's programming was actually carried over the entire system.

Although the post–World War II structure of Mutual was basically the same as it had always been—a program-sharing cooperative owned by three major stockholders (WGN, WOR, Don Lee) and a few minor stockholders (including the New England–based Colonial regional network)—major changes in the broadcasting landscape and in Mutual's corporate structure loomed on the horizon.

After decades of development, television was finally ready for its commercial launch immediately following World War II. Although CBS, NBC, ABC, and an electronics firm named DuMont all announced plans for television networks, Mutual's stand on the matter was ambivalent. Although WGN, WOR, Don Lee, and some Mutual affiliates were getting into television, MBS announced in 1948 that it would "leave actual video operations to its stockholder stations." The decision was made that these stockholder stations might provide programming to MBS affiliates, but that such programming, for the time being, would be outside of Mutual. Although he assumed that MBS would eventually become the fourth television network, MBS president Edgar Kobak stated, "I have a hunch that a few years from now survival may be difficult and one way to survive is to be careful now. That's what we at Mutual are doing." Survival for Mutual would indeed soon become difficult, and the decision not to actively develop a television arm would be one of the major contributing factors.

In 1943 General Tire and Rubber bought the Colonial network, giving it a small stake in MBS. In 1950 it expanded its broadcast holdings with the acquisition of the Don Lee network (giving the company 38 percent ownership of MBS), and its purchase of WOR the following year made it the controlling partner in Mutual, with 58 percent of the stock. Under the new corporate name of General Teleradio, the company acquired the remainder of Mutual's outstanding stock to become the sole owner of MBS. General Teleradio bought RKO-Radio Pictures in the mid-1950s. General's interest in the studio was solely to obtain its backlog of old theatrical movies as programming for General's growing roster of independent television stations. General had no intention of getting into the theatrical film business or of using RKO's studios as a production center for a possible MBS television network. Content to run its television outlets as independent stations, the newly renamed RKO-General immediately liquidated the film studio and, in 1957, sold MBS to oil tycoon Armand Hammer. The company would pass through the hands of five more owners in the next three years.

In the fall of 1958, Mutual was sold again, this time to Hal Roach Studios. A venerable producer of theatrical short comedies in the 1920s and 1930s, Roach had become a major television producer in the 1950s. By the late 1950s, however, the company was in the throes of serious financial reversals, compelling it to accept a buyout offer from a businessman named Alexander Guterma. Guterma immediately announced plans to combine the Roach operation and MBS into a broadcasting powerhouse of both radio and television networks, the latter of which would be supplied with programming from the Roach Studios. Within months, however, the Guterma empire collapsed under allegations of stock fraud.

In February 1959, MBS was sold by the Hal Roach Studios to recording executive Malcolm Smith. Smith sold MBS within months to a new set of owners, the McCarthy-Ferguson Group, who entered the network into bankruptcy reorganization.

During the transition from Smith to McCarthy-Ferguson, a final peculiar twist to the Roach-Guterma era emerged. In an effort to save his flagging business empire, Alexander Guterma had accepted $750,000 in January 1959 from the dictator of the Dominican Republic, Generalissimo Rafael Trujillo, in exchange for up to 425 minutes per month of favorable coverage of the Dominican Republic on Mutual radio news broadcasts. Negative reports on the Dominican Republic would not appear on MBS. The arrangement was reported to federal authorities by the new MBS management, who had found themselves accosted by agents of the Dominican Republic demanding performance on the deal or return of the money. In addition to his problems with the Securities and Exchange Commission, Guterma found himself tried and convicted of failing to register himself as an agent of a foreign principal.

In April 1960, McCarthy-Ferguson sold the network to the giant manufacturing company 3M, which was seeking to diversify into new fields. By the 1960s, however, network radio had become little more than a news-delivery service. As a corollary enterprise to the news division of a television network, a radio news service could return a profit. Despite the fact that it could boast the largest number of affiliates of any radio network, over 500 in 1967, with no television operation to share the costs of news gathering Mutual was locked in a terminal slide toward oblivion.

Despite the inevitability of its demise, MBS lasted considerably longer than most industry analysts expected (and longer than the pioneering NBC Radio Network). The end for Mutual came on 18 April 1999. Its final owner was the Westwood One radio group, which had bought Mutual from Amway in 1985. Shortly before Mutual's demise, Westwood One had turned most management decisions over to CBS Radio, which saw Mutual as redundant to other services offered by both Westwood One and CBS. The last Mutual stations were offered affiliation with Westwood One's CNN Radio operation to replace the departed 65-year-old Mutual network.

RICHARD WARD

See also American Broadcasting Company; CKLW; Columbia Broadcasting System; National Broadcasting Company; Network Monopoly Probe; Westwood One; WGN; WLW; WOR

Further Reading

"After 50 Years the Feeling's Still Mutual," *Broadcasting* (10 September 1984)

Crater, Rufus, "What Happens to MBS?" *Broadcasting-Telecasting* (15 October 1951)

Federal Communications Commission, *Report on Chain Broadcasting,* Washington, D.C.: Government Printing Office, 1941; reprint, New York: Arno Press, 1974

"First 12 Years of Mutual Network: A Study in 'Operations Grass Roots'" *Variety* (23 October 1946)

"Grand Jury Indicts Guterma Trio: Charged with Selling MBS as Dominican Propaganda Vehicle," *Broadcasting* (7 September 1959)

The Money Faces of Mutual, New York: Mutual Radio Networks, 1974

"New Giant Growing in Radio-TV?" *Broadcasting* (15 September 1958)

Robertson, Bruce, "Mutual Reaches Its 20th Birthday," *Broadcasting-Telecasting* (27 September 1954)

Robinson, Thomas Porter, *Radio Networks and the Federal Government,* New York: Columbia University Press, 1943

N

Narrowcasting

Narrowcasting is the process of identifying or selecting a specific portion of the overall radio audience and designing a station's programming to attract and retain that audience. Other related terms include *audience fragmentation, target audience, listener segmentation, niche audience,* and *format-specific, cultural-specific,* or *audience-specific programming.*

Narrowcasting stands in contrast to the older word *broadcasting,* which was borrowed from the agricultural industry. To a farmer, broadcasting means to sow seeds as widely as possible throughout a field; in radio, stations transmit their signal widely throughout their coverage area. In the early decades of radio, stations designed their programming to meet the needs of the largest, widest possible audience. However, stations could not always meet everyone's needs adequately, and portions of the population were left underserved or neglected by programming intended for an aggregate audience.

During the 1930s and 1940s, some independent, non–network stations in larger U.S. markets sold air time to African-American or non English-language programmers. In addition, some country and folk listeners could find programs to meet their needs in the various barn dance and jamboree shows around the country, although few stations featured around-the-clock music for rural listeners.

After World War II, the number of radio stations on the air increased dramatically, increasing the pressure on each outlet to find programs that would appeal to at least some listeners. Managers of newer stations who were willing to forgo the "golden age" approach to programming focused their programs to attract specific audiences; as a result, African-American and country stations first appeared in the late 1940s and early 1950s. During the same period, Top 40 programmers reached out to teens, and Middle-of-the-Road stations attracted older audiences: thus the first true formats were born.

The practice of narrowcasting is of particular interest to advertisers, because even though audiences are typically divided by listening characteristics, listener segmentation also results in buyer segmentation. When special audiences are targeted for specific products, the result is a more efficient use of advertising dollars. Teens are more likely to buy acne medicine and soft drinks. Mature audiences are more prone to invest in luxury cars, health care products, or mutual funds. This improved efficiency can be illustrated by the question, "Would you rather stand on the street and try to sell hot dogs to all the people passing by or would you rather talk to ten hungry people?" On the street, a salesperson will meet many people, but not all will be prepared to purchase. On the other hand, ten hungry people may find themselves quite interested in the prospect of a hot dog, if the salesperson can only locate them.

Over time, the practice of narrowcasting has enhanced the partnership between stations and advertisers. *Broadcasting and Cable Yearbook* in 2000 recognized 70 different radio formats in use. These various narrowcast formats divide the total audience into listener groups according to the demographic characteristics of age, gender, culture, or income. The station programmer's job is to design a total package of program elements, including music, news, IDs, liners, and public service announcements, to attract and retain their specific audience. The advertiser's task is to match audiences with products. A successful link of programs, audiences, and advertisers results in a more efficient and, in the long run, more economical effort enabling stations to deliver audiences to advertisers.

CHARLES F. GANZERT

See also Formats; Programming Strategies and Processes

Further Reading

Fornatale, Peter, and Joshua E. Mills, *Radio in the Television Age,* Woodstock, New York: Overlook Press, 1980

Ganzert, Charles, "Platter Chatter and the Pancake Impresarios: The Re-Invention of Radio in the Age of Television, 1946–1959," Ph.D. diss., Ohio University, 1992

Hall, Claude, and Barbara Hall, *This Business of Radio Programming: A Comprehensive Look at Modern Programming Techniques Used throughout the Radio World,* New York: Billboard, 1977

MacFarland, David, *The Development of the Top 40 Radio Format,* New York: Arno Press, 1979

Routt, Edd, James McGrath, and Frederick Weiss, *The Radio Format Conundrum,* New York: Hastings House, 1978

National Association of Broadcasters

U.S. Broadcast Trade Organization

The National Association of Broadcasters (NAB) is the primary trade association of the American broadcasting industry. The NAB represents the industry before the Federal Communications Commission (FCC), Congress, and other government entities and takes a proactive role in acquainting the public with the importance of radio and television communications.

Members of the association set policies and make decisions on industry-wide matters through a board of directors composed of radio and television broadcasters elected by fellow members. This joint board is subdivided into a radio and a television board, each with its own chair. The joint board also has a chairman. NAB is overseen by a full-time president.

NAB has an extensive committee structure that enables it to draw on the specialized knowledge of its members in dealing with industry causes and in making recommendations to the board of directors. These committees are composed of representatives of individual stations, broadcast groups, and the networks. Active member support and participation are the basis for NAB decisions and activities.

According to its charter, NAB operates "to foster and promote the development of the arts of aural and visual broadcasting in all forms; to protect its members in every lawful and proper manner from injustices and unjust exactions; to do all things necessary and proper to encourage and promote customs and practices which will strengthen and maintain the broadcasting industry to the end that it may best serve the people."

Origins

The early history of NAB is closely tied to the issue of using recorded music in early radio broadcasts of the 1920s. At that time, broadcasters freely used phonograph records without compensating the artists involved, in spite of a 1917 court decision that upheld the right of creative artists to license their products under provisions of the 1909 copyright act. By early 1922, the declining sale of phonograph records in the face of radio broadcasting caused the American Society of Composers, Authors, and Publishers (ASCAP) to look for ways to recover lost royalties directly from broadcasters. ASCAP, founded in 1914, provided the means for artists to license and copyright their creative efforts.

In April 1922 ASCAP determined that the radio reproduction of copyrighted songs fell under the "public performance for profit" portion of the copyright law and that the copyright owners were entitled to compensation from the broadcasters. ASCAP notified all broadcast stations of their intention to collect royalties for their members, but the announcement was largely ignored by the fledgling industry.

After finding little success pursuing its aims within the broadcast industry as a whole, ASCAP decided specifically to move against Westinghouse, General Electric (GE), the Radio Corporation of America (RCA), and a few other giants of the new industry. ASCAP called for a conference to discuss the issue, but it was once again ignored. The broadcasters agreed to a meeting only after ASCAP threatened to sue them for copyright infringements if they did not meet.

At the meeting on 20 September 1922, ASCAP presented as its major concern artists' rights to royalty payments, while broadcasters expressed their desire not to pay royalties. Of major importance to the broadcasters was the payment of performers; at that time, many musicians performed on radio free for the exposure, and broadcasters felt that if they paid some, they would have to pay them all. The broadcasters told ASCAP that they would go on the air with "The Old Gray Mare, She Ain't What She Used to Be" rather then pay ASCAP for the privilege of broadcasting the latest ASCAP licensed hit, "My Bromo-Seltzer Bride." The meeting ended without resolution.

Organization by Zenith's McDonald

The second ASCAP-broadcaster conference occurred one month later, on 25 October 1922. At this meeting, broadcasters expressed sympathy for the artists' position but stated that they could not afford to pay for the music if it would lead to paying all composers, individual performers, and orchestras. ASCAP responded by filing a suit and notifying all broadcasters that it was revoking all temporary licenses for broadcast of ASCAP members' music. Additionally, ASCAP established a rate schedule that fixed fees for the use of music at $250 to $5,000 per year per station depending on the size of the station's audience (determined by location, wattage, and profits). The arbitrary nature of the ASCAP action caused a small group of broadcasters, organized by "Commander" Eugene F. McDonald, Jr., founder-president of Zenith Radio Corporation (and station WJAZ), to meet in Chicago in early 1923 to form an organization to oppose ASCAP; this organization would become the National Association of Broadcasters.

The group, all pioneer broadcasters, moved quickly from a discussion of ASCAP to the need for a regulatory body for radio similar to the Interstate Commerce Commission. At this initial meeting of concerned broadcasters, McDonald first used the name "Federal Communications Commission" for such a group. They also considered rules and regulations that they felt should apply to a free enterprise system for radio.

Shortly after this meeting, McDonald, who knew little of the music business, called on his friend and business colleague Thomas Pletcher, president of the QRS Music Company, for advice. He told Pletcher that his group of broadcasters believed that authors and composers should be paid directly for their contributions, rather then through ASCAP, but that they did not know how to proceed with the organization. Pletcher suggested that the group hire Paul B. Klugh (a knowledgeable, recently retired music roll manufacturer) as secretary of the new organization. McDonald successfully recruited Klugh, who assumed the position of executive chairman. McDonald embarked on a campaign to persuade RCA, GE, Westinghouse, and American Telephone and Telegraph (ATT) to join the fledgling group, but he was initially unsuccessful.

The actual organizational meeting for the NAB was held in the studios of WDAP (Chicago) on 25 and 26 April 1923, with 54 representatives of various radio constituencies. Representatives of ASCAP presented their positions and, after discussion, left the meeting. A committee was then formed to propose the structure for an organization that would carry out the aims of the broadcasters. The committee consisted of McDonald (WJAZ), T. Donnelley (WDAP), J.E. Jenkins (WDAP), W.S. Hedges (WMAQ), P.B. Klugh, R.M. Johnson (Alabama Power and Light Company), and George Lewis (Crosley Manufacturing Company). The group, known as "the Committee of Seven," reported on 26 April 1923, and their recommendations were accepted: the association was to be known as The National Association of Broadcasters, NAB's offices would be established in New York City, and they would employ a managing director. Paul Klugh was selected as the managing director by a unanimous vote. Two other meetings were held in 1923 with the main subject of discussion being the development of an NAB "music bureau" of copyright-free music.

The first annual meeting of the National Association of Broadcasters was held in conjunction with the annual National Radio Show held in New York on 11 October 1923. This meeting, called to order by Chairman Klugh in the Commodore Hotel in New York City, resulted in the election of the first real officers of the NAB. McDonald was elected president. A number of addresses were presented, and the group received a list of holdings in the NAB Music Bureau. A discussion of legislation plans was also undertaken. The group wanted to accomplish two goals with their legislation: (1) music copyright revision and (2) modernization of the 1912 Radio Act. During the meeting, McDonald conducted a test of the size of the audience at WJAZ by asking listeners to send in paid telegrams acknowledging their reception. The audience was estimated to be 400,000, based on receiving 4,284 telegrams in four hours. The public relations coup resulted in considerable publicity for the young broadcasters group.

In conjunction with the 1923 Chicago Radio Show in November, McDonald and his good friend Thorne Donnelley of WDAP, both officers of the NAB, along with Chicago station KYW, conducted an audience census to determine music preference in the Chicago listening area. For 12 days the stations requested listeners to write in telling what they most desired to hear. The three stations received a total of 263,410 pieces of mail, with WJAZ receiving 170,699; WDAP, 54,811; and KYW, 37,900. It was estimated that not more than 1 in 50 listeners would respond, which suggested that the three stations were being heard by an audience of more than 13 million.

In 1924 ASCAP attempted to flex its muscle and thus created a situation that brought the music copyright problem before Congress. The Edgewater Beach Hotel (home of WJAZ) had always paid a fee to ASCAP for the music used in its dining room. Because at times this music had also been broadcast live over WJAZ, ASCAP refused to renew the performance license for the music unless a broadcast license was also secured, even though the hotel no longer allowed WJAZ to broadcast the music. The broadcasters determined that this situation was a good legal test case. The fact that McDonald's WJAZ was involved and that he was also president of the NAB undoubtedly was an important factor in choosing this incident for the test case.

On 22 February 1924, Senator C.C. Dill, at the urging of the NAB, introduced a bill to the Senate that amended the Copyright Act of 1909 to make radio performances of copyrighted material essentially legal and royalty-free. A nasty battle ensued, with ASCAP waging a publicity campaign encouraging all musicians to join the fight. NAB, small and new, had little money to fight back, and a plea to broadcasters for financial help brought nothing.

The NAB cause was represented by NAB President McDonald, Executive Secretary Klugh, and Counsel Charles Tuttle. In January 1925, the ASCAP-supported Perkins Bill (H.R. 11258) was introduced in the House. This bill called for massive changes in the copyright law in general and particularly in those portions concerned with radio broadcasting.

Years of debate followed. The NAB endorsed its stand at each succeeding annual conference. McDonald's involvement also continued. As NAB president, he was appointed one of ten members of the Copyright Committee of the Fourth National Radio Conference called by then Commerce Secretary Herbert Hoover in Washington on 9 November 1925. The fight to avoid paying royalties, however, was rapidly being lost, and the NAB and ASCAP entered secret negotiations. McDonald stepped down as president of the NAB in 1926 as he began a battle with Secretary Hoover over frequency allocations, but he continued his strong involvement with NAB for many years, holding a variety of offices.

NAB Radio and Television Codes

The NAB has been involved in voluntary compliance broadcast codes since 1929, when it produced its first "Code of Ethics" in an attempt to preempt the Federal Radio Commission from imposing such a code. In 1939, attempting to avoid FCC action regarding children's programs, the NAB issued a "Radio Code" dealing with profanity and limits on commercial time in children's programming. NAB issued a guide for broadcasters in 1942 covering security in wartime broadcasting. Although compliance with these codes was voluntary, many NAB members subscribed to them, and the various NAB radio and television codes existed until they were dropped in 1982 after a court case found a portion of the code unconstitutional. Today the NAB operates under a "Statement of Principles" dealing with program content.

The NAB expanded greatly in 1938 and has grown with the broadcasting industry, incorporating other groups in the industry. The FM Broadcaster Association became a department of NAB in 1945, and in 1951 the Television Broadcasters Association also merged with NAB. Springing from NAB membership in 1959, a new and independent National Association of FM Broadcasters met for the first time; by 1984, the 2000-member group, then named the National Radio Broadcasters Association, merged with NAB. NAB also accepts associate members in fields allied to broadcasting.

The Modern NAB

Although much of the association's focus since the 1950s has been on television, cable, and newer media, radio continues to occupy NAB lobbying and developmental efforts. NAB championed FM in the 1960s and 1970s, publishing a monthly newsletter tracing industry developments. In the early 1980s, NAB fought strenuously and eventually successfully to beat back an FCC proposal to reduce AM channels from 10 kilohertz to 9 kilohertz, parallel to much of the rest of the world. The association argued that such a move would increase interference and make many push-button radios obsolete. At the same time, NAB supported efforts to create AM stereo and to select a technical standard for the service.

In the late 1990s, NAB radio interests focused on digital radio and on developing a successful technical standard to allow digital audio broadcasting (DAB) service to begin. At the same time, NAB fought hard against allowing satellite digital services to develop, because they would threaten the local stations represented by the association. And NAB fought against the highly popular introduction of low-power FM stations (LPFM) in 2000, arguing that the potential for hundreds of new stations would greatly increase interference problems. NAB maintains an ongoing educational program on its agenda.

The huge annual four-day NAB conventions, which now attract more than 115,000 attendees, are held every spring (in Las Vegas since the early 1970s) and devote considerable conference time and exhibition space to radio and audio topics. A fall conference focused entirely on radio programming and operations is also held, and radio-related publications are issued regularly. With more than 100 full-time employees housed in its own modern building in Washington, D.C., the NAB has gained a reputation as one of the strongest and most effective lobbies in the nation's capital. Part of this strength comes from the clout inherent in member stations, which provide airtime for political candidates and which will readily call congresspersons to press their views.

HAROLD N. CONES AND JOHN H. BRYANT

See also American Society of Composers, Authors and Publishers; Broadcast Music Incorporated; FM Trade Associations; McDonald, Eugene F.; Trade Associations

Further Reading

Barnouw, Erik, *A History of Broadcasting in the United States*, 3 vols., New York: Oxford University Press, 1966–70; see especially vol. 1, *A Tower in Babel: To 1933*, 1966, and vol. 2, *The Golden Web, 1933 to 1953*, 1968

Cones, Harold, and John Bryant, *Eugene F. McDonald: Communications Pioneer Lost to History*, Record of Proceedings, 3rd International Symposium on Telecommunications History, Washington, D.C.: Independent Telephone Historical Foundation, 1995

Hilliard, Robert L., and Michael C. Keith, *The Broadcast Century: A Biography of American Broadcasting*, Boston: Focal Press, 1992; 2nd edition, Boston and Oxford, 1997

Mackey, David, "The Development of the National Association of Broadcasters," *Journal of Broadcasting* 1 (Fall 1957)
National Association of Broadcasters website, <www.nab.org>
Smith, Frederick, "Fees for Composers—None for Broadcasters," *Radio Age* (February 1923)
Smulyan, Susan, *Selling Radio: The Commercialization of American Broadcasting, 1920–1934*, Washington, D.C.: Smithsonian Institution Press, 1994

National Association of Educational Broadcasters

The National Association of Educational Broadcasters (NAEB) was the oldest professional educational broadcasting organization in the United States, founded as the Association of College and University Stations in 1925. Until its demise in 1981, the NAEB served as the nation's most influential force in the establishment and preservation of an alternative system of noncommercial educational (public) radio stations.

Origins

The historical roots of American public radio extend back at least as far as those of commercial radio broadcasters. Early in the 1900s experimental stations began appearing in electrical engineering departments of universities and colleges across the country, the first being station 9XM (now WHA) at the University of Wisconsin in Madison. Unfortunately, the primary motivation for building many of these stations was limited to the study of technical considerations, without much concern about the programming and service potential of this new electronic medium. By 1926, roughly half of these early stations had gone off the air, but among those remaining, there was a growing interest in exploring educational uses for radio.

Secretary of Commerce Herbert Hoover had already begun a series of annual National Radio Conferences in Washington, D.C., and it was at the fourth of these gatherings, on 12 November 1925, that a group of educational broadcasters created a new organization, the Association of College and University Broadcasting Stations. At first, this fledgling collective of broadcast pioneers was loosely knit and had no specific purpose other than to support and promote radio for educational use. Membership was open to all educational institutions that operated radio stations, but even with annual dues set at only $3.00, less than half of the qualified institutions joined the association during its first few years of existence. Documentation for this period is extremely limited, but it is clear that the members struggled against enormous odds to hold the organization together, as the Great Depression began to bring financial hardships to institutions of higher learning.

The first formal convention of the 25-member association was held on the Ohio State University campus in Columbus in July of 1930, convened in conjunction with the Institute for Education by Radio. Recent licensing actions by the Federal Radio Commission were seen as clearly favoring the commercial use of radio at the expense of educational development. Hence, there was a growing sense of urgency that something needed to be done to stem the tide of lost licenses for educational institutions. Association president Robert Higgy, Director of Ohio State's WOSU, launched a campaign to seek legislation that would reserve a portion of the radio spectrum exclusively for noncommercial educational use. Association members joined with other educational radio advocates, including the National Advisory Council on Radio in Education and the National Committee on Education by Radio, in that pursuit over the next several years. However, when the Communications Act was signed into law on 19 June 1934, the provision for which they had fought so hard had been deleted. Instead, Congress specified that the newly created Federal Communications Commission would study the matter of nonprofit allocations to educational institutions and report back on their findings.

Reorganization and Name Change

When the association members gathered on 10 September 1934, in Kansas City, Missouri, for their annual meeting, there was a renewed determination to influence the work of the FCC. A new constitution was adopted and the organization's name was changed to the National Association of Educational Broadcasters. This new label better reflected the interests of existing members, particularly those that were producing educational programs that were being broadcast over commercial radio stations. The NAEB members rededicated themselves to three goals: reserving channels for educational use, establishing a national headquarters, and creating a mechanism for program exchange. Toward this third goal, three committees were established, each to study a specific means of program exchange: shortwave transmission between stations, the establishment of a chain network, and the recording of programs.

The FCC hearings that stemmed from the congressional statute to study the matter failed to result in the desired educational reservations, though the NAEB continued to lobby both the FCC and members of Congress in the years to follow. When the NAEB assembled in Iowa City on 9 September 1935, each of the program exchange committees made their reports. The use of shortwave interconnection was judged impractical, but there was some optimism expressed that federal and state appropriations might enable the creation of a chain network at some future date. However, the only short-term means of program exchange seemed to be to record programs and exchange them among stations through the U.S. mail service. The goal of establishing a national office was still well beyond the members' reach.

The NAEB continued to meet annually through 1938 but suspended its regular meeting schedule during the years 1939–41. Throughout this period the association achieved modest accomplishments, including purchasing of sound transcription equipment, beginning the publication of a regular newsletter, creating of a radio script exchange, and successfully lobbying the FCC to reserve designated "curricular channels" in 1938, though the authorization of FM broadcasting would render this initial lobbying victory moot. The continued efforts by the NAEB culminated in the FCC's acceptance of the reservation principle when the Commission first authorized FM service in 1941, reserving the five lowest channels for noncommercial educational use. Four years later, the FCC shifted the placement of FM broadcasting to its present location, and set aside the lowest 20 channels (88–92 MHz) for educational use.

Although membership in the association during the early 1940s did not increase significantly, the organization gained greater cohesion and confidence as representatives from member stations worked cooperatively on a variety of association initiatives, including FM channel reservations. Efforts were stepped up to get educational institutions to apply for construction permits for FM stations, and discussions about establishing a national headquarters with paid personnel continued to gain momentum. In an attempt to encourage more stations into operation, the NAEB convinced the FCC to liberalize its FM rules by creating a new Class D license in 1948 that allowed stations to broadcast with as little power as 10 watts, thus greatly reducing the costs of transmitter equipment and ongoing operations for the many colleges and universities that wanted to mount student-operated radio stations.

In 1949, the long-envisioned program exchange was formally begun when station WNYC in New York City made five sets of recordings of the *Herald Tribune Forum* series that were mailed among 22 NAEB member stations in a distribution system that came to be known as the bicycle network. Prompted by the success of this bicycle tape network, University of Illinois Dean Wilbur Schramm offered a plan at the 1950 NAEB convention in Lexington, Kentucky, to house the network headquarters on his Urbana, Illinois, campus, with funding generated by a series of grants. The following year the NAEB received a major grant from the W.K. Kellogg Foundation to fund a permanent national headquarters and distribution center at the University of Illinois. A series of other major grants were soon forthcoming, and both the financial posture and the services provided by the NAEB improved significantly. Increases in NAEB membership quickly followed, and by January of 1954 there were 218 members and 78 stations participating in the tape network. The national headquarters staff had expanded to seven full-time employees. After nearly 30 years of struggling for its own survival, the NAEB had become the dominant force in U.S. educational broadcasting.

Organizational Transitions

During the 1950s, the NAEB greatly improved its stature within the educational community, both nationally and internationally. Increased human and financial resources enabled the association to exhibit expanded leadership and professional development for the educational broadcasting establishment. Workshops, seminars, and regional conferences soon complemented the annual convention. In 1956, a sister organization of individual members from a wide range of educational professions—the Association for Education by Radio-Television (AERT)—merged with the NAEB and brought with it a scholarly publication, the *AERT Journal*. The following year, the association began publication of the *NAEB Journal*, in addition to the monthly *NAEB Newsletter* and other occasional reports and monographs. By the 1959 convention in Detroit, Michigan, it was evident that the NAEB's ever-widening vision had outgrown its home in Urbana, Illinois.

From the inception of the national headquarters at the University of Illinois, the NAEB had operated with an elected president from one of the member stations and a full-time executive director located at the national office. It was time to move the national headquarters to Washington, D.C., and to hire a full-time president. On 1 September 1960, the NAEB's new offices opened in the DuPont Circle Office Building at 1346 Connecticut Avenue, NW. The new president who would lead the organization into a new era of educational broadcasting was William G. Harley, a former elected NAEB president and chairman of the board of directors who had gained national prominence as manager for the highly successful WHA-AM-FM-TV stations at the University of Wisconsin.

Harley moved quickly to establish the NAEB as a lobbying force on Capitol Hill while expanding the number of grants that enabled the NAEB to enhance its intellectual position. A new publication, the *Washington Report*, was created to help keep members posted of important developments in the capital city. The NAEB Radio Network took full advantage of the Washington connection by producing a new public affairs show, *Report from Washington,* that was sent by air mail to the Center in Urbana for distribution to member stations. Harley also secured letters pledging support from both presidential hopefuls, Richard M. Nixon and John F. Kennedy, as a way of building bridges with the new administration. And among the first academic projects to come out of this period was the commissioning of Marshall McLuhan to prepare a report on the new media's role in the future of education. The book resulting from that project, *Understanding Media,* remains one of the most influential mass communication publications of the 20th century.

NAEB leaders had long advocated that the association should be more of a professional organization than a trade association in the traditional sense. The influx of individual members brought about by the merger with AERT in 1956 and the growing ranks of members from closed-circuit instructional television facilities were causing growing dissatisfaction with the existing governance structure, which was controlled by radio and television station representatives. At the 1963 NAEB convention in Milwaukee, Wisconsin, the members voted unanimously to reorganize the association into four divisions: Radio Station Division, Television Station Division, Instruction Division, and Individual Member Division. Each was to elect its own board, with the four boards comprising the NAEB Board of Directors. Offices and support staff for each of the units would be established within the national headquarters.

For radio interests, the reorganization was a major step forward. Television representatives had been exercising more and more control over the NAEB since the mid-1950s. Growing numbers of radio representatives wanted a separate organization of their own but knew full well that they did not have the resources to go it alone. The reorganization gave radio the independence it needed to chart a new course, while allowing it to benefit from the largesse of the higher television station dues. The new configuration also afforded the opportunity for a new name—National Educational Radio (NER)—while remaining under the NAEB umbrella. In the spring of 1964, the radio board appointed WUOM (University of Michigan) production manager Jerrold Sandler to be NER's first executive director. The tape exchange network—now officially named the National Educational Radio Network—remained in Urbana.

Under Sandler's guidance, NER acquired the kind of representational and leadership presence in Washington, D.C., that the radio representatives had envisioned. In addition to continuing the program exchange system, the new division became far more active on Capitol Hill, raised major grants for program and research projects, built relationships in the international community, provided a unified voice in professional circles, offered consulting advice to member stations, and distributed grants-in-aid to support special projects. During 1966–67, Sandler contracted with Herbert W. Land Associates to conduct a national study of the status of educational radio in the United States. The resulting report, *The Hidden Medium: Educational Radio,* offered the documentation needed for Sandler to lobby Congress on behalf of radio during the drafting of what had been the Public Television Act of 1967. As a direct result of Sandler's efforts, radio was written into the language during the eleventh hour, and the final legislation was called the Public Broadcasting Act of 1967.

A Forecast of Demise

While passage of this historic legislation dramatically changed educational radio and television for the better, it signaled the beginning of a transformation that would eventually mean an end to the NAEB. The Act created the Corporation for Public Broadcasting (CPB), which was charged with strengthening the newly relabeled *public* radio and television stations in the United States. This mandate led to the creation of the Public Broadcasting Service (PBS) and National Public Radio (NPR). In 1971, NPR began live national network interconnection of member stations, and the National Educational Radio Network soon merged with NPR. For a number of months, NAEB continued to represent radio stations before Congress, but by 1973, the stations had created a new lobbying and public relations organization—the Association of Public Radio Stations (APRS)—and so NAEB ceased its radio representation function. (APTS would later merge with NPR in 1977.)

Just as NPR and APRS took over the functions of NAEB's Radio Division, PBS soon acquired the functions of the Television Division. By the mid-1970s, the NAEB was again forced to undergo a major reorganization and a redirection of its mission solely as a professional organization. When Harley retired in 1975, James A. Fellows became NAEB's last president. Fellows worked tirelessly to revitalize the association and to generate a solid funding base through individual member services. A variety of public telecommunication institutes on such topics as management skills, graphic arts, instructional design, and audience research methods were scheduled throughout the country. The publications program was expanded with the creation of *Public Telecommunications Review,* a research index and reprint series, and later with the *Current* newspaper. Members were served by various professional councils, ranging from broadcast education and research to engineering and management. The annual convention afforded additional professional training opportunities and a career placement center. In short, the NAEB attempted to become the professional standard bearer for the public broadcasting industry.

Despite a herculean effort by Fellows and his ever-shrinking staff, the NAEB could not sustain itself as an organization supported solely by individual members. At its final convention at the Hyatt Regency Hotel in New Orleans on 3 November

1981, NAEB board chairman Robert K. Avery recalled the association's important accomplishments over the preceding 56 years. With the vote taken to declare bankruptcy, he brought down the final gavel and dissolved the organization.

ROBERT K. AVERY

See also Educational Radio to 1967; National Public Radio; Public Broadcasting Act of 1967; Public Radio Since 1967

Further Reading

Avery, Robert K., Paul E. Burrows, and Clara J. Pincus, *Research Index for NAEB Journals: NAEB Journal, Educational Broadcasting Review, Public Telecommunications Review, 1957–1979,* Washington, D.C.: National Association of Educational Broadcasters, 1980

Blakely, Robert J., *To Serve the Public Interest: Educational Broadcasting in the United States,* Syracuse, New York: Syracuse University Press, 1979

Hill, Harold E., and W. Wayne Alford, *NAEB History,* 2 vols., Urbana, Illinois: National Association of Educational Broadcasters, 1966

Witherspoon, John, Roselle Kovitz, Robert K. Avery, and Alan G. Stavitsky, *A History of Public Broadcasting,* Washington, D.C.: Current, 2000

National Association of Educational Broadcasters Tape Network

Early Program Exchange for Educational Broadcasters

The National Association of Educational Broadcasters (NAEB) Tape Network was the first formal agreement among educational broadcasters to allow for the exchange of programs for rebroadcast. This system was important because it provided for the sharing of much-needed program materials between financially strapped broadcast stations across the United States.

Origins

In the 1930s a growing concept of educational broadcasting was developing at a few scattered stations, most of which were loosely affiliated with colleges and universities. Representatives held annual conventions, which led to formation of the Association of College and University Broadcasting Stations, predecessor to what became the NAEB. The group's primary purpose was to persuade the government to set aside radio channels for state, college, and university operated stations. Another goal was to develop a mechanism for program exchange.

These broadcasters watched the successful sharing of programming taking place in commercial radio through the National Broadcasting Company (NBC) Red and Blue networks and the Columbia Broadcasting System (CBS). A network for educational radio was discussed but was dismissed as too costly; an idea for a script exchange met with little enthusiasm. A few stations did exchange scripts, but few of these exchanges resulted in produced programs.

In 1932 NAEB secured a $500 grant from the National Advisory Council on Radio Education and purchased a wire recorder. The device was to be circulated among stations for recording programs for air. In 1949 the director of New York's municipal station WNYC, Seymour Siegel, made five sets of recordings of the *Herald Tribune Forum* and distributed them to 22 NAEB member stations throughout the year. This event marks the start of what was labeled a "bicycle" (mailed tape exchange) network.

Postwar Developments

In 1950 NAEB was able to secure more funding, this time from the Kellogg Foundation. The purpose of the grant was to support a systemized national noncommercial education and culture tape network to serve the growing demand for educational radio programming. The exchange system delivered programming from the British Broadcasting Corporation (BBC), the Canadian Broadcasting Corporation (CBC), and domestic stations and production centers. In its first year, what became the National Educational Radio (NER) network mailed over 500 hours of programming to 52 NAEB member stations coast to coast.

A network headquarters was established in Urbana, Illinois. Programs were duplicated on high-speed equipment. A fee was established for member stations, and within five years, the organization was modestly financed but self-sustaining. By 1967 it was estimated that the network distributed some

35,400 hours (more than 80 million feet of tape) of educational radio programming in the United States on a budget of less than $60,000.

Sample bulletins from the early years of the tape exchange demonstrate diverse program offerings. Program topics included physical sciences ("The Impact of Atomic Energy"), social sciences ("Woman's Role in Society"), mental and physical health ("The Effects of Smoking"), and arts and literature ("The Alabama String Quartet"). Children's programming was a category listed in early bulletins, but specific examples are difficult to find. Networks supplied their affiliates with tapes to be used in local schools as well. Subject matter for kindergarten through 12th grade included science, foreign languages, guidance, language arts, music and art, safety and health, and social studies.

Most of the programs distributed by the tape network in return for the regular affiliation fee consisted of offerings from individual affiliates. Production costs for these programs were generally covered by the local stations. Some stations enhanced programs with modest grants from organizations such as the National Home Library Foundation and the Johnson Foundation.

The network also delivered some special programming to affiliates at no additional cost. For instance, a 30-minute *Special of the Week* produced out of the University of Michigan featured addresses by national and world leaders on public affairs. In addition, an 11-program series by WGBH Boston (*A Chance to Grow*) examined how families dealt with critical changes in their lives. The series featured interviews with families and was produced with the aid of grants and contributions. The government also provided some "no-charge" programs, such as a panel discussion from a conference held by the Selective Service System at the University of Chicago. And finally, the Library of Congress had a special arrangement with the tape network to allow the network to distribute certain readings and lectures given by the library.

Programs were also provided to affiliates from a wide variety of international sources. Regular contributors to the network included the BBC (*Translantic Forum, The World Report,* and *Science Magazine*), Radio Netherlands, Berne, Italian Radio, UNESCO, and the CBC, among others.

While the tape exchange network was still in a growth mode, educational broadcasters were still pushing for a more permanent way to exchange programming—a real network. In September 1965 some 70 NER stations linked together for a historic live interconnection to broadcast three hours of German national election results coverage.

In 1966 at a NER conference (The Wingspread Conference), 70 leaders from the industry, government, the academic community, philanthropy, and the arts came up with a seven-point plan for developing educational radio that included a national production center and the use of communications satellites for transmitting noncommercial programming. A concrete step toward centralized programming was made in March 1967 when NER set up a Public Affairs Bureau in Washington. However, a centralized network was yet to come. During the Carnegie Commission's study of 1965–67, the need for an interconnected educational radio system was again stressed. The Public Broadcasting Act (1967) that followed set the stage for the development of that network. In the years following the passage of the act, educational broadcasters met to plan the network. Finally, on 3 May 1971, the goal of national programming distribution sources for noncommercial radio broadcasters was realized, when National Public Radio premiered its first live show—*All Things Considered.* Other programs and a satellite interconnect to allow members to share programming would follow. Educational broadcasters no longer had to duplicate and mail tapes to deliver shared programs to the public.

PAMELA K. DOYLE

See also Educational Radio to 1967; National Association of Educational Broadcasters; National Public Radio; Public Radio Since 1967

Further Reading

Blakely, Robert J., *To Serve the Public Interest: Educational Broadcasting in the United States,* Syracuse, New York: Syracuse University Press, 1979

Herman W. Land Associates, *The Hidden Medium: A Status Report on Educational Radio in the United States,* New York: Herman W. Land Associates, 1967

Saettler, L. Paul, *A History of Instructional Technology,* New York: McGraw Hill, 1968

Siegel, Seymour N., "Educational Broadcasting Comes of Age," in *Education on the Air: Twenty-First Yearbook of the Institute for Education by Radio,* Columbus: Ohio State University, 1951

Witherspoon, John, Roselle Kovitz, Robert K. Avery, and Alan G. Stavitsky, *A History of Public Broadcasting,* Washington, D.C.: Current, 2000

National Barn Dance

Country Music Variety Program

Although the *Grand Ole Opry* is best remembered because it survived far longer, the WLS *National Barn Dance* before World War II ranked as America's most popular country music program. After being picked up by the National Broadcasting Company (NBC) network in 1933, and sponsored and supported by Miles Laboratories' Alka-Selzer, the WLS *National Barn Dance* became a Saturday night radio network fixture.

The show's long time home was WLS-AM, a pioneering and important station in Chicago, named by owner Sears-Roebuck to herald itself as the "World's Largest Store." First under Sears, and then after 1928 with new owner *Prairie Farmer* magazine, the focus of this clear channel station was always the rural American. Clear channel status made WLS-AM a fixture in homes throughout the Midwestern farm belt.

Originally called the *WLS Barn Dance*, by 1930 the program had expanded to fill WLS's Saturday nights. When the NBC Blue network began on 30 September 1933 to run a portion, sponsored by Miles Laboratories' Alka-Seltzer, the pro-

WLS National Barn Dance Cast, 1944
Courtesy Radio Hall of Fame

gram was renamed the *National Barn Dance*. Though NBC varied the program's length (from 30 to 60 minutes and back, again and again), from 1936 to 1946, the network always penciled the *National Barn Dance* on the schedule starting at 9 P.M. on Saturday nights. Indeed, in 1940 when NBC picked up the *Grand Ole Opry*, the *National Barn Dance* was already a network fixture, signaling to all who paid attention to radio industry trends a growing interest in the "hillbilly" musical form.

The statistics were impressive. For example, on 25 October 1930 nearly 20,000 fans poured into Chicago's massive International Amphitheater for a special performance, and an estimated 10,000 had to be turned away. The popularity peak of the *National Barn Dance* came during World War II, but after the war the *Opry* surpassed its predecessor in ratings, and the shift to Nashville was underway. NBC dropped the *National Barn Dance* in 1946.

But in its heyday, beginning with a move in 1932 to the Loop's Eighth Street Theater, the success of the *National Barn Dance* could literally be seen as crowds lined up for precious Saturday night tickets and regularly filled the theater's 1,200 seats for two shows. Stars included those remembered by country music historians (Bradley Kincaid, Arkie the Arkansas Woodchopper, and Lulu Belle and Scotty) and those who would help define popular culture of the 20th century (Gene Autry).

Bradley Kincaid was one of country music's first popular sellers. He was a student in Chicago in 1926, having moved there from his native Kentucky, when a friend suggested he try out for the hillbilly show on the radio. Kincaid borrowed a guitar, practiced a few ballads he had learned from his family, and soon was a star. His name is usually lost in country music history, but Kincaid helped define the genre as it emerged during the 1920s and 1930s.

Missourian Luther Ossenbrink renamed himself "Arkie, the Arkansas Woodchopper" when he arrived at WLS in the middle of 1929 after some experience on the radio in Kansas City. While playing the fiddle, guitar, or banjo, he sang and told cornball jokes. Sears executives must have loved his favorite song, "A Dollar Down and a Dollar a Week."

Although the husband-and-wife singing duo, Lulu Belle and Scotty (Wiseman), who appeared from 1935 through 1958, may have been the *Barn Dance*'s most enduring act, to the world Gene Autry symbolized the star-making power of the National Barn Dance. Autry appeared during the early 1930s as the "Oklahoma Yodeling Cowboy," but when his "Silver Haired Daddy of Mine" became a hit, he was off to Hollywood and became part of radio legend.

Following NBC's dropping of the *National Barn Dance*, the program reverted to again being a local show until 1949, when the American Broadcasting Companies (ABC) radio network, sponsored by the Phillips Petroleum Company, picked up the show. The ratings on ABC were anemic, and so by the mid-1950s the *National Barn Dance* was a faded memory for everyone except the aging Chicagoans who continued to embrace its radio and television versions. The Eighth Street Theater closed on 31 August 1957, and WLS abandoned it.

In 1960, many of the former *National Barn Dance* regulars appeared on Chicago's WGN-AM under the name *WGN Barn Dance*. But after the Tribune Company, the owner of WGN, syndicated the show for television in the 1960s with limited success, Tribune executives closed the show for good in 1971.

DOUGLAS GOMERY

See also Autrey, Gene; Country Music Format; Grand Ole Opry; WGN; WLS

Cast

Hosts	Hal O'Halloran (pre-network), Joe Kelly (1933–50)
Announcer	Jack Holden
Performers (partial listing)	Bradley Kincaid, Gene Autry, George Gobel, Red Foley, Homer and Jethro, Lulu Belle and Scotty, Louise Massey Mabie and the Westerners, Arkie, the Arkansas Woodchopper, Wilson Sisters, Dolph Hewitt, Hoosier Hotshots, Pat Butrum

Producers

Walter Wade, Peter Lund, Jack Frost

Director

Bill Jones

Programming History

WLS	April 1924–September 1933
NBC-Blue	September 1933–June 1940
NBC-Red	June 1940–September 1946
ABC	March 1949–March 1950

Further Reading

Baker, John Chester, *Farm Broadcasting: The First Sixty Years*, Ames: Iowa State University Press, 1981

Evans, James F., *Prairie Farmer and WLS: The Burridge D. Butler Years*, Urbana: University of Illinois Press, 1969

Malone, Bill C., *Country Music U.S.A.*, Austin: University of Texas Press, 1968; revised edition, 1985

Summers, Harrison Boyd, *A Thirty-Year History of Programs Carried on National Radio Networks in the United States, 1926–1956*, Columbus: Ohio State University, 1958

WLS Magazine (12 April 1949) (special 25th anniversary issue entitled "Stand By")

National Broadcasting Company

While now focused on television, the National Broadcasting Company (NBC) was the first purpose-built national radio network, although it continues as such today in name only. Begun with an emphasis on public service program orientation, NBC became a very profitable commercial venture that helped to dominate—and define—radio's golden age.

Origins

The origins of NBC lie in the extensive political and legal maneuvering of its parent company, the Radio Corporation of America (RCA), in the 1920s. One member of the RCA group, AT&T, found its phone lines could be used not only for remote broadcasts but could also connect stations together in a "chain" or network. AT&T announced the formation of 38 "radio telephone" stations linked by telephone lines, the purpose of which was not to provide programming but rather to "provide the channels through which anyone with whom it makes a contract can send out their own programs," an arrangement that soon became known as "toll broadcasting."

But RCA was operating its own New York City station as well. WJZ began operations as a Westinghouse outlet in 1921 but was purchased by RCA two years later. Unlike AT&T, RCA's interest in the medium was based on its desire to sell more receivers, the assumption being that entertaining programs would result in the sale of more sets. Thus, by 1923 there were two factions battling for control of radio: the "Telephone Group" led by AT&T and Western Electric, and the "Radio Group" consisting of RCA, General Electric (GE), and Westinghouse. AT&T sold its interest in RCA when conflict became inevitable and refused to allow any station aligned with the Radio Group to use its telephone lines to establish a network.

With the threat of government intervention looming, it was left to RCA General Manager David Sarnoff to broker a compromise. He proposed that "all stations of all parties [be put] into a broadcasting company which can be made self-supporting and probably revenue-producing, the telephone company to furnish the wires as needed." This marked the creation of NBC, which began operation in November 1926 with Merlin H. Aylesworth, former managing director of the National Electric Light Association, as president. Although ownership of NBC was originally divided among RCA, GE, and Westinghouse, AT&T profited the most, since it controlled the wires that would eventually connect thousands of stations nationwide (a franchise that would be extended even further with the introduction of television). In short, AT&T got to keep the toll without having to worry about the broadcasting.

Consolidation and Growth

NBC immediately adopted the practice of toll broadcasting by selling studio space and a time slot—"four walls and air" in trade lingo—to interested advertising agencies and their sponsor-clients. RCA also decided to operate two NBC networks, with WEAF as the flagship station of NBC-Red and WJZ anchoring NBC-Blue (the colors apparently derived from either the company's color-coded program charts or the pencil lines AT&T engineers drew to map the wire paths for the two networks).

NBC grew rapidly in size and profitability. In 1927 the network had 48 affiliates, including both the Red and Blue networks, and lost almost $500,000 in net income; by 1932—when RCA assumed complete ownership of the network—it had 85 affiliates and pretax profits of $1.2 million.

As both networks grew, so did the interest of advertising agencies and sponsors, who saw network radio as an increasingly effective way to reach a national audience of consumers. Fortunately for advertisers, the ability to reach a mass audience with radio intersected with an expansion of the American economy in the 1920s. In 1929 Merlin Aylesworth proclaimed that radio was "an open gateway to national markets, to millions of consumers, and to thousands upon thousands of retailers."

Soon, major advertising agencies were enthusiastically embracing the new medium, a revolution that was not dampened by the Great Depression. The NBC schedule in the 1930s was dominated by shows named for their sponsors, such as *The Chase and Sanborn Hour*, *Cliquot Club Eskimos*, and *Maxwell House Showboat*. By the early 1930s, the economic structure that would dominate radio for the next 20 years had emerged. The networks learned to be the middleman, selling time to advertising agencies that produced the commercial shows on behalf of paying sponsors, while their affiliates were responsible for developing a rapport with local audiences.

Throughout the decade, NBC continued to flourish. A sharper distinction between the Red and Blue networks came into focus. NBC-Red was home to the most popular programs, including *Fibber McGee and Molly*, *One Man's Family*, and *Amos 'n' Andy*, as well as such stars as Bob Hope, Jack Benny, and Fred Allen. Not surprisingly, it accounted for most of NBC's profits. NBC-Blue, on the other hand, was somewhat schizophrenic in character, as it was the home of cultural programming of the highest quality—the *NBC Symphony* led by Arturo Toscanini chief among them—but also was the dumping ground of sustained (unsponsored) programming that as often as not placed fourth in the ratings behind NBC-Red, CBS, and the less successful (after 1934) Mutual Broadcasting System. Still, Blue served as something of a loss leader for

RCA, as it was frequently touted by the company as a prestigious public service. Blue also allowed NBC to cultivate a reputation superior to that of CBS, which was always a special consideration for David Sarnoff.

NBC's economic strength derived almost wholly from its affiliate relationships. In 1932 the network initiated a plan to pay its affiliates a fee for every network-originated, sponsored program that the affiliate carried on its schedule. In return, the affiliate agreed to purchase NBC's sustained programs at a rate lower than what the station might typically produce in-house. This a la carte system still allowed the affiliates some freedom to pick and choose among the various network offerings, and although many of the larger stations objected to what they considered an inadequate reimbursement for their time costs, the arrangement allowed NBC to provide nationwide coverage to paying advertisers. However, both NBC and CBS continued to extract more concessions from their affiliates throughout the 1930s, knowing there was no shortage of local stations eager to accept whatever demands the networks might place upon them. As a result, government intervention became almost inevitable.

Report on Chain Broadcasting: 1941

Although the economic structure of commercial broadcasting was consolidated in the early 1930s, its regulatory parameters were slightly more fluid. In March 1938 the Federal Communications Commission (FCC) initiated an investigation into all phases of the broadcast industry, primarily at the instigation of the Mutual Broadcasting System. Mutual complained to the FCC that it was unable to expand into a national operation because NBC and CBS had affiliation agreements with more than 80 percent of the largest radio stations in the country.

The result of that inquiry, the 1941 *Report on Chain Broadcasting,* was highly critical of the network-affiliate relationship, and the regulations derived therein signaled the beginning of an on-going battle between broadcasters and the government over monopoly practices. To NBC and CBS, the rules set by the *Report* threatened to undermine the very structure of the broadcasting industry.

Most crucial among these were the establishment of strict limits on the length of affiliation contracts; the loosening of affiliation ties by allowing stations to broadcast programs from other networks or sources; the power of affiliates to reject network programs if the stations felt the offering was not in the public interest; and, most dramatically, the abolition of the practice of "option time."

This provision struck directly at the heart of network operations, and NBC, CBS, and Madison Avenue howled in protest. Option time was a standard feature of every affiliation contract, giving the network the legal right to preempt a station's schedule for network programming. CBS affiliates agreed to give up their entire broadcast day if the network demanded it; NBC was slightly less stringent, asking for options on eight and a half hours a day, including the profitable 8 to 11 P.M. evening block. The stations were compensated for all hours claimed by the networks at a rate adequate to cover the loss of potential sales to local advertisers (these local rates were set by the network as well, a procedure also abolished by the 1941 regulations). It was a profitable arrangement. The affiliates received popular national programming, relieving them of the chore of local production, and were compensated for their airtime—NBC took in three times more money from time sales than it dispensed in compensation and could guarantee advertisers a national audience.

The networks bitterly denounced the new regulations and brought suit in federal court to stop their implementation, but it was pressure from the business and advertising communities that caused the FCC to revise the chain broadcasting rules in an October 1941 supplemental report. The commission reasoned that while it remained unconvinced by the NBC and CBS contention that option time was indispensable to network operations, "it is clear that some optioning of time by networks in order to clear the same period of time over a number of stations for network programs will operate as a business convenience." As a result, the networks were permitted to maintain control over the 8 to 11 P.M. slot, the most heavily attended and profitable portion of the broadcast day. Thus, the prime-time schedule remained closed to independent and local producers on affiliate schedules.

Of more direct impact to NBC was the regulation that "no license shall be issued to a standard broadcasting station affiliated with a network organization which maintains more than one network." In other words, either Red or Blue had to go. Despite some public grumbling, RCA was not entirely unhappy with this ruling, having long considered the sale of Blue as a possible source of financing for television activities. Still, NBC filed suit in federal court challenging the Chain Broadcasting rules, but in May 1943, the Supreme Court ruled for the FCC in *NBC v the United States.* Five months later, RCA sold the Blue Network to Life Savers magnate Edward J. Noble for $8 million. The sale provided RCA with a cash infusion with which to prepare NBC-TV for an anticipated boom following World War II and led to the formation of the American Broadcasting Company (ABC) in 1945.

NBC Radio and the Emergence of Television

The immediate postwar era was enormously profitable for radio as pent-up consumer demand, combined with a shift from military to domestic manufacturing, unleashed a spectacular buying binge. However, television loomed on the horizon, and the new medium promised to have a dramatic impact on network radio. In 1945, 95 percent of all radio stations were network affiliates; in 1948—the year NBC, CBS, and ABC began seven-day-a-week television broadcasting—the figure was 68 percent and dropping fast.

Ironically, the end of network radio as a source of major entertainment was hastened by the CBS "talent raids" of 1948–49. In order to attract NBC's biggest radio stars, CBS designed a clever finance mechanism that had the practical effect of placing performers under long-term contract to the network. It was an elegant scheme: radio stars (who were otherwise taxed at personal income rates of up to 75 percent) would incorporate themselves and, in turn, license their company to CBS. The amount paid by CBS would be considered a capital gain and taxed at 25 percent. As a result, NBC performers left the network in droves for CBS, starting with *Amos 'n' Andy* creators Freeman Gosden and Charles Correll in September 1948, soon followed by Jack Benny, George Burns and Gracie Allen, Red Skelton, Bing Crosby, and Edgar Bergen.

The talent raids certainly paid off for CBS in terms of income (a profit increase of almost $7 million), more successful programming (12 of the top 15 radio shows in 1949), and tangential publicity. Most important, radio provided both the financial and programming foundation for the network's television operation. Every star brought over in the talent raids eventually appeared on CBS-TV, draining resources away from NBC.

As television continued to attract an increasingly larger audience, major advertisers rapidly left radio and moved their dollars into the new medium. By the early 1950s, the trend toward television was readily apparent. Fewer and fewer network-originated radio shows were made available to affiliates, while the number of programs that were simulcast (broadcast simultaneously on both radio and TV) increased. By the end of the decade, stations across the country began severing their network affiliations to produce their own programming, an action unthinkable during radio's heyday.

In 1960 NBC stopped production on its last remaining daytime radio serial, *True Story*, and thereafter existed almost solely as a news feed to subscribing stations. In 1986 RCA (including NBC) was purchased by GE (in a sense returning to its original owner), and that same year the network formally split its broadcasting divisions. NBC Radio was then sold to radio conglomerate Westwood One, which continues to maintain "NBC Radio Networks" as a separate brand, although in reality the network has no journalistic responsibility for newscasts labeled as "NBC."

MIKE MASHON

See also American Telephone and Telegraph; Blue Network; Network Monopoly Probe; Radio Corporation of America; Sarnoff, David; Talent Raids

Further Reading

Archer, Gleason Leonard, *History of Radio to 1926*, New York: American Historical Society, 1938; reprint, New York: Arno Press, 1971

Archer, Gleason Leonard, *Big Business and Radio*, New York: American Historical Company, 1939; reprint, New York: Arno Press, 1971

Bilby, Kenneth W., *The General: David Sarnoff and the Rise of the Communications Industry*, New York: Harper and Row, 1986

Campbell, Robert, *The Golden Years of Broadcasting: A Celebration of the First 50 Years of Radio and TV on NBC*, New York: Scribner, 1976

Cook, David, "The Birth of the Network: How Westinghouse, GE, AT&T, and RCA Invented the Concept of Advertiser-Supported Broadcasting," *Quarterly Review of Film Studies* 8, no. 3 (Summer 1983)

Federal Communications Commission, *Report on Chain Broadcasting*, Washington, D.C.: GPO, 1941

Federal Communications Commission, *Supplemental Report on Chain Broadcasting*, Washington, D.C.: GPO, 1941

Fifty Years with the NBC Radio Network, New York: NBC, 1976

"The First 50 Years of NBC," *Broadcasting* (21 June 1976)

Hilmes, Michele, *Hollywood and Broadcasting: From Radio to Cable*, Urbana: University of Illinois Press, 1990

National Broadcasting Company, *NBC and You: An Account of the Organization, Operation, and Employee-Company Policies of the N.B.C.: Designed As a Handbook to Aid You in Your Daily Work*, New York: NBC, 1944

National Broadcasting Company, *NBC Program Policies and Working Manual*, New York: NBC, 1944

"NBC 60th Anniversary Issue," *Television/Radio Age* 33 (May 1986)

"NBC: A Documentary," *Sponsor* 20, no. 10 (May 1966)

Sarnoff, David, *Looking Ahead: The Papers of David Sarnoff*, New York: McGraw-Hill, 1968

Smulyan, Susan, *Selling Radio: The Commercialization of American Broadcasting, 1920–1934*, Washington, D.C., Smithsonian Institution Press, 1994

"A Study of the National Broadcasting Company," *The Advertiser* 14, no. 9 (1943)

National Federation of Community Broadcasters

Member Organization of Community Radio Stations and Producers

The National Federation of Community Broadcasters (NFCB) is a membership organization representing more than 200 radio stations in the United States; it provides legal, technical, and logistical support for community-oriented, educational, and noncommercial broadcasters. Under its leadership, the community radio sector of U.S. broadcasting has emerged as a viable service in major urban centers and rural communities across the country.

In 1973 a group of dedicated community broadcasters met in Seattle to consider the possibilities of coordinating their efforts and promoting community radio throughout the United States. Two years later, in the summer of 1975, the National Alternative Radio Konference (NARK) convened in Madison, Wisconsin. The conference participants—an assortment of radio enthusiasts, artists, musicians, and community activists from across the country—resolved to form a national organization that would represent the interests of community broadcasters before the U.S. Congress and federal regulators. Within a matter of months, the newly formed NFCB located its headquarters in the Washington, D.C. apartment of two of the conference organizers, Tom Thomas and Terry Clifford. From these humble beginnings, the NFCB began its lobbying efforts in support of community-oriented radio. From the outset, the NFCB had two goals: to influence national broadcast policy-making and to secure federal grant money to support this new, locally oriented radio service.

Central to the NFCB's mission is enhancing and increasing diversity in radio broadcasting. Throughout its history, the NCFB has placed special emphasis on opening up the airwaves to women, people of color, and other cultural minorities whose voices are largely absent from mainstream media. In this way, the NFCB promotes volunteerism, supports localism, and encourages the development of programs and services specifically designed to address the needs and interests of underserved audiences. For example, community volunteers program music and public-affairs programming on WFHB in Bloomington, Indiana; WVMR is the only broadcast service for people living in the isolated region of Pocahontas County, West Virginia; and member station KBRW provides multilingual programming for native peoples in Alaska's North Slope region.

In addition, the NFCB provides all manner of technical and logistical support for community broadcasters. To that end, the NFCB has developed training materials that outline the procedures for license applications, describe the use of broadcasting equipment, and offer practical suggestions for enlisting local community support. Some of these publications include *The Public Radio Legal Handbook,* a reference guide to broadcast regulations; *Audiocraft,* a textbook on audio production techniques; and the *Volunteer Management Handbook,* which provides useful strategies for securing and maintaining an enthusiastic volunteer base. Crucially, the NFCB also provides the legal and engineering expertise necessary to successfully secure a broadcasting license from the Federal Communications Commission (FCC). Moreover, through its monthly newsletter, *Community Radio News,* and its annual conventions, the NFCB continues to keep its member stations abreast of ongoing policy debates, new funding initiatives, and the latest technological innovations. Finally, the NFCB established the Program Exchange service in recognition of the need for new and existing stations to round out their broadcast schedules. This scheme encouraged community broadcasters across the country to trade tapes produced by member stations as well as programming developed by independent producers. Not only did this service help offset the costs associated with program production, it had the added benefits of creating an informal network between community stations and helping to define community radio's national identity.

Under the auspices of the NFCB, the community radio movement of the late 1970s challenged the conventions of commercial radio, forever changing the landscape of U.S. broadcasting. The NFCB successfully lobbied the U.S. Congress, regulatory bodies, and government funding agencies to support noncommercial broadcasting in general and community broadcasting in particular. Most important, perhaps, in its commitment to community access, control, and participation, the NFCB popularized listener-supported radio. This model, first championed by Lewis Hill and the Pacifica stations, encourages community residents to become involved in every aspect of the local radio station: management, governance, finance, promotion, and production. Like the Pacifica stations and those associated with Lorenzo Milam's KRAB nebula, NFCB member stations seek to enhance radio's role in the civic, cultural, and social life of the community. Unlike commercial radio, which shies away from innovative and controversial programming, NFCB member stations encourage local cultural expression and support community activism.

Over time, as the organization's influence with industry leaders and policy makers grew, the NFCB became firmly entrenched in the Washington establishment. As a result, the NFCB's fortunes became linked to those of National Public Radio (NPR) and the Corporation for Public Broadcasting (CPB). For a time, this relationship proved beneficial to both parties. For example, in 1982 the NFCB organized the first Minority Producers Conference, which was instrumental in diversifying the staff at public and community radio stations across the country. However, the NFCB's relationship with the CPB has been a source of controversy among some community radio advocates. For instance, the NFCB and NPR supported the FCC's 1978 termination of 10-watt Class D stations, a prohibition that has fundamentally altered the character of community radio in the United States and that has, more recently, prompted the rise of the so-called microradio movement.

Over the past decade, community radio stations have grown increasingly dependent upon CPB funds, such as the Community Service Grant, to support their efforts and improve their services. However, the eligibility requirements for these funds were far beyond the means of small stations with modest resources. As a result, these stations were confronted with the unpleasant choice between shutting down or hiring professional staff to generate income, produce or acquire more "polished" programming, and oversee the station's daily operation. This condition seriously undermines community participation and has led to the "professionalization" of community radio. In recent years, some community stations have become little more than supplemental outlets for the nationally produced programming of NPR, effectively consolidating the public radio sector and eliminating community radio's greatest strength: its localism. Furthermore, the NFCB's Healthy Station Project, with its emphasis on attracting upscale demographics to community stations, has been sharply criticized for its chilling effect on community radio's news and public-affairs programming and the attendant homogenization of music and cultural fare.

Following a period of considerable internal unrest and organizational restructuring, the NFCB moved its headquarters to San Francisco in 1995 and began sharing its operation with Western Public Radio, a not-for-profit radio training and production center. In 2002, NFCB relocated again, this time across the bay to Oakland. With new offices, facilities, and staff, NFCB launched exciting new initiatives, most notably the National Youth in Radio Training Project.

KEVIN HOWLEY

See also Community Radio; Corporation for Public Broadcasting; Low-Power Radio/Microradio; Milam, Lorenzo; National Public Radio; Pacifica Foundation; Public Radio Since 1967; Ten-Watt Stations; Trade Associations

Further Reading

Barlow, William, "Community Radio in the U.S.: The Struggle for a Democratic Medium," *Media, Culture, and Society* 10 (1988)

Bekken, Jon, "Community Radio at the Crossroads: Federal Policy and the Professionalization of a Grassroots Medium," in *Seizing the Airwaves: A Free Radio Handbook*, edited by Ronald B. Sakolsky and Steven Dunifer, Edinburgh and San Francisco: AK Press, 1998

Hochheimer, John L., "Organizing Democratic Radio: Issues in Praxis," *Media, Culture, and Society* 15 (1993)

Lewis, Peter M., and Jerry Booth, *The Invisible Medium: Public, Commercial, and Community Radio,* London: Macmillan, 1989; Washington, D.C.: Howard University Press, 1990

Milam, Lorenzo W., *Sex and Broadcasting: A Handbook on Starting a Radio Station for the Community,* 2nd edition, Saratoga, California: Dildo Press, 1972; 4th edition, as *The Original Sex and Broadcasting,* San Diego, California: MHO and MHO, 1988

National Public Radio

U.S. Noncommercial Radio Network

National Public Radio (NPR) is the U.S.'s largest public radio producer and distributor, providing more than 100 hours of news and cultural programming each week to more than 600 member stations. The network is probably best known for its drive-time newsmagazines, *Morning Edition* and *All Things Considered,* but it produces a wide range of radio fare, including music and cultural programs, such as *Performance Today* and *Jazz Profiles,* and the nationwide call-in program *Talk of the Nation.* In the course of its 30-year history, the network has won virtually every major broadcast award and has figured prominently in the nation's political and artistic life.

Origins

Public radio in the United States had its origins in the 1920s, when low-budget community and college stations sprang up around the country. Hundreds of such stations took hold in the early, unregulated days of radio, but not many of these "educational stations" survived the Depression, especially as commercial broadcasters saw radio's potential as a vehicle for mass-market entertainment and lucrative advertising.

The National Association of Educational Broadcasters (NAEB) argued that educational stations could not compete for spectrum space with commercial giants such as the National Broadcasting Company (NBC) and the Columbia Broadcasting System (CBS). In 1945 the NAEB convinced the Federal Communications Commission (FCC) to allot the low end of the radio dial to educational broadcasters, winning 20 FM channels, from 88 to 92 megahertz. The allotment gave educational radio the stability to build up a small but hardy core of stations.

The Public Broadcasting Act

By the early 1960s, educational broadcasters were exploring the possibility of networking to help fill their program days. Don Quayle, the first president of NPR, recalls "there was a general feeling, both in television and in radio, that no single station had the resources to do the quality of programming we wanted." Quayle helped link up educational radio and television networks in the northeastern United States that were among the precursors of NPR.

Those networks lacked the resources to do much more than instructional broadcasting until the mid-1960s, when President Lyndon Johnson called on the Carnegie Foundation to look into the possibility of a federally funded broadcasting system. After a two-year study, the Carnegie Commission firmly backed the idea of federal funding, but it envisioned a far wider focus, distinguishing between educational and *public* broadcasting.

The commission offered 12 recommendations for the new service, including public-affairs programming that sought insights into controversial issues, diversified programming in which minorities were represented, and coverage of contemporary arts and culture. The commission's recommendations dealt only with television, but President Johnson, a successful radio station owner himself, was sympathetic to radio advocates who fought to be included. Critics argued that adding hundreds of weak and needy radio stations to the measure would dilute the federal funding and drag the whole project down. Supporters pointed out that radio could offer more services than television at a far cheaper cost.

Congress passed the Public Broadcasting Act in October 1967 and included radio, but the funding available to the newly formed Corporation for Public Broadcasting (CPB) quickly dwindled from a proposed $20 million to $5 million. Without financial support from the Carnegie Commission, CBS, the Communications Workers of America, and the Ford Foundation, CPB would not have been able to fund much programming.

In 1969 CPB sponsored a conference in San Diego that laid the foundation for National Public Radio, an entity that, unlike its television counterpart, the Public Broadcasting Service, would produce as well as distribute programs. The corporation invited Bill Siemering, then the station manager of WBFO in Buffalo, New York, to help conduct the discussion. Siemering had already articulated a vision for the service in an essay called "Public Radio: Some Essential Ingredients," in which he argued that public radio should be "on the frontier of the contemporary and help create new tastes," but that it also had to meet the information needs of the public.

Don Quayle was chosen as the organization's first president, not long after he had made a presentation to the Ford Foundation showing that he could interconnect all the qualified public radio stations "for less than [New York public station] WNET spent on television." Quayle hired Bill Siemering as the network's first program director and set him to work on what was to be its first regular program, *All Things Considered.*

National Public Radio Goes On the Air

National Public Radio was incorporated on 26 February 1970 with 90 charter stations. The network set up offices and studios in Washington, D.C., closer to the heart of the nation's politics than the big commercial networks, which had long been based in New York. NPR officially went on the air in April 1971, offering live coverage of the Senate hearings on Vietnam. Less than a month later, on 3 May 1971, the network aired its first edition of *All Things Considered.* Siemering, who directed the first program himself, recalls that it got off to a "rocky and exhilarating" start, with host Robert Conley unable to hear the cues in his headphones. But as to content, it fulfilled practically all the elements that Siemering had outlined. The program went on the air as thousands of antiwar demonstrators filled the streets of Washington in what was to be the last major protest of the Vietnam War. Reporter Jeff Kamin brought back tape of the chanting and the sirens as police waded into the crowds of demonstrators and made more than 7,000 arrests. The same program featured a young black woman speaking dreamily about her heroin addiction. As Siemering had promised, the program transmitted "the experience of people and institutions from as widely varying backgrounds and areas as are feasible." It spoke "with many voices and dialects." Siemering says that it also illustrated many of the network's goals: "using sound to tell the story; and presentation of multiple perspectives rather than a single truth."

Over the next few months, the program took shape. Siemering found a distinctive voice and sensibility for the program and the network in Susan Stamberg, who became the first woman to host a daily national newsmagazine. Stamberg was paired with Mike Waters and later with Bob Edwards. The network offset its limited resources and relatively inexperienced staff with a creative, conversational approach that began to gather fans. Producers at the network tried to explore the possibilities of the medium, emulating the sound-rich work of the Canadian Broadcasting Corporation (CBC) in Canada and the German producer Peter Leonhard Braun at *Sender Freies Berlin.*

During the first decade, cultural programming was a strong component of the network's output. Bob Malesky, now the executive producer of NPR's weekend programming, puts it this way: "*All Things Considered* was experimental, while concert programs gave stations the solid base to build a day's programming around, for what was still a largely conservative audience. NPR's first arts program was *Voices in the Wind,* and during its five-year life span (1974–1979), it was second only to *All Things Considered* in station usage." NPR also provided stations with *Folk Festival USA* (1974–79) and *Jazz Alive* (1977–83), both the work of producer Steve Rathe. In 1982 the network launched the *Sunday Show,* a five-hour cultural newsmagazine that won a Peabody Award for its first year but, like *Jazz Alive,* fell victim to the NPR financial crisis in 1983.

Expanding Role

For its first five years, NPR functioned primarily as a production and distribution center for its member stations, which had grown over the years to 190. The network took on a bigger role in 1977, after a merger with the stations' lobbying organization, the Association of Public Radio Stations. Don Quayle had left the network in 1973 to become a senior vice president at CPB, and the merger took place under President Lee Frischnecht. NPR offered member services, including training, management assistance, and help with program promotion. The network also began representing the member stations' interests before Congress and the FCC.

The network gained a great deal more national visibility under its third president, Frank Mankiewicz, who was hired in 1977. Mankiewicz came to the job from a background in freelance journalism and politics. He had been Robert Kennedy's press secretary during the presidential campaign that ended with the senator's assassination in Los Angeles, and he had managed the presidential campaign of Senator George McGovern. He had barely heard of NPR. He recalls that he got the job after promising the board that he "would do whatever was necessary so that people like me will know what NPR is."

Mankiewicz's first coup was getting permission for NPR to do live coverage of the Panama Canal debate in the Senate in 1978. It was the first live broadcast ever from the Senate floor. He chose NPR's Senate reporter, Linda Wertheimer, to anchor the coverage, despite complaints from station managers who worried that a woman would sound "too shrill."

Mankiewicz also fostered the network's longtime goal of offering a morning program service. When *All Things Considered* was first conceived, many public radio stations were not even on the air in the mornings. Bill Siemering says he felt that a new staff creating a new kind of program needed to work together on the day's news, without having to contend with the problems of preparing a program overnight, so the network's flagship program was designed for afternoon drive time. But the network was well aware that radio's biggest audience was in the morning, and NPR was eager to reach it. Mankiewicz also believed that if listeners tuned to an NPR program in the morning, there was a good chance the dial would remain there the rest of the day.

Mankiewicz says he saw his chance to start a morning program when CPB President Henry Loomis offered him a big chunk of money. Loomis pointed out one day that CPB, unlike most federally funded agencies, received its appropriation in a lump sum at the beginning of the fiscal year. That meant that the money accrued a lot of interest in the course of the year, about $4 million. Loomis, a radio fan, asked Mankiewicz what he would do with that money if he got it. "I'd use it to start a morning radio program." Loomis liked the idea and put up the money to start *Morning Edition,* which went on the air in November 1979. After the member stations roundly rejected a pilot version of the program, Bob Edwards was lured from his position as *All Things Considered* cohost with Susan Stamberg to take over *Morning Edition* on a temporary basis. Edwards stayed on, and the program became a hit.

Morning Edition marked an important departure in program style for the network. Whereas *All Things Considered* flowed relatively freely through its 90 minutes, from beginning to middle to end, *Morning Edition* was conceived as a *program service,* structured into rigid segments that member stations could either use or replace with their own local news, weather, and traffic reporting. The success of the *Morning Edition* format led many station managers to lobby for similar changes in the afternoon program, changes that the producers resisted until the mid-1990s.

The demands of providing material for two major programs each day also forced a significant change in the structure of the organization. In the early years, all reporters and producers worked directly for *All Things Considered,* tailoring their work to the eclectic and free-flowing style of that program. The advent of *Morning Edition* meant that the same reporters had to write and produce for the more rigid time constraints of the program service as well. To avoid having reporters pulled by the conflicting demands of the two shows, the news department adopted a "desk" system like those at major newspapers. Editors with specific expertise and experience set up a national desk, a foreign desk, a science desk, and so on. Reporters now worked for NPR news, rather than for a particular show.

In 1980 the network made a striking improvement in its sound quality with the launch of the nation's first satellite-delivered radio distribution system. Overnight, the programs went from the tinny, telephone quality of a 5 kilohertz phone line to the clarity and intimacy of the studio. Mankiewicz likes to joke that he told the technicians to add a bit of static to the broadcast for the first few days, "to lend an air of verisimilitude" to the otherwise too-perfect sound. The satellite also expanded the network's delivery capacity from a single channel to four, allowing stations to choose alternative programs without having to wait for the tapes to arrive in the mail.

The satellite meant that listeners could hear complex, skillfully layered sound without the degradation of telephone lines, and it cleared the way for the golden period of NPR documentaries. In 1980 the network broadcast a 13-part series by former *Voices in the Wind* producer Robert Montiegal called *A Question of Place*. The series used sound to bring to life the work of writers and thinkers such as James Joyce, Michel Foucault, Simone de Beauvoir, and Bertrand Russell. The following year, writer James Reston, Jr., producer Deborah Amos, and host Noah Adams used tape from the last days of the Jonestown religious cult to show how a leader's egomania led to the mass suicides of more than 900 people in Guyana. The network even branched into radio theater, with the series *Earplay, Masterpiece Radio Theater,* and a radio version of the George Lucas hit film *Star Wars*.

Financial Crisis

NPR was riding high and spending freely in 1983. Under Frank Mankiewicz, the network's membership had grown to more than 250 stations, and as he had promised, it was familiar to people like him, people in politics, the arts, and business. When Mankiewicz took over in 1977, the network was dependent on the federal government for 90 percent of its funding. By 1983 corporate underwriting, foundation grants, and other income had brought that figure down to around 50 percent. But the network was not in good financial shape. The network's chief operating officer, Tom Warnock, came to Mankiewicz with the news that NPR might be as much as $2 million in debt. As Warnock delved deeper, the estimate of the debt grew, finally reaching $9 million, a third of the network's budget. Subsequent audits revealed that NPR's accounting was so sloppy that some of the spending could not be tracked. As an example, the network had issued more than 100 American Express cards to its administrators, who ran up hundreds of thousands of dollars in bills for entertainment and travel. Teetering on the edge of bankruptcy, the network slashed department budgets and ultimately had to lay off about 100 people. The staff of the cultural programs department shrank from 33 to 8. Mankiewicz says he could have saved the situation with an ordinary bank loan, but political differences with the CPB's leadership made that impossible. In the end, Mankiewicz resigned.

Ronald Bornstein, the director of telecommunications at the University of Wisconsin, took over as acting president, leading a rescue team of lawyers and accountants from some of the top firms in Washington. For the first and only time in its history, the network went directly to its listeners to help pay off the debt, by including a fund-raising segment in *All Things Considered.* The member stations were wary of allowing the network to raise money from among their contributors, and only about a third of them chose to carry the segments. Even so, the "Drive to Survive" brought in $2.25 million in just three days. The balance of NPR's debt was to be paid off over a three-year period, under the terms of a loan agreement that was hammered out in often-rancorous negotiations with the CPB.

Messy and embarrassing as it was, NPR reported the story of its near-disaster as it happened. That, too, was part of Mankiewicz's legacy. "By that time, NPR had cast itself unambiguously as a news organization, so the standards of news applied to it," recalls Doug Bennet, who took over as president in 1983. Bennet says that what Mankiewicz "really did was to establish NPR in the news niche, and that was genius. Before, there was some news, but nowhere near the scale of investment or the staff." Scott Simon, then the network's Chicago bureau chief, was brought in to report on the financial crisis, interviewing beleaguered staffers and disgusted members of Congress. Bennet found himself taking over at an organization that was tense and distrustful.

Return to Stability

Bennet recalls that during his first one and a half years, he had to cut the budget eight more times, as new deficits were discovered, but he says, "I never believed that NPR would disappear." Bennet's first task was to develop a plan that would reassure the stations and Congress that NPR could be turned around.

The feud with the CPB eventually led the CPB to open up its funding process so that NPR would not be the only recipient. CPB created a Radio Fund, permitting any radio organization to apply for the money. The new funding system was approved without NPR's knowledge, to the fury of station managers. NPR had lost its funding monopoly and faced the prospect of presiding over an ever-diminishing cut of the pie. Doug Bennet recalls sitting in a bar with then-Board Chairman Don Mullally, charting out a radical and risky new funding plan on a bar napkin. The business plan, announced in February 1985, proposed giving all the CPB money to the member stations and letting them use it to acquire programming

Roundup was anchored by Edward R. Murrow in London. Murrow and CBS are largely credited with helping radio to mature into a full-fledged news medium. One notable legacy of Murrow and his team would prove to be the development of "broadcast journalism" as a distinctive syntax of writing and reporting news—conversational and brief writing that incorporates sound from the field, or *actualities,* into the news story. Murrow's actualities often included exploding bombs and screaming sirens.

Some war correspondents were limited to recording their reports before airtime because of government restrictions. But most reporters managed to broadcast their reports live, proving over time that they could do so without breaching military secrets. CBS introduced the term "news analyst" as a replacement for the term "commentator" specifically to avoid the impression that its radio news improperly shaped public opinion.

While the BBC and other European services widely reported the expanding war after 1939, entry of the United States into the war in late 1941 dramatically increased American radio news reporting. From 1940 to 1944, scheduled network news increased by more than 50 percent. By 1944, NBC-Red was offering 1,726 hours of news annually, while CBS was a close second with 1,497 hours. Despite a postwar decline in overall news hours, news still occupied more than 12 percent of network evening airtime.

Improving technology, such as portable recorders and smaller transmitters, provided the means for such dramatic broadcasts as Edward R. Murrow's recording of a bomber's run over Germany, George Hicks' live coverage of the June 1944 D-Day landings in Normandy, and pick-ups of news from distant Pacific island battlefields. Listeners heard the war begin by radio in 1939 (or in the U.S. on an otherwise quiet Sunday in December 1941)—and heard world leaders announce the end of the war in Europe in May 1945 and in the Pacific just three months later. Millions tuned to the Japanese surrender as it was broadcast from the deck of the battleship *Missouri* in Tokyo Bay in September 1945. And radio carried the resulting celebrations in cities around the world.

Adjustment (1950–70)

Despite the growing diffusion of television into American households in the late 1940s and into the 1950s, radio networks continued providing an extensive schedule of news and commentary for years after World War II. Only after the mid-1950s did network news schedules begin to dip sharply, soon to decline to hourly summaries of top stories. News continued to be an important element of most local station schedules.

Networks attempted to redesign their radio news. Network executives decided that the five-minute news summaries supplied on the hour should now be delivered by experienced reporters with recognizable names, rather than by staff announcers. In addition, the networks tried rolling out "variety news programs" that offered a greater emphasis on feature stories. In 1955, NBC began airing *Monitor,* a mixture of news, music, interviews, dramatic sketches, and sports. *Monitor* was hosted by Frank Blair and Hugh Downs; Gene Shalit did occasional film reviews. ABC began *New Sounds,* a weekday evening series patterned after *Monitor.* In 1960, CBS began *Dimension,* a series consisting of five-minute informational inserts on the half hour. For all the network's efforts, however, their audience and network radio news continued to decline in the face of television's increasing viewership.

As radio stations increasingly programmed according to some music format by the mid-1950s, news was made a part of the schedule. Only major-market stations provided more than a news headline service.

In the 1960s, news finally became its own radio format. KFAX in San Francisco adopted the first "All News" format in 1960 with each hour containing 25 minutes of hard news, updated throughout the day. The remaining minutes were filled with sports, business news, and feature stories. But KFAX failed after four months because of a lack of advertising support. The first commercially successful all-news radio station was founded by Gordon McLendon in 1961. McLendon took a rock-and roll radio station in Tijuana, Mexico, changed its format to hard news, and targeted it at listeners in Los Angeles. News was recycled every half hour to coincide with the commuting times of drivers going into and out of Los Angeles. No reporters were used—just hard news from the newswires AP, UPI and the Los Angeles City News Service. In 1965, WINS in New York City became an all-news station and began airing the promotional advertisement that has now become standard to Westinghouse (Group W) news stations: "All news, all the time." WINS used its own reporters and focused heavily on local news, as did KYW in Philadelphia, Pennsylvania, and other stations in cities large enough to support the expensive format.

Radio News Formats

Most aspects of modern radio news vary widely among stations, including the number, scheduling, and length of newscasts, the content of stories broadcast, and their order of presentation. Each of these, in turn, may depend on market size, station format, and any news policy of the owners, especially of multiple outlets. Overall, less news is offered on radio today than was the case a decade or two ago.

While some stations still retain regular newscasts, many music and all-talk outlets focus on feature material. Contemporary Hit Radio (CHR) and Adult Contemporary (AC) stations often still provide brief newscasts, for example; others, such as Country, Easy Listening, and Album-Oriented Rock (AOR), more often run features or have eliminated news, reasoning that not every station in a multi-station market has to provide it. If stations affiliate with a network at all, it is most often to carry their brief national newscasts. Some networks are designed to serve up news to fit within specific radio formats.

Through the 1980s the typical pattern for small- and medium-market music stations was to offer five minutes of news on the hour, usually combining world, national, and local stories with a growing emphasis on the latter. World and national news is most often received as an audio feed from a network or syndicated satellite service or as text from the Associated Press, and may be recast to include a local angle.

On stations still offering news, newscasts range from a minute of headlines to three or four minutes long. Most "stories" are now limited to a sentence or two totaling 10 to 20 seconds—a 60 second story would be very unusual. Stories may contain a sound bite or sound actuality, perhaps bits of an interview. Short newscasts are often devoted to but one or two stories, especially with breaking news. While major-market AM radio stations once offered up to a half hour of news in morning or afternoon drive time, that model has all but disappeared as most stations trim or even eliminate news staffs.

The largest markets typically include an all-news AM station. These often utilize a "news wheel" to format their newscasts. This displays the length, order, and content-type of stories to be broadcast over a half hour or hour cycle, after which the wheel repeats itself. Ideally, each time the news wheel begins again, stories have been refreshed with new information or have been replaced with new stories. The news wheel is especially convenient for listeners who like to know that they can hear a specific type of information (e.g. the weather forecast) by tuning in at exactly :20 and :50 minutes past the hour. As radio news is easier to prepare than reports for TV or newspapers, radio is often quicker at getting breaking news on the air, though internet-based services can be the fastest of all.

Information Sources

Radio stations gather news from a number of sources, including the local newspaper, the telephone, and the field interview. The local newspaper is especially important to smaller stations with a limited budget and news staff. Reporters will rarely admit to using the local newspaper as a primary source of news, instead describing the newspaper as source for obtaining leads to develop news stories. However, reporters at smaller radio stations often do not have the time to gather their own news, so they end up rewriting newspaper stories for their own newscasts.

The telephone is used for conducting interviews with officials and experts—such as politicians, police officers, and coroners—to acquire actualities to be edited later into sound bites. The radio reporter often initiates a phone call from the studio and then records the conversation on tape or computer. FCC regulations specify that reporters must identify themselves as such and name the station they work for and that they must indicate that the station plans to broadcast portions of the interview. All of this has to be done before the reporter begins recording the interview. The telephone is also used to receive live or pre-recorded traffic and weather reports from companies that sell these services to radio stations. *Accuweather* is widely subscribed to for weather information, while *Cellular 1* and *Shadow Traffic* are widely subscribed to for traffic information.

Hand-held cassette and minidisk recorders are used by radio reporters to gather actualities from the field. Hand-held recorders are used to record actualities and then to play them back over a telephone line to the studio.

To gather national news, radio stations use newswires, satellite feeds, and the internet. The main newswire subscribed to by both small-market and major-market radio stations is AP; what is left of the UPI agency is used by a small and dwindling number of stations. Typically only the very largest stations in the U.S. subscribe to the other three big international newswire services—Reuters from England, the French Agency Press (AFP) from France, and ITAR-Tass from Russia. Major-market radio stations normally subscribe to regional and city newswire services as well. Newswire services deliver information mainly through satellite downlinks or conventional phone lines connected to a radio station's computer.

Radio stations also use satellite dishes to receive audio news from satellite news feeds. These feeds can be re-broadcast as self-contained newscasts, or they can be used as actualities to be edited into sound bites. Satellite feeds come down at specific times determined by the satellite service. Satellite feeds are provided by the networks CBS, NPR, and CNN, as well as the news wire service AP. They are normally provided free of charge to radio stations in exchange for pre-selected advertising spots in the radio station's program schedule, which will generate revenue for the satellite service. The remaining available spots will be made available for the local stations to fill with news sponsorships. Television is also used to gather information for radio newscasts. The advent of all-news channels and mostly news channels (e.g., CNN, Fox, MSNBC, ESPN, Weather Channel) has allowed radio reporters to monitor nationally developing news stories constantly.

Increasingly, radio stations are relying on the internet to gather national and international news. Many traditional media organizations such as CNN and NBC as well as non-traditional news organizations provide websites with news and information that can be downloaded as text or audio. The internet provides a cheaper alternative to subscribing to a newswire or satellite, but the boundaries for copyright infringement and source credibility are less clear than for newswire or satellite news information.

Regulation and Deregulation of News

For many years, the FCC's licensing renewal guidelines strongly encouraged stations to provide from 6 to 8 percent of total airtime to news and public affairs. Those rules disappeared in the early 1980s, leaving stations to determine their own journalistic role—if any. As a result, in the past two decades many stations have opted out of any news programming at all. However, most small-market Top 40/Contemporary Hit Radio stations have retained news programming because of a traditional listener base that has expectations for local, community-oriented news with a practical quality, such as local events, traffic reports, and high school and college sports.

Industry consolidation in the 1990s has led to further reductions in radio news programming as part of corporate cost-cutting strategies. Multiple stations with common corporate ownership now routinely obtain news by purchasing national news feeds from independent "outsourcing" companies, such as Metro Network's *MetroSource*.

Significantly, the marketplace guideline of deregulation has led many radio stations to air a new kind of news in their program schedules. Today news has come to be defined not strictly as hard news but also as entertainment-oriented information. Medium-market and major-market radio stations have reinforced this redefinition through the news covered in syndicated talk shows they program. National hosts such as Rush Limbaugh, Don Imus, G. Gordon Liddy, Laura Schlessinger, Howard Stern, as well as many local talk-show hosts now feature political, sexual, or celebrity news in their shows. These and other talk-show hosts routinely deliver news stories and then offer their own opinionated comments, after which listeners are invited to engage in the discussion by calling, faxing, or emailing the program.

ROBERT MCKENZIE

See also All News Format; British Radio Journalism; Canadian Radio News; Commentators; Documentary Programs; Editorializing; Election Coverage; Fairness Doctrine; Fireside Chats; News Agencies; Politics and Radio; Press-Radio War; Public Affairs Programming; United States Presidency and Radio; World War II and U.S. Radio

Further Reading

Bliss, Edward Jr., *Now the News: The Story of Broadcast Journalism*, New York: Columbia University Press, 1991

Charnley, Mitchell V., *News By Radio*, New York: Macmillan, 1948

Greenfield, Thomas Allen, *Radio: A Reference Guide*, New York: Greenwood Press, 1989

Hausman, Carl, *Crafting the News for Electronic Media: Writing, Reporting and Producing*, Belmont, California: Wadsworth, 1992

Mayeux, Peter E., *Broadcast News: Writing and Reporting*, Dubuque, Iowa: Wm. C. Brown, 1991; 2nd edition, Prospect Heights, Illinois: Waveland Press, 2000

McKenzie, Robert, *How Selected Pennsylvania Radio Stations Use Technology to Gather Information for News Programs: A Field Study*, PhD. diss., Pennsylvania State University, August 1990

Sterling, Christopher H., and John M. Kittross, *Stay Tuned: A History of American Broadcasting*, 3rd edition, Mahwah, New Jersey: Lawrence Erlbaum, 2002

White, Paul W., *News on the Air*, New York: Harcourt Brace, 1947

News Agencies

For most of their history, radio stations have relied on national sources for much of their news. Other than the radio network news divisions, the prime source for national and international news has usually been one or more of the news agencies (also called news or wire services or press associations) such as the Associated Press (AP), United Press International (UPI), or Reuters. In recent years, the news agencies have provided considerable regional and local news as well. As more radio stations reduce their news staffs (or eliminate them entirely), "rip 'n' read" newscasts based entirely on news agency copy have once again become common.

Cooperative news gathering began in the United States with the formation of the Associated Press by a number of New York daily newspapers in 1848. Although reorganized several times, the AP has always been a cooperative rather than a profit-seeking venture. The United Press (UP) was begun as a profit-seeking affiliate of the Scripps newspaper chain in 1907, and the International News Service (INS) appeared as another commercial venture in 1909, controlled by the Hearst newspapers. Each of the three was based on service to newspapers, and newspaper-based board members controlled their operation. Radio's arrival and demand for service created big questions for the agencies.

Associated Press

The oldest and largest news agency had the most trouble deciding how to handle the new medium. As early as 1922, when few stations offered news, the AP warned its member newspapers not to allow use of AP news reports on their own radio stations. But in 1924 and 1928, AP did allow election returns to be broadcast. Starting in 1933, AP adopted a policy of allowing radio use of AP news stories only for events of "transcendent importance," and this continued until 1941, long after its competitors were serving broadcasters.

In 1941 AP initiated a radio wire—a news service written for use on the air, as opposed to the traditional service for newspapers designed to be read. Dubbed "Circuit 7760," it operated 24 hours a day under the direction of Oliver Gramling. Within a year AP was serving 200 (of about 750 total) stations in 120 cities, with 110 stations on its broadcast wire payroll. After the war, some 450 stations were elected to associate membership (an important status within the AP cooperative organization, as radio now had more of a voice in management decisions). By the early 1950s, AP was providing some 75,000 words every 24 hours, written for audio reading, and usually condensed and rephrased from the main newspaper service.

AP news for radio was usually provided in the shape of ready-to-use newscasts of different lengths. This led many smaller stations, lacking their own news staffs, to simply have an announcer assemble a news program from the news agency wire stories ("rip 'n' read"). As more stations developed popular music formats, many relied on this practice for their entire news operation. Now dubbed the AP Radio Network, the agency launched an audio service with actualities (sound recordings from the field) for stations to use in their own newscasts in 1974. Just five years later, the AP Broadcast Wire was said to be the longest leased telecommunication circuit in the world.

In 1980 the AP broadcast service became the first radio network in the world to use a communications satellite. Just four years later, AP owned its own satellite transponder, making it the first news organization to do so. In the meantime, AP had shifted its broadcast operations from New York to Washington, D.C. A decade later, AP was serving just over half the commercial radio stations in the nation with four focused services designed to better serve varied radio formats: AP NewsPower, AP DriveTime, AP NewsTalk, and AP Specialty Wires. To these was added AP All-News Radio in 1994, a 24-hour service of "full packed" radio newscasts that served more than 70 stations by 2000, with another 750 taking news feeds.

United Press and International News Service

The story of UPI and radio is more complex and begins with its two commercial predecessors: the United Press and the International News Service.

The United Press provided a 1924 general election hook-up using WEAF in New York as the base station. UP's president Karl Bickel argued strongly in favor of serving radio in the 1920s and saw the new medium as an exciting development. But his newspaper-dominated board of directors prevented such a service until 1935 when he resigned owing to ill-health at age 53, leaving his successor Hugh Baillie to bring UP service to radio station subscribers. In the meantime, a number of radio stations owned by UP client newspapers had been using UP reports on the air, despite news agency policy banning such practices. In 1943, UP published a *United Press Radio News Style Book,* an indicator of the growing importance of radio to the commercial news agency. By the early 1950s, UP was providing about 70,000 words per day to its radio subscribers.

The smaller and weaker International News Service began a "radio-script" service providing radio material 40 times a week in addition to its regular print service. In the face of AP expansion and success in luring away newspaper clients, however, the weaker INS and UP agreed to merge to form UPI in 1958.

United Press International

With the merger, UPI, under Scripps control, began to provide its client radio stations with audio reports to use in their own news programs. By 1965 clients were receiving about 65 voice stories a day from the UPI radio center in Chicago. Nine years later they were also receiving 20 full newscasts a day with inputs from London and Hong Kong. In 1977 UPI was serving about 900 client stations (almost twice the number reached by the AP). UPI also moved to distribute its radio news service by satellite in the early 1980s.

But UPI was in deep financial trouble. Although intended from the start as a commercial affiliate of a for-profit newspaper company, the agency had fairly consistently lost money. By the 1970s it was rapidly losing newspaper clients to the larger and better-financed AP and began to focus more on its radio station business.

In 1982, after a two year effort, Scripps sold the company it had founded nearly three-quarters of a century earlier. The sale led to two decades of drastic decline under several successive owners and two separate declarations of bankruptcy. UPI declined from about 1,800 employees at the time of the Scripps sale to less than 200 in mid-2000 when a Unification Church affiliate took control of the remains of the Washington-based news agency.

The UPI radio network was now the central part of the now much smaller news service. It offered 24 hours of fully produced programs for use at the top of each hour plus an actuality service for stations to use in their own newscasts (called "Selectnews," it began in 1992). While UPI reached some 2,000 stations in 1994 and radio accounted for half of the agency's income three years later, UPI had to end its radio service on 19 August 1999 for lack of funds, competition from newer news sources, and a drastic drop-off in the number of client stations (to just 400). Its final words:

> This is the final broadcast from UPI Radio. United Press International is getting out of the broadcast news business and has sold its contracts to Associated Press Radio. For those of us suddenly out of work, it's been fun. We feel UPI Radio has done its job well overall, even as we struggled with fewer and fewer resources. So we sign off now with smiles, memories, a few tears . . . but no regrets.

Other Services

Although most stations relied upon their network affiliation (if any), a specialized news service, or AP for world and national news, competing news sources had existed even in the early years of radio. Several entities, for example, had developed to serve radio in the wake of the brief 1930s "war" that limited availability of AP, UP, and INS news feeds. The agencies created the Press-Radio Bureau to combine service from the three in special radio reports, and it lasted until 1938. The Yankee News Service served the New England regional network's affiliates, the Continental Radio News Service based in Washington, and the Radio News Association from Los Angeles also began operation in 1934. Transradio Press Service began at the same time, largely built with former Columbia Broadcasting System (CBS) news people, aimed at serving non-network radio stations. Transradio survived beyond the 1930s (to December 1951, when it closed for lack of sufficient station clients) because stations could again obtain their news from the traditional news agencies.

London-based Reuters made its first radio agreement with the then-commercial BBC in 1922, although with provisions that no news would be broadcast before 7 P.M. in the evening, thus protecting the circulation of evening London dailies. Continuing negotiations in the mid-1920s allowed the BBC to cover current events as they were happening, and by 1929 the now government-chartered BBC received the full Reuters news wire for use in its news programs. Only in 1972 did Reuters begin its first voice news service for local stations in Britain and the United States. By the 1990s the firm had refocused on financial reporting and information and no longer served radio stations.

The expansion (and by the late 1990s, the consolidation) of the radio business contributed to a variety of other radio news services including Unistar (which carried CNN Radio), Capnews, the Business Radio Network, and the USA Radio Network. This expansion of syndicated news sources paralleled the decline of individual local station news efforts; instead of supporting local station news, these services were increasingly replacing local efforts.

CHRISTOPHER H. STERLING

See also News; Press-Radio War; Yankee Network

Further Reading

Bickel, Karl August, *New Empires: The Newspaper and the Radio,* Philadelphia, Pennsylvania: Lippincott, 1930

Bliss, Edward, Jr., *Now the News: The Story of Broadcast Journalism,* New York: Columbia University Press, 1991

Boyd-Barrett, Oliver, *The International News Agencies,* London: Constable, and Beverly Hills: Sage, 1980

Jackaway, Gwenyth L., *Media at War: Radio's Challenge to the Newspapers, 1924–1939,* Westport, Connecticut: Praeger, 1995

Moore, Herbert, et al., *"More News—After This . . . ": The Untold Story of Transradio Press,* Warrenton, Virginia: Sun Dial, 1983

Morris, Joe Alex, *Deadline Every Minute: The Story of the United Press,* Garden City, New York: Doubleday, 1957

"News Agencies and Radio Broadcasting," in *News Agencies, Their Structure and Operation,* Paris: Unesco, 1953

Nielsen. *See* A.C. Nielsen Company

Nightingale, Earl 1921–1989

U.S. Radio Commentator and Actor

With a sonorous voice characteristic of his English surname, combined with his writing and verbal styling, Earl Nightingale led a distinguished career as a radio inspirational speaker and actor. He was the voice of radio serial hero *Sky King,* and with his five-minute daily program *Our Changing World* lasting for almost 30 years, Nightingale created one of the longest-running syndicated programs in the history of radio. The innovative talk show he wrote and hosted for WGN was one of the most heavily sponsored shows in radio. Venturing beyond traditional broadcasting, Nightingale took his radio writing and speaking skills into new media, producing history's best-selling non-entertainment recording, *The Strangest Secret.* In addition, the Nightingale-Conant audio publishing and syndication corporation he co-founded pioneered motivational recordings.

Origins

Nightingale was born in March 1921 in Los Angeles, California. Encouraged by his mother, he became a good reader by the time he entered kindergarten. His passion for reading became the foundation of his writing and gave him the ideas he was later to broadcast. Seeing his family in the depth of the Depression in 1933, 12-year-old Nightingale became intensely curious about what separated the "haves" from the "have-nots." He began looking for answers in his local library. Vowing to find a way to become financially independent by the age of 35, his search for answers became the defining quest of his life and career. It also contributed to his personal library of over 6,000 books.

Nightingale joined the U.S. Marines in 1938 and was one of only 12 Marines to survive the Japanese bombing of the battleship *U.S.S. Arizona* at Pearl Harbor in 1941. While he was a Marine instructor at Jacksonville, North Carolina, in 1945, Nightingale worked part-time at a local radio station and quickly discovered his gift for broadcasting. After leaving the service, he became a broadcaster for KTAR in Phoenix, Arizona. Three years later, he was an announcer and news commentator with the Columbia Broadcasting System (CBS) in Chicago. Having been with CBS for only a year, Nightingale joined the staff of WGN in Chicago as a writer-producer in 1950. From 1950 until 1954, his radio role as the dashing radio adventurer *Sky King* was broadcast nationally.

Inspirational Radio

While portraying *Sky King,* Nightingale created and hosted a 90-minute daily talk show at WGN that soon became, in his words, "one of the most heavily-sponsored programs in the industry"—due in large measure to Nightingale's remarkable ability to sell advertisers. His guest-interview "talk" format, according to some, served as a prototype for talk show formats popular today. The success of the radio talk show that Nightingale wrote, hosted, and sold led to a television version on WGN-TV.

In 1956 the *Chicago Daily Tribune* reported that Nightingale's childhood dream had been realized:

> An intense young radio commentator will retire tomorrow at the age of 35, on a life income he estimates will be from $30,000 to $50,000 a year. He is Earl Nightingale, who heads three corporations built up out of his radio work, and who will say farewell tomorrow night to his WGN listeners who have heard him ten times a week since 1950 (Hughes, 1956).

While meeting the demands of his daily radio and television programs at WGN, Nightingale's entrepreneurial drive had enabled him to develop several sales firms, one of which was a nationally ranked life insurance agency. He also had formed his own firm, Earl Nightingale, Incorporated, through which he bought radio time from WGN. He once estimated that he had worked 12 to 14 hours a day from 1944 until 1956 to achieve his determined childhood goal of early retirement.

Nightingale had also been in demand as a public speaker while he was at WGN. One of his platform messages, "The Strangest Secret," was a compilation of key ideas from his reading about success, wealth, and achievement. On hearing Nightingale's talk, the president of the Pure Oil Company urged him to record it; during the same month he retired from WGN, Nightingale made the recording.

In 1959, Nightingale returned to work to write and record his own daily syndicated radio program, *Our Changing World.* In five years the program was syndicated to the largest number of stations in the history of broadcasting to that time. As Nightingale described his approach, "[W]e take the refined knowledge that has been promulgated by the great thinkers of our time . . . and winnow from it that which we feel is vital to the average person and then put it into language that he can easily understand." With 7,000 radio commentaries recorded, *Our Changing World* lasted for almost 30 years, making it one of the longest-running syndicated radio programs in history.

In 1960, Nightingale and Lloyd Conant formed the Nightingale-Conant Corporation around the success of *The Strangest Secret* and the syndicated *Our Changing World. Standard and Poor's Register of Corporations, Directors, and Execu-*

Earl Nightingale
Courtesy Nightingale-Conant Corporation

tives described the Nightingale-Conant Corporation in January 2000 as having 250 employees, having annual sales of over $30 million, and producing "motivational, educational and communication programs, cassette tapes, recordings and films, radio and television programs, graphic arts." Fellow radio commentator Paul Harvey described Nightingale as "the dean of self-development."

In 1986 Earl Nightingale was inducted into the National Association of Broadcasting Hall of Fame. This radio personality, with an extraordinary genius for touching an audience through a microphone, died in early 1989.

EDGAR B. WYCOFF

See also WGN

Earl Clifford Nightingale. Born in Los Angeles, California, 12 March 1921. Son of Albert Victor and Gladys Fae (Hamer) Nightingale; served in U.S. Marine Corps, 1938–46; broadcaster, KTAR, Phoenix, Arizona, 1946–49; announcer, CBS, Chicago, Illinois 1949–50; formed Earl Nightingale, Inc., 1950; wrote, produced, and hosted *The Earl Nightingale Show* on WGN-Chicago, 1950–56; host, *Our Changing World,* 1959–89; voice of *Sky King* in series from WGN, 1950–54; co-founded radio syndication and motivational firm Nightingale-Conant, 1960. Inducted into National Association of Broadcasters Radio Hall of Fame, 1986; received Napoleon Hill Foundation Gold Medal Award for Literary Excellence, Gold Record from Columbia Records, and Golden Gavel Award from Toastmasters International. Died in Scottsdale, Arizona, 25 March 1989.

Radio Series

1950–54 *Sky King*
1950–56 *The Earl Nightingale Show*
1959–89 *Our Changing World*

Publications

This Is Earl Nightingale, 1969
Earl Nightingale's Greatest Discovery: The Strangest Secret Revisited, 1987

Further Reading

Broadcasting: The Businessweekly of Television and Radio (21 February 1972)
Cruz, Georgina, "Nightingales Migrate to Broward," *Fort Lauderdale News and Sun-Sentinel* (29 March 1956)
Harmon, Jim, *The Great Radio Heroes,* Garden City, New York: Doubleday, 1967
Hughes, Frank, "Success at 35: Retirement at $30,000 a Year," *Chicago Daily Tribune* (29 March 1956)
"He Hit Pay Dirt by 35," *Parade Magazine* (11 January 1959)
Wycoff, Edgar Byron, "An Analysis of Earl Nightingale's Audio Cassette Counseling Programs: A Case Study in the Rhetoric of Success Motivation," Ph.D. diss., Florida State University, 1974

Non-English-Language Radio in the United States

In the United States, the radio boom of 1923 coincided roughly with the legislated end of the largest wave of international immigration in modern history. Some 20 million immigrants arrived on the shores of the United States between 1871 and 1920, some settling permanently, some returning home, and others traveling that route a number of times. The result was a proliferation of people and communities who spoke German, Polish, Spanish, Yiddish, Italian, and other languages. At the end of the 19th century, newspapers were the primary vehicle for the dissemination of information, with countless daily papers from every ethnic and political angle being published every day. As cities grew and the networks of newspaper distribution had not yet expanded, radio became a vital alternative to the printed word.

As early as 1926, foreign-language markets were being identified as potential profit centers. Despite indications by the advertising industry, major networks such as the National Broadcasting Company (NBC) and soon the Columbia Broadcasting System (CBS) were loath to incorporate foreign-language programming into their schedules. (However, for two years beginning in 1928, NBC briefly broadcast *Der Tog*, a Yiddish-language program.) Yet by 1964 more than 340 radio and television stations broadcast non-English-language programs in everything from Italian to Navajo. The majority of these stations were local, low-wattage stations without access to national networks. More recently the number of non-English broadcasts has increased exponentially with the popularization of web-based broadcasting. The development of non-English programming through alternative formats (local versus network and web versus traditional radio) is not coincidental but speaks of a particular power relationship between language and radio.

Throughout radio's history, the number of non-English stations in the United States has been quite significant, which suggests a considerable listening audience despite its traditional exclusion from market studies. As NBC and CBS rapidly grew to dominate the national networked radio dial, numerous local radio stations, generally with a broadcast power between 100 and 500 watts, began to spring up in urban areas. Almost always, these stations rented portions of their broadcast day to different community groups who wished to broadcast. Two of the most prominent examples are New York's WEVD and Chicago's WCFL. Founded in 1927, WEVD was owned by the Debs Radio Trust and broadcast programs in at least four languages, while WCFL hosted broadcasting in no fewer than 11. In 1924 WOAI in San Antonio, Texas, aired its first Spanish language broadcast, and Cleveland's WJAY initiated a weekly Polish-language program beginning in 1926.

Following the Federal Radio Commission's reorganization of the radio dial in November 1928, the majority of the stations that carried non-English-language programming found themselves relegated to the low and high frequency margins of the broadcast spectrum. Additionally, the FRC forced many of these stations to share frequencies and therefore also divide up the broadcast day. Broadcasting from the margin and on power that typically ranged from 250 to 1,000 watts, these stations cobbled together whatever broadcasting they could, usually comprised of a loose coalition of multi-lingual programs, performers, advertising agents, and sponsors. Few if any of these stations could choose to broadcast in only one language. The only significant exception to this rule was organized during the late 1930s, when New York-based station WOV organized 15 east-coast stations into the International Broadcasting Corporation which served as an Italian-only network, serving an audience of nearly 3 million listeners.

With the organization of the Federal Radio Commission in 1927 and amid the growing concern about the "decency" of radio programs, Section 29 of the Radio Act of 1927 sought to regulate U.S. airwaves by providing that "whoever utters any obscene, indecent, or profane language by means of radio communication shall be fined not more than $10,000 or imprisoned not more than two years." As it was impossible to listen to every broadcast nationwide, the FRC mandated that broadcasts be recorded if a listener had filed a complaint against that program. In the beginning, these recordings were made on glass plates (78-rpm records could not hold long enough segments, and magnetic recording tape had not yet been invented). This concern about the "decency" of language contributed to a general suspicion about non-English broadcasts and led to additional federal policing and harassment of such programs. Of course, it was also true that non-English broadcasts could elude the surveying ear of the FRC because often FRC monitors could not understand their content.

If non-English-language programs were considered a marginal segment of radio broadcasting in the United States, the incorporation of non-standard English speakers into English-language programs fueled the popular imagination. During the Depression years, the networks established themselves nationally via the appeal of ethno-comedies such as *The Goldbergs* and *Amos 'n' Andy.* These two programs, as some of the first to reach national audiences, drew significantly on cultural, linguistic, and dialectical differences for their humor. In the case of *The Goldbergs,* the common problems of language acquisition (mispronunciation, spoonerisms, malapropisms, etc.) and accent were the source of a great deal of the humor that Molly Berg wrote into the program. *Amos 'n' Andy* drew on a much older tradition of minstrelsy (and played on its racial stereotypes), but the particular challenge of putting blackface on radio turned the emphasis from appearance to dialect as the primary signifier of difference. Even though the Goldbergs were clearly on the path toward becoming ordinary Americans, whereas Amos and Andy were depicted as unassimilable, in both cases mastery of English was highlighted as the key determinant of mainstream acceptance. These programs clearly appealed to English-speaking audiences as they poked fun at members of non-traditional-English-speaking population groups.

Although culturally and linguistically marginal, non-English-language programs occupied a substantial amount of the radio dial. Statistics for non-English broadcasts during the Depression era are scant, but one source reports that nearly 200 stations out of a total of 850 broadcast non-English-language programs for some part of the day.

In a 1941 anthology entitled *Radio Research* (edited by Paul Lazarsfeld and Frank Stanton), sociologists Arnheim and Bayne published a survey of non-English-language broadcasts. They reported the presence of German, Italian, Yiddish, Polish, Lithuanian, and Spanish broadcasts. However, there were almost certainly Greek, Croatian, Russian, Hungarian, Bulgarian, Romanian, Chinese, Japanese, and Gaelic broadcasts at the time, as well. Although Arnheim and Bayne were primarily interested in the content of a typical broadcast day, their study is the first organized examination of ethnic radio in the United States, representing an early effort to include non-English-speaking audiences in radio market research.

Limitations

As soon as the United States entered World War II, the Federal Communications Commission (FCC) and the Foreign Language Division of the Office of Facts and Figures began investigations of all major East Coast stations (as well as others farther west) that broadcast foreign-language programs. Stations in New York, Philadelphia, Boston, and Chicago were targeted, and Alan Cranston, then chief of the foreign language division, recommended that certain broadcasters be "barred from the air immediately." Following the removal of a handful of German and Italian broadcasters and a general decline of foreign-language broadcasts, in March 1942 Cranston initiated new programs in German and Italian. The programs, entitled *Uncle Sam Speaks*, were designed to encourage ethnic listeners to join the war effort through volunteering or taking jobs in the defense industry.

Despite this overall reduction in U.S. foreign-language broadcasting, the World War II era also marked the emergence of substantial Spanish-language broadcasting, which has grown exponentially since that time, with about 500 stations including Spanish-language programming by 1980. In the postwar years, broadcasters in Yiddish actively involved themselves in reuniting Jewish refugees with their families by broadcasting names of people who were looking for family members. Despite these brief highlights, the 1940s were devastating to non-English-language broadcasting, with the exception of Spanish broadcasts, which managed steady growth. Not coincidentally, the postwar years also witnessed the near-total domination of radio by the networks.

With the rise of McCarthyism in the 1950s, radio stations that previously housed foreign-language programs began dropping them from their rosters, fearing that broadcasters from Eastern Europe might use their airtime to spread communist propaganda. These station owners were responding to demographic changes, as well; first-generation groups began to give way to their English-speaking children. As language and residence patterns changed, so did cultural tastes, and programming once valued for its cultural specificity began to sound old-fashioned. However, this era should not be seen as the end of foreign-language radio, but rather as a reflection of changing immigration and settlement patterns and changes in the cultural preferences of many European immigrants. Thus, as European-language broadcasts decreased, a sizable immigration from Asia (most significantly from the Philippines, India, and Korea) gave birth to new broadcast options. And Spanish-language broadcasting continued its growth.

Since the late 1960s, non-English-language radio in the United States has seen a massive growth in both the overall number of stations, as well as the size and impact of audiences. Stations broadcasting primarily in Spanish, Korean, and Chinese have multiplied in conjunction with the growth of immigrant populations in primarily urban areas. What distinguishes the growth of non-English-language radio programming in the second half of the century from that of the first half is the development of single-language stations that are able to compete in larger metropolitan markets. Whereas network interests choked off the development of single-language radio stations during radio's golden age, the virtual dominance of radio by local interests has opened the door for radio stations that target a particular ethno-linguistic population in a particular area.

Online Radio

Recently, with the popularity of the internet and the increasing availability of web-based broadcasts, non-English "radio" broadcasts are flourishing. For example, www.live-radio.net contains a listing of online broadcasts from radio stations all over the globe in virtually every language imaginable. The two primary interfaces from accessing online media also include simple ways of locating and accessing online broadcasts of all kinds. No longer restricted by the narrow spectrum of radio frequencies, broadcasters can reach audiences of size and scope never before imaginable. At the same time, audiences can tune in to a wider variety of programs originating from more locations, and broadcasting in more languages than has ever before been possible.

With the spread of online broadcast technology and an FCC ruling in 2000 to create a class of low-wattage stations, the future of non-English-language broadcasting is bright, if not in traditional broadcasting. Insofar as non-English-language programs have long been on the margin of mainstream broadcasting, they have also often been in the vanguard of broadcast practices, conventions, and styles. Thus the sheer number of non-English web-based broadcasts should come as no surprise. Their proliferation indicates that the future of radio is wide open, a form of expression that cannot be limited by traditional broadcast practices or geographical location. It also heralds a return to the origins of non-English-language broadcasting in the United States, which were rooted in the needs and preferences of immigrants from other countries.

ARI KELMAN

See also Canadian Radio and Multiculturalism; Hispanic Radio; Internet Radio; Jewish Radio; Native American Radio; Stereotypes on Radio; WCFL; WEVD

Further Reading

Arnheim, R., and M. Bayme, "Foreign Language Broadcasts over Local American Stations," in *Radio Research,* edited by Paul Felix Lazarsfeld and Fred Stanton, New York: Duell, Sloan, and Pearce, 1941; reprint, New York: Arno Press, 1971

Grame, Theodore C., *Ethnic Broadcasting in the United States,* Washington, D.C.: American Folklife Center, Library of Congress, 1980

Gutiérrez, Felix F., and Jorge Reina Schement, *Spanish Language Radio in the Southwestern United States,* Austin: Center for Mexican American Studies, University of Texas, 1979

Hilmes, Michele, *Radio Voices: American Broadcasting, 1922–1952,* Minneapolis: University of Minnesota Press, 1997

Horton, Gerd, "Unity on the Air? Fifth Columnists and Foreign Language Broadcasting in the United States during World War II," *Ethnic Forum* 13, no. 1 (1993)

Keith, Michael C., *Signals in the Air: Native Broadcasting in America,* Westport, Connecticut: Praeger, 1995

Klopf, Donald, and John Highlander, "Foreign Language Broadcasting in the Western States" *Western Speech* 29, no. 4 (Fall 1965)

Migala, Józef, *Polish Radio Broadcasting in the United States,* Boulder, Colorado: East European Monographs, 1987

North American Regional Broadcasting Agreement

Sharing Frequencies among the U.S., Canada, Mexico, and the Caribbean

First placed into effect early in 1941, the North American Regional Broadcasting Agreement (NARBA) treaty doled out radio channels to Canada, Cuba, Mexico, and the United States. Renewed after extensive negotiations in 1960, it remained for years the basis for cooperative regulation and interference reduction among these countries until it was replaced by bilateral agreements and treaties affecting all of the Western Hemisphere.

As the number of radio stations in North America grew, countries neighboring the United States felt increasingly squeezed out of valuable medium wave frequencies used for AM radio broadcasting. Naturally the potential for trouble was greatest along the northern and southern borders of the United States, where American stations could—and did—cause interference to outlets in other countries, and vice versa. Given the larger U.S. population and expanding radio industry, American broadcasters sought the lion's share of available frequencies, leaving little to be shared by Canada, Mexico, and Cuba.

In 1937 representatives of the four nations met in Havana and hammered out the gist of a proposed frequency-sharing treaty. The task was not easy: given the great distances covered by AM signals, the work was complex and occasionally contentious. Nevertheless, the treaty was ratified by each nation and entered force on 29 March 1941. At the time, there were about 750 broadcast stations on the air in the United States, most of which had to shift their frequency (some only slightly), primarily to clear some radio channels for expanded use in Mexico. But for the remainder of the decade, the four nations were able to license stations in accordance with NARBA, thus greatly reducing potential interference problems.

The treaty expired in March 1949 after initial attempts to renew it failed. The demise of NARBA occurred despite considerable effort and controversy. The key problem was the dramatic expansion of American broadcasting—to 2,127 stations when NARBA expired—and thus the greater (and steadily expanding) use of spectrum by the U.S. radio industry. At the same time, driven in part by understandable nationalism, governments in the neighboring countries felt they were (again) getting the short end of the frequency stick, because the number of their stations had increased as well.

Cuba was the loudest complainer, even in the late 1940s. It demanded use of more frequencies and threatened to not ratify a NARBA renewal if it did not get them. Likewise, Mexico moved to protect some of the border stations serving American audiences. Complicating matters was the addition of new negotiating players: Haiti, the Dominican Republic, and Britain (on behalf of the Bahamas and Jamaica). Still, after several rounds of engineering work and diplomatic negotiation, a draft of a second NARBA treaty was initialed in 1950. But because of continuing negotiation problems—and rising pressure from big clear channel broadcasters fearful of losing some of their coverage as well as from small daytime-only stations hoping for longer broadcast hours (both groups felt they had given up enough already)—American ratification was delayed.

Between 1953 and 1962, a second NARBA was ratified by Cuba, Canada, the Dominican Republic, and the Bahamas. In 1960 the U.S. Congress ratified it as well, placing the treaty into force despite the lack of agreement.from Mexico, Jamaica, and Haiti. Bilateral agreements with Mexico in 1969–70 and again in 1986 (as well as with Canada in 1984), served to keep the lid on potential interference problems and allowed many American daytime-only stations operating on Mexican or Canadian clear channels to begin operations before local sunrise and sometimes to extend operating hours into the evening with greater power. These agreements remain in force as long as the three nations agree.

Politics, always potent in international agreements, became central in dealing with the island nation of Cuba. By the late 1960s, the Castro regime in Cuba was informally ignoring the 1960 NARBA treaty. In 1981 Cuba formally abrogated the agreement and began to build stations that went well beyond the agreement in terms of power and frequencies used. When the United States began propaganda broadcasts into Cuba over Radio Martí, the Castro government retaliated by building high-power transmitters that caused considerable interference, especially with stations in the American South and Midwest. The Federal Communications Commission began to make case-by-case decisions allowing the affected stations to increase their own power—in essence recognizing exactly the kind of "radio war" the original NARBA treaty was designed to prevent.

As for the other nations in the region, the second NARBA treaty was eventually superseded by various Region II (Western Hemisphere) radio broadcasting agreements established under the auspices of the International Telecommunication Union in a series of regional radio conferences.

CHRISTOPHER H. STERLING

See also Border Radio; Canada; Cuba; Frequency Allocation; International Telecommunication Union; Licensing; Mexico; Radio Martí

Further Reading

North American Regional Broadcasting Agreement: Final Protocol to the Agreement, Resolutions, and Recommendations, Washington, D.C.: International Telecommunication Union, 1950

Rankin, Forney Anderson, *Who Gets the Air? The U.S. Broadcaster in World Affairs,* Washington, D.C.: National Association of Broadcasters, 1949

Nostalgia Radio

Broadcast audio programming has been collected by both institutions and interested individuals and most recently has become a battleground of interests. Central to the ability to collect such programs, of course, was the development of a means to make permanent recordings. Broadcasters from the beginning of radio needed some way to record. Initially the only means for recording was cutting a disc. Western Electric had developed the 16-inch, 33 1/3-rpm recording disc for use as the sound tracks for early 1920 "talkies." Sound-on-Film (SOF) optical tracks did not follow until later. The same medium was used in radio to record programs, usually for archiving, but sometimes for program syndication to multiple markets. Those Electrical Transcriptions (ETs) that survived were preserved by sponsors, their advertising agencies, the talent on the shows, and some broadcast engineers.

Beginnings of Programming Collecting

The Library of Congress began to collect and preserve some programming in 1949 in its role as the U.S. copyright depository. The National Archives also collected and preserved programming from governmental sources and increasingly received donated event and news materials from stations and networks. Institutional archives are as variable as the institutions preserving the available material. Funding difficulties led the UCLA Film and Television Archive to concentrate on their film collection and discontinue the development of its radio archive, which consisted of 50,000 ETs and 10,000 tapes of radio dating from 1933 to 1983 (which the Archive still retains). The Milo Ryan Phonoarchive at the University of Washington, obtained from radio station KIRO, consisted of CBS programming from 1938 to 1962. These ETs were subsequently transferred to the National Archives. Material became available as people gained access to more ETs as radio stations began disposing of their stored material and donations were made to institutions. A large source of material was recordings that had been made for Armed Forces Radio to bring radio programs to U.S. troops during World War II. Those ETs that survived and a few network and syndicated discs comprised the basis for collectors in the sixties, when radio as it had been was almost gone. From these and other sources ETs were transferred to audiotape.

Serious collecting of radio programs by individuals was the result of the introduction of home reel-to-reel tape recording decks to the consumer market around 1950. Small groups formed to exchange material, information, and sources on both the East and West coasts. A number of clubs began to trade tapes, the earliest being the Radio Collectors of America. Some of these grew into large organizations of members who gathered and traded from their shared collections, such as the North American Radio Archive (NARA) and the Society to Preserve and Encourage Radio Drama, Variety and Comedy (SPERDVAC).

In 1954 Charles Michelson developed a rebroadcast market by obtaining an umbrella agreement to license *The Shadow* to individual radio stations, long-playing record manufacturers, and producers of home-enjoyment tapes. The first aggressively marketed private dealer was J. David Goldin, a former engineer at CBS, NBC, and Mutual, who formed "Radio Yesteryear" and an album subsidiary, "Radiola," in the late 1960s. Michelson began to send "cease and desist" letters to collectors selling any of the series he had licensed.

Newsletters about radio program collecting began to circulate in the late 1960s. The most influential, which set the standard, was "Radio Dial" by the *Radio Historical Society of America* founded by Charles Ingersoll. Carrying on the tradition, the leading newsletter today is "Hello Again" by Jay Hickerson, which began publication in 1970 and tied together more than 100 of the most active collectors at that time. Today, more than 160 active collectors comprise the mass of privately collected broadcast material available, but thousands of other collectors maintain some program recordings. No one knows how many shows survive, but 150,000 or more are documented as existing in Jay Hickerson's *Ultimate History of Network Radio Programming,* which is an attempt to catalog every radio program currently circulating. The publications of *Radio's Golden Age* in 1966 and its updated revision as *The Big Broadcast 1920–1950* in 1973 by Frank Buxton and Gary Owen also increased interest in old radio programs.

Despite the interest of individual private collectors and the growth of institutional archives, the preservation of radio programming faces a crisis stemming from a combination of concerns. The most basic problem is the increasing rate of disposal and destruction of material. The way programs have been recorded—electrical transcription to tape formats—poses problems for preservationists. As transcription turntables disappear and reel-to-reel tape recorders are replaced with cassette recorders (and cassette recorders replaced with CD players), the means for playing the available material is lost or exists only in museums. The need to transfer the older formats into new formats is a time consuming and costly process. Many radio programs have been made available over the worldwide web in the downloadable MP3 format, free to anyone with a computer and an internet connection. Some collectors sell home-recorded CDs on their own websites, with as many as 50 or more shows on a single disk. There are numerous sites on the internet dealing with Old Time Radio (OTR).

Copyrighting and Collecting

Another problem is one of copyright ownership and control. As the nostalgia market for old radio programming has developed, copyright owners became more interested in protecting their copyrights. Ownership of many programs is very complex and depends upon contracts with directors, writers, performers, and rights holders of music and other materials used in the broadcasts. Private collectors who charge for duplication or sell programs are more susceptible to copyright problems than are institutions. Under certain conditions specified in the copyright law, libraries and other archives are authorized to photocopy or make other reproductions for research and teaching. However, Congress, through changes in the current copyright law, has placed most old radio programs under copyright even though they aired 75 years ago. Although sound recordings could not be copyrighted until 1972, the underlying script could be copyrighted as an "unpublished